Microbiology: Advancements and Applications

Microbiology: Advancements and Applications

Edited by Drew Farmer

SYRAWOOD
PUBLISHING HOUSE

New York

Published by Syrawood Publishing House,
750 Third Avenue, 9th Floor,
New York, NY 10017, USA
www.syrawoodpublishinghouse.com

Microbiology: Advancements and Applications
Edited by Drew Farmer

© 2019 Syrawood Publishing House

International Standard Book Number: 978-1-68286-838-6 (Hardback)

Cataloging-in-Publication Data

Microbiology : advancements and applications / edited by Drew Farmer.
    p. cm.
Includes bibliographical references and index.
ISBN 978-1-68286-838-6
1. Microbiology. 2. Biochemistry. 3. Microorganisms. 4. Biology. I. Farmer, Drew.
QR41.2 .M53 2019
579--dc23

TABLE OF CONTENTS

PREFACE

In my initial years as a student, I used to run to the library at every possible instance to grab a book and learn something new. Books were my primary source of knowledge and I would not have come such a long way without all that I learnt from them. Thus, when I was approached to edit this book; I became understandably nostalgic. It was an absolute honor to be considered worthy of guiding the current generation as well as those to come. I put all my knowledge and hard work into making this book most beneficial for its readers.

The study of acellular, unicellular and multicellular microorganisms is termed as microbiology. It has a number of significant sub-disciplines, which include parasitology, virology, bacteriology and mycology, among others. Culture, staining and microscopy are some traditional techniques of this field. Modern microbiology often relies on molecular biology tools like DNA sequence based identification for bacterial identification. Microbes can be used to produce biotechnologically significant enzymes such as reporter genes and Taq polymerase for use in molecular biology techniques and other genetic systems. Bacteria are useful in the industrial production of amino acids and for the synthesis of antibiotics. Symbiotic microbial communities help animal and human health by aiding digestion, suppressing pathogenic microbes and producing beneficial amino acids and vitamins. This book provides comprehensive insights into the field of microbiology. It traces the progress of this field and highlights some of its key advancements and applications. Students, researchers, experts and all associated with this field will benefit alike from this book.

I wish to thank my publisher for supporting me at every step. I would also like to thank all the authors who have contributed their researches in this book. I hope this book will be a valuable contribution to the progress of the field.

Editor

Ciliate diversity and distribution patterns in the sediments of a seamount and adjacent abyssal plains in the tropical Western Pacific Ocean

Feng Zhao[1,2], Sabine Filker[2], Thorsten Stoeck[3] and Kuidong Xu[1,4]*

Abstract

Background: Benthic ciliates and the environmental factors shaping their distribution are far from being completely understood. Likewise, deep-sea systems are amongst the least understood ecosystems on Earth. In this study, using high-throughput DNA sequencing, we investigated the diversity and community composition of benthic ciliates in different sediment layers of a seamount and an adjacent abyssal plain in the tropical Western Pacific Ocean with water depths ranging between 813 m and 4566 m. Statistical analyses were used to assess shifts in ciliate communities across vertical sediment gradients and water depth.

Results: Nine out of 12 ciliate classes were detected in the different sediment samples, with Litostomatea accounting for the most diverse group, followed by Plagiopylea and Oligohymenophorea. The novelty of ciliate genetic diversity was extremely high, with a mean similarity of 93.25% to previously described sequences. On a sediment depth gradient, ciliate community structure was more similar within the upper sediment layers (0-1 and 9-10 cm) compared to the lower sediment layers (19-20 and 29-30 cm) at each site. Some unknown ciliate taxa which were absent from the surface sediments were found in deeper sediments layers. On a water depth gradient, the proportion of unique OTUs was between 42.2% and 54.3%, and that of OTUs shared by all sites around 14%. However, alpha diversity of the different ciliate communities was relatively stable in the surface layers along the water depth gradient, and about 78% of the ciliate OTUs retrieved from the surface layer of the shallowest site were shared with the surface layers of sites deeper than 3800 m. Correlation analyses did not reveal any significant effects of measured environmental factors on ciliate community composition and structure.

Conclusions: We revealed an obvious variation in ciliate community along a sediment depth gradient in the seamount and the adjacent abyssal plain and showed that water depth is a less important factor shaping ciliate distribution in deep-sea sediments unlike observed for benthic ciliates in shallow seafloors. Additionally, an extremely high genetic novelty of ciliate diversity was found in these habitats, which points to a hot spot for the discovery of new ciliate species.

Keywords: Deep sea, Seamount, Protozoa, Depth gradient, Vertical distribution, Ciliate

* Correspondence: kxu@qdio.ac.cn
[1]Department of Marine Organism Taxonomy and Phylogeny, Institute of Oceanology, Chinese Academy of Sciences, 7 Nanhai Road, Qingdao 266071, People's Republic of China
[4]University of Chinese Academy of Sciences, Beijing 100049, China
Full list of author information is available at the end of the article

Background

Deep-sea systems (> 1000 m) cover more than 65% of the Earth's surface and fulfill a range of key ecosystem functions [1]. Yet, they are amongst the least understood ecosystems on Earth [2]. Among various deep-sea habitats, seamounts are widespread and prominent features of the world's underwater landscape [3, 4]. Their specific topography creates distinct habitats, characterized by particular hydrography and substrate types, which influence the diversity of the benthos [5]. So far, most biological research related to seamounts focused on patterns of mega- and macrobenthic biodiversity and their biogeography [4]. Information about the diversity and distribution of benthic protists from seamount sediments are scarce, with the exception of benthic foraminifera, whose seamount communities have been described by several authors using morphological criteria [6]. In the past decade, molecular techniques and sequencing the 18S rDNA as a taxonomic marker were used to investigate the diversity of protists in deep-sea sediments. These studies did not only reveal high protistan diversities in various deep-sea habitats [7, 8], but also numerous undescribed taxa and even several early branching eukaryotic lineages [9]. In some of these habitats, ciliates were the predominant and often the most diverse microeukaryotic group [10–13]. However, knowledge on their distribution patterns and the major factors enforcing their dispersal in deep-sea sediments are far from complete.

In a microscopy study, water depth emerged as an important factor structuring the distribution of benthic protists, including ciliates, in the deep sea [14]. However, because protists are not highly abundant in the deep sea [15], it is not surprising that microscopy studies miss a large proportion of the deep-sea protist diversity. Accordingly, using molecular techniques, more than 125 ciliate OTUs were detected in less than 1 g of abyssal sediments [7]. This exceeded the number of ciliate OTUs detected in the coastal sediments [16]. Thus, it is reasonable to assume that a large proportion of ciliate diversity in the deep-sea is still unknown to science.

Moreover, studies of benthic protists mainly focused on the top 1-2 cm surface layer of deep-sea sediments [7]. However, it is well known that different ciliate species are not equally distributed along a vertical sediment gradient, mostly owed to oxygen, organic carbon and grain-size gradients [17–20]. Thus, some unknown ciliate taxa which are absent from the surface sediments are likely to be found in deeper sediment layers.

In this study, we collected sediment samples from four sites in the tropical Western Pacific Ocean with water depths ranging between 813 m and 4566 m as well as from different sediment layers. This sampling strategy was to maximize the number of different deep-sea habitats and thus, the proportion of novel ciliate diversity to be found. These habitats included a seamount and the adjacent abyssal plain. Statistical analyses were then used to assess shifts in ciliate communities across vertical sediment gradients and water depth.

Results

Environmental parameters of sampling sites

For each sediment sample, the median grain size, the proportion of each sediment component and the total organic carbon (TOC) content were analyzed. The median grain size (mgs) of sediments differed notably among the different layers and sites and ranged between $4.5 * 10^{-13}$ µm (DS2.2) and 799.4 µm (S-B.1; Table 1). The mgs of surface sediments was highest at all

Table 1 Environmental parameters of each sample (S = sand, St = silt, C = clay)

	Layer (cm)	Depth (m)	Longitude	Latitude	Total organic Carbon (%)	Median grain Size (µm)	Sand (%)	Silt (%)	Clay (%)	Type of sediments
S-M.1	0-1 cm	813	137°45.3'E	8°52.2'N	0.10	236.2	88.7	9.0	2.3	S
S-M.2	9-10 cm				0.14	73.8	53.8	30.8	15.4	S-St
S-M.3	19-20 cm				0.53	45.2	39.2	42.9	17.9	St-S
S-M.4	29-30 cm				0.06	42.4	39.4	44.8	15.8	St-S
S-B.1	0-1 cm	3812	137°54.4'E	8°47.0'N	0.63	799.4	62.7	23.3	9.1	S
S-B.2	9-10 cm				0.26	8.4	5.7	62.5	31.8	C-St
S-B.3	19-20 cm				0.41	25.3	26.1	53.8	20.2	S-C-St
DS1.1	0-1 cm	4042	134°50.1'E	10°0.3 'N	0.52	21.6	15.4	67.9	16.7	C-St
DS1.2	9-10 cm				0.42	33.7	39.7	55.3	5.1	S-St
DS1.3	19-20 cm				0.82	85.2	63.6	33.3	3.1	St-S
DS2.1	0-1 cm	4566	136 °0.2'E	9 °0.3'N	7.14	18.1	3.3	78.3	18.5	C-St
DS2.2	9-10 cm				7.95	$4.5 * 10^{-13}$	0.0	75.3	24.7	C-St

sampling sites, but site DS1. Here, the highest mgs was detected at 20 cm depth. Sediment composition was different for each sampling site (Table 1). The TOC content of the sediments at sites S-M, S-B and DS1 varied between 0.1% (S-M.1) and 0.82% (DS1.3). The TOC content at site DS2 was, in contrast, more than 8 times higher than at the other sampling sites (e.g. DS2.2: 7.95%; Table 1).

Overview of the sequencing data
For the 12 sediment samples analyzed in this study, we obtained a total of 640,860 high-quality ciliate V4 sequences, which clustered into 104 distinct operational taxonomic units (OTUs) based on a 97% sequence similarity. The number of reads per sample varied between 25,169 (DS1.2) and 137,990 (S-B.2), with an average of 56,820 reads. The total number of OTUs obtained from each sample and used in downstream analyses ranged between 12 (S-B.2) and 55 (S-M.3), with an average of 33 OTUs (Table 2). Rarefaction analyses indicated near-saturated sampling for all samples except of S-B.1 (Additional file 1).

Distribution of ciliate OTUs along water depth gradients and sediment layers
Observed benthic ciliate alpha diversity varied remarkably among the different sites and sediment depths. The effective number of species was on average highest at sampling site S-M and lowest at site S-B (S-M: 13 ± 1.9; DS2: 9.9 ± 2.4; DS1: 3.1 ± 0.8; S-B: 2.8 ± 1.1), whereas OTU richness was on average highest at site S-M and lowest at DS1 (S-M: 46.3 ± 7.1; DS2: 41 ± 0.0; S-B: 21.7 ± 12.9, DS1: 21.3 ± 7.5; Table 2). The number of OTUs in surface sediments along water depth was relatively stable and ranged from 33 to 41. Except of sampling site S-M, the effective number of species and the OTU richness were higher in the surface sediment samples than in the deeper layers (Table 2).

About 24% of the ciliate OTUs detected in this study were unique to one sample. No single OTU was present in all the 12 samples, whereas 20 OTUs (19%) were

detected at more than six sampling sites. Abundant OTUs in the dataset accounted for 20% (DS2.2) to 45% (S-M.2), whereas rare OTUs contributed 26% (S-M.2) to 69% (DS1.2; Table 2).

At a local scale (i.e. sediment depth gradient), the major proportion of OTUs was unique to one sample and accounted for 30.1% (S-M) to 75% (S-B), whereas OTUs shared by 4 (S-M), 3 (S-B, DS1) or 2 (DS2) samples only contributed 1.9% (S-B) to 46.4% (DS2) to the ciliate OTU distribution (Fig. 1 a-d). Also on a water depth gradient, the proportion of unique OTUs was between 42.2% (Fig. 1 f) and 54.3% (Fig. 1 e) and that of shared OTUs (by 4 sites) only around 14%. 29 OTUs (78%) of the surface layer at the shallow site S-M were shared with the surface layers at sites deeper than 3800 m (Fig. 1 e).

Partitioning of diversity revealed that the ciliate community structure was more similar within the upper sediment layers (0-1 and 9-10 cm) compared to the lower sediment layers (19-20 and 29-30 cm) at each site. Ciliate communities in the upper sediment layers (0-1 and 9-10 cm) of sites DS1, DS2 and S-M grouped together, respectively, with a high mean Jaccard similarity indicating a high stability of these clusters (Fig. 2 a). Ciliate communities from site S-B are distinct to the other ciliate communities and do not group into the clustering pattern (Fig. 2 a). Taking sequence abundances into account, this pattern is confirmed, although the overall grouping of the samples is slightly different (Additional file 2 a).

Ciliate diversity, i.e. OTU richness and effective number of species, and ciliate community composition in the surface sediments were not significantly correlated with any of the measured environmental parameters (Additional file 3). Additionally, changes of ciliate community composition in the deep-sea sediments were not significantly related to any combination of the environmental factors. The Mantel test showed that the pairwise community dissimilarity had no significant correlation with the geographic distance at a 350 km distance scale ($r = -0.432$, $p = 0.83$).

Table 2 Sample statistics including number of sequences and OTUs per sample, alpha diversity estimates and proportions of rare and abundant OTUs

	S-M.1	S-M.2	S-M.3	S-M.4	S-B.1	S-B.2	S-B.3	DS2.1	DS2.2	DS1.1	DS1.2	DS1.3
No. of filtered ciliate sequences	48,426	46,003	30,233	39,786	78,685	137,990	86,461	40,417	29,913	27,525	25,169	50,252
No. of ciliate OTUs (97%)	37	41	55	51	41	12	13	41	41	33	17	16
Effective no. of species	10.3	14.1	15.3	12.4	4.3	1.9	2.2	12.3	7.6	3.8	2.0	3.7
No. of rare OTUs	15	11	22	15	24	7	6	13	14	15	11	9
Proportion of rare OTUs [%]	41	26	40	29	60	58	46	32	33	47	69	56
No. of abundant OTUs	15	19	19	11	9	3	5	15	8	7	6	7
Proportion of abundant OTUs [%]	41	45	35	22	23	25	38	37	20	22	38	44

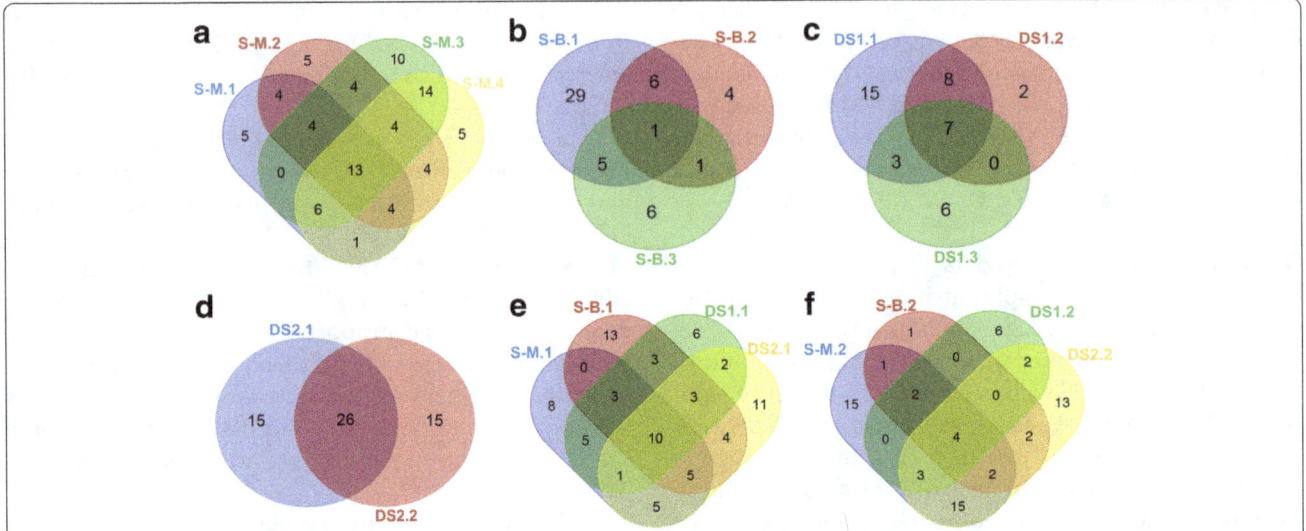

Fig. 1 Venn diagrams displaying the number of unique and shared OTUs in different layers at sites S-M (**a**), S-B (**b**), DS1 (**c**) and DS2 (**d**), as well as in the surface layers (**e**) and in the layers of 9-10 cm depth (**f**) of different sites

Taxonomic classification of ciliate OTUs

Ciliate OTUs were assigned to 9 (out of 12) ciliate classes (Fig. 2 b; Additional file 2 b), originating from 40 families and 51 genera (Additional file 4). The classes Armophorea, Cariacotrichea and Colpoda were not detected. Most OTUs belonged to the class Litostomatea (on average 23.3% of total OTUs at each site). The Litostomatea accounted for up to 33.3% (S-M.4: 17 OTUs) of the total ciliate OTU diversity in the investigated sediment samples. However, litostomatean sequences contributed only 4.7% to the total ciliate sequence abundance (up to 15.6% in sample S-M.2; Additional file 2 b). All of these OTUs were affiliated to 8 known genera. Within the Litostomatea, sequences related to *Phialina*, *Loxophyllum* and *Litonotus*

were most abundant (Additional file 4). No litostomatean sequences could be retrieved from sample DS1.3.

Plagiopylea were the second most diverse ciliate group, contributing on average 18.6% to the OTU diversity at each site, but 47.8% to the total sequence abundance. Plagiopylean ciliates were detected at 11 of 12 sites, with the exception of site S-B.3. The proportion of plagiopylean OTUs varied between 0 (S-B.3) and 26.8% (DS2.1: 11 OTUs) and the proportion of sequences between 0 (S-B.3) and 78% (S-B.2) (Fig. 2 b, Additional file 2 b). All OTUs/sequences were related to the anaerobic ciliate *Epalxella* (Additional file 4). The sequence similarities of these OTUs to *Epalxella* ranged between 85% and 94%.

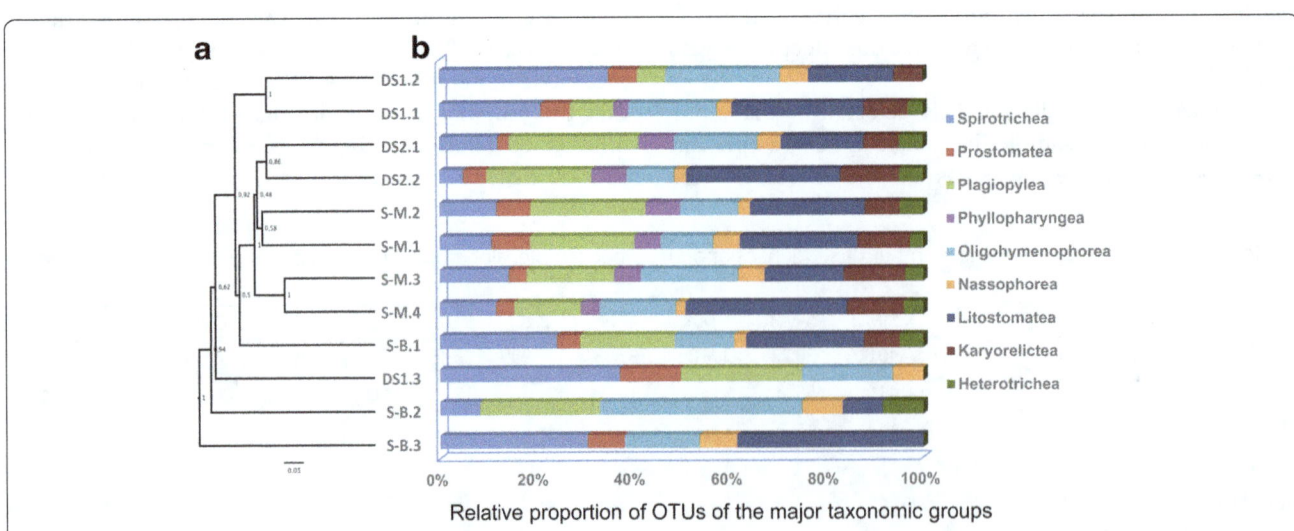

Fig. 2 UPGMA clustering analysis based on the incidence Jaccard index (**a**), and the relative proportion of OTUs (**b**) belonging to the major taxonomic groups of ciliates detected in the 12 sediment samples

Oligohymenophorean and spirotrichean OTUs accounted for 16% on average of all detected OTUs at each site, respectively. The proportion of OTUs affiliated to Oligohymenophorea ranged between 4.9% (S-B.3: 2 OTUs) and 20% (S-M.3: 11 OTUs) (Fig. 2 b). Oligohymenophorean sequences accounted for 35.4% of all ciliate sequences, with the lowest proportion of sequences detected in sample DS1.3 (3%) and the highest in sample S-B.3 (90%) (Additional file 2 b). All oligohymenophorean OTUs/sequences were assigned to 10 known genera and one unclassified taxon. Within this class, sequences related to *Pseudocyclidium*, *Trichodina* and *Pleuronema* were most abundant. The most cosmopolitan oligohymenophorean OTU, which could be detected in 11 of 12 samples, was closely related to *Pleuronema setigerum* with the similarity of 99% (Additional file 4).

The proportion of spirotrichean OTUs ranged between 4.9% (DS2.2) and 37.5% (S-M.3) in the different benthic ciliate communities, all of these OTUs were affiliated to 15 known genera (Additional file 4). Spirotrichean sequences contributed 8.9% to the total number of ciliate sequences, the sequence proportion accounted for up to 35.6% in sample DS1.3 (Additional file 2 b). Within this class, sequences related to the genera *Amphisiella*, *Apokeronopsis* and *Aspidisca* were most abundant. Ten of 21 OTUs, which were observed in six or more samples, belonged to Spirotrichea (Additional file 4), of which five were affiliated to the order Urostylida. The OTUs affiliated

to oligotrich ciliates of *Novistrombidium orientale* and *Parallelostrombidium obesum* were detected in the downcore sediments only (Additional file 4).

Classes Karyorelictea (on average 8.8% of total number of OTUs in each sample), Prostomatea (5.3%), Phyllopharyngea (4.3%), Nassophorea (4%) and Heterotrichea (3.8%) were relatively rare (Fig. 2 b, Additional file 2 b) at all sampling sites.

Degree of novel diversity

The observation of numerous low-identity (< 97%) OTUs in our dataset points to a high genetic novelty within the deep-sea habitats. The mean similarity of all detected ciliate OTUs to previously described sequences was only 93.25%. 73% of OTUs had a sequence similarity of less than 97% to reference sequences and 25% of the OTUs had an identity match of less than 90%. The mean similarities for each sample were relative stable and ranged between 92.86% (S-M.1) and 94.55% (DS1.1). The lowest overall mean sequence similarities, i.e. the highest degree of novel diversity occurred within the class Plagiopylea (89%). All of the 18 plagiopylean OTUs had a sequence similarity < 95%. The highest overall mean sequence similarities were observed within the classes Spirotrichea and Karyorelictea (95.7%, respectively; Fig. 3). Eleven of 23 spirotrichean OTUs and 4 of 8 karyorelictean OTUs had a sequence similarity of equal or more than 97%.

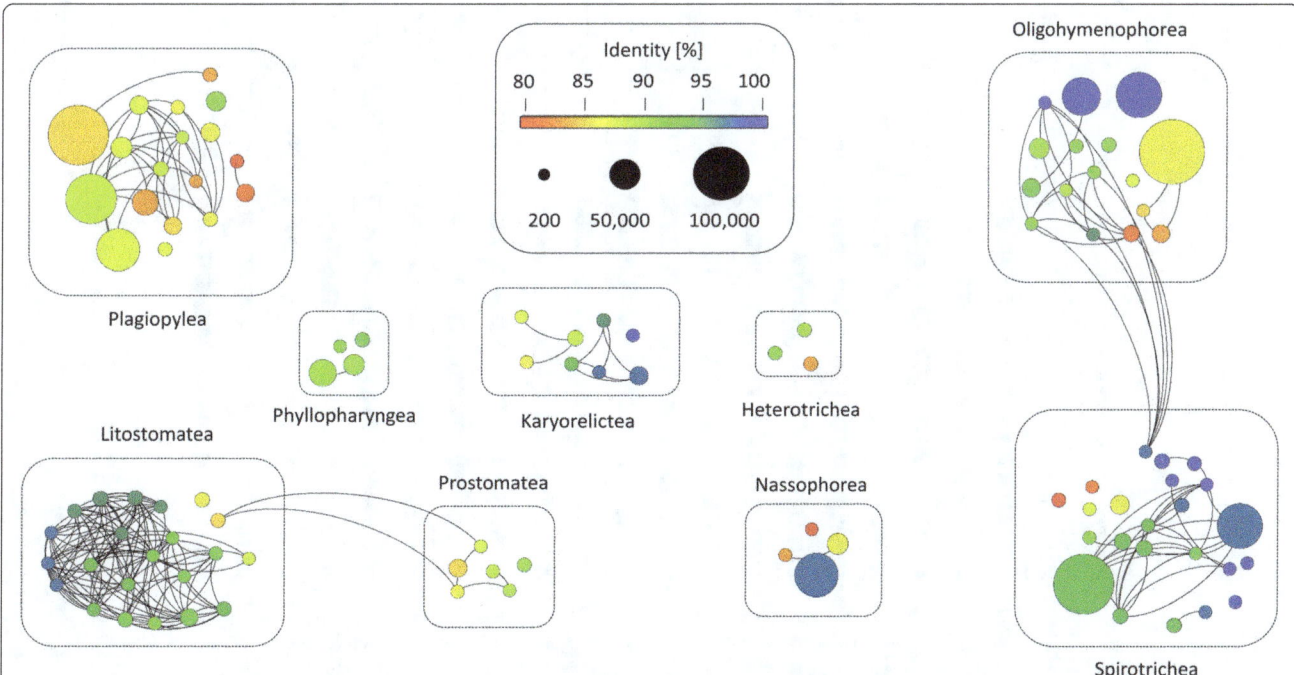

Fig. 3 Analyses of the novel diversity within the ciliate communities in the sediments of the seamount and adjacent deep-sea plain. Each node represents one OTU, the different colors indicate the sequence similarity to a deposited reference sequence, and the size of the node indicates the number of V4 sequences. An edge weight (sequence similarity) of 90% was chosen to discriminate between the different OTUs

Comparison of benthic ciliate communities from different marine habitats

OTU richness peaked in the continental shelf samples (mean OTU richness: 105.8 ± 44), followed by the intertidal samples (mean OTU richness: 93.3 ± 25) and the hydrothermal vent samples (mean OTU richness: 40 ± 21). Alpha diversity was lowest in the seamount and abyssal plain samples (mean OTU richness: 34 ± 4). Changes in alpha diversity were significant between ciliate communities from the intertidal zone ($p = 0.01$) or continental shelf ($p = 0.01$), respectively, and the communities of the seamount and abyssal plain. No significant difference was detected between the OTU richness of the seamount and abyssal plain communities and the hydrothermal vent communities.

The proportion of spirotrichean OTUs was highest in the communities from the intertidal zone and continental shelf (Fig. 4). Almost no Plagiopylea ciliate was detected in the coastal samples, whereas Plagiopylea was one of the most diverse ciliate group in the seamount, abyssal plain, and hydrothermal vent samples (Fig. 4). The proportion of oligohymenophorean OTUs was the highest in the communities of the hydrothermal vent samples. UPGMA clustering based on the Sorensen-Dice similarity coefficient between the ciliate communities revealed two large clusters, separating intertidal and offshore communities from the deep-sea seamount, abyssal plain and hydrothermal vent communities (Fig 4). ANOSIM analyses confirmed this observation showing a significant disparity among different habitat types ($r = 0.92$; $p = 0.001$).

Discussion

Ciliate diversity and their distribution along the water depth gradient

Our study revealed the diversity of ciliates along a wide water depth gradient (813-4566 m) in the sediments

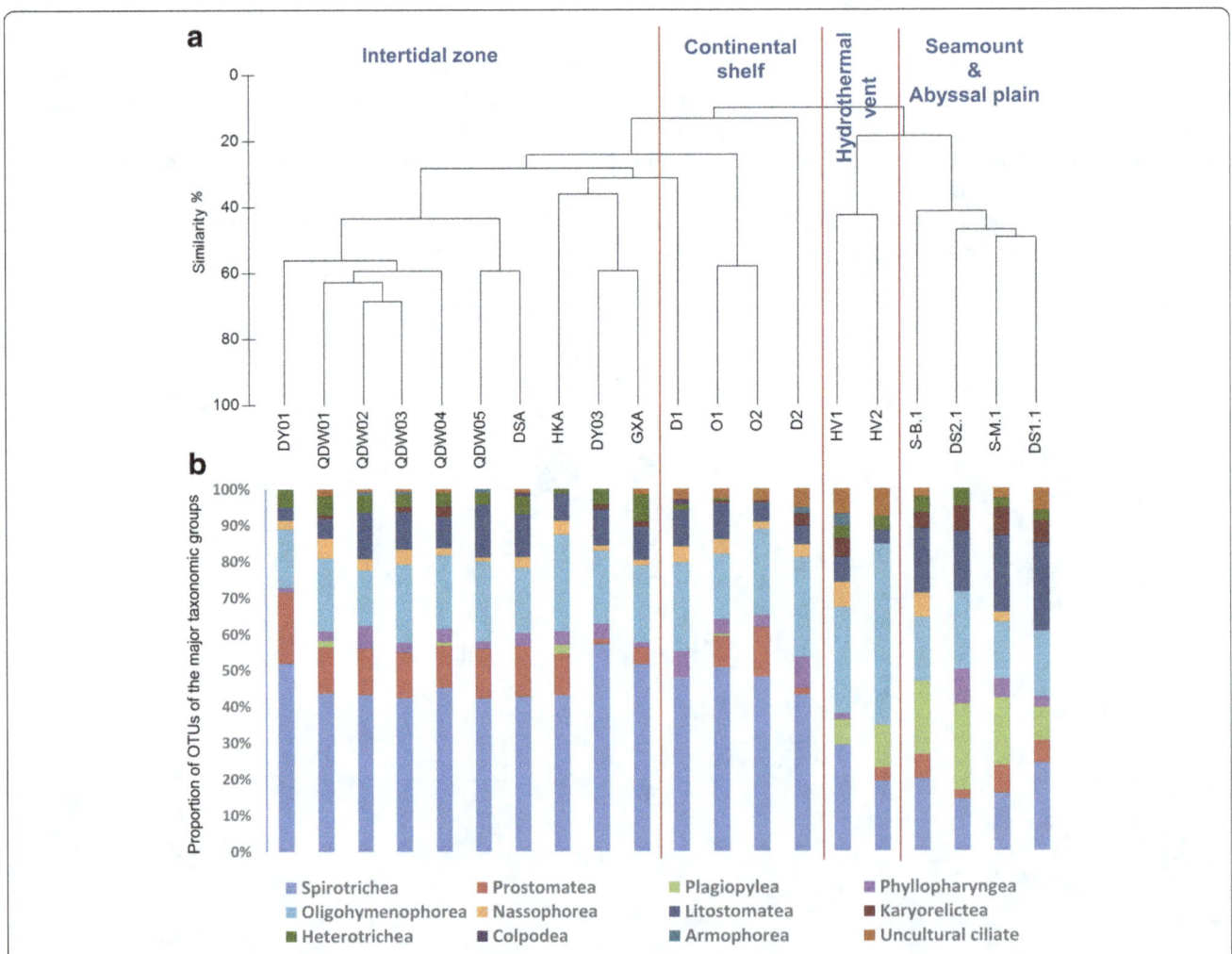

Fig. 4 UPGMA clustering analysis based on the Sorensen-Dice values (**a**), and the relative proportion of OTUs (**b**) belonging to the major taxonomic groups of ciliates detected from the surface sediments of the intertidal zone and the continental shelf of China sea area, and the hydrothermal vent in Okinawa trough, as well as the seamount and abyssal plain in the Western Pacific Ocean

from a seamount and an adjacent abyssal plain in the Western Pacific Ocean. Although sample saturation was reached for all but one sample, we found a notably lower number of distinct ciliate OTUs (33-41 OTUs at each site) in the surface sediment samples analyzed in this study compared to a previous molecular diversity survey of protists in deep-sea sediments (68-172 ciliate OTUs at each site) [7]. This could be partially explained by methodological differences, such as the use of another barcode marker (V9 vs. V4 region of SSU rDNA) and/or the deletion of singletons (OTUs containing only one sequence) in our study, which aimed to improve specificity at the cost of a possible loss in sensitivity. Instead, our results corroborate well with a microscopy study [14], suggesting a lower diversity of ciliates in deep-sea sediments compared to that in sediments from an intertidal zone and a continental shelf (e.g. 58-115 OTUs in the Yellow Sea and the East China Sea [21, 22]; 77-365 OTUs in the European coastal regions [23]). However, if water depth is deep enough, e.g. 813 m or more than 3800 m, it seems that water depth has little to no effect on the benthic ciliate OTU richness. This is different from the conclusion of the investigation in coastal sediments, where water depth overrides both geographic distance and environmental heterogeneity in determining the community composition of protists [16]. Likewise, ciliate biomass and the number of species showed a significantly negative correlation with water depth in the offshore sediments of the Yellow Sea [24]. However, little variation was observed in microeukaryotic communities from abyssal depths ranging between 5033 m and 5655 m [25]. Likewise, using the data provided by the study of Pawlowski et al. for water depths ranging between 2292 and 6326 m, we could not detect a significant correlation between the benthic ciliate OTU richness and water depth [7]. This indicates that water depth is likely more important on influencing the distribution of microbial eukaryotes in coastal sediments than in deep-sea sediments.

In our study, other factors than water depth might be more relevant for the structuring of ciliates in deep-sea sediments, e.g. oxygen saturation, which is a decisive variable for shaping protist community in shallow water sediments [17–20]. Unfortunately, this parameter could not be measured in our study.

Sediment patchiness at the deep-sea bottom is extremely high [26], possibly explaining the high dissimilarity of ciliate communities between all samples and the low number of shared OTUs between samples. Since rarefaction analyses confirm near-saturated sampling profiles for all but one sample, undersampling is no reason for the observed differences in ciliate communities.

Previous studies mainly focused on high taxonomic levels, such as the phylum level, when investigating the proportion of ciliates in a protist community (e.g. [7]). However, there is no detailed information on the assemblage of benthic ciliates in deep sea environments, except of the deep-sea hydrothermal vents [27] and cold seeps [28]. In our study, we, for the first time, revealed that the pattern in the sediments of a seamount and the adjacent abyssal plain is quite different from that observed in intertidal zones and offshore areas, where spirotrichs are predominant in terms of relative abundance and species richness [20–22, 29]. By contrast, in the sediments of the seamount and the abyssal plain, Plagiopylea ciliates were obviously abundant and diverse. The high sequence contribution implies that they might have contributed a high biomass in the ciliate community in these habitats [30]. Plagiopylea ciliates have generally been found in high sulfide, anoxic sediments [31]. Previous studies indicate that these ciliates are abundant in sediments from the hydrothermal vents [22, 27]. Our study, therefore, extends the knowledge on the distribution of Plagiopylea ciliates, and reveals their potential importance in the benthic ciliate communities of seamounts and abyssal plains. However, some ciliate genera, like *Pleuronema*, which contains more than 20 morphotypes reported from various geographical locations of intertidal and coastal zones [32], were found to be widely distributed along the pronounced range of water depths, revealing their cosmopolitan character.

Regarding the proportion of rare OTUs in the sediments of a seamount and adjacent abyssal plain, there was no obvious shift along the water depth gradient. The proportion of rare OTUs detected in our study was obviously lower than what is known from intertidal and coastal sediments [21]. Rare taxa can be considered as a seed bank of genetic resources, and are hypothesized to include ecologically redundant taxa that could increase in abundance following environmental perturbation to maintain continuous ecosystem functioning [33, 34]. Compared to coastal zones, deep-sea benthic habitats are relatively stable environments, explaining the observed lower proportion of rare taxa.

Vertical distribution of ciliates in the deep-sea sediments

In intertidal sediments, previous studies of benthic protists have shown that the abundance and species richness of protozoa were highest at the top 1 to 2 cm [17, 18]. Likewise, ciliate abundance and biomass in the 0-2 cm layers of coastal sediments accounted for 77% and 81%, respectively [24]. Several species, however, were also found to live below 5 cm, even in 8 cm sediment depth [17, 18]. Previous studies showed that the respective proportions of ciliate abundance were about 23% in the 3–8-cm layers [24]. Our study detected a high diversity of ciliate sequences in 20 cm sediment depth. When working with DNA, this is certainly no proof for the presence of living and active ciliate species, as DNA can be preserved in marine sediments over time. However, ciliates are very well adapted to a variety of different environment types, being able to not only tolerate anoxia, high salt or hydrogen sulfide concentrations or high pressure but also combinations of those [22, 27, 28]. It, therefore, would not be surprising if ciliates evolved to live in this water and sediment depths, especially in the presence of large bacterial populations in these deep sediment spheres [35, 36], which could serve as food source for heterotrophic ciliates.

The observation of OTUs affiliated to oligotrich ciliates in the deep layer of the deep-sea sediments confirmed a fraction of ciliate DNA sequences which likely do not belong to living organisms, but represent either extracellular DNA or cysts of planktonic species. The deep-sea floor, therefore, appears as a global DNA repository, which preserves molecular information of organisms living in the sediments, as well as in the overlying water column [37]. In contrast, OTUs, which are found in RNA surveys, were more likely active species [38, 39]. Thus, RNA surveys should provide a better representation of in situ protist biomass and diversity. Clearly, in the future, the analyses of deep-sea RNA will be helpful to identify metabolically active organisms.

High novelty of ciliate genetic diversity

A high degree of genetic novelty was uncovered within the ciliates from the sediments of a seamount and the adjacent abyssal plain, although ciliates are considered as the best-known group within the protists. The application of high throughput DNA sequencing contributed substantially to the detection of a broader protist diversity [7, 40]. The extremely high molecular diversity largely exceeded the one detected by morphological methods. Thus, there might be some artifacts resulting from limitations of the techniques. Previous studies have indicated that protists are well known for having extremely high rDNA copy numbers [41, 42], which might result in a high polymorphism of SSU rDNA in a single cell and lead to the overestimation of molecular diversity in environmental samples. Gong et al. [41] showed that the minimal similarity between two copies in a single ciliate cell was 99.1%. The commonly used cutoff of 97% for creating OTUs based on the ciliate V4 fragment of their 18S rDNA is very well analyzed and established and excludes the effect of intragenomic variations [23, 43–46].

Moreover, there is still an obvious gap between the obtained sequences and the reference databases, which are used for taxonomic assignment of the sequences. The

ciliate class Plagiopylea is such an example. Only one species, i.e. *Epalxella antiquorum*, within the order Odontostomatida and a total of eight plagiopylean species had their 18S rDNA sequences deposited in the NCBI database (searched on July 4th, 2017). The lack of reference sequences in the database resulted in all plagiopylean OTUs having a sequence similarity of less than 95% to the closest reference sequence. Additionally, even the most abundant plagiopylean OTU had only a sequence similarity of less than 90% to the reference sequences. More efforts in the isolation, cultivation and description of protists are necessary to link the environmental sequences to the real protist inventory [23, 47]. The design of novel species-specific primers and probes based on the retrieved sequences will also help to identify the target species by molecular techniques [48, 49]. In addition, different ciliate species were not equally distributed along a vertical sediment gradient and a large proportion of unknown ciliate diversity was found in deeper sediment layers. In the future, a more adequate sampling strategy is needed, which could maximize the number of different deep-sea habitats, and thus, detect more new ciliate species.

Conclusions

Our data point to obvious variations in ciliate communities along a sediment depth gradient in a seamount and its adjacent abyssal plain in the tropical Western Pacific Ocean and reflect the heterogeneity and diversity of benthic habitats. Water depth occurred as a less important factor impacting ciliate distribution in deep-sea sediments unlike observed for benthic ciliates in shallow seafloor habitats. Furthermore, our results indicate that ciliate diversity is not as well known as previously assumed and further efforts have to be made towards the identification of ciliate species. In this respect, deep-sea sediments appear as a hotspot for capturing and investigating novel ciliate species.

Methods

Study sites and sampling

Sediment samples were collected from four sites in the tropical Western Pacific Ocean (Fig. 5). Among these, site S-M is located at the middle of a seamount adjacent to the Yap Trench at a water depth of 813 m, and site S-B at the foot of the seamount at a water depth of 3812 m (Table 1). Sediment samples of S-M and S-B were collected using a push-corer operated by the remotely operated vehicle (ROV) "Discovery" onboard the *R/V KEXUE* in December 2014. Sampling sites DS1 and DS2 are located in the adjacent abyssal plain with water depths of 4042 m and 4566 m, respectively (Fig. 5). Sediment samples of DS1 and DS2 were taken using a 0.25-m^2 modified Gray-O'Hara box corer in

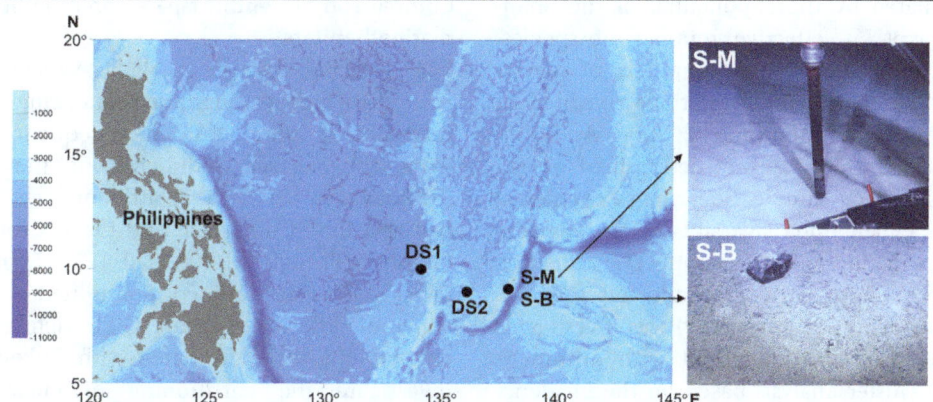

Fig. 5 Location of the four sampling sites in the tropical Western Pacific Ocean. Site S-M is located in the middle of a seamount adjacent to the Yap Trench at a water depth of 813 m, and site S-B at the foot of the seamount at a water depth of 3812 m. Sampling sites DS1 and DS2 are both located in the adjacent deep-sea plain with depths of 4042 m and 4566 m, respectively. Images on the right illustrate the different sediment types observed at the seamount

December 2014. Sites S-M and S-B are about 20 km distant from each other, and about 200 km apart from site DS2 and about 350 km apart from site DS1. Site DS1 is about 170 km distant from site DS2.

Four layers of the sediment cores (0-1 cm, 9-10 cm, 19-20 cm and 29-30 cm; labelled as 1, 2, 3 and 4, respectively) were sampled at site S-M, three layers (0-1 cm, 9-10 cm and 19-20 cm; labelled as 1, 2 and 3, respectively) at sites S-B and DS1, and two layers (0-1 cm, 9-10 cm; labelled as 1 and 2, respectively) at site DS2.

For each site and each layer, three replicate samples, each containing about 15 g sediments, were taken. Samples were stored at –80 °C for DNA extraction, at –20 °C for examination of total organic carbon (TOC), and at 4 °C for grain size measurements. TOC was determined in a Vario TOC cube (Elementar, Germany). Grain size analyses were performed using a Laser Diffraction Particle Size Analyzer (Cilas 940 L).

DNA extraction, PCR amplification and high-throughput sequencing

Environmental DNA was extracted from 0.3 g sediment of each replicate sample (a total of 3 DNA samples for each sediment sample) using the Power Soil DNA isolation kit (MoBio Laboratories, USA) according to the manufacturer's protocol. Ciliate sequences were amplified by a nested PCR approach [50], using ciliate-specific 18S rRNA gene primers in the first reaction. The second PCR reaction employed a primer set specific for the hypervariable V4 region, which was included in the first PCR product. To minimize PCR errors, we used the PrimeSTAR GXL DNA High Fidelity Polymerase (TAKARA BIO INC., Japan). Three products from each sample were pooled.

Sequencing libraries were constructed using the NEB Next® Ultra™ DNA Library Prep Kit for Illumina (NEB, USA). Quality of the libraries was assessed with an Agilent Bioanalyzer 2100 system. Finally, libraries were sequenced on an Illumina MiSeq platform, generating 300-bp paired-end reads. The ciliate sequence reads have been deposited at the National Center for Biotechnology Information (NCBI) Sequence Read Archive under the accession number SRP101585.

Sequence data processing

Paired-end reads were merged and then filtered. Quality filtering of the raw sequences was performed according to the QIIME quality control process [51]. Afterwards, a set of unique sequences was identified and the number of occurrences for each sequence was recorded by UPARSE software, followed by discarding all singleton sequences [52]. A sequence similarity of 97% was used to delineate ciliate OTUs [43–46] by the UPARSE-OTU algorithm. Representative sequences from each OTU were extracted and subjected to the basic local alignment search tool (BLAST) analyses against the Silva database (v. 123) as implemented in the QIIME pipeline (v. 1.9.0) [51].

In each sample, OTUs were defined as rare when their number was equal to or less than 0.1% of all the sequences in the sample, and abundant, when their number was equal to or more than 1% of all the sequences in the sample [34, 53].

Statistical analysis

Statistical analyses were conducted in R using the vegan and fossil packages, unless stated otherwise. Rarefaction analyses were conducted in order to investigate the degree of sample saturation. Alpha diversity for each sample was calculated by the exponential of the Shannon index H′ (exp(H′), effective number of species) [54]. After conversion, the effective numbers of species lets us avoid the serious misinterpretations spawned by the nonlinearity of the Shannon index H′ [54]. Additionally, the ciliate OTU richness was determined as the number of OTUs in each sample. Spearman's coefficient was used to relate ciliate alpha diversity and the environmental parameters, including water depth, TOC, the median grain size and the proportions of sand, silt and clay.

Partitioning of diversity (beta diversity) was investigated by an unweighted pair-group method with arithmetic means (UPGMA) cluster analysis based on the incidence Jaccard index. Stability in cluster analysis was evaluated by bootstrap resampling (100 times) using the clusterboot function in the fpc package of R [55]. Here, the Jaccard similarities of the original clusters to the most similar clusters in the resampled data were computed. The mean over these similarities was then used as an index of the stability of a cluster. Generally, "highly stable" clusters should yield average Jaccard similarities of 0.85 and above. Clusters with a Jaccard similarity value smaller or equal to 0.5 were considered as a "dissolved cluster" [56]. Prior to alpha and beta diversity analyses, the OTU table was randomly subsampled to the lowest number of sequences present in a sample (n = 25,169).

RALATE function in PRIMER v6 (Plymouth Marine Laboratory, UK) was applied to calculate Spearman's correlation coefficients between the ciliate community composition and the environmental parameters. The BIOENV function in PRIMER v6 was used to calculate the correlation between the ciliate community and the environmental similarity matrix to determine the variables that best explain variation in the ciliate community.

The significance of the relationship between community Bray-Curtis dissimilarity and geographical distance (distance in kilometers between pairs of sites) within the detected distance-decay extent was assessed by a Mantel test for each data set. The Mantel tests were performed using the vegan package in R [57].

Identification of novel diversity

Identification of novel diversity followed the description of Filker et al. 2015 [47]. Briefly, representative sequences of all detected OTUs were aligned with Seaview v.4.6.1 using "clusto" [58]. Based on these alignments, pairwise similarities were calculated (custom script) and used for network construction in R ("igraph" package) [59]. The resulting network was visualized and modified with GEPHI v.0.9.1. according to the OTU taxonomic affiliation and the Blast Hit values [60]. In the network, two nodes were connected by an edge if they shared a sequence similarity of at least 90%.

Comparison of benthic ciliate communities from different marine habitats

We collected a set of publicly available data that were related to high throughput sequencing of the ciliate V4 fragment of 18S rDNA. These sequences were obtained from the surface sediments of the intertidal zone (SPR068269, [21]), the continental shelf of the Yellow Sea and East China Sea, and the hydrothermal vents in Okinawa Trough (SPR064020, [22]). Sequence data were processed as the protocol mentioned above. The significance of differences in the OTU richness among different habitats were estimated with Tukey's HSD test. To investigate the partitioning of ciliate diversity, the abundance-based OTU table was transformed into a presence/absence table prior to calculating Sorensen-Dice indices. Sorensen-Dice values were then used for UPGMA cluster analyses. The ANOSIM function in PRIMER v6 was used to test the differences among different habitats. ANOSIM provides an R-statistic to evaluate the dissimilarity of groups, thus, groups are dissimilar if R-statistic is close to 1. Prior to alpha and beta diversity analyses, the OTU table was randomly subsampled to the lowest number of sequences present in a sample (n = 22,000).

Additional files

Additional file 1: Rarefaction curves for all samples under study.

Additional file 2: UPGMA clustering analysis based on the Bray-Curtis similarity coefficient (a) and the relative proportion of sequences (b) related to the major taxonomic groups of ciliates detected in the 12 sediment samples.

Additional file 3: Spearman's rank correlation coefficients between ciliate alpha diversity in the surface layer sediments and the environmental parameters. P-value <0.05 are considered as significant.

Additional file 4: OTU-table displaying, for each sample, the number of sequences per OTU, the BLAST similarity of the representative sequence of each OTU to the reference sequence and the taxonomic assignment of the representative sequence.

Abbreviations
BLAST: Basic local alignment search tool; MGS: Median grain size; OTU: Operational taxonomic unit; ROV: Remotely operated vehicle; TOC: Total organic carbon; UPGMA: Unweighted pair-group method with arithmetic means

Acknowledgments
We thank the WPOS sample center of Institute of Oceanology, Chinese Academy of Sciences and RV KEXUE for providing the sediments samples. Thanks are extended to Dr. Junlong Zhang for his help in making the figure of sampling sites.

Funding
This work was supported by the Strategic Priority Research Program of the Chinese Academy of Sciences (No. XDA11030201), the National Basic Research Program of China (973 Program) (No. 2015CB755902), the Scientific and Technological Innovation Project Financially Supported by Qingdao National Laboratory for Marine Science and Technology (No. 2016ASKJ14) and the China Scholarship Council (CSC, No. 201604910395).

Authors' contributions

FZ was the main contributor to the experimental work, analysis of the sequence data and drafted the manuscript. KX contributed with the design of the study. SF participated in the analysis of sequencing data and interpretation of the data. SF, KX and TS performed a critical revision of the manuscript. All authors approved the final version of the article.

Competing interests

The authors declare that they have no competing interests.

Author details

[1]Department of Marine Organism Taxonomy and Phylogeny, Institute of Oceanology, Chinese Academy of Sciences, 7 Nanhai Road, Qingdao 266071, People's Republic of China. [2]Department of Molecular Ecology, University of Kaiserslautern, 67663 Kaiserslautern, Germany. [3]Department of Ecology, University of Kaiserslautern, 67663 Kaiserslautern, Germany. [4]University of Chinese Academy of Sciences, Beijing 100049, China.

References

1. Danovaro R. Extending the approaches of biodiversity and ecosystem functioning to the deep ocean. In: Solan M, Aspden RJ, Paterson DM, editors. Marine biodiversity and ecosystem functioning: frameworks, methodologies, and integration. Oxford: Oxford University Press; 2012. p. 115–26.

2. Creer S, Sinniger F. Cosmopolitanism of microbial eukaryotes in the global deep seas. Mol Ecol. 2012;21:1033–5.

3. Pitcher TJ, Morato T, Hart PJB, Clark MR, Haggan N, Santos RS. Seamounts: ecology, fisheries, and conservation. Oxford: Blackwell; 2007.

4. Clark MR, Rowden AA, Schlacher T, Williams A, Consalvey M, Stocks KI, Rogers AD, O'Hara TD, White M, Shank TM, Hall-Spencer JM. The ecology of seamounts: structure, function, and human impacts. Annu Rev Mar Sci. 2010;2:253–78.

5. Ramirez-Llodra E, Brandt A, Danovaro R, De Mol B, Escobar E, German CR, Levin LA, Martinez Arbizu P, Menot L, Buhl-Mortensen P, Narayanaswamy BE, Smith CR, Tittensor DP, Tyler PA, Vanreusel A, Vecchione M. Deep, diverse and definitely different: unique attributes of the world's largest ecosystem. Biogeosciences. 2010;7:2851–99.

6. Stefanoudis PV, Bett BJ, Gooday AJ. Abyssal hills: influence of topography on benthic foraminiferal assemblages. Prog Oceanogr. 2016;148:44–55.

7. Pawlowski J, Christen R, Lecroq B, Bachar D, Shahbazkia HR, Amaral-Zettler L, Guillou L. Eukaryotic richness in the abyss: insights from pyrotag sequencing. PLoS One. 2011;6:4.

8. Bik HM, Sung W, De Ley P, Baldwin JG, Sharma J, Rocha-Olivares A, Thomas WK. Metagenetic community analysis of microbial eukaryotes illuminates biogeographic patterns in deep-sea and shallow water sediments. Mol Ecol. 2012;21:1048–59.

9. Edgcomb VP, Kysela DT, Teske A, Gomez AD, Sogin ML. Benthic eukaryotic diversity in the Guaymas Basin hydrothermal vent environment. Proc Natl Acad Sci U S A. 2002;99:7658–62.

10. López-García P, Philippe H, Gail F, Moreira D. Autochthonous eukaryotic diversity in hydrothermal sediment and experimental microcolonizers at the mid-Atlantic ridge. Proc Natl Acad Sci U S A. 2003;100:697–702.

11. López-García P, Vereshchaka A, Moreira D. Eukaryotic diversity associated with carbonates and fluid-seawater interface in lost City hydrothermal field. Environ Microbiol. 2007;9:546–54.

12. Sauvadet AL, Gobet A, Guillou L. Comparative analysis between protist communities from the deep-sea pelagic ecosystem and specific deep hydrothermal habitats. Environ Microbiol. 2010;12:2946–64.

13. Urich T, Lanzen A, Stokke R, Pedersen RB, Bayer C, Thorseth IH, Schleper C, Steen IH, Ovreas L. Microbial community structure and functioning in marine sediments associated with diffuse hydrothermal venting assessed by integrated meta-omics. Environ Microbiol. 2014;16:2699–710.

14. Hausmann K, Hulsmann N, Polianski I, Schade S, Weitere M. Composition of benthic protozoan communities along a depth transect in the eastern Mediterranean Sea. Deep-Sea Res I. 2002;49:1959 70.

15. Pachiadaki MG, Taylor C, Oikonomou A, Yakimov MM, Stoeck T, Edgcomb V. In situ grazing experiments apply new technology to gain insights into deep-sea microbial food webs. Deep Sea Res II. 2016;129:223–31.

16. Gong J, Shi F, Ma B, Dong J, Pachiadaki M, Zhang X, Edgcomb VP. Depth shapes alpha- and beta-diversities of microbial eukaryotes in surficial sediments of coastal ecosystems. Environ Microbiol. 2015;17:3722–37.

17. Fenchel T. The ecology of marine microbenthos. IV structure and function of the benthic ecosystem, its chemical and physical factors and the microfauna communities with special reference to the ciliated protozoa. Ophelia. 1969;6:1–182.

18. Berninger UG, Epstein SS. Vertical distribution of benthic ciliates in response to the oxygen concentration in an intertidal North Sea sediment. Aquat Microb Ecol. 1995;9:229–36.

19. Carey PC. Marine interstitial ciliates: an illustrated key. New York: Chapman and Hall; 1992.

20. Hamels I, Muylaert K, Sabbe K, Vyverman W. Contrasting dynamics of ciliate communities in sandy and silty sediments of an estuarine intertidal flat. Eur J Protistol. 2005;41:241–50.

21. Zhao F, Xu K. Distribution of ciliates in intertidal sediments across geographic distances: a molecular view. Protist. 2017;168:172–81.

22. Zhao F, Xu K. Molecular diversity and distribution pattern of ciliates in sediments from deep-sea hydrothermal vents in the Okinawa trough and adjacent sea areas. Deep-Sea Res I. 2016;116:22–32.

23. Forster D, Dunthorn M, Mahe F, Dolan JR, Audic S, Bass D, Bittner L, Boutte C, Christen R, Claverie JM, Decelle J, Edvardsen B, Egge E, Eikrem W, Gobet A, Kooistra W, Logares R, Massana R, Montresor M, Not F, Ogata H, Pawlowski J, Pernice MC, Romac S, Shalchian-Tabrizi K, Simon N, Richards TA, Santini S, Sarno D, Siano R, Vaulot D, Wincker P, Zingone A, de Vargas C, Stoeck T. Benthic protists: the under-charted majority. FEMS Microbiol Ecol. 2016;92:8.

24. Meng Z, Xu K, Dai R, Lei Y. Ciliate community structure, diversity and trophic role in offshore sediments from the Yellow Sea. Eur J Protistol. 2012;48:73–84.

25. Scheckenbach F, Hausmann K, Wylezich C, Weitere M, Arndt H. Large-scale patterns in biodiversity of microbial eukaryotes from the abyssal sea floor. Proc Natl Acad Sci U S A. 2010;107:115–20.

26. Lejzerowicz F, Esling P, Pawlowski J. Patchiness of deep-sea benthic foraminifera across the Southern Ocean: insights from high-throughput DNA sequencing. Deep-Sea Res II. 2014;108:17–26.

27. Coyne KJ, Countway PD, Pilditch CA, Lee CK, Caron DA, Cary SC. Diversity and distributional patterns of ciliates in Guaymas Basin hydrothermal vent sediments. J Eukaryot Microbiol. 2013;60:433–47.

28. Takishita K, Kakizoe N, Yoshida T, Maruyama T. Molecular evidence that phylogenetically diverged ciliates are active in microbial mats of deep-sea cold-seep sediment. J Eukaryot Microbiol. 2010;57:76–86.

29. Du Y, Xu K, Warren A, Lei Y, Dai R. Benthic ciliate and meiofaunal communities in two contrasting habitats of an intertidal estuarine wetland. J Sea Res. 2012;70:50–63.

30. Fu R, Gong J. Single cell analysis linking ribosomal (r)DNA and rRNA copy numbers to cell size and growth rate provides insights into molecular protistan ecology. J Eukaryot Microbiol. 2017; 10.1111/jeu.12425.

31. Esteban G, Finlay BJ, Embley TM. New species double the diversity of anaerobic ciliates in a Spanish lake. FEMS Microbiol Lett. 1993;109:93–100.

32. Wang Y, Song W, Warren A, Al-Rasheid KAS, Al-Quraishy SA, Al-Farraj SA, Hu X, Pan H. Descriptions of two new marine scuticociliates, Pleuronema sinica n. Sp and P. wilberti n. Sp (Ciliophora: Scuticociliatida), from the Yellow Sea, China. Eur J Protistol. 2009;45:29–37.

33. Caron DA, Countway PD. Hypotheses on the role of the protistan rare biosphere in a changing world. Aquat Microb Ecol. 2009;57:227–38.

34. Pedrós-Alió C. The rare bacterial biosphere. Annu Rev Mar Sci. 2012;4:449–66.

35. Kallmeyer J, Pockalny R, Adhikari RR, Smith DC, D'Hondt S. Global distribution of microbial abundance and biomass in subseafloor sediment. Proc Natl Acad Sci U S A. 2012;109:16213–6.

36. Parkes RJ, Cragg BA, Wellsbury P. Recent studies on bacterial populations and processes in subseafloor sediments: a review. Hydrogeol J. 2000;8:11–28.

37. Coolen MJL, Saenz JP, Giosan L, Trowbridge NY, Dimitrov P, Dimitrov D, Eglinton TI. DNA and lipid molecular stratigraphic records of haptophyte succession in the Black Sea during the Holocene. Earth Planet Sci Lett. 2009; 284:610–21.

38. Stoeck T, Zuendorf A, Breiner HW, Behnke A. A molecular approach to identify active microbes in environmental eukaryote clone libraries. Microb Ecol. 2007;53:328–39.

39. Jones SE, Lennon JT. Dormancy contributes to the maintenance of microbial diversity. Proc Natl Acad Sci U S A. 2010;107:5881–6.

40. Stoeck T, Bass D, Nebel M, Christen R, Jones MDM, Breiner HW, Richards TA. Multiple marker parallel tag environmental DNA sequencing reveals a highly complex eukaryotic community in marine anoxic water. Mol Ecol. 2010;19:21–31.

41. Gong J, Dong J, Liu XH, Massana R. Extremely high copy numbers and polymorphisms of the rDNA operon estimated from single cell analysis of Oligotrich and Peritrich ciliates. Protist. 2013;164:369–79.

42. Weber AAT, Pawlowski J. Wide occurrence of SSU rDNA intragenomic polymorphism in foraminifera and its implications for molecular species identification. Protist. 2014;165:645–61.

43. Balzano S, Abs E, Leterme SC. Protist diversity along a salinity gradient in a coastal lagoon. Aquat Microb Ecol. 2015;74:263–77.

44. Massana R, Gobet A, Audic S, Bass D, Bittner L, Boutte C, Chambouvet A, Christen R, Claverie JM, Decelle J, Dolan JR, Dunthorn M, Edvardsen B, Forn I, Forster D, Guillou L, Jaillon O, Kooistra W, Logares R, Mahe F, Not F, Ogata H, Pawlowski J, Pernice MC, Probert I, Romac S, Richards T, Santini S, Shalchian-Tabrizi K, Siano R, Simon N, Stoeck T, Vaulot D, Zingone A, de Vargas C. Marine protist diversity in European coastal waters and sediments as revealed by high-throughput sequencing. Environ Microbiol. 2015;17: 4035–49.

45. Nebel M, Pfabel C, Stock A, Dunthorn M, Stoeck T. Delimiting operational taxonomic units for assessing ciliate environmental diversity using small-subunit rRNA gene sequences. Environ Microbiol Reports. 2011;3:154–8.

46. Dunthorn M, Klier J, Bunge J, Stoeck T. Comparing the hyper-variable V4 and V9 regions of the small subunit rDNA for assessment of ciliate environmental diversity. J Eukaryot Microbiol. 2012;59:185–7.

47. Filker S, Gimmler A, Dunthorn M, Mahe F, Stoeck T. Deep sequencing uncovers protistan plankton diversity in the Portuguese Ria Formosa solar saltern ponds. Extremophiles. 2015;19:283–95.

48. Orsi W, Edgcomb V, Faria J, Foissner W, Fowle WH, Hohmann T, Suarez P, Taylor C, Taylor GT, Vd'acny P, Epstein SS. Class Cariacotrichea, a novel ciliate taxon from the anoxic Cariaco Basin. Venezuela Int J Syst Evol Microbiol. 2012;62:1425–33.

49. Gimmler A, Stoeck T. Mining environmental high-throughput sequence data sets to identify divergent amplicon clusters for phylogenetic reconstruction and morphotype visualization. Environ Microbiol Reports. 2015;7:679–86.

50. Stock A, Edgcomb V, Orsi W, Filker S, Breiner HW, Yakimov MM, Stoeck T. Evidence for isolated evolution of deep-sea ciliate communities through geological separation and environmental selection. BMC Microbiol. 2013;13:150.

51. Caporaso JG, Kuczynski J, Stombaugh J, Bittinger K, Bushman FD, Costello EK, Fierer N, Pena AG, Goodrich JK, Gordon JI, Huttley GA, Kelley ST, Knights D, Koenig JE, Ley RE, Lozupone CA, McDonald D, Muegge BD, Pirrung M, Reeder J, Sevinsky JR, Tumbaugh PJ, Walters WA, Widmann J, Yatsunenko T, Zaneveld J, Knight R. QIIME allows analysis of high-throughput community sequencing data. Nat Methods. 2010;7:335–6.

52. Edgar RC. UPARSE: highly accurate OTU sequences from microbial amplicon reads. Nat Methods. 2013;10:996–8.

53. Fuhrman JA. Microbial community structure and its functional implications. Nature. 2009;459:193–9.

54. Jost L. Partitioning diversity into independent alpha and beta components. Ecology. 2007;88:2427–39.

55. Hennig C. Cluster-wise assessment of cluster stability. Comput Stat Data An. 2007;52:258–71.

56. Hennig C. Dissolution point and isolation robustness: robustness criteria for general cluster analysis methods. J Multivar Anal. 2008;99:1154–76.

57. Oksanen J, Guillaume BF, Kindt R, Legendre P, O'Hara RB, Simpson GL, Solymos P, Stevens MHH, Wagner H. Vegan: community ecology package. R package version 1.17-3. 2010. http://CRAN.R-project.org/package=vegan

58. Galtier N, Gouy M, Gautier C. SEAVIEW and PHYLO_WIN: two graphic tools for sequence alignment and molecular phylogeny. Comput Appl Biosci. 1996;12:543.

59. Csárdi G, Nepusz T. The igraph software package for complex network research. Inter J Complex Syst. 2006;1695:1–9.

60. Bastian M, Heymann S, Jacomy M. Gephi: an open source software for exploring and manipulating networks. ICWSM. 2009;8:361–2.

2

Antimicrobial structure activity relationship of five anthraquinones of emodine type isolated from *Vismia laurentii*

Gislaine Aurelie Kemegne[1], Pierre Mkounga[2], Jean Justin Essia Ngang[1], Sylvain Leroy Sado Kamdem[1*] and Augustin Ephrem Nkengfack[2]

Abstract

Background: Antimicrobial activity of anthraquinone compounds of emodine type has been reported by many authors. These compounds are found in *Vismia laurentii* (Clusiaceae), a plant used in traditional pharmacopoeia for treatment of microbial infections among others affections. The continuous identification of new compounds has raised the problem of the relation between the structure and antimicrobial properties.

Results: The yeast growth kinetics parameters were not influenced by the pH variation as it was the case for the other tested bacteria. Fungicidal activities were noted for all molecules while only few of them had bactericidal activities, mostly on Gram positive bacteria. Mathematical model establishing a quantitative relationship between physicochemical properties of molecules and their fungicidal activities were obtained for *Candida albicans* and showed that physicochemical properties impacting on antifungal activity were polarizability, partition coefficient, molecular weight and hydrogen bond acceptor.

Conclusions: This work demonstrated that the presence of a long aliphatic chain methoxy group substituted in position two of the emodine structure increased the antibacterial properties of the studied compounds. Moreover this antimicrobial property depends on the pH of the environment, and specifically on the polarizability and number of hydrogen bond acceptors of the compound.

Keywords: Anthraquinones, Emodine, Antimicrobial activity, Physicochemical property, Structure-activity relationship

Background

Plants belonging to *Vismia* genus have been studied since 1979 [1] because of their biological activity due to secondary metabolites that they contain. Different parts of *Vismia laurentii* are used in the traditional pharmacopoeia in the treatment of different affections including microbial infections [2]. Previous chemical assessments carried out on this plant have resulted in the isolation and characterization of a great number of secondary metabolites, the most significant belonging to the xanthone, benzophenone and quinone classes. Quinones especially

anthraquinones, present numerous biological activities such as antiprotozoa [1, 3], antituberculous, fungicidal [4], antioxidant [5], cytotoxic and antitumor activities [6]. Xanthones and anthraquinones are known to bind irreversibly with nucleophilic amino acids in proteins, often leading to the inactivation of proteins and loss of function [7]. The rarity of plant diseases in *V. laurentii* is explained by the development of a natural defense system resulting in the synthesis of a multitude of antimicrobial molecules, which enable them to fight effectively against the pathogenic microbes [8, 9].

Anthraquinones are divided into two types: alizarin and emodine [10]. The alizarin type is used as natural dye in the textile industry [11], while the emodine type was formerly used as a like laxative compound [12–14]. Many studies have reported on antimicrobial activity

---
* Correspondence: sadosylvain@hotmail.com
[1]Department of Microbiology, Faculty of Science, University of Yaoundé I, P.O. Box 812, Yaoundé, Cameroon
Full list of author information is available at the end of the article

of anthraquinone compounds of the emodine type [2, 15, 16]. Their variety and the continuuos discovery of new emodine derivate molecules always call in question, the specific properties of the most antimicrobial effective compounds. Moreover, the number of molecules extracted from the biological and/or potentially existing systems is by far higher than the capacity of analysis of their biological properties.

Facing these limitations, a solution consists of building models which allows for correlating the activity to structure within a family of compounds, hereby increasing the effectiveness of high throughput screening [17]. On the basis of their physical and chemical properties, the antimicrobial activity of natural substances can be predicted in order to have information on biomechanism, gain time of bio-prospection of new molecules and to study their use in the sectors of the production of antiseptics, disinfectants and drugs [18, 19]. The structure activity relationship (SAR) or the quantitative structure activity relationship (QSAR) offers approaches which could be useful to predict these antimicrobial activities according to the physical and chemical properties of the molecules concerned. This approach could constitute a first stage of molecules screening and thus making it possible to reduce the number of compounds to be tested in the laboratory.

The purpose of this study was hence to establish a relationship between antimicrobial activities and physicochemical properties of some anthraquinone molecules of emodine type isolated from *Vismia laurentii*.

Methods
Plant material and purification
The roots and leaves of *Vismia laurentii* De Wild were collected in March 2004 in Mbalmayo, located in the Center Region of the Republic of Cameroon and identified by Mr Nana (plant taxonomist) of the *National Herbarium* of *Cameroon*, Yaounde. A voucher specimen (N° 1882/SRFK) documenting the collection was deposited.

The extraction and purification were carried out according to [1, 2]. Briefly, air dried powder of the roots of *Vismia laurentii* (2 kg) was extracted exhaustively at room temperature with methanol (8 L) for 48 h by maceration. The suspension was filtered and the filtrate was concentrated on reduced pressure to give 100 g of brown residue. This residue was subjected to flash chromatography on silica gel (Merck, 230–400 mesh), eluted with the gradient polarity of cyclohexane and ethyl acetate to give 5 fractions labelled : A (20 g; cyclohexane), B (35 g; cyclohexane/ethyl acetate 4:1), C (18 g; cyclohexane/ethyl acetate 1:1) and D (10 g ethyl acetate). Fraction B, which according to the works of [1, 2] could contain most of the emodine type compounds based on the solvent polarity used, was further subjected to

column chromatography on silica gel (Merck 70–230 mesh) and eluted with cyclohexane/ethyl acetate mixture of increasing polarity. One hundred fractions of 100 mL each were collected and analysed by TLC using the mixture of cyclohexane/dichloromethane (7:3) as mobile phase. Fractions 1–25, eluted with cyclohexane afforded three compounds which were identified as: 3-geranyloxyemodine (300 mg); compound A, friedelin (25 mg) and stigmasterol (35 mg). Fractions 27–47, eluted with the mixture of cyclohexane/ethyl acetate (9:1) gave 1.3 g of brown residue which was subjected to further column chromatography to yield laurentixanthone (25 mg), 3-methoxyemodine (25 mg); compound C and compound E bivismiaquinone (40 mg). Fraction A eluted with cyclohexane/ethyl acetate (4:1), gave after repetitive column chromatography, kampherol (16 mg), laurentixanthone A (50 mg), 1,7- dihydroxy xanthone (18 mg), vismiaquinone B (50 mg); compound B,, 2-isoprenyl-3-methoxyemodine (22 mg); compound D. The chemical structure of each isolated compound was established on the basis of their NMR spectra (one and two dimensions) [2, 20] and data recorded on BRUKER DRX-400 instrument.

Physicochemical properties determination
The Compound polarity (Rf) was assessed by Thin Layer Chromatography method [21, 22]. The number of hydrogen bond acceptors (HA) and donors (HD) were assessed by calculations with available equations [1, 23]. Partition coefficient ($LogK_{O/W}$), water solubility (S_W), superficial tension ($S\_tens°$) and polarizability (Polarz) properties were obtained by using the following predicting softwares: SMILES Translator and Structure File Generator, ACDLABS and EPIWEB version 4.1.

Data set
Six (06) microorganisms consisting of three Gram positive (*Bacillus cereus* ATCC 11966, *Listeria monocytogenes* 56 Lγ and *Staphylococcus aureus* NCTC 10652), two Gram negative (*Escherichia coli* 555, *Salmonella enteritidis* 155A) and one yeast of the species *Candida albicans* were tested for their sensitivity to 5 emodine derived compounds: 3-geranyloxyemodine, vismiaquinone B, 3-methoxyemodine, 2-isoprenyl-3-methoxyemodine and bivismiaquinone (Fig. 1). Microorganisms were obtained from copies stored at −80 °C and subcultured twice in Brain Heart Infusion broth at 37 °C for bacteria and 25 °C for yeast.

Growth kinetic of microorganisms and antimicrobial activity of selected compounds
The microbial counting was performed by dilution and seeding method on Mueller Hinton agar medium (Oxoid, Basingstoke, UK) for bacteria [24] and microscope direct

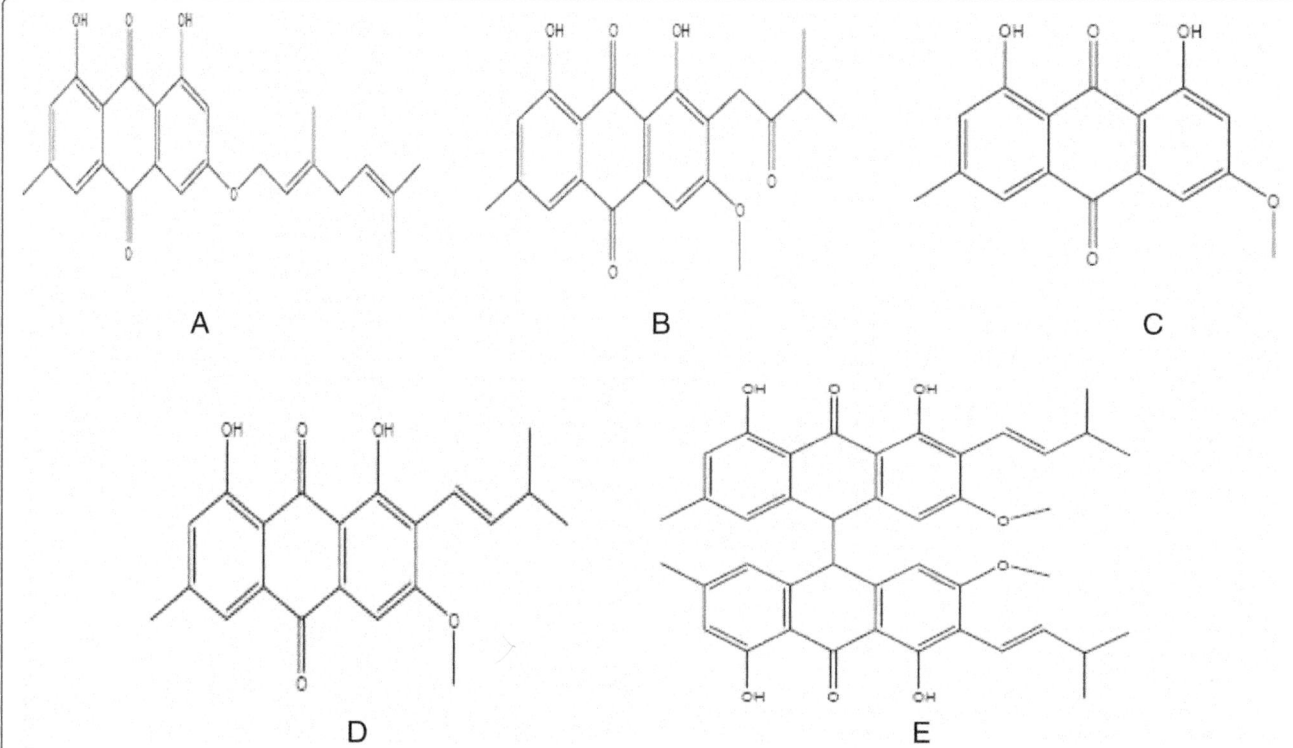

Fig. 1 Chemical structures of chemical compounds used in this work. **a** 3-geranyloxyemodine, **b** Vismiaquinone B, **c** 3-methoxyemodine, **d** 2-isoprenyl-3-methoxyemodine, **e** Bivismiaquinone

counting for yeast [25] using Mueller Hinton broth medium (Oxoid, Basingstoke, UK). Antimicrobial activity was performed by macrodilution method in liquid medium for the MBC/MFC (Minimal Bactericidal Concentration/Minimal Fungicidal Concentration) according to [26].

Statistical analysis

The Quantitative structure activity relationship was established by regression analysis using Statistica.7 of Statsoft.

Results

Microorganisms growth kinetics

In order to assess the impact of pH on the antimicrobial properties of the tested compounds, growth kinetics of the microorganisms were first obtained in those conditions and are presented for each microorganisms at pH5 and pH7 in Fig. 2. This kinetics showed that pH variation of the medium does not affect the lag and growth rate of *Candida albicans* while for the other strains, these parameters are affected. In general, it can be observed that the lag was increased and the growth rate reduced when pH was 5 compared to pH7, independently on the bacteria strain. Moreover, *Listeria monocytogenes* a Gram positive strain grew to higher final cell load notwithstanding their slow growth rates.

Minimal bactericidal (MBC) and fungicidal (MFC) concentration

Sensitivity test reveals that reference molecules (gentamicin for bacteria and nystatin for yeast) are more active than the tested molecules which were more active on *Candida albicans*. Gram negative bacteria were less sensitive than Gram positive bacteria (Table 1).

In general, the sensitivity of the strains to all the compounds tested decreased with increase in pH. As exception to this rule, compounds C and D were more active at pH7 than at pH5 and 6. While almost all the compounds were active on *Candida albicans*, this was not the case for bacteria strains.

Compounds physicochemical properties

Bivismiaquinone (E) had the highest partition coefficient while 3-methoxyemodine (C) had the lowest. These results were confirmed by the water solubility property of the compounds. In fact, compound C had the highest water solubility coefficient. Regarding the superficial tension, which is the tendency of a compound to contract due to internal forces and resist external forces, it was noticed that 3-geranyloxyemodine (A) had the lowest superficial tension (52) while 3-methoxyemodine had the highest (63.7). Compound C also had the lowest polarizability while compound E had the highest. Regarding the electron bond donors (HD) and electron

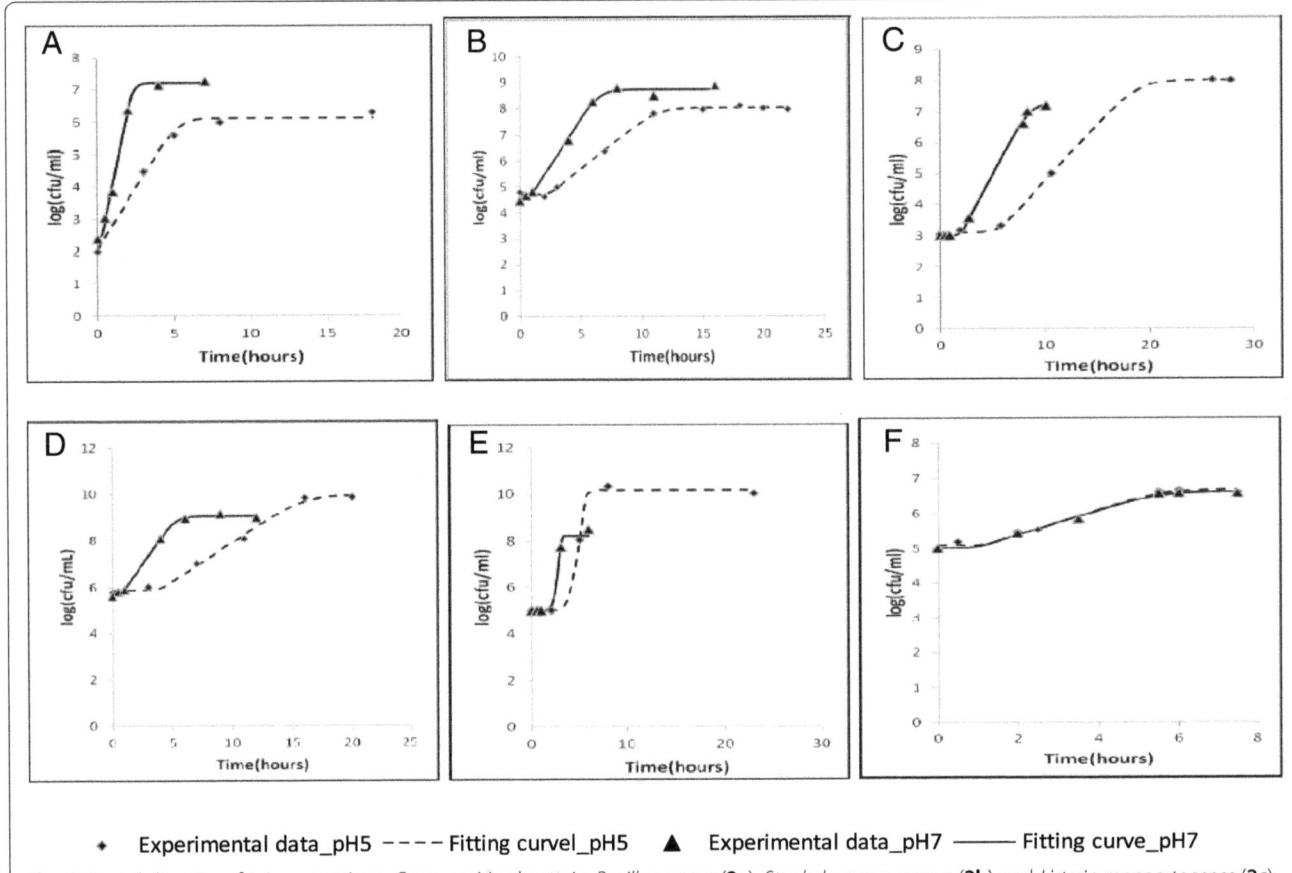

◆ Experimental data_pH5 – – – – Fitting curvel_pH5   ▲ Experimental data_pH7 ———— Fitting curve_pH7

Fig. 2 Growth kinetics of microorganisms: Gram positive bacteria: *Bacillus cereus* (**2a**), *Staphylococcus aureus* (**2b**) and *Listeria monocytogenes* (**2c**); Gram negative bacteria: *Escherichia coli* (**2d**) and *Salmonella enteritidis* (**2e**); *Candida albicans* (**2f**) in BHI medium at pH5 and pH7

bond acceptors (HA), bivismiaquinones (E) proved to be quite different from the other compounds by demonstrating an HD of 4 and HA of 8. This was also the case for the insaturation number (IN) where compound E was far more unsaturated than the other compounds. The polarity degree also confirmed the Log $K_{o/w}$ and Sw results, indicating that compound C was the most polar (Table 2).

Structure-activity relationship

Quantitative relationship between structure and activity of molecules (QSAR) was possible only for data obtained with *Candida albicans*. In other to perform this, the MIC (Minimum Inhibitory Concentration) at pH7 for compound C was assessed by increasing the maximum concentration limit tested and found to be 3000 ppm. First, all the data for *Candida albicans* were merged together irrespective of the pH and a regression analysis was performed in order to obtain a quadratic polynomial equation. Unfortunately, no significant result was obtained. After hypothesizing that the data set could not be sufficient to explain the effect of pH on the CMF, we decided to split the regression analysis of each pH. Equations 1, 2 and 3 report the polynomial equation

indicating the relationship between statistically significant physicochemical parameters and the antifungal potential (Log MFC) of the anthraquinones studied. QSAR Models obtained for the activity of tested molecules on *Candida albicans* can thus be written in the following equations where the increase of Log MFC indicates a reduction of the antifungal activity of the compound.

$$\text{Log (MFC) at pH7} = 1.671 + 0.488 * \text{LogK}_{o/w} + 0.774 * \text{HA} - 0.147 * \text{Polarz} \quad (1)$$

$$\text{Log (MFC) at pH6} = 2.410 - 0.077 * \text{MW} + 1.321 * \text{HA} + 0.537 * \text{Polarz} \quad (2)$$

$$\text{Log (MFC) at pH5} = 2.197 - 0.092 * \text{MW} + 1.620 * \text{HA} + 0.642 * \text{Polarz} \quad (3)$$

At the different pH, the common independent variables that significantly affected the Log (MFC) were the polarizability (tendency of a compound to negatively charge itself and hence be easily modified in the presence of positively

Table 1 Minimal bactericidal (MBC) and fungicidal (MFC) concentration (ppm) of compounds tested against selected microorganisms

		Compounds						
		A	B	C	D	E	N	G
Yeast								
Candida albicans	pH 5	150	600	600	1200	550	25	
	pH 6	300	600	1200	1200	1100	50	
	pH 7	1200	1200	/	1200	1100	25	
Gram-positive bacteria								
Listeria monocytogenes	pH 5	600	/	/	/	/		25
	pH 6	600	/	/	/	/		6.25
	pH 7	1200	1200	/	/	/		25
Staphylococcus aureus	pH 5	600	1200	1200	/	1100		75
	pH 6	600	/	1200	/	/		< 3.12
	pH 7	1200	/	75	150	/		< 3.12
Bacillus cereus	pH 5	300	/	/	/	/		< 3.12
	pH 6	600	/	/	/	/		< 3.12
	pH 7	600	/	/	300	/		< 3.12
Gram-negative bacteria								
Salmonella enteritidis	pH 5	/	/	1200	/	/		600
	pH 6	/	/	/	/	/		600
	pH 7	/	/	/	/	/		< 3.12
Escherichia coli	pH 5	/	/	1200	/	/		12.50
	pH 6	/	/	/	/	/		6.25
	pH 7	/	/	/	/	/		25

A = 3-geranyloxyemodine, *B* = vismiaquinone B, *C* = 3-methoxyemodine, *D* = 2-isoprenyl-3-methoxyemodine, *E* = bivismiaquinone, *G* = gentamicin (antibacterial), *N* = nystatin (antifungicidal), <= inferior,/= not active at concentration ≤ 1200 ppm

Table 2 Physicochemical properties of compounds tested in this work

QSAR independent variables[a]	Tested compounds				
	A	B	C	D	E
Log $K_{O/W}$	8.67	4.950	4.570	6.790	13.450
S_W (mg/L)	$3.25^a 10^{-5}$	0.083	0.563	0.003	$4.93^a 10^{-11}$
S_tens°	52	58.600	63.700	57.800	59.300
Polarz	45.760	38.600	29.310	39.260	77.650
MW (g/mol)	406.17	362.00	284.06	352.00	674.28
HD	2	2	2	2	4
HA	5	6	5	5	8
R_f	0.80	0.76	0.86	0.80	0.60
IN	13	12	11	12	23

Log $K_{O/W}$ partition coefficient, *S_W* water solubility, *S_tens* superficial tension, *Polarz* polarizability, *MW* molecular weight, *HD* hydrogen bond donor, *HA* hydrogen bond acceptor, *R_f* polarity degree, *IN* insaturation number
[a]The Compound polarity (Rf) was assessed by Thin Layer Chromatography method. The number of hydrogen bond acceptors (HA) and donors (HD) were assessed by calculations with literature available equations. Partition coefficient (LogK$_{O/W}$), water solubility (S$_W$), superficial tension (S_tens°) and polarizability (Polarz) properties were obtained by using the following predicting softwares: SMILES Translator and Structure File Generator, ACDLABS and EPIWEB version 4.1

charged compounds) and the number of hydrogen acceptors (HA). As the pH of the inactivation medium decreased, the effect of those two variables on the fungicidal properties increased as it can be observed from the coefficients. At pH7, the octanol/water partition (Log $K_{o/w}$) which describe the affinity of a compound to lipid or water phases, had a significative impact on the fungicidal property while at pH6 and 5, it was the molecular weight (Table 3).

The quantitative structure activity relationship (QSAR) on bacteria could not be performed because of the low activity of the compounds. However, it was possible to propose a structure activity relationship (SAR). In fact, by observing the results, it can be noticed that 3-geranyloxyemodine (A) was bactericidal to the three Gram positive bacteria strains with activity increasing with decreasing pH while 3-methoxyemodine (C) was active only on *Staphylococcus aureus* with activity decreasing with pH. On the other hand 2-isoprenyl-3-methoxyemodine (D) was active only at pH7 and only on *Staphylococcus aureus* and *Bacillus cereus*. In terms of decreasing antimicrobial activity, the tested compounds can be classified in the following way: 3-geranyloxyemodine (A) > 3-

Table 3 Statistically significant parameters obtained by multiple regression analysis, and by correlations between experimental and calculated values

Parameters	pH 7				pH 6				pH 5			
	coef ± errSt	IC$_f$ 95%	IC$_s$95%	P	coef ± errSt	IC$_f$ 95%	IC$_s$95%	P	coef ± errSt	IC$_f$ 95%	IC$_s$95%	P
Const	1.671 ± 0.072	1.531	1.813	0	2.410 ± 0.156	2.104	2.716	0	2.197 ± 0.206	1.792	2.601	0
Log K$_{o/w}$	0.488 ± 0.020	0.448	0.528	0	NS	-	-	-	NS	-	-	-
MW (mol/g)	NS	-	-	-	−0.077 ± 0.009	−0.094	−0.059	0	0.092 ± 0.012	−0.116	−0.069	0
HA	0.774 ± 0.031	0.714	0.835	0	1.321 ± 0.142	1.043	1.598	0	1.620 ± 0.187	1.254	1.987	0
Polarz	−0.147 ± 0.006	−0.158	−0.135	0	0.537 ± 0.065	0.409	0.665	0	0.642 ± 0.086	0.473	0.810	0
R^2 Model	0.995				0.943				0.936			
SSE Model	0.001				0.015				0.026			

coef ± errSt parameters coefficient, *P* probability of kindness of the equation, *Const* constancy, *MW* molecular weight, *Log K$_{o/w}$(or Log P)* partition coefficient, *Polarz* polarizability, *HA* hydrogen bond acceptor, *NS* non significative, *IC$_f$ 95% and IC$_s$ 95%* inferior (IC$_f$) and superior (IC$_s$) confidence interval, correlation values: R^2 model and SSE model

methoxyemodine (C) > 2-isoprenyl-3-methoxyemodine (D) > vismiaquinone (B) > bisvismiaquinone (E). It can be noticed that compound E is an association of two molecules of compound D by the presence of a ketone group on the isoprenyl substitution in position 2. This substitution in position 2 is the difference between compound C and compounds B and D. Finally, compound A differs from compound C by the length of the aliphatic chain of the methoxy substitution. It can hence be assumed that steric effect, weight and the presence of substitutions in position 2 of emodine derivatives is detrimental to their bactericidal activity while increase in the aliphatic chain length of the methoxy substitution in position 6 is beneficial to the antibacterial activity of these emodine derived anthraquinones.

Discussion

The differences observed during the growth kinetics at pH5 and 7 of the tested strains can be associated to the different nature of their cell walls. In fact, the effect of pH that is mostly described on the cell internal pH [27] is most important on bacteria than on fungi. Moreover, homeostasis regulations that sometimes involve ATP dependent processes may also include cell membrane lipid modifications to reduce fluidity [28]. Fungi resistance to pH also depends on their high cell wall thickness and composition mainly made of 80–90% glucomannoproteines, glucanes and chitins; this last compound being higher in *Candida albicans* with respect to other yeast species [29].

The emodine derived compounds tested in this work were highly colored and hence permitted only evaluation of the MBC and MFC. The microbiocidal concentrations observed for the different compounds can be explained by the compound interference with the cell wall, the membrane, nucleic acid and enzymes [30, 31]. The presence of an external membrane on the Gram negative bacteria can explain the difference of sensibility observed between the

two groups of bacteria. Kosanić and Ranković [32] suggested that the cell wall structure and composition of bacteria and fungi could account for the different sensitivity to antimicrobial compounds. On the other hand, Gram positive bacteria and *Candida albicans* cells have their cell walls exposed, and compounds that can interact with these cell walls should have a long aliphatic chain to help disorder the cell wall. This is the case of compound A compared to compounds C and D. Sikkema et al. [33] observed that saturated alcanes had very low antimicrobial activity, while [34] have demonstrated that unsaturated aldehydes had more antimicrobial activity than saturated ones. This explains the difference between compounds D and B. Moreover, another possible mechanism was proposed by [35] who demonstrated that the antimicrobial activity of emodine on *Helicobacter pylori* was also due to the interference with saturated and unsaturated fatty acid elongation by inhibiting the β-hydroxyacyl-Acp dehydratase (HpFabZ).

Regarding the physicochemical properties and their relation with the antimicrobial activity of the compounds tested, different methods as described before were used to assess them. The molecular weight (MW) which is an indication of a compound's steric effect is also one of the key parameters of compounds with pharmacological properties (Lipinski et al., [23]). This author also proposes the ideal molecular weight to be lower than 500 g. The numbers of electron donors (HD) and acceptors (HA) indicate the capacity of a compound to form hydrogen bonds with the cell membrane compounds. Moreover, the partition coefficient (Log K$_{o/w}$) gives information on the lipophilicity of the compound and hence its capacity to distribute itself in the membrane lipidic phase and the aqueous environmental or cytoplasmic phase. Regarding the degree of polarization, it is the tendency of a compound to negatively charge itself and hence be easily modified in the presence of positively charged compounds. The surface tension on the other hand represents the internal attractive forces of a compound that limits bond creation with other compounds. The QSAR obtained for *Candida albicans* indicated that HA and Polarz were the common significant variables of the equations for each pH. Both physicochemical parameters are associated to the compound capacity to bind to other molecules by forming bonds with positively charged atoms. In fact the growing importance of HA and Polarz in the equations as pH decreases (protons concentration increase) in the environment confirm this. At pH7, the concentration of protons is lower and hence the increase of the Polarz negatively affects the fungicidal property. The fact that the antifungal property of the compounds depended on the pH, and that the pH did not affect the growth of *Candida albicans* may suggest that the effect of pH on the compounds antimicrobial effect is associated with compound modification.

The logical structure activity relationship deduced for the bacteria strains tested indicates that the substitution in position 2 of the emodine compound is detrimental for the antibacterial activity of these compounds while the insaturation of the substitute isimportant for this activity. Moreover the increase of the aliphatic chain length of the methoxy substitute in position 3 increases the lipophilicity of the compound. The antimicrobial property which is increased by the lipophilicity of the compound is reduced as the compound molecular weight increases. In fact the lipophilicity associated to the Log (Ko/w) denotes the capacity to integrate the cell wall and membrane capacity which may be slowed by the steric effect of the compound, revealed here by the molecular weight.

Conclusion

In conclusion, the present work has demonstrated the antifungal and antibacterial properties of some anthraquinones of emodine type isolated from *Vismia laurentii*. This antimicrobial property is increased by the presence of a long aliphatic chain methoxy group substituted in position 2 of the emodine structure. The mathematical equations produced demonstrate that QSAR can contributeto understanding the diversity of compound antimicrobial activity present in plant extracts.

Abbreviations

coef ± errSt: Parameters Coefficient; Const: Constancy; G: Gentamicin; HA: Hydrogen bond acceptor; HD: Hydrogen bond donor; IC$_f$ 95%: inferior Confidence Interval; IC$_s$ 95%: superior Confidence Interval; IN: Insaturation Number; Log K$_{O/W}$(or Log P): Partition coefficient; MBC: Minimum Bactericidal Concentration; MFC: Minimum Fungicidal Concentration; MIC: Minimum Inhibitory Concentration; MW: Molecular weight; MW: Molecular weight; N: Nystatin; NS: Non significant; P: Probability of kindness of the equation; Polarz: Polarizability; QSAR: Quantitative Structure Activity Relationship; R$_f$: Polarity degree; S_tens°: Superficial tension; SAR: Structure Activity Relationship; S$_W$: Water solubility

Acknowledgements

The Authors acknowledge the contribution of the University of Yaoundé I for supporting this research with the facilities needed for the good functioning of the different laboratories involved in this research.
The Authors also acknowledge the funding for the modernization of research in the Ministry of Higher Education that was provided to some of the authors to booster their research activities.

Funding

No direct funding was received for this work.

Authors' contributions

GAK: contributed in designing the experimental plan. Did the bench work on antimicrobial activity assessment and contributed in statistical assessment of the QSAR as well as writing of the paper. PM: did the extraction, purification and identification of the molecules, and contributed in the writing of the paper. JJEN: contributed in designing the experimental plan, and contributed in the writing of the paper. SLSK: contributed in designing the experimental plan, performed predictions of the compounds physicochemical properties and the statistical assessment of the QSAR and coordinated the writing of the paper. AEN: supervised the extraction, purification of the molecules and coordinated the identification of the molecules, and contributed in the writing of the paper. All authors read and approved the final manuscript.

Competing interests

The authors declare that they have no competing interests.

Author details

[1]Department of Microbiology, Faculty of Science, University of Yaoundé I, P.O. Box 812, Yaoundé, Cameroon. [2]Department of Organic Chemistry, Faculty of Science, University of Yaoundé I, P.O. Box 812, Yaoundé, Cameroon.

References

1. Nguemeving JR. Etudes de métabolites secondaires de deux plantes médicinales Camerounaises du genre Vismia : Vismia laurentii et Vismia guineensis (clusiaceae) : structures chimiques-transformations chimiques-activités antimicrobiennes. Thèse de Doctorat de l'Université de Yaoundé I, Cameroun. 2008. p. (49–78), (140–146)/379.
2. Kuete V, Nguemeving JR, Penlap BV, Azebaze AGB, Etoa F-X, Meyer M, Bodo B, Nkengfack AE. Antimicrobial activity of the methanolic extracts and compound from *Vismia laurentii* De Wild (Guttiferae). J Ethnopharmacol. 2007;109:372–9.
3. Mbwanbo ZH, Apers S, Moshi MJ, Kapingu MC, Van Miert S, Claeys M, Brun R, Cos P, Pieters L, Vlietinck A. Anthraquinoids compounds with antiprotozoal activity from *Vismia orientalis*. Planta Med. 2004;70:706–10.
4. Kanokmedhakul K, Kanokmedhakul S, Phatchana R. Biological activity of anthraquinones and triterpenes from *Prismatomeris fragrans*. J Ethnopharmacol. 2005;100:284–8.
5. Jasril LNH, Mooi LY, Abdullah MA, Sukari MA, Ali AM. Antitumor promoting and antioxidant activities of anthraquinones isolated from the cell suspension culture of *Morinda elliptica*. Asia Pac J Mol Biol. 2003;11:3–7.
6. Cassinelli G, Geroni C, Botta B, Monache GD, Monache FD. Cytotoxic and antitumor activity of vismiones isolated from vismieae. J Nat Prod. 1986;49:929–31.

7.   Stern JL, Hagerman AE, Steinberg PD, Mason PK. Phorotannin-protein interactions. J Chem Ecol. 1996;22:1887–99.

8.   Jones JDG, Dangl JL. The plant immune system. Nature. 2006;444:323–9.

9.   Gibbons S. Phytochemicals for bacterial resistance - strengths, weaknesses and opportunities. Planta Med. 2008;74:594–602.

10. Mohanlall V, Steenkamp P, Odhav B. Isolation and characterization of anthraquinone derivatives from *Ceratotheca triloba* (Bernh.) Hook f. J Med Plant Res. 2011;5:3132–41.

11. Derksen GCH, Naayer M, van Beek TA, Capelle A, Haaksman IK, van Doren HA, de Groot AE. Chemical and enzymatic hydrolysis of anthraquinone glycosides from Madder roots. Phytochem Anal. 2003;14:137–44.

12. Izhaki I. Emodin–asecondary metabolite with multiple ecological functions in higher plants. New Phytol. 2002;155:205–17.

13. Akihiro M, Mayumi H, Yosuke S. Emodin has a cytotoxic activity against human multiple Myeloma as a Janus-Activated Kinase 2 Inhibitor. Mol Cancer Ther. 2007;6:987–94.

14. Shu-Chun H, Jing-Gung C. Anticancer potential of emodin. Biomedicine. 2012;2:108–16.

15. Tala MF, Krohn K, Hussain H, Kouam SF, Wabo HK, Tane P, Schulz B, Qunxiu H. Laurentixanthone C: A new antifungal and algicidal xanthone from stem bark of *Vismia laurentii*. Z Naturforsch. 2007;62b:565–8.

16. Mbaveng AT, Kuete V, Nguemeving JR, Penlap BV, Nkengfack AE, Meyer JJM, Lall N, Krohn K. Antimicrobial activity of the extracts and compounds obtained from *Vismia guineensis* (Guttiferae). Asian J Trad Med. 2008;3(6):211–23.

17. Arkadiusz ZD, Tomasz A, Jorge G. Computational methods in developing Quantitative Structure-Activity Relationships (QSAR): a review. Comb Chem High Throughput Screen. 2006;9:213–28. 213.

18. Lessigiarska I. Development of structure-activity relationships forpharmacological endpoints relevant to European union legislation, Thèse de Doctorat. England: Liverpool John Moores University; 2006. p. 292.

19. van de Waterbeemd H, Gifford E. *ADMET* in silico modelling: towards prediction paradise? Nat Rev Drug Discov. 2003;2(3):192–204.

20. Do Carmo M, Miraglia M, Mesquita AAL, de Jesus M, Varejão C, Gottlieb OR, Gottlieb HE. Anthraquinones from *Vismia species*. Phytochemistry. 1981;20(8):2041–2.

21. Tarride I, Desarnaud JC. Réaliser une chromatographie sur couche mince. AIX-Marseille; 2015. p. 1

22. Magna J-Y. La Chromatographie sur couche mince (CCM). GPL Ghostscript 8.70; 2011. p. 1

23. Lipinski CA, Lombardo F, Dominy BW, Feeney PJ. Experimental and computational approaches to estimate solubility and permeability in drug discovery and development settings. Adv Drug Deliv Rev. 1997;23:3–25.

24. Larcher C. Techniques de Dénombrement; 2012. p. 7.

25. Harlé J. Utiliser une cellule de Malassez. Fiche technique. Méthodes & Techniques (Sciences de la Vie et de la Terre); 2009. p. 1.

26. Perumal S, Suthagar P, Lee WC, Roziahanim M, Surash R. Determination of minimum inhibitory concentration of *Euphorbia hirta* (L.) Extracts by Tetrazolium Microplate Assay. J Nat Prod. 2012;5:68–76.

27. Hall HK, Karem KL, Foster JW. Molecular responses of microbes to environmental pH stress. Adv Microbiol Physiol. 1995;37:229–64.

28. Beales N. Adaptation of microorganisms to cold temperatures, weak acid preservatives, low pH, and osmotic stress: A Review. Comp Rev Food Sci & Food Saf. 2004;3:1–20.

29. Kuwahara T, Kaneda S, Shimono K, Inoue Y. Growth of microorganisms in total parenteral nutrition solutions without lipid. Int J Med Sci. 2010;7:43–7.

30. Lambert HP, O'Grady FW. Antibiotic and chemotherapy. 6th ed. Edinburgh: Churchill-Livingstone; 1992.

31. Tenover FC. Mechanisms of antimicrobial resistance in bacteria. Am J Med. 2006;119(6A):S3–S10.

32. Kosanić M, Ranković B. Screening of antimicrobial activity of some lichen species in vitro. Kragujevac J Sci. 2010;32:65–72.

33. Sikkema J, de Bont JAM, Poolman B. Mechanisms of membrane toxicity of hydrocarbons. Microbiol Rev. 1995;59:201–22.

34. Giuseppe B, Lagana MG, Trombetta D, Arena S, Nostro A, Uccella N, Mazzanti G, Saija A. In vitro antibacterial activity of some aliphatic aldehydes from *Olea europaea* L. FEMS Microbiol Lett. 2001;198:9–13.

35. Chen J, Liang Z, Yu Z, Haitao Z, Jiamu D, Jianping D, Yuewei G, Hualiang J, Xu S. Emodin Targets the β-Hydroxyacyl-acyl Carrier Protein Dehydratase from *Helicobacter pylori*: Enzymatic inhibition assay with crystal structural and thermodynamic characterization. BMC Microbiol. 2009;9:91.

3

Isolation, characterization and transcriptome analysis of a novel Antarctic *Aspergillus sydowii* strain MS-19 as a potential lignocellulosic enzyme source

Bailin Cong* ⓘ, Nengfei Wang, Shenghao Liu, Feng Liu, Xiaofei Yin and Jihong Shen

Abstract

Background: With the growing demand for fossil fuels and the severe energy crisis, lignocellulose is widely regarded as a promising cost-effective renewable resource for ethanol production, and the use of lignocellulose residues as raw material is remarkable. Polar organisms have important value in scientific research and development for their novelty, uniqueness and diversity.

Results: In this study, a fungus *Aspergillus sydowii* MS-19, with the potential for lignocellulose degradation was screened out and isolated from an Antarctic region. The growth profile of *Aspergillus sydowii* MS-19 was measured, revealing that *Aspergillus sydowii* MS-19 could utilize lignin as a sole carbon source. Its ability to synthesize low-temperature lignin peroxidase (Lip) and manganese peroxidase (Mnp) enzymes was verified, and the properties of these enzymes were also investigated. High-throughput sequencing was employed to identify and characterize the transcriptome of *Aspergillus sydowii* MS-19. Carbohydrate-Active Enzymes (CAZyme)-annotated genes in *Aspergillus sydowii* MS-19 were compared with those in the brown-rot fungus representative species, *Postia placenta* and *Penicillium decumbens*. There were 701CAZymes annotated in *Aspergillus sydowii* MS-19, including 17 cellulases and 19 feruloyl esterases related to lignocellulose-degradation. Remarkably, one sequence annotated as laccase was obtained, which can degrade lignin. Three peroxidase sequences sharing a similar structure with typical lignin peroxidase and manganese peroxidase were also found and annotated as haem-binding peroxidase, glutathione peroxidase and catalase-peroxidase.

Conclusions: In this study, the fungus *Aspergillus sydowii* MS-19 was isolated and shown to synthesize low-temperature lignin-degrading enzymes: lignin peroxidase (Lip) and manganese peroxidase (Mnp). These findings provide useful information to improve our understanding of low-temperature lignocellulosic enzyme production by polar microorganisms and to facilitate research and applications of the novel Antarctic *Aspergillus sydowii* strain MS-19 as a potential lignocellulosic enzyme source.

Keywords: Polar organisms, *Aspergillus sydowii*, Transcriptome, Lignocellulose degradation, Low temperature enzyme, Lignin peroxidase, Manganese peroxidase

* Correspondence: biolin@fio.org.cn
The First Institute of Oceanography, State Oceanic Administration, Qingdao 266061, People's Republic of China

Background

With the growing demand for fossil fuel and the severe energy crisis, lignocellulose is widely regarded as a promising, cost-effective renewable resource for bioethanol production, and the use of lignocellulose residues as a raw material has become remarkable [1–3]. However, there are numerous technological obstacles to the degradation of lignocellulose. Lignin is important organic matter that is widely present in the plant cell wall. Together with cellulose and hemicellulose, lignin forms the main component of the plant skeleton, representing the second most abundant organic regenerative resource after cellulose on earth. Since lignin and cellulose are cross-linked and lignin has complex physical and chemical properties, it represents the restrictive factor for the utilization of lignocelluloses. To effectively utilize cellulose and hemicellulose from lignocellulose raw materials, it is essential to release them from lignin bonds. Cellulose is composed of D-glucose with beta-1 and 4 glycosides, while lignin is a natural amorphous high-molecular-weight polymer. Because of the complex structure of lignin, there are no conclusive assessments of its structure to date, but a consensus has developed that the basic structure of lignin consists of phenyl propane units [4].

In their natural environments, only a small number of microorganisms are capable of degrading lignin. Lignin can be successfully implemented not only by pure cultures of particular microorganisms but also by the application of a variety of lignocellulolytic species and some non-lignocellulolytic microbes that work synergistically to break down the tough lignocellulosic structure [5–7]. The complete degradation of lignin results from the co-operation of fungi, bacteria and actinomycetes, among which fungi play the most important role. Fungi enter wood materials through hyphae while secreting extracellular enzymes that attack cellulose in the plant cell wall, resulting in the depolymerization and dissolution of lignin and cellulose. According to the type of decay caused in different lignocellulose components, the fungi can be divided into white-rot fungi, brown-rot fungi and soft-rot fungi [8]. The essence of lignin degradation consists of an oxidative process, with almost equal importance of phenol oxidase conduction. It is generally believed that lignin degradation mainly depends on four enzymes that are secreted by white-rot fungi [9]: Lac (laccase, EC 1.10.3.2) [10], LiP (lignin-peroxidase, EC 1.11.1.14) [11], Mnp (manganese peroxidase, EC 1.11.1.16) [12] and VP (versatile peroxidase, EC 1.11.1.16) [13]. Some lignin-degrading fungi do not secrete laccase, including *Phanerochete chrysosporium*, which indicates that laccase is not necessary for the degradation of lignin but could be involved in the process coordinating with other peroxidases [14]. Recently, several *Aspergillus* fungi have

been shown to produce such enzymes, and additional enzymes are indispensable for complete degradation [15, 16]. Cellulase and hemicellulose are also required during lignin degradation, most of which can be categorized into the glycoside hydrolase (GH) families and carbohydrate esterase (GE) families in the Carbohydrate-Active Enzymes database (CAZy). Apart from the four above-described peroxidases, other particular enzymes also participate in or have a certain impact on lignin degradation, including cellobiose dehydrogenase (CDH,EC 1.1.99.18), glyoxal oxidase (GLOX, EC1.2.3.5), aryl alcohol oxidase (AAO,EC 1.1.3.7), glucose 1-oxidase (EC 1.1.3.4),phenol oxidase, and catalase, among others. Elena Fernández-Fueyo et al. conducted a genome analysis of *Ceriporiopsis subvermispora* and screened out all peroxidases related to lignin degradation. Their results suggests that the Lip and VP genes are not present in this strain, but two other enzymes with similar functions were identified [17]. This unexpected finding may imply that lignin degradation mechanisms vary among species.

The production of bioethanol requires the degradation of cellulose and lignin to glucose, followed by the fermentation of glucose by yeast. The temperature required for yeast fermentation is 30 °C. Low-temperature enzymes are defined as those with an optimum temperature of approximately 35 °C while maintaining a certain catalytic efficiency at 0 °C compared to the most optimum reaction temperature for cellulose ranging from 45 °C to 65 °C. Hence, if low-temperature cellulase and lignin degradation enzymes could be obtained, synchronized fermentation of the two procedures could be achieved, which would greatly simplify the production process of bioethanol and reduce costs. Although large amounts of research have been published on cellulase and lignin-degrading enzymes, few studies have investigated low-temperature lignocellulose degradation enzymes. Cecil w. Forsberg et al. reported low-temperature glucanase from the rumen thermophilic anaerobic bacteria, *Fibrobacter succinogenes* S85 [18], and another low-temperature cellulose, CelG, from the Antarctic marine thermophilic bacteria, *Pseudoalteromonas haloplanktis*, was also discovered [19]. To date, no reports on low-temperature lignin degradation enzymes have been identified.

High-throughput sequencing of transcriptomes (RNA-Seq) has provided new routes to study the genetic and functional information stored within any organism at an unprecedented scale and speed. These advances greatly facilitate functional transcriptome research in species with limited genetic resources, including many "non-model" organisms with substantial ecological or evolutionary importance [20]. Most genomics studies of lignin-degrading fungi have focused on white-rot fungi, brown-rot fungi belonging to the Basidiomycota and filamentous fungi (trichoderma, neurospora, penicillium, among others) belonging to the Ascomycota. The most widely studied white-rot fungi is *Phanerochete chrysosporium*, the genome sequence of which was published in 2004 [21]. Analyses of its genome sequence and subsequently of its transcriptional and secretary proteins have provided ample information [22–24]. Analyses of the transcriptome suggested that 545 genes or proteins were significantly altered during the lignin degradation process. Some proteins contain signal peptide and carbohydrate (CBM) domains, which may be related to the degradation of lignocelluloses. Martinez D et al. analysed the genome, transcriptome and secretome of *Postia placenta*, the most well-studied brown-rot fungus. They discovered three groups of peroxidase (LiP, Mnp and VP) and laccasein the fungal genome [25], in accordance with the inability of brown-rot fungi to degrade lignin.

This study aimed to isolate, identify and perform a transcriptome analysis of novel strains of fungi with the potential to degrade the lignocellulosic biomass isolated from the Antarctic Pole, to screen for filamentous fungi capable of producing low-temperature lignin enzymes, and to obtain lignin degradation-related enzymes through RNA-seq, with the goal of gaining insight into the mechanism underlying lignin degradation in fungi and providing a potential lignocellulosic enzyme source for industrial production.

Methods

Sample collection and isolation of fungi
Soil, macro-algal rot and sediment samples were collected from Ardley Island - near Fildes Peninsula, Antarctica, during the Chinese 27th Antarctic Scientific Expedition. All samples were placed in sterilized plastic bags or flasks and transported to the laboratory at 4 °C for microorganism isolation. One gram of each sample was placed in a 50-mL sterile centrifuge tube containing 10 mL of sterile distilled water and shaken at 120 rpm overnight. The tube was maintained stably overnight. Next, 10^{-1} and 10^{-2} serial dilutions of each sample suspension were spread as 0.1-mL aliquots on plates containing potato dextrose agar (PDA). The PDA plates were incubated at 12 °C for 1–2 weeks, and distinct colonies were picked and sub-cultured for further analysis.

Phylogenetic analysis
The total DNA from 15 native fungal isolates was extracted according to the method of Gonzalez-Mendoza et al. with minor modifications [26]. Each DNA sample was amplified by PCR with Taq DNA polymerase following the manufacturer's instructions (Tiangen, Beijing, China). Next, 2 μL DNA was used as PCR template. The primers used for amplification were ITS 1 forward (TCCGTAGGTGAACCTGCGG) and ITS 4 reverse (TCCTCCGCTTATTGATATGC). PCR amplification was performed using the following protocol: 94 °C for 5 min (1 cycle), 94 °C for 30 s, 55 °C for 30 s and 72 °C for 40 s (30 cycles), and 72 °C for 10 min(1 cycle). To confirm the quality of the PCR, the amplification products were run on 0.8% Tris acetate EDTA agarose gels, and bands were visualized by staining with ethidium bromide. The PCR products were purified using a universal DNA purification kit (Tiangen, Beijing, China), and the amplified ITS regions were sequenced (Jimei, Shanghai, China) and submitted to GenBank.

The similarities of the 15 native isolate sequences with other known species were investigated by comparisons with sequence data in the National Center for Biotechnology Information (NCBI) database using the BLASTN programme. The phylogenetic analysis was based on BioEdit multiple alignment with sequences from their closest relatives and from common fungi in the Antarctic. A phylogenetic tree based on the ITS region was constructed using MEGA5.1 software with the neighbour-joining method, and the statistical analysis utilized bootstrapping with 1000 replications.

Enzyme assays
Fresh spores of the 15 native fungi were filtered through three layers of sterile gauze and inoculated into 150-mL flasks containing 50 mL of optimized medium for lignocellulosic enzyme production (glucose 10 g/L, ammonium sulphate 0.2 g/L, KH_2PO_4 2 g/L, $MgSO_4 \cdot 7H_2O$ 0.5 g/L, $CaCl_2$ 0.01 g/L, Vitamin B_1 1.0 mg/L). The flasks were incubated at 12 °C and shaken at 120 rpm. Supernatant was collected at 2d, 3d, 4d, 5d, 7d and 10d after inoculation, centrifuged at 5000 rpm to obtain cell-free samples and specific enzymatic activities were measured as follows.

Lignin peroxidase (Lip) activity was measured by the oxidation of Azure B in a reaction mixture consisting of 32 μM Azure B, 100 μM H_2O_2 and 50 mM sodium citrate buffer at pH 4.5 in a final volume of 1 mL. Azure B oxidation was monitored at an absorbance of 651 nm. The unit of enzymatic activity (U) was defined as an OD value reduction of 0.1 OD per mL of supernatant in 1 min, and its linear reaction time can be extended to 20 min [27].

Laccase (Lac) activity was measured using a colorimetric assay based on the oxidation of 2,2′-azino-bis(3-ethylbenzothiazoline-6-sulphonic acid (ABTS). The reaction mixture consisted of 2 mL of ABTS (0.5 mM), 1 mL of sample and 0.1 mMHAc-NaAc buffer at pH 5 and 25 °C in a final volume of 3 mL. The oxidation of ABTS was monitored from 0 to 200 s at an absorbance of 420 nm. The unit of enzymatic activity (U) was defined as 1 μM of ABTS oxidized per mL of supernatant in 1 min ($\varepsilon = 3.6 * 10^4$ M^{-1} cm^{-1}) [28].

Manganese peroxidase (Mnp) activity was measured by the oxidation of guaiacol in a reaction mixture added in turn to 2.9 mL of phosphate buffer (50 mM), 1 mL of H_2O_2 (2%), 1 mL of guaiacol (50 mM) and 0.1 mL of fermentation broth. The enzyme solution was boiled for 5 min and used as a control. The reaction system was incubated in a 34 °C water bath for 3 min immediately after addition of the fermentation broth, quickly diluted one-fold, and then monitored for phenol red oxidation at an absorbance of 465 nm once per minute a total of five times. The unit of enzymatic activity (U) was defined as $\triangle OD465/t_{min} \times 1000$.

To evaluate the effects of different temperatures and pH values on the enzymatic activity, Lip and Mnp activities were measured from 0 to 60 °C and in buffers with different pH values: Na_2HPO_4-KH_2PO_4 buffer (pH 6.0–7.5), Tris-HCl buffer (pH 7.5–9.0) and Na_2CO_3-$NaHCO_3$ (pH 9.0–11.0). All enzymatic activities were assessed in triplicate using a GE healthcare NanoVue spectrophotometer.

Growth measurements of *Aspergillus sydowii* MS-19
The growth of *Aspergillus sydowii* MS-19 in fermentation medium (glucose 10 g/L, ammonium sulphate 0.2 g/L, KH_2PO_4 2 g/L, $MgSO_4 \cdot 7H_2O$ 0.5 g/L, $CaCl_2$ 0.01 g/L, Vitamin B_1 1.0 mg/L) [29] was followed based on measurement of the dry weight of mycelia: to evaluate the effect of temperature on *Aspergillus sydowii* MS-19 growth, fermentation broth was filtered using a moderate speed qualitative filter, and mycelia were washed three times with sterile distilled water and dried at 80 °C for 3 h. Fresh spores were inoculated at a 2% inoculum concentration in six 150-mL flasks containing 50 mL of fermentation medium. The flasks were incubated at 0, 3, 10, 20, 37, and 45 °C shaken at 120 rpm for 10 days. The dry weights of the mycelia were measured once daily from day two after inoculation. To evaluate the effect of pH on *Aspergillus sydowii* MS-19 growth, fresh spores were inoculated at a 2% inoculum concentration in six 150-mL flasks containing 50 mL of fermentation medium. The flasks were incubated at pH 4, 5, 6, 7, 8 and 9 and shaken at 120 rpm at an optimized temperature (20 °C). The dry weights of the mycelia were measured upon initiation of the logarithmic growth period beginning at 7 d.

To assess the effects of lignin as the sole carbon source on growth, fermentation medium was replaced by medium containing lignin as the sole carbon source (lignin 0.3 g/L, K_2HPO_4 1.0 g/L, NaCl 0.5 g/L, $MgSO_4 \cdot 7H_2O$ 0.3 g/L, $NaNO_3$ 2.5 g/L, $CaCl_2$ 0.1 g/L, $FeCl_3$ 0.01 g/L, pH 7.0).

RNA extraction and transcriptome sequencing
The *Aspergillus sydowii* MS-19 strain was grown in fermentation medium with shaking for 1 week. Mycelia were harvested by centrifugation and ground in liquid nitrogen. Total RNA samples were isolated using a standard TRIzol method, eluted in RNase-free water and stored at –80 °C until further use. Two replicates were performed for library preparation. The integrity of the total RNA was checked on an agarose gel, and its quantity and purity were determined using NanoVue (GE). mRNA was enriched from total RNA using oligo T (dT) beads, and broken into short fragments by the addition of fragmentation buffer.

cDNA was then synthesized using these fragments as template, purified, end-repaired and dA-tailed, and then ligated to sequencing adaptors. The samples were gel size-selected for the 150-bp fragment size. Size-selected adaptor-ligated cDNA was purified with an AgencourtAMPure kit (Beckman Coulter, CA, USA) and used as template for PCR amplification to create the cDNA library. The purified library were profiled using the Agilent Bio analyser and sequenced using the PE100 strategy on the IlluminaHiSeq 2000 platform to yield paired-end reads.

Assembly and annotation of the transcriptome
Raw reads from the library were filtered to remove low-quality reads as well as adapters and poly-A/T-containing reads. The resulting clean reads were assembled to produce unigenes using the short reads assembling programme Velvet (v1.2.08) [30] and Osase (v0.2.08) [31] with default parameters. Potentially contaminated sequences were removed using BLAST.

For functional annotations, all unigene sequences were searched against Nr, eggNOG, and KEGG using BLASTX to extract predicted coding region sequences with high sequence similarity to the given unigenes along with their protein function annotations. We use the blastx with cutoff of similarity > = 80%, coverage > = 80% and evalue <1e-5. In addition, the gene ontology (GO) term for each unigene and GO enrichment analysis was obtained using the Blast2GO programme. Pathway annotation of unigenes was performed according to the KEGG database mapping method. Unigenes were aligned to the KOG database to predict and classify possible functions.

CAZyme annotation of the *Aspergillus sydowii* MS-19 transcriptome was conducted by searching against the dbCAN database [32], which is a network resource for automatic annotation of carbohydrate active enzymes based on CAZyme tag domains for any submitted protein dataset. For each CAZyme family, a tag domain was defined by combining the search results for the conserved domain database (CDD) and published literature, and a corresponding Hidden Markov Model (HMM) was constructed. Bioedit software (v7.2.5) was used to assign genes with particular activities, including the enzyme commission (EC) number annotations.

Results

Fungal isolation

Six compost types were subjected to screening for the isolation and characterization of fungi. A total of 168 isolates were obtained, and their species and genera were determined based on their colony morphology and the microscopic characteristics of the spore apparatus, spore stalks and spores. Fifteen representative isolates with clear differences were selected for further analysis (Table 1). The phylogenetic tree constructed based on the ITS sequences of the 15 native isolates indicated that five isolates were members of the genus *Penicillium*, two isolates were closely related to *Pseudeurotium, Geomyces* and *Cladosporium*, and one isolate matched *Bionectria, Aspergillus, Aureobasidium* and an unclassified *Onygenales* (Additional file 1: Figure S1). The Antarctic region exhibited a rich diversity of fungal species.

Neighbour-joining tree showing the relationship between the ITS sequences from 15 Antarctic native isolates and their closest relatives as well as common fungi in the Antarctic. The bootstrap values for the neighbour-joining analysis with 1000 replicates are shown on the branches. The scale bar represents 0.05 substitutions per amino acid site.

Aspergillus sydowii MS-19 is able to produce low-temperature Lip and Mnp enzymes

To test the ability of the 15 representative fungi to synthesize lignin-degrading enzymes, the strains were grown in liquid medium supplemented with the corresponding substrates. The activities of Lip, Lac and Mnp were measured. Four isolates, PU-01, SS-04, MS-05 and MS-19, showed Lip and Mnp activities (Table 2). Lac activity was not detected. Since MS-19 exhibited the highest activities of both Lip and Mnp enzymes, it was chosen as our target fungal strain in this study.

Lip and Mnp activities were measured at different temperatures and pH values to identify the optimum condition for enzymes produced by MS-19. Lip and Mnp maintained a certain activity at 0 °C and gradually increased with temperature, showing the highest activity at 30 °C and declining sharply above this temperature (Fig. 1a). Since low-temperature enzymesare defined to have an optimal temperature of approximately 35 °C and to maintain a certain catalytic efficiency at 0 °C, the Lip and Mnp enzymes synthesized by MS-19 were considered low-temperature enzymes. The optimal pH for Lip and Mnp was 3.0 and 4.5, respectively. Lip was found to have sufficiently high activity between pH 2.0–4.0, while Mnp showed high activity between pH 4.0–6.0 (Fig. 1b).

BLAST searches demonstrated that MS-19 was highly similar to *Aspergillus* (Additional file 1: Figure S1). To further identify the phylogeny of MS-19, a phylogenetic tree was constructed based on the ITS sequences of MS-19 and its 15 closest relatives. MS-19 displayed 99% identity to *Aspergillus sydowii* (Fig. 2a). Combined with the morphological and microscopic characteristics of MS-19 (Fig. 2b), the genus of the MS-19 isolate was identified as *Aspergillus. sydowii*.

Table 1 Sampling information and the representative fungal isolates

Sample name	Longitude and latitude	Number of fungi isolated	Species of fungi isolated	Representative isolates and their species	GenBank accession No.
Freshwater lake sediments (FS)	58°54'W, 62° 11'S	11	*Penicillium* (5), *Geomyces* (4), *Cladosporium* (2)	FS-03 *Cladosporium cladosporioides*	JX139700
Peak umber (PU)	58°59'W, 62° 11'S	14	*Penicillium* (4), *Geomyces* (5), *Aureobasidium* (5)	PU-01 *Aureobasidium pullulans*	JX675048
				PU-05 *Penicillium*	JX139707
Hill soil (HS)	58°57'W, 62° 13'S	23	*Penicillium* (6), *Pseudeurotium* (5), *Geomyces* (7), *Cladosporium* (2), *Aureobasidium* (3)	HS-07 *Penicillium commune*	JX139703
				HS-11 *Pseudeurotium*	JX139704
Snow sediments (SS)	57°57'W, 62° 13'S	35	*Pseudeurotium* (11), *Geomyces* (6), unclassified *Onygenales* (4), *Bionectria* (3), *Penicillium* (9), *Cladosporium* (1), *Aureobasidium* (1)	SS-04 *Cladosporium*	JX675049
				SS-08 unclassified *Onygenales*	JX139708
				SS-10 *Geomyces*	JX139709
				SS-13 *Penicillium chrysogenum*	JX139710
Macroalgae sediments (MS)	58°58'W, 62° 13' S	58	*Pseudeurotium* (9), *Geomyces* (7), unclassified *Onygenales* (4), *Bionectria* (6), *Penicillium* (12), *Aspergillus* (5), *Cladosporium* (8), *Aureobasidium* (7)	MS-02 *Penicillium chrysogenum*	JX139706
				MS-05 *Bionectria ochroleuca*	JX675045
				MS-17 *Pseudeurotium*	JX139705
				MS-19 *Aspergillus sydowii*	JX675047
Freshwater lake water (FW)	57°54'W, 62° 11' S	27	*Pseudeurotium* (6), *Geomyces* (4), *Penicillium* (5), *Aspergillus* (3), *Cladosporium* (3), *Aureobasidium* (6)	FW-04 *Penicillium polonicum*	JX139701
				FW-13 *Geomyces*	JX139702

Table 2 Lignin-degrading enzymatic activities of the fungal isolates

Genus	Isolate No.	Highest enzymatic activity detected (U/L)	
		Lip	Mnp
Aureobasidium	PU-01	126.5	113.8
Cladosporium	SS-04	136.8	126.4
Bionectria	MS-05	154.2	132.6
Aspergillus	MS-19	182.6	159.7

Aspergillus sydowii MS-19 could grow in the temperature range from 3 to 37 °C. Its growth was observed at temperatures as low as 3 °C. The growth rate was maximal at 20 °C and pH 7 (Fig. 2c and d). A white mycelial pellet could be observed after 7 days of culturing at 20 °C, demonstrating that *Aspergillus sydowii* MS-19 could utilize lignin as a sole carbon source (Fig. 2e).

Summary of the RNA-seq dataset

To obtain an overview of the *Aspergillus sydowii* MS-19 transcriptome, the poly (A)-enriched mRNA sample was subjected to high-throughput IlluminaHiSeq sequencing, resulting in 18,453,231 reads with an average length of 101 nt. Assembler Velvet and Oases was employed to complete the assembly and cluster of the *Aspergillus sydowii* MS-19 transcriptome using default parameters. From the Velvet assembly were obtained 72,387 contigs with a total length of 23,862,098 nt. Oases continued to generate longer transcripts, resulting in 27,600 transcripts with a length of 21,976,433 nt. Additional sequence cluster analyses were conducted among all transcript sequences, generating 11,269 unigenes with a mean size of 1130 nt (Table 3).

GO and KOG classification

GO assignments were used to classify the unigene functions of *Aspergillus sydowii* MS-19. Based on the sequence homology, 8199 unigenes were categorized into 55 functional groups. In terms of biological process, the majority of the unigenes were involved in "hydrolase activity" (3143 members), "transferase activity" (2345 members) and "transport" (1918 members). For the cellular component, the majority of the unigenes were involved in "membrane" (2393 members), "nucleus" (1760 members) and "ribosome" (479 members). The investigation of molecular functions revealed that most unigenes were involved in "DNA binding" (1522 members), "RNA binding" (621 members) and "structural molecule activity" (515 members) (Fig. 3). To further evaluate the completeness of the transcriptome and the effectiveness of the annotation process, the annotated sequences were screened for genes involved in KOG classifications. In total, among17767 nr hits, 11,192 sequences had KOG classifications. Among the 25 KOG categories, the cluster for "function unknown" represented the largest group (3,039members), followed by "general function prediction" (1064 members) and "transcription" (818 members). The following categories represented the smallest groups: Extracellular structures (3 members); Cell motility (6 members) and Nuclear structure (31 members) (Fig. 4). To identify the biological pathways that were active in *Aspergillus sydowii* MS-19, we mapped all the unigenes to the reference canonical pathways in KEGG, and we found that a total of 3533 sequences could be assigned to 39KEGG pathways. The most representative pathways by the unigenes were "metabolism pathways" (2229 members), "genetic information processing" (1158 members), "cellular processes" (552 members), Organismal Systems (522 members) and "environmental information processing" (196 members) (Fig. 5). These annotations provide a valuable resource for investigating specific processes, functions and pathways in *Aspergillus sydowii* MS-19.

GO has three ontologies: molecular function, cellular component and biological process, indicating the GO functional classification annotation and the number of unigenes in each category. The x-axis indicates the term description, and the y-axis indicates the number of genes annotated in the current group (GO terms with less than 50 annotated genes are not shown).

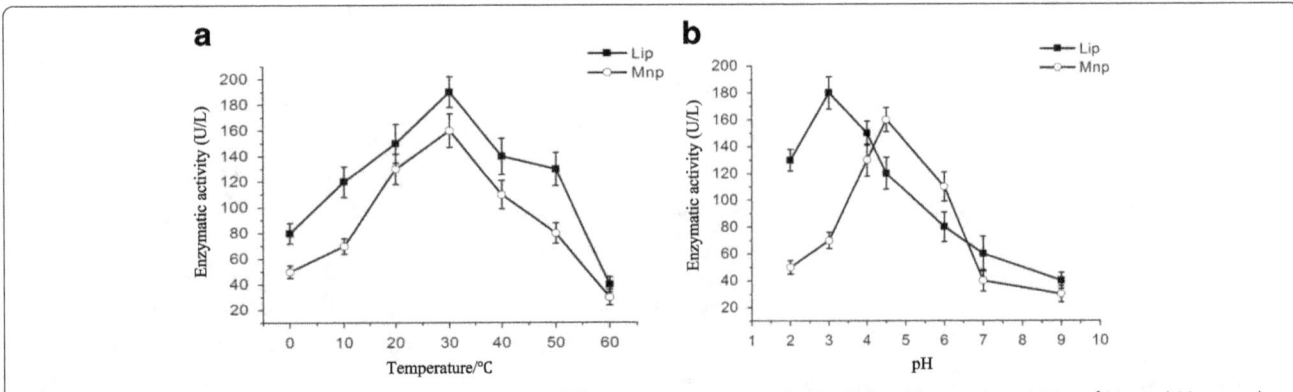

Fig. 1 Enzymatic assay of Lip and Mnp produced by MS-19 at different temperatures and pH values. **a** Enzymatic activities of Lip and Mnp synthesized by MS-19 at different temperatures. **b** Enzymatic activities of Lip and Mnp synthesized by MS-19 at different pH values

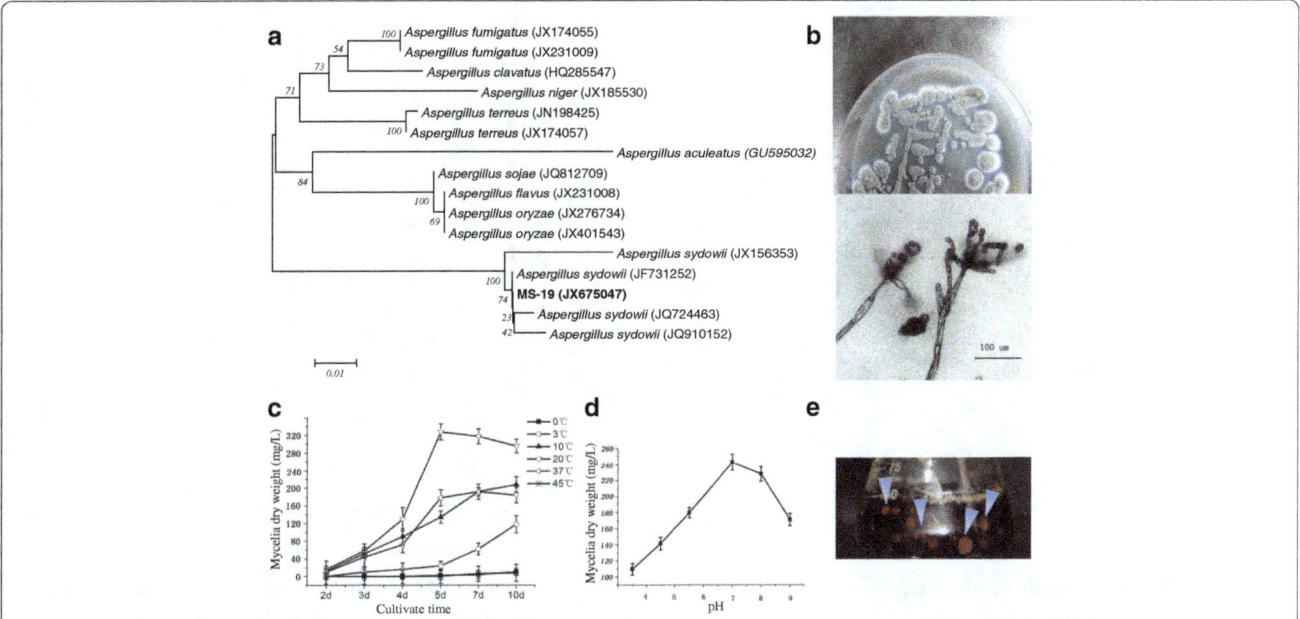

Fig. 2 Phylogeny and growth profile of Aspergillus sydowii MS-19. **a** Neighbour-joining tree showing the relationship between ITS sequences from MS-19 and its 15 closest relatives. Bootstrap values for the neighbour-joining analysis with 1000 replicates are shown on the branches. The scale bar represents 0.01 substitutions per amino acid site. **b** The morphological and microscopic characteristics of Aspergillus sydowii MS-19. **c** Time course of Aspergillus sydowii MS-19 growth at different temperatures. **d** Growth of Aspergillus sydowii MS-19 at different initial pH values. **e** Mycelial pellets were observed when Aspergillus sydowii MS-19 was cultured in medium with lignin as the sole carbon source. Blue arrowheads indicate white mycelial pellets in growth medium

The KEGG pathway annotation provides a mapping of the transcriptomic dataset to the KEGG pathway maps for biological interpretation of higher-level systemic functions. It indicates the KEGG pathway classification annotation and the percentage of each pathway.

Table 3 Output statistics for the sequencing and assembly

Sample	Aspergillus sydowii MS-19
Raw Reads	28,267,863
Clean Reads	18,453,231
Read size (nt/read)	101
Total nucleotides (nt)	3,690,646,200
Contigs	
Number of contigs	72,387
Mean size of contigs	330
Length of all contigs (nt)	23,862,098
Transcripts	
Number of transcripts	27,600
Mean size of transcripts	796
Length of all transcripts (nt)	21,976,434
Unigenes	
Number of unigenes	11,269
Mean size of unigenes	1130
Length of all unigenes	12,735,577

Since our main concern was material degradation, we focused on sequences that participate in carbohydrate metabolism and xenobiotic biodegradation pathways (Additional file 2: Table S2). Most fungi adopt the Embden-Meyerhof-Parnas pathway (EMP) for carbohydrate metabolism, while a few fungi such as red yeast utilize the pentose phosphate pathway (HMP). Annotation of the *Aspergillus sydowii* MS-19 transcriptome revealed enzymes that play important roles in the EMP, such as hexokinase, glucose-6-phosphate isomerase, fructose-phosphate kinase phosphatases, glyceraldehyde-3-phosphate dehydrogenase, pyruvate kinase, acetyl coenzyme synthetase and 6-phosphate glucose isomerase, indicating that the EMP is the main carbohydrate metabolism pathway in *Aspergillus sydowii* MS-19.

Regarding the annotated unigenes associated with xenobiotic biodegradation, some enzymes are capable of metabolizing the benzene ring structure. These enzymes may contribute to lignin degradation, which is a class of polymer composed of phenyl propane as the structural unit. Relevant benzene ring-metabolizing enzymes are as follows: 2-haloacid dehalogenase, aldehyde dehydrogenase and S-(hydroxymethyl) glutathione dehydrogenase related to the degradation of chloroalkane and chloroalkene; catechol 1,2-dioxygenase and carboxymethylenebutenolidase related to the degradation of toluene and chlorobenzene; phenylacetate 2-hydroxylase, fumarylacetoacetase, fumarylacetoacetase, amidase, nitrilase and 3-hydroxyphenylacetate 6-hydroxylase

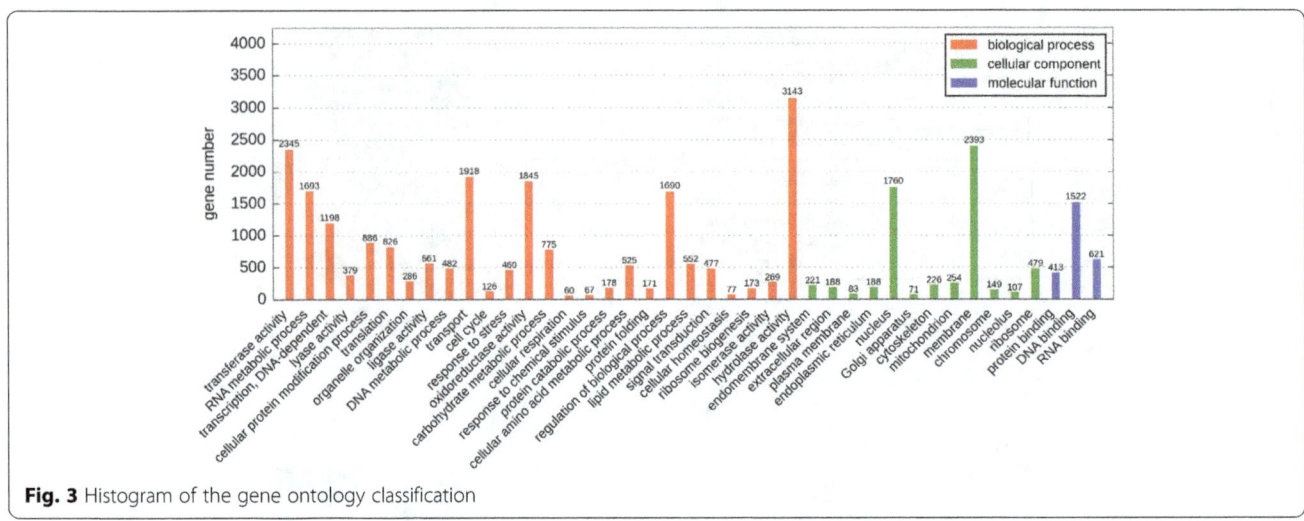

Fig. 3 Histogram of the gene ontology classification

related to the degradation of styrene; salicylate hydroxylase related to the degradation of dioxin;glutathione S-transferase; and S-(hydroxymethyl) glutathione dehydrogenase in the cytochrome P450 superfamily related to xenobiotic metabolism, among others. These annotations indicate that *Aspergillus sydowii* MS-19 not only has the ability to metabolize xenobiotics and environmental pollutants but also may degrade lignin.

CAZyme expression profiles

Using the CAZy database, a total of 701 unigenes were annotated, including 355 glycoside hydrolase (GH), 208 glycosyltransferase (GT), 9 polysaccharide lyase (PL), 92 carbohydrate esterase (CE) and 37 carbohydrate combined structure domain (CBM) proteins (Fig. 6a). The most representative families were GT41 (38 members), CE10 (31 members) and GH5 (24 members).

There were 37 genes encoding proteins containing the cellulose-binding domain (CBM), which may assist in lignocellulose-degrading enzyme attachment to the surface of cellulose and thus facilitate co-degradation of the natural cellulose-hemicellulose network.

CAZyme-annotated genes in *Aspergillus sydowii* MS-19 and the brown-rot fungi representative species, *Postia placenta* and *Penicillium decumbens*, were compared (Fig. 6, b). The total number of CAZymes and the number of CAZymes in the GH and GT classes of *Aspergillus sydowii* MS-19 were much higher than in the other two.

The CAZymes in *Aspergillus sydowii* MS-19 related to lignocellulose degradation are listed in Table 4. There were 17 cellulase (eight cellobiohydrolase and nine endo-1,3-beta-glucanase) and 19 feruloyl esterase-encoding genes. In addition, 13 chitinase genes were annotated, and the chitin-degrading ability of *Aspergillus sydowii* MS-19 is subject to verification.

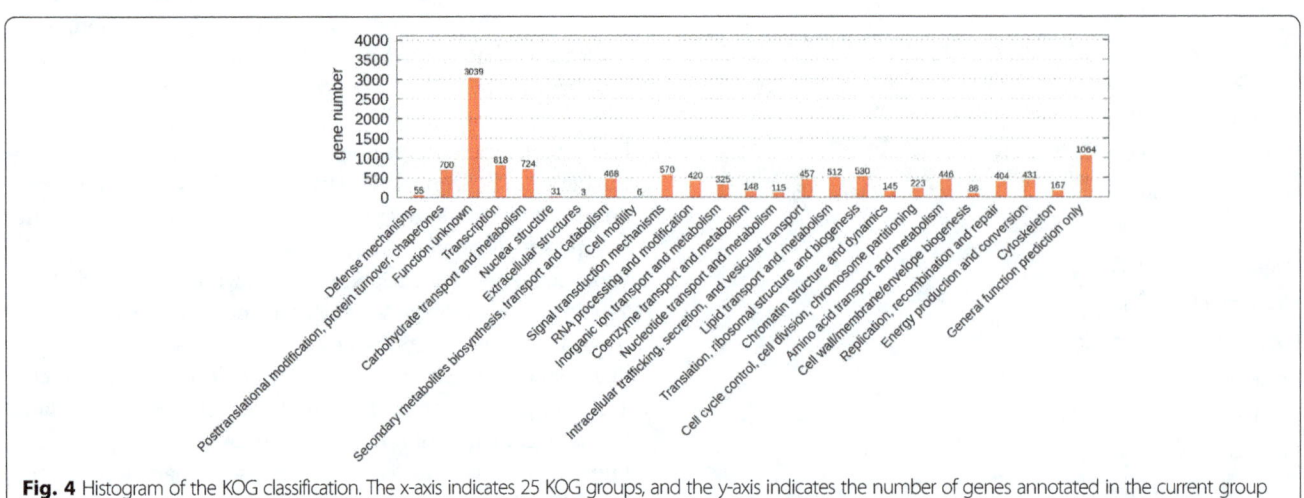

Fig. 4 Histogram of the KOG classification. The x-axis indicates 25 KOG groups, and the y-axis indicates the number of genes annotated in the current group

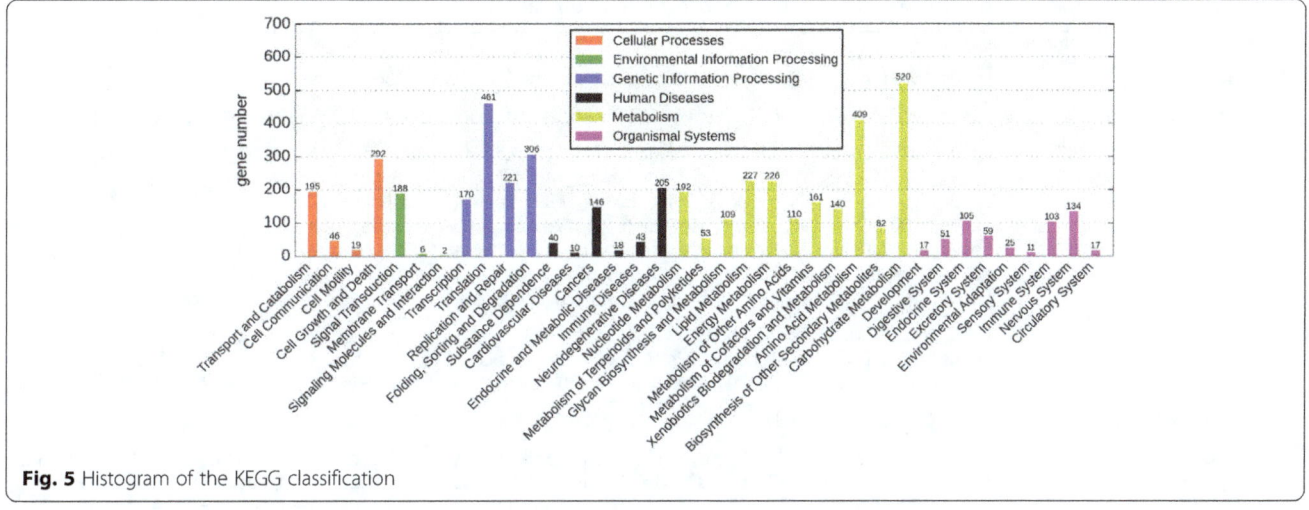

Fig. 5 Histogram of the KEGG classification

Remarkably, one sequence annotated as laccase, which can degrade lignin, was found at Locus_6288. The BLAST results showed that the nucleic acid sequence homology between this sequence and the laccase sequences from *Coccidioides immitis* (EAS35187), *Grosmannia clavigera* (EFX01145), *Aspergillus kawachii* (GAA87354), and *Aspergillus niger* (XP_001389525) was 52%, 40%, 54% and 53%, respectively (Fig. 7).

Three peroxidase sequences sharing a structure similar to typical lignin peroxidase and manganese peroxidase were found and annotated as haem-binding peroxidase, glutathione peroxidase and catalase-peroxidase. Among them, the haem-binding peroxidase annotated sequence (Locus_2354) displayed the highest similarity, and the BLAST results showed a nucleic acid sequence homology of 75% between this sequence and *Aspergillus niger* CBS 513.88 haem-binding peroxidase (Fig. 8). It is

speculated that this gene may have different names but similar functionalities.

Discussion

Polar microorganisms have important value in scientific research as well as in application development since they provide novelty, uniqueness and diversity. An increasing number of people are now focused on the study and application of low-temperature enzymes. In recent years, due to the shortage of energy sources, utilization of lignocellulosic material to produce liquid fuels and chemicals has become an important alternative approach for supporting the sustainable development of human society. Low-temperature lignocellulosic enzymes also show great potential applications in industrial production [33]. In this study, we isolated and identified a fungus, *Aspergillus sydowii* MS-19, from the Antarctic region that is able to produce the low-temperature enzymes, Lip and Mnp. We also adapted high-throughput sequencing technology to characterize the transcriptome features and genes encoding CAZymes in *Aspergillus sydowii* MS-19. This work will provide useful information to facilitate our understanding of low-temperature lignocellulosic enzyme production by polar microorganisms, as well as further research and the application of a novel Antarctic *Aspergillus sydowii* strain MS-19 as a potential source of lignocellulosic enzymes.

RNA-seq reads of *Aspergillus sydowii* MS-19 were assembled and subsequently clustered into 11,269 unigenes, among which most of the sequences were annotated against three databases. The Blast2GO framework was used to analyse the annotation results based on transcriptomic research. Among the unigenes, 3.82% of 11,269 accounted for catabolism processes involved in lignin degradation and glycometabolism. Most of the unigenes generated products inside the cell, whereas a few generated extracellular products, potentially due to the different

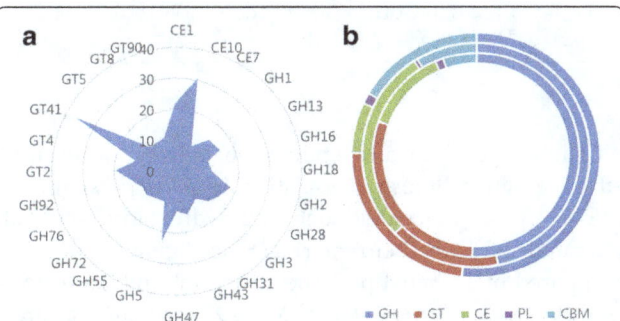

Fig. 6 Comparison of the numbers of unigenes belonging to CAZy families (**a**) in *Aspergillus sydowii* MS-19 and (**b**) in *Aspergillus sydowii* MS-19, *Postia placenta* and *Penicillium decumbens*. **a** The numbers of unigenes belonging to particular CAZy families in *Aspergillus sydowii* MS-19, only CAZy families containing 10 or more unigenes are shown. **b** Distribution of various CAZymes in *Aspergillus sydowii* MS-19 (outer ring), *Postia placenta* (middle ring) and *Penicillium decumbens* (outer ring). GH, glycoside hydrolase; GT, glycosyltransferase; CE, carbohydrate esterase; PL, polysaccharide lyase; CBM, carbohydrate-binding module

Table 4 CAZymes in *Aspergillus sydowii* MS-19 related to lignocellulose degradation

CAZyme	Number	Locus
cellobiohydrolase	8	Locus_1390,Locus_2944,Locus_4414,Locus_6027,Locus_7420,Locus_15666,Locus_22469,Locus_24089
endo-1,3-beta-glucanase	9	Locus 526,Locus 1086,Locus 1179, Locus 2141,Locus 3549,Locus 7936,Locus 21,439,Locus 22,695,Locus 24,357
β-glucosidase	57	Locus_10259, Locus_10547, Locus_10915, Locus_1108, Locus_11132, Locus_114, Locus_11458, Locus_12029, Locus_13235, Locus_13894, Locus_13917, Locus_14024, Locus_14178, Locus_14313, Locus_14583, Locus_14893, Locus_1560, Locus_16303, Locus_16488, Locus_16896, Locus_16926, Locus_17760, Locus_17789, Locus_17844, Locus_1900, Locus_1912, Locus_1939, Locus_19428, Locus_20804, Locus_2085, Locus_21268, Locus_21724, Locus_21959, Locus_22387, Locus_23309, Locus_23358, Locus_23698, Locus_25281, Locus_26343, Locus_27, Locus_2734, Locus_3147, Locus_3781, Locus_3913, Locus_4347, Locus_4646, Locus_4820, Locus_5135, Locus_5202, Locus_5954, Locus_6125, Locus_6476, Locus_6499, Locus_7420, Locus_8287, Locus_931, Locus_9711
α- glucosidase	24	Locus_11276, Locus_12706, Locus_13235, Locus_14736, Locus_18, Locus_18186, Locus_18361, Locus_19137, Locus_19676, Locus_2131, Locus_22493, Locus_23154, Locus_24653, Locus_3331, Locus_354, Locus_3592, Locus_4621, Locus_5377, Locus_5565, Locus_5981, Locus_8096, Locus_856, Locus_8769, Locus_9406
endo-1,4-beta-xylanase	2	Locus_21348,Locus_14631
beta-xylosidase	37	Locus_10259, Locus_11943, Locus_13192, Locus_13650, Locus_13917, Locus_14313, Locus_14583, Locus_15712, Locus_16267, Locus_16303, Locus_16896, Locus_17760, Locus_18132, Locus_1844, Locus_1912, Locus_1915, Locus_2085, Locus_21472, Locus_21724, Locus_23358, Locus_25098, Locus_25281, Locus_3147, Locus_3913, Locus_4820, Locus_5135, Locus_5202, Locus_5954, Locus_6125, Locus_6476, Locus_7051, Locus_8287, Locus_8971, Locus_9039, Locus_931, Locus_9711, Locus_9815
beta-mannosidase	42	Locus_10547, Locus_10915, Locus_1108, Locus_11132, Locus_114, Locus_11458, Locus_11470, Locus_12029, Locus_1319,Locus_13235, Locus_13828, Locus_13894, Locus_14024, Locus_14178, Locus_14381, Locus_14823, Locus_14893, Locus_1560, Locus_16488, Locus_16926, Locus_17112, Locus_17789, Locus_17844, Locus_1862, Locus_1900, Locus_19428, Locus_20143, Locus_20804, Locus_20846, Locus_21268, Locus_21959, Locus_21986, Locus_22387, Locus_23309, Locus_23698, Locus_2531, Locus_26343, Locus_3619, Locus_4347, Locus_4646, Locus_6354, Locus_6499
galactosidase	63	Locus_10547, Locus_1086, Locus_10915, Locus_1108, Locus_11372, Locus_11458, Locus_11470, Locus_1179, Locus_11943, Locus_12029, Locus_1319, Locus_13192, Locus_13235, Locus_13565, Locus_13650, Locus_13828, Locus_13915, Locus_14024, Locus_14381, Locus_14823, Locus_15680, Locus_15712, Locus_16267, Locus_16926, Locus_17112, Locus_17184, Locus_17789, Locus_17844, Locus_18132, Locus_1844, Locus_1862, Locus_1915, Locus_20143, Locus_20804, Locus_20846, Locus_21268, Locus_2141, Locus_21,439, Locus_21465, Locus_21472, Locus_21986, Locus_22270, Locus_22,695, Locus_24,357, Locus_24863, Locus_25098, Locus_2531, Locus_26343, Locus_3549, Locus_3619, Locus_3930, Locus_4347, Locus_4646, Locus_526, Locus_6354, Locus_6561,Locus_7051, Locus_7512, Locus_7936, Locus_8060, Locus_8971, Locus_9039, Locus_9815
feruloyl esterase	19	Locus_10175, Locus_11024, Locus_11435, Locus_12982, Locus_17024, Locus_17982, Locus_18274, Locus_2117, Locus_3617, Locus_365, Locus_4119, Locus_4156, Locus_5080, Locus_5575, Locus_5779, Locus_6122, Locus_7841, Locus_8559, Locus_858
pectate lyase	6	Locus_10429, Locus_13318, Locus_16762, Locus_17762, Locus_19898, Locus_4613
acetylesterase	6	Locus_15519, Locus_15609, Locus_18146, Locus_18619, Locus_20150, Locus_21291
chitinase	13	Locus_1056, Locus_12689, Locus_12831, Locus_14176, Locus_1429, Locus_16340, Locus_1636, Locus_1648, Locus_18421, Locus_21977, Locus_2206, Locus_3516, Locus_483

cultivating conditions. In many lignocellulose-degrading microorganisms, lignocellulosic enzyme secretion issignificantly enhanced following the addition of particular inducers were added [34, 35]. Among the MFs (molecular functions) of the annotated sequences, 10.48%, 14.12%, 1.14% were responsible for DNA-binding, ion binding and enzyme binding, respectively, which may be related to gene transcription regulation and protein-folding transportation, and can be used for further studies of the synthesis and regulatory mechanisms of lignocellulosic enzymes [36]. The largest portion of annotated unigenes against the KEGG database was associated with metabolic pathways. Enzymes involved in these pathways could metabolize cellulose, hemicellulose, and starch, and they were also carbohydrate-active enzymes.

Cellobiohydrolase and endo-1,4-beta-xylanase ensured effective lignocellulose degradation in *Aspergillus sydowii* MS-19 [37, 38]. Enzymes such as 2-hydrochloric acid halide can degrade the benzene ring. Since lignin is a polymer composed of a phenyl propane structural unit, its expression in *Aspergillus sydowii* MS-19 suggested a great potential ability to degrade xenobiotic and environmental pollutants, even lignin [39]. Mariana et al. have reported that *Aspergillus sydowii* Gc12 isolated from sponge can secrete peroxidase to catalyse loop-opening of benzyl glycidyl ether (a lignin monomer analogue) [40].

An interesting sequence located in Locus_6288 was presumed to perform a similar function to laccase. The sequence identity between Locus_6288 in *Aspergillus sydowii* MS-19 and laccase in *Aspergillus niger* ATCC

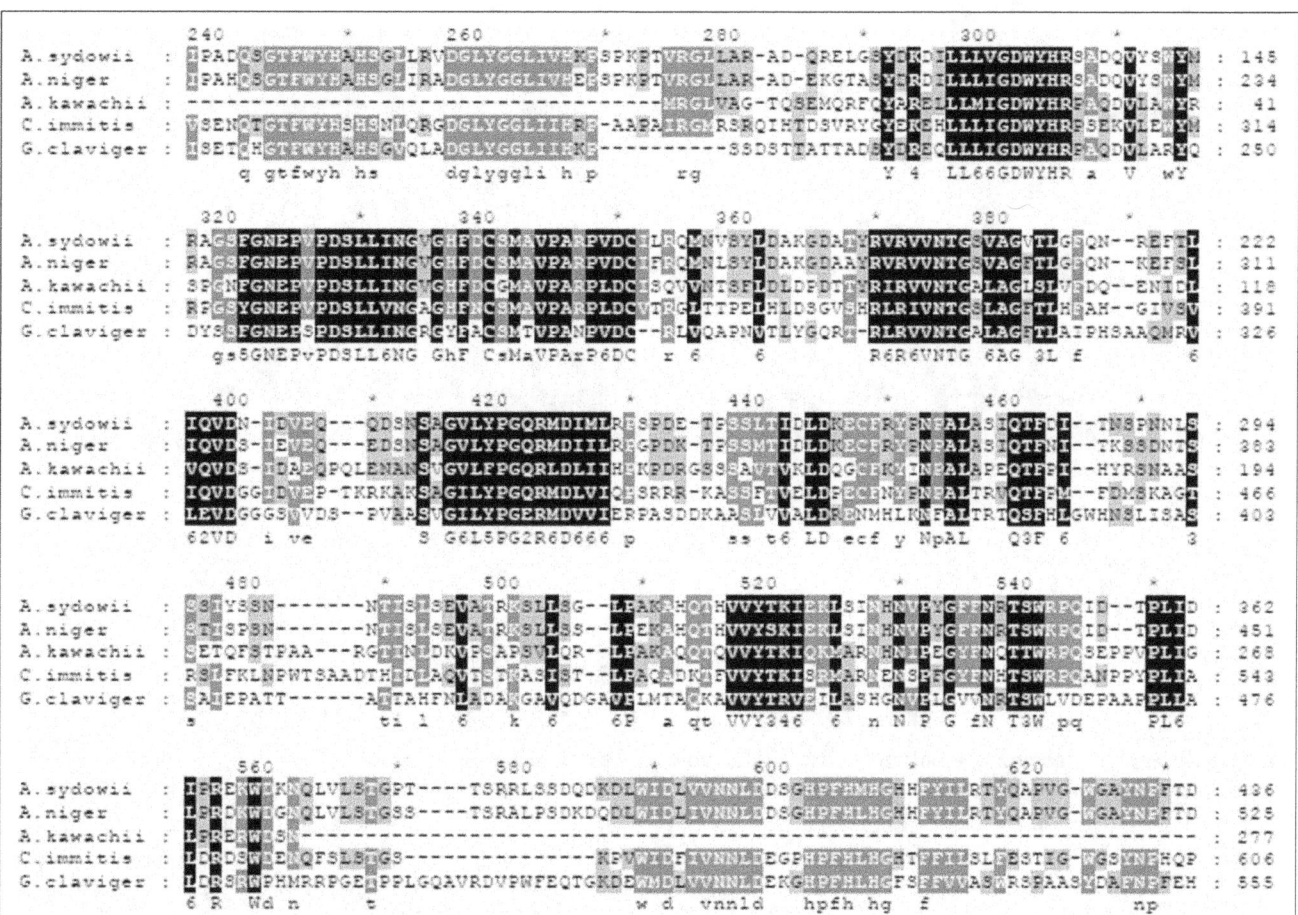

Fig. 7 Alignment of the amino acid sequence of *Aspergillus sydowii* MS-19 (A.sydowii) laccase-like gene with the laccase sequences from four other fungal species. The species selected for the alignment were *Coccidioides immitis* (EAS35187, C.immitis), *Grosmannia clavigera* (EFX01145, G claviger), *Aspergillus kawachii* (GAA87354, A.kawachii), and *Aspergillus niger* (XP_001389525, A.niger)

1015 reached 53%. However, the enzyme activity assay of *Aspergillus sydowii* MS-19 in this study did not detect laccase activity. These contradictory results may be due to the small amount and low activity of laccase secreted extracellularly. The laccase activity of *Aspergillus sydowii* MS-19 could by further analysed by increasing the enzyme concentration.

We also observed that three sequences, annotated as haem-binding peroxidase, glutathione peroxidase and catalase-peroxidase, were similar to typical lignin peroxidase and manganese peroxidases, all of which had haem binding sites based on a multi-alignment analysis. This result suggests the presence of peroxidase sequences with different names that perform similar functions in *Aspergillus sydowii* MS-19. Elena Fernández-Fueyo et al. [17] screened all peroxidase-encoding genes associated with lignin degradation in *Ceriporiopsis subvermispora*. Instead of the LiP and VP gene, they found two other peroxidases with similar functions. There is a relatively conserved tryptophan residue exposed outside the protein tertiary structure in the two peroxidases, which

plays an important role in electron transportation during lignin degradation. Whether enzymes encoded by haem-binding peroxidase genes participate in *Aspergillus sydowii* MS-19 lignin degradation remain to be elucidated.

At present, *Phanerochete chrysosporium* is the most comprehensive studied. It shows the best enzyme activity in ligninase research field. According to the unit definition and the measurement method of ligninase proposed by Archibald [41] and Gleen [42], the optimized Lip and Mnp activities of *Phanerochete chrysosporium* may reach 300 ~ 1000 U/L, and the optimal enzyme activity temperature is as high as 40 ~ 45 °C. The novel Antarctic *Aspergillus sydowii* MS-19 we obtained in this study has low temperature ligninase activity, its optimal enzyme activity is 30 °C. The maximum enzyme activity of Lip and Mnp are 182.6 U/ L and 159.7 U/L for *Aspergillus sydowii* MS-19, which are lower than that of *Phanerochete chrysosporium*. This can be attributed to the most suitable conditions for enzyme production, influencing factors and mutagenesis for *Phanerochete chrysosporium* that have been studied

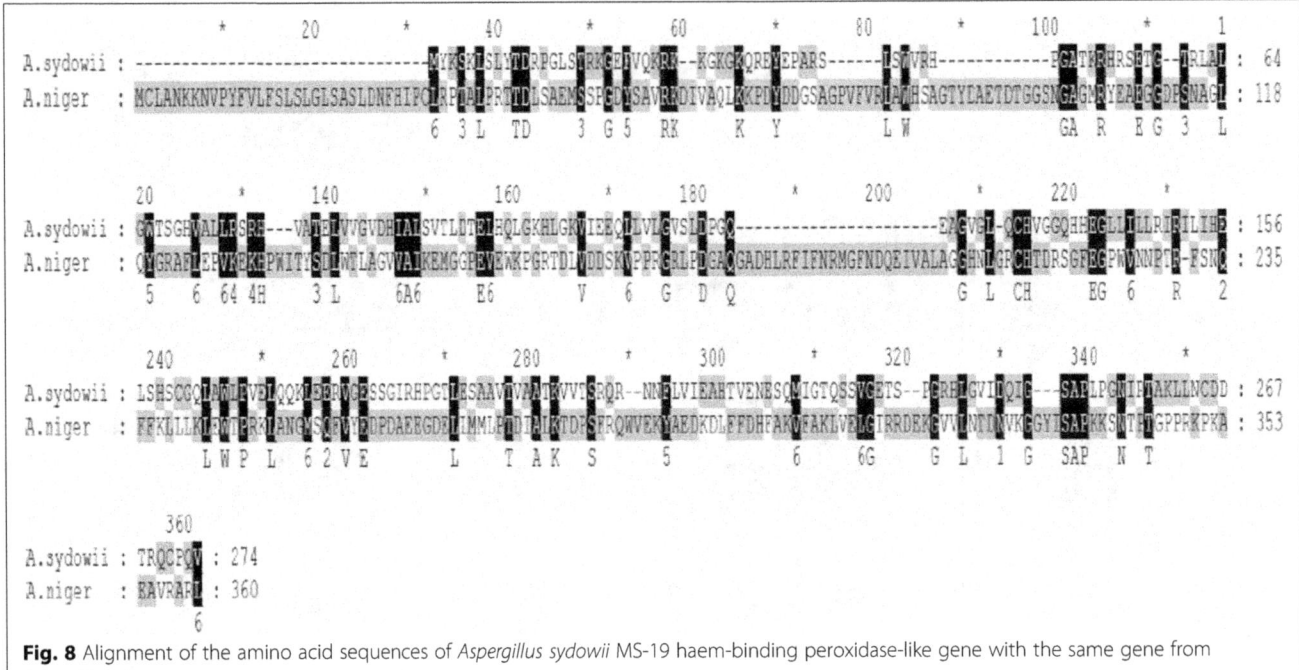

Fig. 8 Alignment of the amino acid sequences of *Aspergillus sydowii* MS-19 haem-binding peroxidase-like gene with the same gene from *Aspergillus niger*

thoroughly and higher enzyme activity is obtained. Low temperature enzyme such as *Aspergillus sydowii* MS-19 has advantages over *Phanerochete chrysosporium* in bio energy applications field in our study. The production of bio ethanol requires degradation of cellulose and lignin to glucose, followed by the fermentation of glucose by yeast. The reaction temperature of conventional cellulase and ligninase are between 45 ~ 65 °C, while the temperature required for yeast fermentation is 30 °C. Thus the two processes are separate, and with the increase of reaction time, the concentration of glucose is feedback inhibited. It not only affects the production efficiency, but also greatly increases the cost of bioethanol production. Hence, if low temperature cellulase and lignin degradation enzymes is obtained, synchronization fermentation of the two procedures would achieve. This will greatly simplify the production process of bio ethanol with lower cost. Excess glucose can also be used for yeast fermentation medium. Synchronization fermentation can also reduce the concentration of glucose, remove feedback inhibition and improve the yield of ethanol. We have obtained the low temperature cellulase (optimal enzyme activity temperature 30 °C). Meanwhile, the isolated strain in this study has the ability of producing low temperature ligninase. For further research, these results will provide a solid theoretical basis for the industrial production of bioethanol, and provide a new way of production. In a word, because of the huge difference of reaction temperature, there is no direct comparability between the strains of *Aspergillus sydowii* MS-19 and industrialized strains.

Conclusions

In the present investigation, a fungus, *Aspergillus sydowii* MS-19, with the potential for lignocellulose degradation was screened out and isolated from the Antarctic region. We measured the growth profile of *Aspergillus sydowii* MS-19 and found that it could utilize lignin as a sole carbon source. We also verified its ability to synthesize low-temperature lignin peroxidase (Lip) and manganese peroxidase (Mnp) enzymes, the properties of which were also investigated. High-throughput sequencing was employed to identify and characterize the transcriptome of *Aspergillus sydowii* MS-19. Carbohydrate-Active Enzymes (CAZyme)-annotated genes in *Aspergillus sydowii* MS-19 were compared with those in the brown-rot fungi representative species, *Postia placenta* and *Penicillium decumbens*. One sequence annotated as laccase, which can degrade lignin, and three peroxidase sequences sharing a similar structure with typical lignin peroxidase and manganese peroxidase were obtained. The novel Antarctic *Aspergillus sydowii* strain MS-19 could be utilized as a potential source of lignocellulosic enzymes.

Additional files

Additional file 1: Figure S1. Phylogeny of Antarctic fungal isolates. Neighbour-joining tree showing the relationship between the ITS sequences from 15 Antarctic native isolates and their closest relatives as well as common fungi in Antarctic. The bootstrap values of the neighbor-joining analysis with 1000 replications are shown on the branches. The scale bar represents 0.05 substitutions per amino acid site. The isolated strains can be classified into 4 classes: *Eurotiomycetes,*

Leotiomycetes, Sordariomycetes and Dothideomycetes. All the 18S/ITS sequences of isolated fungus have been submitted to GenBank and the GenBank accession No. can be available in the brackets.

Additional file 2: Table S2. Annotated unigenes associated with carbohydrate metabolism and xenobiotic biodegradation. Annotated unigenes, pathways and corresponding members number associated with carbohydrate metabolism and xenobiotic biodegradation after the CAZyme annotation of *A. sydowii* MS-19 transcriptome.

Abbreviations
CAZyme: Carbohydrate-active enzymes; CBM: Carbohydrate-binding modules; CE: Carbohydrate esterase; GH: Glycoside hydrolase; GO: Gene ontology; GT: Glycosyltransferase; ITS: Internal transcribed spacer; KEGG: Kyoto encyclopaedia of genes and genomes; Lac: Laccase; Lip: Lignin peroxidase; Mnp: Manganese peroxidase; PL: Polysaccharide lyase; VP: Versatile peroxidase

Acknowledgements
The author would like to thank Xiaomo Li, Sihui Zhu for the assistance of bioinformatics analysis and Huanghao Yang, Xian Chen for the assistance of biosensor technology.

Funding
This work was financially supported by Chinese National Natural Science Foundation Grant No. 41006102, Chinese Polar Environment Comprehensive Investigation & Assessment Programs (Grant nos. CHINARE2015–01-06 and CHINARE 2015–02-01), the Agriculture Guiding Project of Fujian Provincial Department of Science and Technology Grant No. 2016N0016.

Authors' contributions
BC designed and supervised the experiments, wrote and revised the manuscript. NW collected soil samples, SL conducted enzyme activity assay, FL prepared nuclear acid material for RNA-seq. XY and BC contributed to data analysis and interpretation. JS participated in discussion of results. All authors read and approved the final manuscript.

Competing interests
The authors declare that they have no competing interests.

References
1. Zaldivar J, Nielsen J, Olsson L. Fuel ethanol production from lignocellulose: a challenge for metabolic engineering and process integration. Appl Microbiol Biotechnol. 2001;56(1–2):17–34.
2. Dashtban M, Schraft H, Qin W. Fungal bioconversion of lignocellulosic residues; opportunities & perspectives. Int J Biol Sci. 2009;5(6):578–95.
3. Sun Y, Cheng J. Hydrolysis of lignocellulosic materials for ethanol production: a review. Bioresour Technol. 2002;83(1):1–11.
4. Del Rio JC, Marques G, Rencoret J, Martinez AT, Gutierrez A. Occurrence of naturally acetylated lignin units. J Agric Food Chem. 2007;55(14):5461–8.
5. Lynd LR, Weimer PJ, van Zyl WH, Pretorius IS. Microbial cellulose utilization: fundamentals and biotechnology. Microbiol Mol Biol Rev. 2002;66(3):506–77. table of contents
6. JVRE L, Howard S. Lignocellulose biotechnology: issues of bioconversion and enzyme production. Afr J Biotechnol. 2004;2:602–19.
7. Sanchez C. Lignocellulosic residues: biodegradation and bioconversion by fungi. Biotechnol Adv. 2009;27(2):185–94.
8. Buswell JA, Eriksson KE, Gupta JK, Hamp SG, Nordh I. Vanillic acid metabolism by selected soft-rot, brown-rot, and white-rot fungi. Arch Microbiol. 1982;131(4):366–74.
9. Heidelberg SB. Degradation of Plant Cell Wall Polymers by Fungi. Berlin Heidelberg: Springer; 2007.
10. Kirk TK, Tien M. Lignin-degrading enzyme fromPhanerochaete chrysosporium. Appl. Biochem. Biotechnol. 1984;9(4):317–8.
11. Bao W, ., Fukushima Y,., Jensen KA, Moen MA, Hammel KE: Oxidative degradation of non-phenolic lignin during lipid peroxidation by fungal manganese peroxidase. FEBS Lett. FEBS Lett 1994, 354(3):297-300.
12. Bourbonnais R, Paice MG. Oxidation of non-phenolic substrates: An expanded role for laccase in lignin biodegradation. FEBS Lett. 1990;267(1):99–102.
13. Rodakiewicz-Nowak J, Jarosz-Wilkołazka A, Luterek J. Catalytic activity of versatile peroxidase from Bjerkandera fumosa in aqueous solutions of water–miscible organic solvents. Appl. Catal. A Gen. 2006;308(4):56–61.
14. Edwards SL, Raag R, Wariishi H, Gold MH, Poulos TL. Crystal structure of lignin peroxidase. Proc Natl Acad Sci. 1993;90(2):750–4.
15. MartíNez AT. Molecular biology and structure-function of lignin-degrading heme peroxidases. Enzym. Microb. Technol. 2002;30(01):425–44.
16. Martínez AT, Speranza M, Ruiz-Dueñas FJ, Ferreira P, Camarero S, Guillén F, Martínez MJ, Gutiérrez A, JCD R. Biodegradation of lignocellulosics: microbial, chemical, and enzymatic aspects of the fungal attack of lignin. Int Microbiol Off J Span Soc Microbiol. Int Microbiol. 2005;8:195–204.
17. Fernández-Fueyo E, Ruiz-Dueñas FJ, Miki Y, Martínez MJ, Hammel KE, Martínez AT. Lignin-degrading peroxidases from genome of selective ligninolytic fungus Ceriporiopsis subvermispora. J Biol Chem. 2012;287(20):16903–16.
18. Akila G, Chandra TS. A novel cold-tolerant Clostridium strain PXYL1 isolated from a psychrophilic cattle manure digester that secretes thermolabile xylanase and cellulase. FEMS Microbiol Lett. 2003;219(1):63–7.
19. Violot S, Haser R, Sonan G, Georlette D, Feller G, Aghajari N. Expression, purification, crystallization and preliminary X-ray crystallographic studies of a psychrophilic cellulase from Pseudoalteromonas haloplanktis. Acta Crystallogr Sect D. 2003;59(7):1256–8.
20. Haas BJ, Papanicolaou A, Yassour M, Grabherr M, Blood PD, Bowden J, Couger MB, Eccles D, Li B, Lieber M, et al. De novo transcript sequence reconstruction from RNA-seq using the Trinity platform for reference generation and analysis. Nat Protoc. 2013;8(8):1494–512.
21. Martinez D, Larrondo LF, Putnam N, Gelpke MDS, Huang K, Chapman J, Helfenbein KG, Ramaiya P, Detter JC, Larimer F. Erratum: Genome sequence of the lignocellulose degrading fungus Phanerochaete chrysosporium strain RP78. Nat Biotechnol. 2004;22(6):695–700.
22. Vanden Wymelenberg A, Gaskell J, Mozuch M, Kersten P, Sabat G, Martinez D, Cullen D. Transcriptome and Secretome Analyses of Phanerochaete chrysosporium Reveal Complex Patterns of Gene Expression. Appl. Environ. Microbiol. 2009;75(12):4058–68.
23. Kersten P, Cullen D. Extracellular oxidative systems of the lignin-degrading Basidiomycete Phanerochaete chrysosporium. Fungal Genet. Biol. 2007;44(2):77–87.
24. Sato S, Liu F, Koc H, Ming T. Expression analysis of extracellular proteins from Phanerochaete chrysosporium grown on different liquid and solid substrates. Microbiology. 2007;153(Pt 9):3023–33.
25. Diego M, Jean C, Ingo M, David H, Monika S, Kubicek CP, Patricia F, Ruiz-Duenas FJ, Martinez AT, Phil K. Genome, transcriptome, and secretome analysis of wood decay fungus Postia placenta supports unique mechanisms of lignocellulose conversion. Proc Natl Acad Sci U S A. 2009;106(6):1954–9.
26. González-Mendoza D, Argumedo-Delira R, Morales-Trejo A, Pulido-Herrera A, Cervantes-Díaz L, Grimaldo-Juarez O, Alarcón A: A rapid method for isolation of total DNA from pathogenic filamentous plant fungi. Genetics and molecular research. 2010;9(1):162-6.
27. Archibald FS. A new assay for lignin-type peroxidases employing the dye azure B. Appl Environ Microbiol. 1992;58(9):3110–6.

28. Saparrat MC, Martinez MJ, Cabello MN, Arambarri AM. Screening for ligninolytic enzymes in autochthonous fungal strains from Argentina isolated from different substrata. Rev. Iberoam. Micol. 2002;19(3):181–5.

29. Chen H, Tan Z, Feng K, Xie M, Wang Z, Gong D. Optimization of Fermenting Culture Medium for Cellulase Production from Trichoderma by Response Surface Method. Liquor-Making Science & Technology. 2011;

30. Zerbino DR, Birney E. Velvet: algorithms for de novo short read assembly using de Bruijn graphs. Genome Res. 2008;18(5):821–9.

31. Schulz MH, Zerbino DR, Vingron M, Birney E. Oases: robust de novo RNA-seq assembly across the dynamic range of expression levels. Bioinformatics (Oxford, England). 2012;28(8):1086–92.

32. Yin Y, Mao X, Yang J, Chen X, Mao F, Xu Y. dbCAN: a web resource for automated carbohydrate-active enzyme annotation. Nucleic Acids Res. 2012; 40(Web Server issue):W445–51.

33. Masran R, Zanirun Z, Bahrin EK, Ibrahim MF, Yee PL, Abd-Aziz S. Harnessing the potential of ligninolytic enzymes for lignocellulosic biomass pretreatment. Appl Microbiol Biotechnol. 2016;100(12):5231–46.

34. Black GW, Rixon JE, Clarke JH, Hazlewood GP, Theodorou MK, Morris P, Gilbert HJ. Evidence that linker sequences and cellulose-binding domains enhance the activity of hemicellulases against complex substrates. Biochem J. 1996;319(Pt 2):515–20.

35. Faison BD, Kirk TK. Factors Involved in the Regulation of a Ligninase Activity in Phanerochaete chrysosporium. Appl Environ Microbiol. 1985;49(2):299–304.

36. Yoon J, Maruyama JI, Kitamoto K. Disruption of ten protease genes in the filamentous fungus Aspergillus oryzae highly improves production of heterologous proteins. Appl. Microbiol. Biotechnol. 2011;89(3):747–59.

37. Bansal P, Hall M, Realff MJ, Lee JH, Bommarius AS. Modeling cellulase kinetics on lignocellulosic substrates. Biotechnol Adv. 2009;27(6):833–48.

38. Collins T, Gerday C, Feller G. Xylanases, xylanase families and extremophilic xylanases. FEMS Microbiol Rev. 2005;29(1):3–23.

39. de Vries RP, Visser J. Aspergillus enzymes involved in degradation of plant cell wall polysaccharides. Microbiol Mol Biol Rev. 2001;65(4):497–522.

40. Martins MP, Mouad AM, Boschini L, Regali Seleghim MH, Sette LD, Meleiro Porto AL: Marine fungi Aspergillus sydowii and Trichoderma sp. catalyze the hydrolysis of benzyl glycidyl ether. Mar Biotechnol 2011, 13(2):314-320(317).

41. Archibald FS. A new assay for lignin-type peroxidases employing the dye azure B. Appl. Environ. Microbiol. 1992;58(58):3110–6.

42. Glenn JK, Gold MH. Purification and characterization of an extracellular Mn(II)-dependent peroxidase from the lignin-degrading basidiomycete, Phanerochaete chrysosporium. Arch Biochem Biophys. 1985;242(2):329–41.

4

Probiotic bacteria prevent *Salmonella*–induced suppression of lymphoproliferation in mice by an immunomodulatory mechanism

R. Doug Wagner* and Shemedia J. Johnson

Abstract

Background: *Salmonella enterica* infections often exhibit a form of immune evasion. We previously observed that probiotic bacteria could prevent inhibition of lymphoproliferation and apoptosis responses of T cells associated with *S. enterica* infections in orally challenged mice.

Results: In this study, changes in expression of genes related to lymphocyte activation in mucosa-associated lymphoid tissues (MALT) of mice orally infected with *S. enterica* with and without treatment with probiotic bacteria were evaluated. Probiotic bacteria increased expression of mRNA for clusters of differentiation antigen 2 (*Cd2*), protein tyrosine phosphatase receptor type C (*Ptprc*), and Toll-like receptor 6 (*Tlr6*) genes related to T and B cell activation in mouse intestinal tissue. The probiotic bacteria were also associated with reduced mRNA expression of a group of genes (*RelB, Myd88, Iκκα, Jun, Irak2*) related to nuclear factor of kappa light chains enhancer in B cells (NF-κB) signal transduction pathway-regulated cytokine responses. Probiotic bacteria were also associated with reduced mRNA expression of apoptotic genes (*Casp2, Casp12, Dad1, Akt1, Bad*) that suggest high avidity lymphocyte sparing. Reduced CD2 immunostaining in mesenteric lymph nodes (MLN) was suggestive of reduced lymphocyte activation in probiotic-treated mice. Reduced immunostaining of TLR6 in MALT of probiotic-treated, *S. enterica*-infected mice suggests that diminished innate immune sensitivity to *S. enterica* antigens is associated with preventing lymphocyte deletion.

Conclusions: The results of this study are consistent with prevention of *S. enterica*-induced deletion of lymphocytes by the influence of probiotic bacteria in mucosal lymphoid tissues of mice.

Keywords: Probiotic, Gene expression, Signal transduction genes, *Salmonella*, Apoptosis

Background

Salmonella enterica is known to evade host defenses and induce persistent infections [1]. *Salmonella* sp. persistence is also aided by a virulence mechanism that causes clonal deletion of high-avidity CD4$^+$ T cells [2]. Furthermore, *Salmonella* sp. induces an inflammatory response and secretes a factor that blocks macrophage and dendritic cell migration [3]. The intestinal microbiota can modulate the pathogenesis and host evasion mechanisms of *S. enterica*; for example, commensal *E. coli* improved

* Correspondence: doug.wagner@fda.hhs.gov
Microbiology Division, National Center for Toxicological Research, U.S. Food and Drug Administration, 3900 NCTR Rd, Jefferson, AR 72079, USA

the host response to *Salmonella* sp. infection by an immune system regulatory effect [4]. The mechanisms of mucosal immune system regulation are only partially understood and the roles of microbes in this regulation are even less well understood. Addition of probiotic bacteria to the indigenous intestinal microbiota also affects mucosal immune regulation.

In a previous study, we observed that bacteria from a commercial probiotic product modified the host immune response to *S. enterica* by enhancing the proliferative response of spleen lymphocytes to *S. enterica* antigens. We also observed that the T and B cell enriched splenocyte fractions had reduced activities of apoptosis-related cysteine peptidase or caspase (CASP) 3 and CASP7 [5]. Caspases are involved in the signaling pathway that directs programmed cell death (apoptosis) in activated immune cells. These results were observed in mice that were not previously immunized to *S. enterica*, suggesting that the innate pathways of immune activation were involved. T cells that have been activated, but lacking secondary signals, activate an intrinsic caspase pathway that is arrested by signals from B- cell CLL/lymphoma 2 (BCL) 2 and its related proteins. The probiotic bacteria may be providing analogs to these secondary signals that maintain the survival of activated lymphocytes.

The present study sought to determine whether the secondary signals are mediated by intermediary antigen-presenting cells (APC), such as dendritic cells and intestinal epithelial cells or by direct contact to lymphocytes. APC would signal to lymphocytes with cytokines to affect a change in their activation and apoptotic status. Expression of these and other signaling molecules in mucosal tissues of probiotic-treated and untreated *S. enterica*-infected mice were compared. A direct T cell response to microbial antigens would have to occur by a yet unrevealed mechanism or possibly by the same pathways that function in the APC. Intestinal epithelial cells are also involved in T cell activation by production of pro-inflammatory cytokines in the presence of commensal or probiotic bacteria [6].

Methods
Microorganisms
A blend of bacteria derived from a commercial probiotic product, which contained: *Lactobacillus reuteri, Lactobacillus rhamnosus, Lactobacillus acidophilus, Lactobacillus casei, Lactobacillus gasseri, Bifidobacterium thermophilus, Bifidobacterium longum,* and *Bifidobacterium adolescentis* was used, as previously described [5]. These isolates are available from the corresponding author upon reasonable request. The products were cultured for isolation of component bacteria on de Man Rogosa Sharpe (MRS) agar (REMEL Laboratories) and *Bifidobacterium* agars (Anaerobe Systems) incubated anaerobically for 48 h at 37 °C. Bacterial isolates were identified by their cellular fatty acid methyl ester compositions (Microbial ID, Inc.) and by their 16S rRNA sequences with the MicroSeq 16S rRNA sequence assays (Applied Biosystems, Inc.). Rapid identifications were made with the Biolog Microbial Identification System. Serial dilution plate counts on MRS or *Bifidobacterium* agar were used to quantify the components of the probiotic products. Probiotic products were suspended in drinking water at a concentration of 5×10^6 CFU/mouse and administered to mice with a feeding tube, as previously described [5]. *Salmonella enterica* Serotype Cubana, originally isolated from poultry [5] was grown on Trypticase Soy agar with 5% sheep blood

(REMEL) or in Trypticase Soy broth at 37 °C in an atmosphere of 5% CO_2 and air. The *Salmonella* isolate was originally characterized as Serotype Typhimurium and subsequently determined to be serotype G2 (serovar Cubana) by the Arkansas Regional Laboratory of the Food and Drug Administration (FDA), Jefferson, AR. The isolate is available upon reasonable request from the corresponding author.

Mice
A total of 4 male and 4 female 8 week-old human microbiota-associated [5] and 4 male and 4 female 8 week-old defined-microbiota BALB/c mice (Charles River) were used in this study with the approval of the Institutional Animal Care and Use Committee of the National Center for Toxicological Research. Sterile water and NIH-31 mouse chow (Purina) were supplied ad libitum to the mice. The probiotic bacteria were administered with a feeding tube to mice 7 days before oral challenge with *S. enterica*. Isolator sterility was assessed with weekly swab cultures on Trypticase soy blood agar plates (REMEL). The cultures were incubated at 37 °C in 95% air, 5% CO_2 atmosphere overnight and the plates were assessed for bacterial growth. The mice were fed a NIH-31 diet sterilized in the transfer box as previously described [5].

Experimental design
Control mice were human microbiota-associated in our laboratory, as previously described [5] or specific pathogen free from the vendor and fecal samples were cultured to assure their status. Treatment mice were colonized with the probiotic bacteria blend and fecal samples were collected from the mice 1 day later to culture for the presence of the probiotic bacteria. At 7 days after probiotic treatment, 4 probiotic-treated mice and 4 untreated mice were each orally inoculated with 2×10^8 colony forming units (CFU) of *Salmonella enterica* using a feeding tube. Seven days later, the mice were euthanized and MALT consisting of intestinal lamina propria, Peyer's patches, and mesenteric lymph nodes were excised for analysis. Spleen cells were isolated and incubated with *S. enterica* antigens and mitogens to evaluate lymphoproliferative effects, as previously described [5]. Total RNA samples were extracted from the tissues and cDNA were generated, from which quantitative real-time-polymerase chain reaction (qRT-PCR) analyses were run to quantify mRNA expressed from genes of the mucosal immune system in probiotic-treated and untreated mice. The data were analyzed to identify cell type markers and intracellular signaling molecules involved in the host response to *S. enterica* that were affected by the probiotic bacteria. Control mice and probiotic-bacteria-treated BALB/c mice were orally challenged

with *S. enterica* and pathway-focused gene expression profiles were generated from qRT-PCR expression arrays, as described below, to compare signal transduction in MALT from defined-microbiota mice treated with or without probiotic bacteria and orally challenged with *S. enterica*.

The experiment was repeated on a group of 8 specific pathogen free mice to obtain intestinal lamina propria, Peyer's patches, and mesenteric lymph nodes for immunolocalization by an immunohistochemical method using antibodies from Santa Cruz Biotechnology. Immunolocalization of CD21$^+$ B cells and dendritic cells, CD2$^+$ T lymphocytes, and detection of PTPRC, TLR6, and v-rel avian reticuloendotheliosis viral oncogene homolog B (RELB) cellular expression was used to evaluate the probiotic effects that were observed on mRNA expression.

Antigen preparations

Antigens were prepared from crude lysates of *S. enterica* for in vitro activation and apoptosis assays of lymphocytes collected from the spleens of mice from the experiments, as previously described [5]. Briefly, the entire volume of a 500 ml log phase broth culture of bacteria was centrifuged at 2000 × *g* for 15 min. The bacterial pellet was washed three times with an equal volume of PBS and centrifuged again. The final bacterial pellet was suspended in 10 ml of PBS and passed through a French pressure cell (SLM/AMINCO) at 15,000 lb./in^2 to disrupt the bacteria. The disrupted bacteria were centrifuged at 2000 × *g* and the protein content of the supernatant was determined by the bicinchoninic acid protein assay (Pierce Chemical Co.) to express antigen mass as mg protein, and used as the antigens for lymphocyte proliferation assays.

Lymphocyte proliferation assay

Lymphocytes from the spleens of mice treated with probiotics after *S. enterica* challenge were assayed for proliferative responses to *S. enterica* antigens, as previously described [5]. Lymphocyte proliferation assays were performed with the CellTiter Aqueous 96 assay (Promega, Corp.). Lymphocytes from the spleens of experimentally treated mice were prepared and incubated at a density of 5 × 10^5 cells/well of a 96-well culture plate in RPMI medium (Thermo Fisher Scientific, Inc.) containing *S. enterica* antigens. Antigens were added to 3 wells with spleen cells at a concentration of 10 μg whole cell lysate protein antigen preparation per well. Antigens were incubated with the cells 56 h at 37 °C in a humidified 5% CO$_2$ incubator before testing for lymphocyte proliferation. The formation of proliferating clonal clusters was also verified microscopically. The proliferation of lymphocytes in response to the antigens was measured as absorbance of reduced 3-(4,5-dimethylthiazol-2-yl)-5-(3-carboxymethoxyphenyl)-2-(4-sulfonyl)-2H–tetrazolium,

inner salt (MTS) at 490 nm, which was measured with a plate reader (Applied Biosystems). The average of three wells per sample was used to determine the mean ± standard error of the mean (SEM) Abs$_{490}$ for three mice per group. Proliferative responses of lymphocytes to antigens were compared as % increases in MTS absorbance as a result of the effects of probiotics.

Apoptosis assay

Lymphocytes from the spleens of the mice 7 days after probiotic treatment and *S. enterica* challenge were analyzed for activation of caspases 3 and 7 (Apo-ONE, Promega Corp.). Each assay well of a Nunc F16 black Maxisorp 96 well fluorescent assay plate (Thermo Fisher Scientific, Inc.) contained 50 μl of cell suspension in Roswell Park Memorial Institute (RPMI) 1640 medium at a cell concentration of 2 × 10^5 cells/ml to which was added 10 μg antigen preparation per well. After 56 h incubation at 37 °C in a humidified 5% CO$_2$ incubator, 100 μl of working substrate solution was added to each well. The plate was rotated at 300 rpm for 30 min at room temperature. Fluorescence intensities were measured in a plate reader (Applied Biosystems) set with filters for an excitation wavelength of 485 nm and an emission wavelength of 530 nm at 30 min intervals until the rates of increase reached a plateau state. Endpoint fluorescence intensities from each treatment group were compared as the % change in relative fluorescence intensities resulting from probiotic inhibition of CASP3/7activation.

qRT-PCR array profiling of signaling pathway and cytokine genes

RT^2 Profiler™ PCR arrays from Qiagen Bioscience were used to assess expression of mRNA for 327 genes involved in the host response to bacteria in the mucosal immune tissues of the mouse GI tract. Arrays for genes involved in apoptosis (PAMM-012 Mouse Apoptosis Array), for genes involved in NF-κB activation (PAMM-025 Mouse NF-κB Array), and for mouse T and B cell activation markers (PAMM-053) were used according to the manufacturer's instructions. Total cellular RNA from Peyer's patches, mesenteric lymph nodes, and lamina propria from 8 mice were isolated using ArrayGrade total RNA isolation kits (Qiagen, Inc.). The RNA samples were treated with DNAse-1 (Thermo Fisher Scientific, Inc.) reverse-transcribed with the RT^2 PCR Array first strand kit and the resulting cDNA was analyzed by real-time PCR for detection in a BioRad Pci,Q5 instrument. The housekeeping genes used in the study were: beta glucuronidase, hypoxanthine phosphoribosyltransferase 1, heat shock protein 90 alpha family class B member 1, glyceraldehyde-3-phophate dehydrogenase, and beta actin. Results of the PCR array experiment were analyzed with the Excel™ (Microsoft Corp.) template provided by

Qiagen, Inc. to determine the key signal transduction pathways and immune system cells involved in the probiotic effects.

Immunolocalization of responses to probiotic bacteria in the murine MALT

The gene expression profiling was used to indicate specific mouse immune system genes that respond to the effect of probiotic bacteria in response to *S. enterica*. Two-color immunohistochemistry was used to determine the cell types involved with specific gene product markers and their tissue locations. Toxicologic Pathology Associates (Jefferson, Arkansas) prepared frozen sections of the MALT tissues from 8 mice and conducted the immunolocalization on those sections for cell markers and specific intracellular signal transduction gene products. Antibodies labelled with horseradish peroxidase specific for CD21, CD2, PTPRC, TLR6, and RELB were purchased from Santa Cruz Biologics. Densitometry of the stained areas in Peyer's patches, lamina propria of intestinal villi, or the cortical and paracortical regions of mesenteric lymph nodes were measured by the Positive Pixel Count Algorithm from Aperio Technologies (Leica Biosystems).

Analysis of data

Evaluation of statistically significant differences between the results from treatment groups and control groups were determined with Repeated Measures Analysis of Variance and Bonferroni's post tests using Prism v.6.0 software (GraphPad Software). Numerical count data were $\log_{10}$ transformed prior to statistical analysis to make the data better fit a normal distribution. Statistical significance was defined at $P < 0.05$.

Results

Persistence of probiotic bacteria and *Salmonella* sp. in the murine GI tracts

In our previous study [5], colonization of germfree mice by probiotic bacteria was easily verified by microbial culture from feces, but the mice in this study had conventional microbiota. The presence of *L. reuteri*, which is not a component of the BALB/c mouse colony microbiota (Charles River) in the defined-microbiota mice in the present study was confirmed by culture on MRS agar and microbial identification of colonies from feces at 7 days after oral feeding of 8.0 $\log_{10}$ CFU of the probiotic bacterial mixture. Infection of mice was confirmed by recovery of *S. enterica* from the mice 7 days after oral challenge (Table 1).

S. enterica was recovered from one of the probiotic-fed mice, and all of the control mice 7 days after oral *Salmonella* challenge. Table 1 also shows numbers of lactic acid bacteria that were recovered from mice 7 days after oral *S. enterica* challenge and grown on MRS agar.

Table 1 Numbers of bacteria recovered from experimental mice

Animal[a]	Probiotic-treated	Salmonella detected (CFU/mL)[b]	No. total lactobacilli (CFU/g)[c]
1	-	2.2×10^6	5.0×10^{10}
2	-	6.7×10^6	3.3×10^{10}
3	-	1.6×10^7	4.8×10^{10}
4	-	2.8×10^6	4.2×10^{10}
5	+	0	6.9×10^{11}
6	+	0	4.2×10^{11}
7	+	0	4.4×10^{11}
8	+	7.6×10^6	5.2×10^{11}

[a]Mice 1–4 were non-treated with probiotic lactic acid bacteria. Mice 5–8 were orally challenged with 1×10^8 CFU probiotic bacteria mixture. Seven days later, all mice were orally challenged with 1×10^8 CFU/mL *S. enterica*. [b]The number of *S. enterica* isolated from feces of mice 7 days after challenge were enumerated on SS agar plates. [c]The numbers of lactic acid bacteria were enumerated on MRS agar plates and reported as CFU/g feces 7 days after *S. enterica* challenge

Probiotic bacteria prevented immunosuppression of T cells in mice to *S. enterica* antigens

Lymphocytes from the spleens of BALB/c mice challenged with *S. enterica* only for 7 days did not proliferate in response to the B-cell mitogen lipopolysaccharide (LPS) or to the T cell mitogen concanavalin-A, nor did they respond to soluble antigens from *S. enterica* (Fig. 1a). This lack of responsiveness was observed under similar conditions in our previous report [5]. Decreased lymphoproliferation of splenocytes in mice treated with probiotics was observed in the control treatment and LPS treatment groups. Splenocytes from mice that were fed the mixture of probiotic bacteria prior to *S. enterica* challenge had significant proliferative responses to concanavalin-A and the *S. enterica* antigens (Fig. 1a), showing a preventative effect of the probiotic bacteria on *S. enterica*-induced immunosuppression. The lymphoproliferative responses to *S. enterica* LPS were not significantly changed by probiotic bacteria compared with the *S. enterica*-treated control mice.

The probiotic bacteria altered the activation of cellular apoptosis in lymphocytes, as reported previously [5]. In the present study, activation of CASP3 and CASP7, which mediate lymphocyte apoptosis, was significantly increased in *S. enterica*-infected BALB/c mice, compared to the response to LPS or concanavalin-A, but was strongly induced by *S. enterica* antigens (Fig. 1b). The mice that were fed probiotic bacteria before *S. enterica* challenge did not have a significant increase in lymphocyte CASP3 and CASP7 activation by *S. enterica* antigens (Fig. 1b), which suggests that strong *S. enterica* induction of T cell apoptotic responses was suppressed by the probiotics.

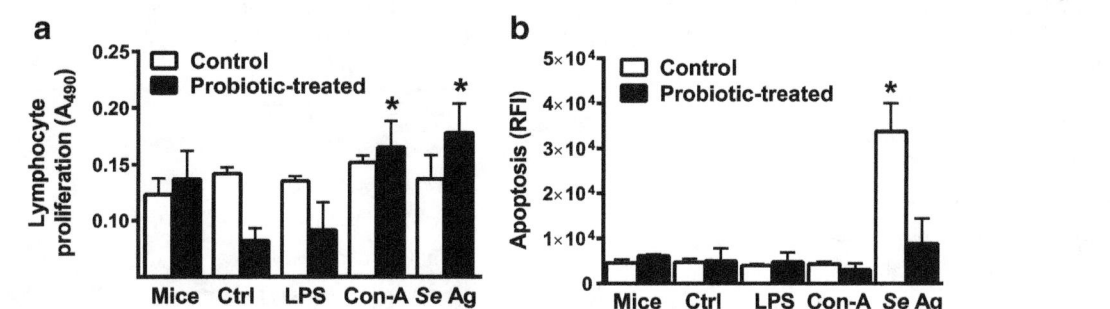

Fig. 1 Lymphocyte proliferation and apoptosis responses. Proliferation of splenocytes from uninfected mice (Mice), *S. enterica*-infected BALB/c mice (Ctrl), or *S. enterica*-infected, probiotic-treated mice to LPS (LPS), concanavalin-A (Con-A), or soluble *S. enterica* antigens (*Se* Ag) was measured as change in absorbance of MTS at 490 nm (**a**). Apoptosis was measured as activation of caspases 3 and 7 in splenocytes from uninfected mice (Mice), *S. enterica*-infected mice (Ctrl), or *S. enterica*-infected, probiotic-treated mice to LPS (LPS), concanavalin-A (Con-A), or soluble *S. enterica* antigens (*Se* Ag), detected as relative fluorescence intensity (RFI) of caspase substrate, were compared (**b**). The *asterisks* indicate statistically significant differences between control (Ctrl) and treated groups by ANOVA, $P < 0.05$

Probiotic bacteria altered mRNA expression of genes involved in responses to *S. enterica* infection

The presence of probiotic bacteria induced changes in expression of immune response genes and cellular locations in the MALT of mice infected with *S. enterica*. There was a general increase in T cell function and a decrease in inflammatory responses, as evidenced by increased expression of T cell activation genes and apoptosis-related genes and decreased gene expression of NF-κB signal transduction pathway components and proinflammatory cytokines, respectively (Table 2). Probiotic lactobacilli had significant effects on the expression of mRNA from genes involved in B and T cell activation, signal transduction by the NF-κB pathway, and apoptosis, when the mice were challenged with *S. enterica*. Expression of the mRNA for the lymphokines interleukin (IL) -12 p40 (*Il12p40*) and *Il4* were reduced, while *Il10* mRNA was significantly increased by the presence of probiotic lactobacilli (Table 2). Other genes involved in lymphocyte activation were also affected by probiotic bacteria. Nearly 3-fold induction of mRNA for *Cd2*, and *Ptprc* genes (Table 2) indicates activation and differentiation of B and T cells. Other components of the T and B cell activation signal-transduction pathways were also induced (Table 2). Since activated splenocytes proliferated in response to concanavalin-A, but not *Salmonella* LPS (Fig. 1a), it appears that mostly T cell activation accounts for the increased expression of the lymphocyte activation genes listed in Table 2. This conclusion is supported by the result that expression of the mRNA for *Tlr4*, the protein of which is a receptor for *S. enterica* lipopolysaccharides on APC, was not affected by probiotic bacteria (Table 2). Decreased expression of *Tlr1* and increased expression of *Tlr6* in MALT suggests that probiotic bacteria modified the spectrum of bacterial lipoproteins

that could be detected by APC in the MALT. RNA for *Cd21* was not detected in the *S. enterica* challenged mice that were not treated with probiotic bacteria. It was present in samples of MALT from the probiotic-treated mice, consistent with the possibility that probiotics prevented the loss of antigen-specific lymphocytes at sites of infection.

Additional effects of the probiotic lactobacilli were observed in genes that relate to the NF-κB signal transduction pathway that controls expression of cytokine genes in response to receptor detection of microbial antigens. mRNA expression of several signaling protein genes CASP8 and FADD like apoptosis regulator (*Cflar*), *Casp8*, receptor TNFRSF-interacting serine threonine kinase 1 (*Ripk1*), TNF receptor associated factor 2 (*Traf2*), *Bcl10*, and *Bcl2*) was slightly reduced, but not to the 2-fold change level of significance (Table 2). Significant reduction ($P < 0.05$) in mRNA expression of *RelB*, myeloid differentiation primary response gene 88 (*Myd88*), ring finger protein 7 (*Rnf7*), inhibitor of light polypeptide gene enhancer in B cell kinase alpha (*Iκκα*), and Map/Erk kinase kinase (*Mekk*) are indicative of a response that suppresses transduction of proinflammatory signals in lymphocytes (Table 2). The increased mRNA expression of interleukin-*1 receptor-associated kinase 2* (*Irak2*) and v-jun sarcoma virus 17 oncogene homolog (*Jun*) induced by probiotic bacteria in Table 2 suggest differential regulation of these genes that may function simultaneously in other signal transduction pathways.

Activation of T and B lymphocytes to produce cytokines is linked to activation of programmed cell death responses in these cells. In this study, expression of mRNA for genes involved in apoptosis was reduced by the presence of probiotic bacteria (Table 2). Multiple signaling pathways of apoptosis induction and suppression function simultaneously to regulate this important process. We observed that probiotic bacteria were associated with suppression of

Table 2 Gene Expression Effects of Probiotic Bacteria on *S. enterica*-Challenged Mice (Fold Change in mRNA Expression vs. Control)

Gene-coded protein/Function	Fold Change[a]
B and T cell activation genes[b]	
Cd2/Induces IFN-γ, interacts with PTPRC	+ 2.9
Cd21 (Cr2)/B cell complement receptor	+ 3.0
Hells/Helicase - Cell growth, DNA methylation	+ 2.5
Hsp90aa1/AKT signal transduction pathway, protein chaperon	+ 2.3
Il4/TH-2 cytokine (IL-4)	- 2.6
Il10/TH-2 cytokine (IL-10)	+ 3.6
Il12b/TH-1 cytokine (IL-12p40)	- 2.7
Impdh/Cell growth	+ 2.5
Nkx2.3/Lymphocyte cellular differentiation	+ 2.3
Prkcd/B-cell signal transduction and apoptosis	+ 2.2
Prlr/Prolactin receptor	+ 2.1
Ptprc/(CD45/B220) cell growth/differentiation	+ 3.0
Tlr1/Proinflammatory receptor for microbes	- 2.5
Tlr6/Proinflammatory receptor for microbes	+ 2.3
Vav1/T and B cell activation via JNK and P38	+ 2.1
Wwp1/E3 ubiquitin protein ligase	+ 2.4
Zap70/T cell receptor cofactor	- 2.4
Nf-κb signal transduction genes[c]	
Bcl2/Anti-apoptotic signaling	- 1.1
Bcl2l10/Pro-apoptotic, pro-inflammatory signaling	- 1.8
Cflar/Pro-apoptotic signaling	- 1.2
Egr1/B-cell receptor signaling	- 2.1
Fasl/Intercellular signaling, pro-apoptotic	- 3.3
Fos/Intracellular signaling, pro-apoptotic	- 3.1
Icam1/Leukocyte movement, signaling	- 2.3
Ikκa (Chuk)/Pro-inflammatory activator of NF-κB	- 2.1
Irak2/Induces NFκB nuclear translocation	+ 2.3
Jun/(AP-1) B-cell and T-cell receptor signaling	+ 2.1
Mekk/(Map3k1) Pro-inflammatory	- 2.2
Myd88/Pro-apoptotic and inflammatory adapter	- 3.8
Nfkbia/Blocks NFκB nuclear translocation	- 2.1
Relb/NFκB cofactor activation in B-cells	- 3.0
Ripk1/Pro-inflammatory, apoptosis regulatory	- 1.1
Rnf7/IκB ubiquitination – pro-apoptotic, pro-inflammatory	- 3.8
Traf2/Pro-inflammatory, apoptosis regulatory	- 1.7
Apoptosis genes[d]	
Akt1/Anti-apoptotic	- 2.1
Bad/Anti- apoptotic	- 2.2
Card10/Proinflammatory via NFκB, pro-apoptotic	- 4.3
Casp2/Pro-apoptotic	- 2.3

Table 2 Gene Expression Effects of Probiotic Bacteria on *S. enterica*-Challenged Mice (Fold Change in mRNA Expression vs. Control) *(Continued)*

Casp12/Pro-apoptotic	- 2.8
Dad1/Possible apoptosis inhibitor	- 3.1
Fadd/Pro-apoptotic	- 2.2
Pak7/Protein kinase induces proliferation/ anti-apoptotic	- 4.2
Tnf/Pro-apoptotic cytokine	- 2.4

[a]The mean (n = 4/group) fold change in qRT-PCR threshold cycles (C_T) between tissue RNA samples amplified from probiotic-treated versus untreated control mice were calculated as $2^{(-\Delta\Delta CT)}$, where ΔC_T is the housekeeping gene-normalized average C_T and $\Delta\Delta C_T$ is the $\Delta C_{T(treatment)}$- $\Delta C_{T(control)}$. Positive values indicate increased expression and negative values indicate reduced expression
[b]Results from Qiagen PAMM-053 Mouse T and B Cell Activation Array
[c]Results from Qiagen PAMM-025 Mouse NF-κB Array
[d]Results from Qiagen PAMM-012 Mouse Apoptosis Array

caspase recruitment domain protein 10 (*Card10*), tumor necrosis factor alpha (*Tnf*), and fas-associated death domain protein (*Fadd*) mRNA expression, which are involved in initial steps of the apoptosis activation cascade and also reduction of mRNA expression of *Casp2* and *Casp12*, which are involved in the activation of enzymes involved in apoptosis (Table 2). Some genes involved in suppression of apoptosis cascade activation were also suppressed, including: defender against cell death 1 (*Dad1*), p21-activated kinase 7 (*Pak7*), v-akt murine thymoma viral oncogene homolog 1 (*Akt1*), *Bcl10*, and bcl2-associated agonist of cell death (*Bad*) (Table 2). These seemingly contradictory findings illustrate the precise balance needed in feedback regulation mechanisms for apoptosis in antigen-activated lymphocytes.

Probiotic bacteria changed the tissue distribution of cellular activation markers

In addition to mRNA expression effects of probiotic bacteria, tissue distributions of several proteins were investigated. Immunohistochemical staining of MALT and mesenteric lymph node tissues was used to trace the tissue locations of cells with changes in production of CD21, CD2, PTPRC, RELB, and TLR6 proteins within the MALT tissues. Immunolocalization helped to interpret the qPCR results in terms of probiotic effects on host responses to *S. enterica* and these results will be presented in the following narrative that includes the observations and discussion of their significance in the context of T cells and antigen-presenting cells responding to *S. enterica* in the MALT and movement of these cells between MALT and MLN.

The CD21 protein is expressed on B lymphocytes and dendritic cells, which can be differentiated morphologically in MALT and MLN. The germinal centers of Peyer's patches (Fig. 2a, b) and secondary follicles of MLN (Fig. 2c, d) contained numerous cells with CD21

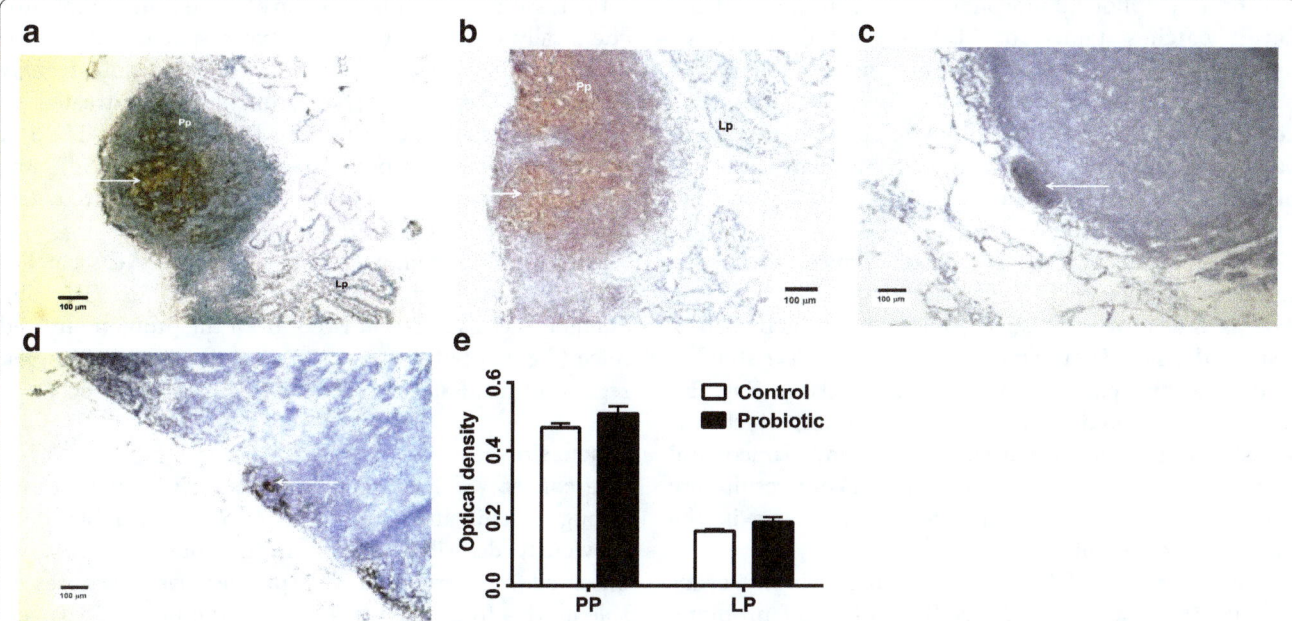

Fig. 2 Immunohistochemical staining of CD21in Peyer's Patches and MLN. The germinal centers of Peyer's patches (Pp) (**a**, control mice; **b**, probiotic-treated mice) and MLN (**c**, control mice; **d**, probiotic-treated mice) contained numerous cells with CD21 (*arrows*) that have lymphocyte morphology. Images 10× magnification. There were no significant differences in optical density of stained cells in probiotic-treated Peyer's patches (Pp) and lamina propria (Lp) compared to control *S. enterica*-challenged Peyer's patches and lamina propria by ANOVA, *P* < 0.05 (**e**)

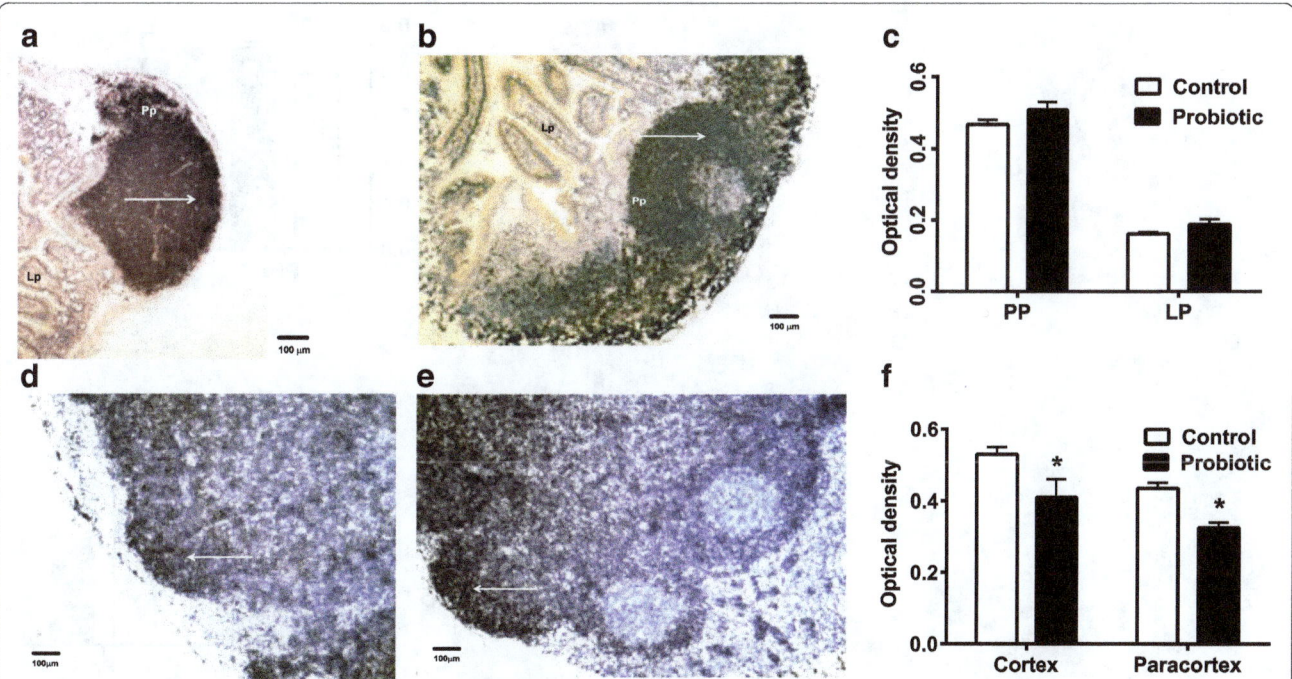

Fig. 3 Immunohistochemical staining of CD2 in Peyer's Patches and MLN. Peyer's Patches (Pp) of *S. enterica*-infected mice were highly reactive with heavy follicular CD2- staining (arrows), especially in probiotic-treated mice (**a**, control mice; **b**, probiotic-treated mice). There were no significant differences in staining between control and treated Peyer's patches or lamina propria (Lp) (**c**). CD2 staining of cortical and paracortical lymph node tissue (**d**, control mice; **e**, probiotic-treated mice) in probiotic-treated *Salmonella*-infected mice were different than control mice. Images 10× magnification. Statistically significant differences in staining of cortical and paracortical regions of the lymph nodes between control and probiotic-treated mice are shown at the asterisk, *P* < 0.05 (**f**)

that have lymphocyte morphology. Densitometry of the Peyer's patches stained for CD21 suggested no change in the numbers of cells expressing this marker in the probiotic-treated group, (P = 0.055) by ANOVA (Fig. 2e). The increased CD21 mRNA expression in Peyer's patches of the mice with probiotic treatment (Table 2) did not appear to change the numbers of CD21-producing cells in the Peyer's patches.

Peyer's patches of *S. enterica*-infected mice were highly reactive with heavy follicular CD2- staining (Fig. 3a). This is indicative of a strong migration of activated T cells to the site. There was reduced diffuse CD2 staining of cortical and paracortical lymph node tissue (Fig. 3e) in probiotic-treated *S. enterica*-infected mice, which was statistically significant between cortical and paracortical regions of the lymph nodes (Fig. 3f). These results are consistent with reduced inflammatory response in the probiotic-treated mice.

The infiltration of PTPRC-expressing cells in Peyer's Patches (Fig. 4a, b) and MLN (Fig. 4d, e) of probiotic treated and control *S. enterica*-infected mice in the two tissues appeared similar. Comparisons of staining densities of the replicates in these tissues of both treatment groups also showed no significant difference (Fig. 4c, f).

Expression of TLR6 was investigated immunohistochemically because its mRNA expression was increased in the probiotic-treated mice. TLR6 was not strongly stained in Peyer's patches from probiotic-treated *S. enterica*-infected mice (Fig. 5b). The amount of TLR6 in cortical areas of probiotic-treated MLN (Fig. 5d), was significantly reduced, compared with untreated *S. enterica*-challenged mice (Fig. 5e).

RELB cortical staining was apparent in Peyer's patches of probiotic-treated *S. enterica*-infected mice (Fig. 6b). Staining was also seen in MLN from the probiotic-treated mice (Fig. 6e) but was not significantly different between regions of the tissues or between treatments (Fig. 6c, f).

Discussion

The concept of T cell depletion by early induction of a strong inflammatory response by *S. enterica* has been previously described [2, 3]. In the present study, we observed recovery of T cell proliferative responses in splenocytes from mice treated with the probiotic bacterial blend prior to *S. enterica* infection, which is consistent with the concept that T cell depletion was averted. In a previous report, an immunosuppressive effect of *S. enterica* infection was observed on splenic lymphoproliferative responses in gnotobiotic BALB/c mice [5].

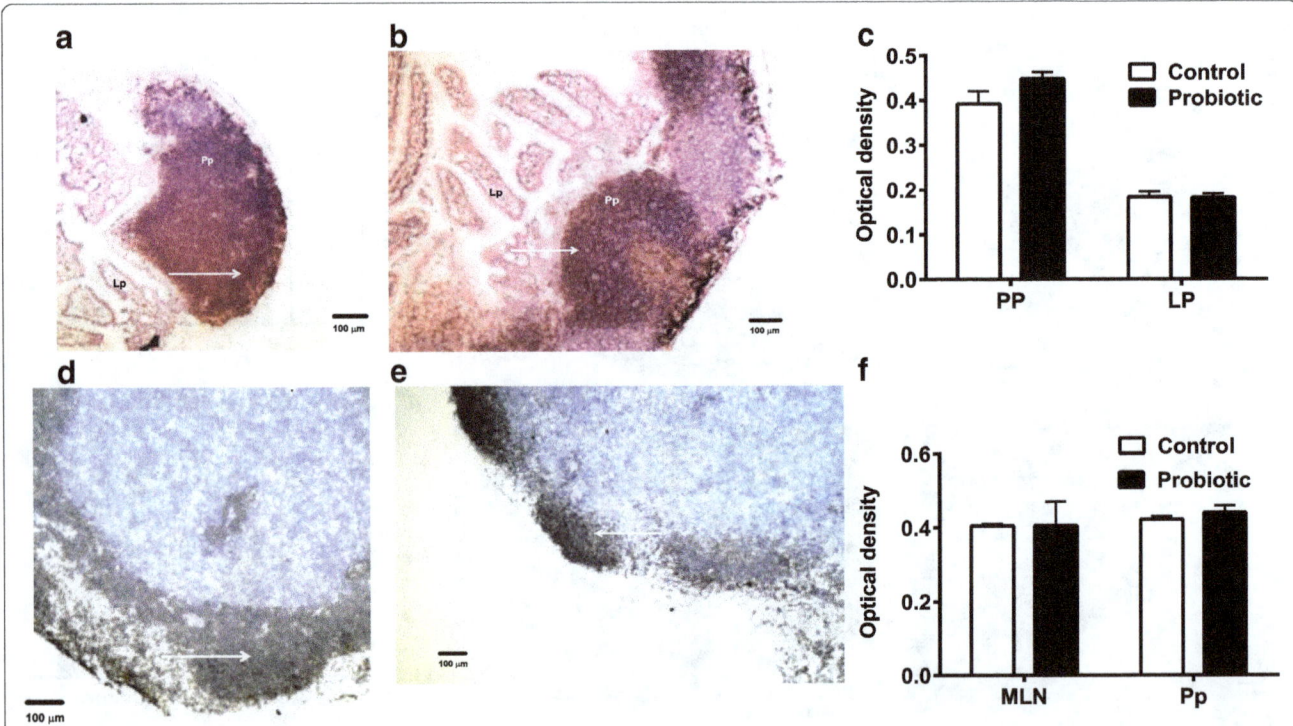

Fig. 4 PTPRC production shown by immunohistochemistry. Peyer's patches (Pp) and lamina propria (Lp) (**a**, control mice; **b**, probiotic-treated mice) appeared to have the same amount of PTPRC staining (arrows). There were no significant differences measured by densitometry (**c**). MLN (**d**, control mice; **e**, probiotic-treated mice) of probiotic-treated and *S. enterica*-infected mice also showed similar amounts of PTPRC staining in germinal centers. No significant differences in PTPRC staining between control and treated mice were confirmed by densitometry (**f**). PTPRC staining is shown at the arrows

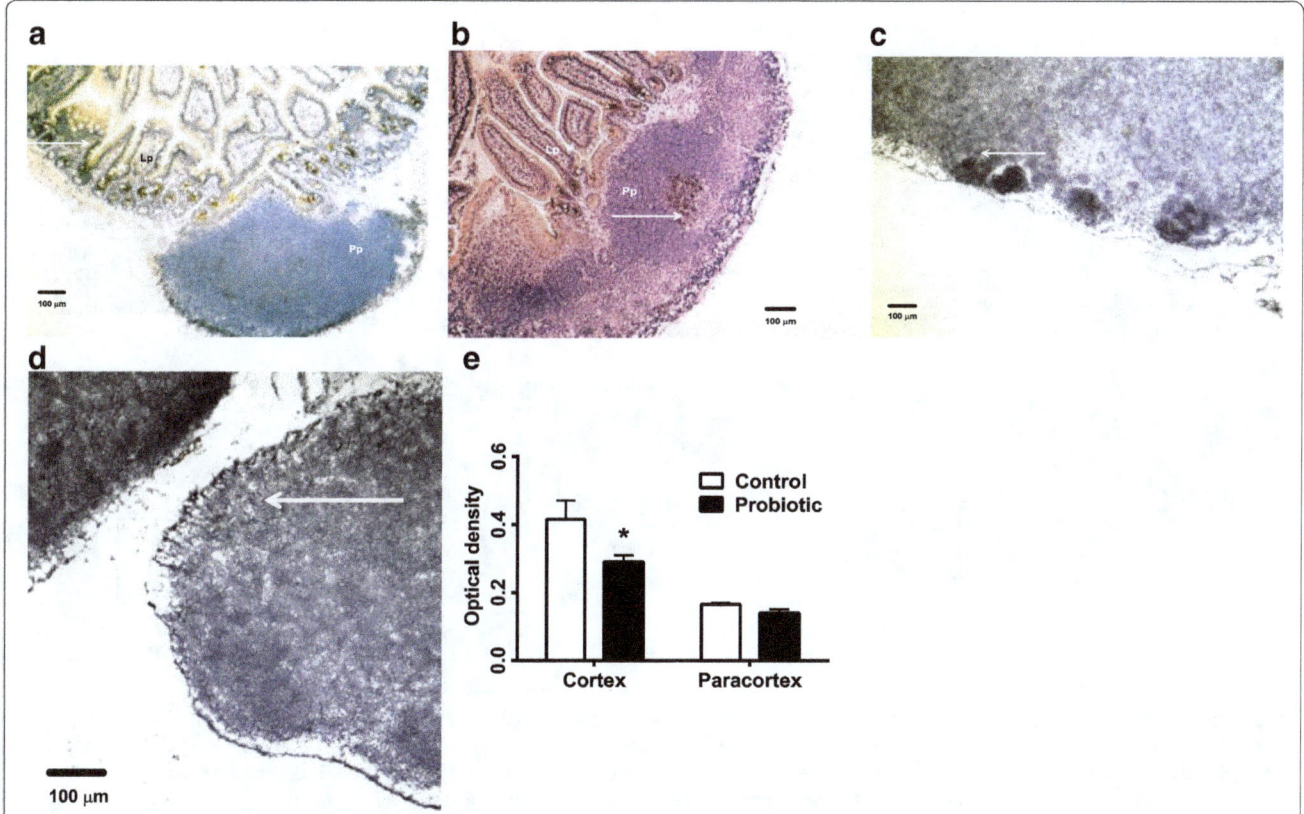

Fig. 5 Immunohistochemical staining of TLR6. Cells were stained for TLR6 (*arrows*) in Peyer's Patches (**a**, control mice; **b**, probiotic-treated mice) and MLN (**c**, control mice; **d**, probiotic-treated mice) from *S. enterica*-infected and probiotic-treated *S. enterica*-infected mice. The amount of TLR6 in cortical areas of probiotic-treated MLN was significantly reduced, $P < 0.05$ by ANOVA, compared with untreated *S. enterica*-challenged mice (**e**)

Treatment of the mice with probiotic bacteria prevented the immunosuppressive effects. There are several immunosuppressive mechanisms used by virulent *S. enterica* to evade host responses, including elimination of high-avidity antigen-specific T cells [2], activation of suppressive regulatory T cells [3], and phagolysosomal escape in macrophages and dendritic cells [1, 7].

In the present study, the probiotic bacteria suppressed basal and LPS-induced splenic lymphoproliferation. A growing body of literature is reporting this characteristic of probiotic lactobacilli with inhibition of LPS-stimulated lymphocyte proliferation [8], inhibition of Concanavalin-A-stimulated proliferation [9], and anti-CD3ε-stimulated splenocyte proliferation [10]. It is consistent with the concept that pathogen-induced lymphocyte clonal deletion can be blocked by bacterial induction of mechanisms that inhibit proliferation. Cell surface receptors for commensal or probiotic bacteria may activate pathways that prevent proliferation and apoptosis of lymphocytes.

Bacterial activation of a pathway that inhibits lymphocyte activation and apoptotic caspase activity most likely occurs through the recognition of bacterial surface glycoproteins by TLR1, TLR2, or TLR6 on APC [11]. TLR4 (detects lipopolysaccharides-LPS) and TLR5 (detects

flagellin) are also involved in the T cell activation response to *S. enterica* by APC [11]. In our present study, gene expression profiling qRT-PCR panels and immunohistochemistry were used to observe changes in expression of genes associated with lymphocyte activation and apoptosis and signal transduction from Toll-like receptors through the NF-κB and MAPK signal transduction pathways that showed evidence for a T cell-sparing effect from the influence of probiotic bacteria. Activation and signaling responses by lymphocytes are regulated by the NF-κB transcription factor and signal transduction pathway [12]. Some of the intracellular molecules that activate NF-κB include: CFLAR [13], CASP8 [13], conserved helix-loop-helix ubiquitous kinase (CHUK/IκκA) [14], RIPK1 [15], TRAF2 [15], and BCL10 [16].

Here we show evidence for a probiotic mechanism that prevents *S. enterica* immune evasion by T cell deletion through a mechanism of reduced inflammatory response early during the infection in BALB/c mice. The probiotic bacteria modulated expression of genes related to inflammation and T cell apoptosis, including some of the Toll-like receptors involved in *S. enterica*-induced inflammation. In this present study, we observed increased *Tlr6* mRNA in MALT of probiotic-treated mice

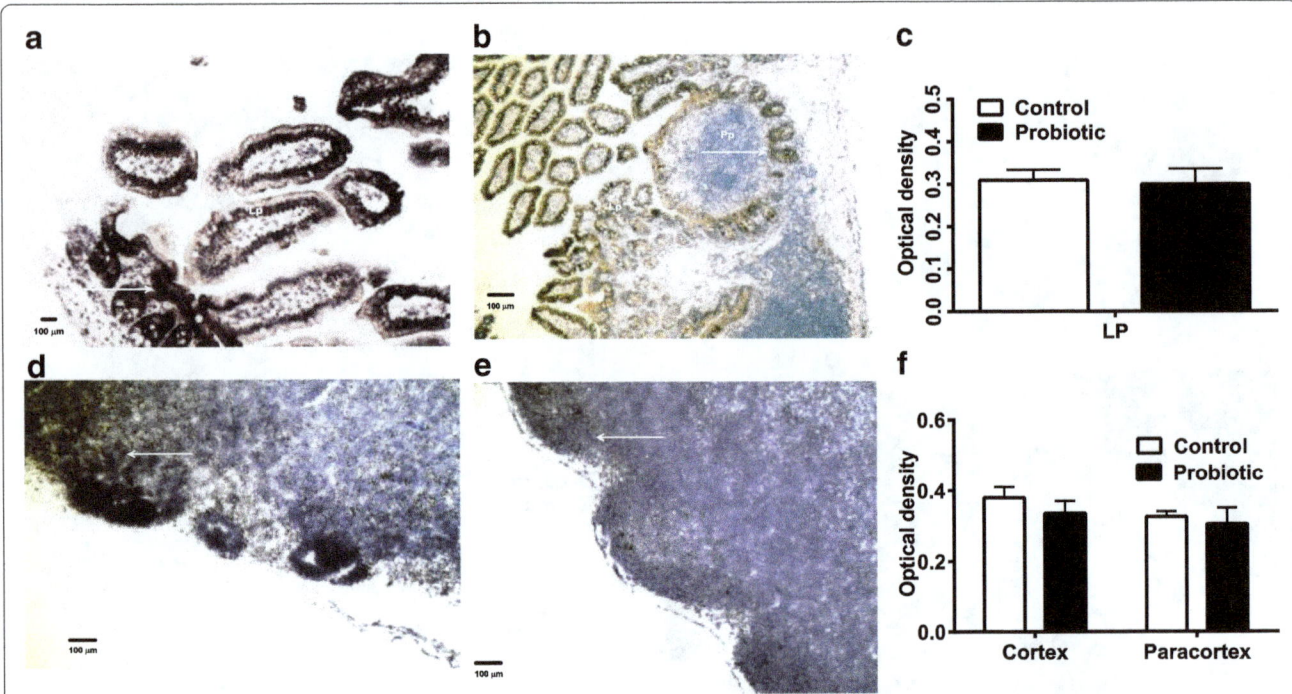

Fig. 6 Immunohistochemical staining of RELB. The Peyer's patches and lamina propria (**a**, control mice; **b**, probiotic-treated mice) and MLN of probiotic-treated *S. enterica*-infected mice (**d**, control mice; **e**, probiotic-treated mice) had RELB expressed in cells (*arrows*) with lymphocyte and epithelial cell morphologies. There were no significant differences in staining of lamina propria (Lp) of control or probiotic-treated mice (**c**). Staining in MLN from the probiotic-treated mice was not significantly different from controls, $P < 0.05$ by ANOVA (**f**)

compared with *S. enterica*-treated mice. This agrees with evidence that probiotic *Lactobacillus plantarum* induced *Tlr6* mRNA expression in intestinal epithelial cells of cyclophosphamide-suppressed mice [17]. We also observed that there was not an increase in TLR6 expressing cells in Peyer's patches, suggesting a role for post-transcriptional regulation of TLR6 on the surface of lymphocytes in our mice. It is possible that TLR2 and TLR6 heterodimer activation can be inhibited by probiotics through activation of peptide regulators, as shown with TLR6 transmembrane domain peptides in vitro [18]. The latter paper suggested this inhibition is specific to TLR2, but the work was in vitro and may not show how the process may affect TLR2/TLR6 heterodimer surface expression in cells in vivo. Another possibility for this in the present study is that the different microbiota used in the human microbiota-associated mice versus the conventional mice may not have supported the probiotic effect.

Since we saw probiotic bacteria could reverse the suppression of splenic lymphocyte proliferation to *S. enterica* antigens, we looked at changes in expression of genes associated with T cell activation in mice treated with probiotic bacteria. Initial contacts by intestinal epithelial cells or professional phagocytes with *S. enterica* surface molecules that have pathogen-associated molecular patterns activate receptors that initiate signal transduction within the cell. These signals initiate mRNA

transcription and protein synthesis of cytokines that are secreted into the extracellular environment and recruit inflammatory cells. When dendritic cells encounter *S. enterica* antigens in the intestinal tissues, they process the antigens for presentation to lymphocytes and migrate to MLN [19]. In the MLN, the dendritic cells activate antigen-specific T cells that are selected by stromal cells for expression of α4β7 integrin and CCR9, which is essential for their migration along a gradient of CCL25 and other chemokines to the intestinal MALT [20–22].

Clues to the probiotic mechanism that blocks *S. enterica*-induced immunosuppression might be seen in changes of expression of genes involved in recruitment of immune responses to the MALT and MLN of mice protected by the bacteria from *S. enterica* infection. MLN are essential for resistance to *S. enterica* infections, but are also the sites of long term persistence of infections [1]. The MLN hold *S. enterica*-infected dendritic cells and limit the dissemination of the infection by arresting DC migration [23]. Still, some of the bacteria can escape the MLN because host cell death can be induced by *S. enterica*, especially Serovar Typhimurium [1].

Some probiotic mechanisms do not involve immune and inflammatory cell migration. For example, cell-free spent media from probiotic *Bifidobacterium bifidum* inhibits growth of *S. enterica* [24]. These authors suggest that *Bifidobacter*-derived factors interfere with expression of *S.*

enterica virulence genes encoded on the *Salmonella* pathogenicity islands 1 and 2. Some effects of commensal and probiotic bacteria appear to occur at the mucosal surface in epithelial cells. Some surface layer proteins of *Lactobacillus acidophilus* inhibit CASP3 responses to *Salmonella* sp. in IEC cell lines [25] thus, reducing apoptotic death and loss of epithelial barrier functions to *Salmonella* sp. invasion. Clearly, evidence exists for the combination of microbial interactions for colonization resistance to *S. enterica* infections and influences on the innate and adaptive immune responses can be attributed to probiotic bacteria.

Conclusions

Despite all the efforts of public health and agricultural programs to reduce acquisition of *Salmonella* sp. infections, they persist as a major health issue. One reason for this is the asymptomatic carrier state of some individuals who have recovered from acute infections. Persistence of a carrier state is facilitated by immune system evasion mechanisms inherent to the genus. Strains of bacteria in the intestinal microbiota and found in probiotic products can antagonize *Salmonella* sp. infections, perhaps by inhibition of these inherent immune system evasion mechanisms. In this study, the presence of probiotic bacteria in the intestinal tracts of mice during *S. enterica* infections prevented loss of splenocyte proliferation and decreased splenocyte apoptosis. Gene expression data showed changes that suggest lymphocyte activation, survival. And immune cell homing functions were restored, This was related to an apparent ability of the probiotic bacteria to limit exhaustive lymphocyte proliferation and apoptosis. The migration of lymphocytes and dendritic cells in intestinal mucosal tissues was restored by the presence of the probiotic bacteria. These results are consistent with the theory that *S. enterica* induced clonal deletion of lymphocytes that was inhibited by the presence of probiotic bacteria.

Abbreviations

AKT1: v-akt murine thymoma viral oncogene homolog 1; APC: Antigen presenting cell; BAD: bcl2-associated agonist of cell death; BCL2: B- cell CLL/lymphoma 2; CARD10: Caspase recruitment domain protein 10; CASP2: Apoptosis-related cysteine peptidase 2; CD2: Clusters of differentiation antigen 2; CFLAR: CASP8 and FADD like apoptosis regulator; CFU: Colony forming unit; DAD1: Defender against cell death 1; FADD: Fas-associated death domain protein; IL4: Interleukin-4; IRAK2: Interleukin-1 receptor-associated kinase 2; IκκB: Inhibitor of light polypeptide gene enhancer in B cell kinase alpha; JUN: v-jun sarcoma virus 17 oncogene homolog; LPS: Lipopolysaccharide; MALT: Mucosa associated lymphoid tissue; MEKK: Map/Erk kinase kinase; MLN: Mesenteric lymph node; MRS: de Man Rogosa Sharpe; MTS: 3-(4,5-dimethylthiazol-2-yl)-5-(3-carboxymethoxyphenyl)-2-(4-sulfonyl)-2H–tetrazolium, inner salt; MYD88: Myeloid differentiation primary response gene 88; NF-κB: Nuclear factor of kappa light chains enhancer in B cells; PAK7: p21-activated kinase 7; PTPRC: Protein tyrosine phosphatase receptor type C; qRT-PCR: quantitative real-time polymerase chain reaction; RELB: v-Rel avian reticuloendotheliosis viral oncogene homolog B; RIPK1: Receptor TNFRSF-interacting serine threonine kinase 1; RNF7: Ring finger protein 7; RPMI 1640: Roswell Park Memorial Institute 1640; SEM: Standard error of the mean; TLR6: Toll-like receptor 6; TNF: Tumor necrosis factor alpha; TRAF2: TNF receptor associated factor 2

Acknowledgements

The authors would like to thank Dr. Sangeeta Khare and Dr. Steven Foley for critical review of the manuscript. This work was supported by funds from the Food and Drug Administration. The opinions expressed in this manuscript are the authors' and do not necessarily reflect the position of the Food and Drug Administration.

Funding

This project was solely supported by US Public Health Service funds.

Authors' contributions

SJJ maintained the animals and conducted laboratory analyses. RDW conceived of the study, conducted the microbial challenges and euthanasia of the animals, and wrote the manuscript. Both authors read and approved the final manuscript.

Competing interests

The authors declare that they have no competing interests.

References

1. Ruby T, McLaughlin L, Gopinath S, Monack D. *Salmonella's* long-term relationship with its host. FEMS Microbiol Rev. 2012;36:600–15.
2. Ertelt JM, Johanns TM, Mysz MA, Nanton MR, Rowe JH, Aguilera MN, et al. Selective culling of high avidity antigen –specific CD4+ T cells after virulent *Salmonella* infection. Immunol. 2011;134:487–97.
3. Monack DM. *Salmonella* persistence and transmission strategies. Curr Opin Microbiol. 2012;15:100–7.
4. Lima-Filho JVM, Viera LQ, Arantes RME, Nicoli JR. Effect of the *Escherichia coli* EMO strain on experimental infection by *Salmonella enterica* serovar Typhimurium in gnotobiotic mice. Braz J Med Biol Res. 2004;37:1005–13.
5. Wagner RD, Johnson SJ, Kurniasih-Rubin D. Probiotic bacteria are antagonistic to *Salmonella enterica* and *Campylobacter jejuni* and influence host lymphocyte responses in human microbiota-associated immunodeficient and immunocompetent mice. Mol Nutr Food Res. 2009; 53:377–88.
6. Bahrami B, Macfarlane S, Macfarlane GT. Induction of cytokine formation by human intestinal bacteria in gut epithelial cell lines. J Appl Microbiol. 2010; 110:353–63.
7. Bueno SM, Riquime S, Riedel CA, Kalergis AM. Mechanisms used by virulent *Salmonella* to impair dendritic cell function and evade adaptive immunity. Immunol. 2012;137:28–36.
8. Hosoya T, Sakai F, Yamashita M, Shiozaki T, Endo T, Ukibe K, et al. *Lactobacillus helveticus* SBT2171 inhibits lymphocyte proliferation by regulation of the JNK signaling pathway. PLoS One. 2014;9:e108360.
9. Li C-Y, Lin H-C, Lai C-H, Lu JJ-Y, Wu S-F, Fang S-H. Immunomodulatory effects of *Lactobacillus* and *Bifidobacterium* on both murine and human mitogen-activated T cells. Int Arch Allergy Immunol. 2011;156:128–36.
10. Yoshida A, Yamada K, Yamazaki Y, Sashihari T, Ikegami S, Shimizu M, et al. Immunology. 2011;133:442–51.
11. Harris G, KuoLee R, Chen W. Role of Toll-like receptors in health and diseases of gastrointestinal tract. World J Gastroenterol. 2006;12:2149–60.
12. Weil R, Israël A. Deciphering the pathway from the TCR to NF-κB. Cell Death Differ. 2006;13:826–33.
13. Budd RC, Yeh W-C, Tschopp J. cFLIP regulation of lymphocyte activation and development. Nat Rev Immunol. 2006;6:196–204.
14. Arnold R, Brenner D, Becker M, Frey CR, Krammer PH. How T lymphocytes switch between life and death. Eur J Immunol. 2006;36:1654–8.
15. Vallabhapurapu S, Karin M. Regulation and function of NF-κB transcription factors in the immune system. Annu Rev Immunol. 2009;27:693–733.
16. Su H, Bidère N, Zheng L, Cubre A, Sakai K, Dale J, et al. Requirement for caspase-8 in NF-κB activation by antigen receptor. Science. 2005;307:1465–8.

17. Xie J, Nie S, Yu Q, Yin J, Xiong T, Gong D, et al. *Lactobacillus plantarum* NCU116 attenuates cyclophosphamide-induced immunosuppression and regulates Th17/Treg cell immune responses in mice. J Agric Food Chem. 2016;64:1291–7.

18. Fink A, Reuven EM, Arnusch CJ, Shmuel-Galia L, Antonovsky N, Shai Y. Assembly of the TLR2/6 transmembrane domains is essential for activation and is a target for prevention of sepsis. J Immunol. 2013;190:6410–22.

19. Willard-Mack CL. Normal structure, function, and histology of lymph nodes. Toxicol Pathol. 2006;34:1533–601.

20. Malhotra D, Fletcher AL, Turley SJ. Stromal and hematopoietic cells in secondary lymphoid organs: partners in immunity. Immunol Rev. 2012;251:160–76.

21. Molenaar R, Greuter M, van der Marel APJ, Roozendaal R, Martin SF, Edele F, et al. Lymph node stromal cells support dendritic cell-induced gut-homing of T cells. J Immunol. 2009;183:6395–402.

22. Hammerschmidt SI, Ahrendt M, Bode U, Wahl B, Kremmer E, Förster R, et al. Stromal mesenteric lymph node cells are essential for the generation of gut-homing T cells in vivo. J Exp Med. 2008;205:2483–90.

23. Voedisch S, Koenecke C, David S, Herbrand H, Förster R, Rhen M, et al. Mesenteric lymph nodes confine dendritic cell-mediated dissemination of *Salmonella enterica* Serovar Typhimurium and limit systemic disease in mice. Infect Immun. 2009;77:3170–80.

24. Bayoumi MA, Griffiths MW. Probiotics down-regulate genes in *Salmonella enterica* Serovar Typhimurium pathogenicity islands 1 and 2. J Food Protect. 2010;73:452–60.

25. Li P, Yin Y, Yu Q, Yang Q. *Lactobacillus acidophilus* S-layer protein-mediated inhibition of *Salmonella*-induced apoptosis in Caco-2 cells. Biochem Biophys Res Comm. 2011;409:142–7.

5

Polynucleotide phosphorylase is implicated in homologous recombination and DNA repair in *Escherichia coli*

Thomas Carzaniga[1,2], Giulia Sbarufatti[1,3], Federica Briani[1] and Gianni Dehò[1*] (ID)

Abstract

Background: Polynucleotide phosphorylase (PNPase, encoded by *pnp*) is generally thought of as an enzyme dedicated to RNA metabolism. The pleiotropic effects of PNPase deficiency is imputed to altered processing and turnover of mRNAs and small RNAs, which in turn leads to aberrant gene expression. However, it has long since been known that this enzyme may also catalyze template-independent polymerization of dNDPs into ssDNA and the reverse phosphorolytic reaction. Recently, PNPase has been implicated in DNA recombination, repair, mutagenesis and resistance to genotoxic agents in diverse bacterial species, raising the possibility that PNPase may directly, rather than through control of gene expression, participate in these processes.

Results: In this work we present evidence that in *Escherichia coli* PNPase enhances both homologous recombination upon P1 transduction and error prone DNA repair of double strand breaks induced by zeocin, a radiomimetic agent. Homologous recombination does not require PNPase phosphorolytic activity and is modulated by its RNA binding domains whereas error prone DNA repair of zeocin-induced DNA damage is dependent on PNPase catalytic activity and cannot be suppressed by overexpression of RNase II, the other major enzyme (encoded by *rnb*) implicated in exonucleolytic RNA degradation. Moreover, *E. coli pnp* mutants are more sensitive than the wild type to zeocin. This phenotype depends on PNPase phosphorolytic activity and is suppressed by *rnb*, thus suggesting that zeocin detoxification may largely depend on RNA turnover.

Conclusions: Our data suggest that PNPase may participate both directly and indirectly through regulation of gene expression to several aspects of DNA metabolism such as recombination, DNA repair and resistance to genotoxic agents.

Keywords: Polynucleotide phosphorylase, Genetic recombination, DNA repair, Mutagenesis, RNA metabolism

Background

Polynucleotide phosphorylase (PNPase, polyribonucleotide nucleotidyltransferase, EC 2.7.7.8), an enzyme widely conserved in Bacteria and in eukaryotic organelles of bacterial origin, reversibly catalyses the 3′-to-5′ phosphorolysis of polyribonucleotides, releasing nucleoside diphosphates (NDPs) and the reverse template-independent 5′-to-3′ polymerization of nucleoside diphosphates, releasing inorganic phosphate (Pi) [1, 2]. The original interest for the RNA polymerizing activity of this enzyme [3] was superseded by the J. Hurwitz discovery of DNA-dependent RNA polymerase (reviewed by [4]) and RNA degradation was

since thought of as the main in vivo activity of PNPase [5]. Its RNA polymerizing activity has also been implicated in PNPase-dependent RNA decay, as in *Escherichia coli* polyadenylation and heteropolymeric tailing of RNA 3′-ends performed by polyadenylpolymerase (PAP) and PNPase, respectively, target bacterial RNAs to degradation [6–8]. A wealth of evidence accumulated in the last decades indicates that the key role of PNPase in vivo is to modulate the abundance of a number of mRNAs and small RNAs (sRNAs), and thus expression of many genes (reviewed by [2]).

Remarkably, PNPase can catalyse both DNA phosphorolysis and template independent synthesis of DNA from dNDPs [9–15]. The latter enzymatic property was exploited in the early era of molecular biology for the synthesis of oligodeoxyribonucleotides but was not generally considered to play a role in vivo. However, features that link this

---

* Correspondence: gianni.deho@unimi.it

[1]Dipartimento di Bioscienze, Università degli Studi di Milano, via Celoria 26, Milan 20133, Italy

Full list of author information is available at the end of the article

enzyme to DNA metabolism have emerged. For example, PNPase deficient *E. coli* mutants are more sensitive to UV [16] and exhibit a lower mutation frequency [17]. Such phenotypes have been thought of as a direct or indirect result of PNPase RNA-degrading activity; however, the recent observation that in *Bacillus subtilis* not only PNPase is implicated in DNA repair but also is part of the RecN repair complex [13, 14, 16, 17], raises the possibility that its DNA degradative and/or polymerizing activities may be implicated in DNA recombination, repair and mutagenesis.

Homologous recombination (HR) was first identified as a mechanism that assorts genes on homologous chromosomes at meiosis, thus contributing to the generation of genetic variability at the population level. Genetic studies on fungi, which led to the pioneering Holliday's model, implicated HR process in mismatch repair, whereas studies on recombination and DNA repair deficient mutants in *Escherichia coli* readily highlighted that the two phenomena are inextricably intertwined and that multiple pathways and mechanisms are implicated in these processes [18]. Finally, as DNA damage, in particular double strand breaks (DSB) generated during progression of the replicative fork, represent an obstacle to completion of chromosome replication, it was recognised that recombination and DSB repair (DSBR) are essential for cell viability. It is now widely accepted that HR is a housekeeping process implicated in the maintenance of genome integrity both during and after DNA replication [18].

Given the pivotal role of recombination and DNA repair for cell survival and homeostasis, it is not surprising that such processes cross talk with other central cellular pathways. In this work we show that PNPase, an enzyme typically implicated in RNA turnover, may play a role both in transduction-mediated HR and in error prone DNA repair, thus providing further evidence for the connections between RNA and DNA metabolism.

Results

Transduction frequency is reduced in *E. coli* Δpnp mutant

PNPase has recently been implicated in recombination and DNA repair in *Bacillus subtilis* [14]. We tested the potential involvement of PNPase in homologous recombination in *E. coli* by assessing the frequency of P1-mediated generalized transduction in the presence and absence of PNPase. Isogenic wild type and Δ*pnp* strains auxotrophic for tryptophan (Δ*trpE::kan*) or leucine (Δ*leuA::kan*) were infected with a lysate of P1 HFT phage grown in the prototrophic strain C-1a and prototrophic transductants were selected on M9 glucose agar minimal medium, as described in Materials and Methods. All prototrophic transductants tested turned out to be Kan^S, thus indicating that recombination had occurred at the homologous loci in both wild type and Δ*pnp* srains. The results of these experiments reported in Table 1 show that

for both independent auxotrophy markers transduction frequency was 5–6 folds lower in the *pnp* mutant.

Recombination is only one of the steps required for transduction and it could be argued that PNPase may affect efficiency of upstream events such as phage adsorption and/or DNA injection. However, P1 adsorption was comparable in wild type and *pnp*$^-$ strains (Table 1) whereas P1 efficiency of plating (e.o.p.) was slightly lower (0.79), with a smaller plaque size, in the two Δ*pnp* recipients than in their isogenic *pnp*$^+$ strains. Even assuming that the lower e.o.p. exclusively depends on less efficient DNA injection rather than on downstream steps in phage growth cycle (as suggested by the smaller plaque size), it does not account for the 5–6 fold reduction in transduction efficiency. It seems thus possible that PNPase contributes to some extent to the efficiency of recombination of transduced DNA.

PNPase is composed of a catalytic core and a C-terminal region composed of two RNA binding domains (RBDs) KH and S1 [19–21]. To test whether either or both catalytic and RNA binding activities were implicated in transduction efficiency, we ectopically complemented the above Δ*pnp* strains with plasmids expressing the wild type PNPase, PNPase mutants lacking either or both the KH and S1 domains [22, 23], and a PNPase with a point mutation in the catalytic sites (PNPase^{S438A}) known to abolish phosphorolytic activity [24, 25]. As shown in Table 2, transduction efficiency in the non-complemented Δ*pnp* strain (harbouring the empty vector) and in strains expressing PNPase lacking either RBD was from about 3–5 fold lower than in the *pnp*$^+$-complemented strain, whereas the catalytically inactive mutation *pnp*^{S438A} marginally (if at all) affected recombination (1.2–1.7 fold decrease). Surprisingly, the lack of both KH and S1 domains strongly reduced (25–45 fold) the recovery of prototrophic transductants. It thus appears that nucleic-acid binding activity is required to regulate transduction efficiency and that the presence of an enzyme devoid of both RBDs impairs recombination more than the mere lack of PNPase.

E. coli Δpnp is more sensitive to the genotoxic agent zeocin

The substrates of transduction-mediated recombination are the circular chromosome and a homologous linear dsDNA fragment [26], likely implicating the RecBCD-dependent recombination pathway [27], which is promoted by free dsDNA ends [28]. We thus addressed whether PNPase also participates in repair processes of DSBs induced by zeocin (phleomycin D1), a glycopeptide of the bleomycin/phleomycin antibiotic family, known to cause DSBs in vivo [29].

Firstly, we measured zeocin cytotoxicity on PNPase mutants by plating Δ*pnp* strains non complemented or

Table 1 P1 transduction in Δpnp strains

Transduced marker	trpE				leuA			
Chromosomal pnp allele	Adsorbed phage[a]	m.o.i.[b]	TF[c] (×10^7)	TF fraction[d]	Adsorbed phage[a]	m.o.i.[b]	TF[c] (×10^7)	TF fraction[d]
wild type	0.98 (±0.01)	5.87×10^{-2}	37.0 (±1.4)	1.00	0.98 (±0.01)	5.66×10^{-2}	23.0 (±1.2)	1.00
Δpnp-751	0.97 (±0.01)	5.84×10^{-2}	6.9 (±0.28)	0.19	0.97 (±0.01)	5.81×10^{-2}	3.7 (±0.17)	0.16

[a]adsorbed phage is input phage - unadsorbed phage as assayed 20 min post infection
[b]multiplicity of infection (m.o.i.) is the ratio of phage to bacterial cells
[c]transduction frequency (TF) is the ratio of transductants to adsorbed phage. Average and standard deviation of three experiments are reported
[d]ratio to TF of wild type strain

complemented with ectopically expressed wild type and mutant pnp alleles in the presence of increasing concentration of the drug. As shown in Fig. 1, the Δpnp mutant was slightly more sensitive than the isogenic pnp$^+$ strain. Complementation by the wild type allele expressed from plasmid pAZ101, increased the resistance to zeocin to a higher level than the wild type parental strain (Fig. 1a), possibly due to a copy-number effect.

We also tested whether Δpnp mutant complementation for zeocin sensitivity required PNPase catalytic and/or RNA binding activities. As shown in Fig. 1a, i) the zeocin survival curves of the two strains expressing pnp mutants lacking a single RBD were similar to that of the wild type-complemented strain; ii) zeocin survival of the strain ectopically complemented with a pnp allele lacking both KH and S1 domains was intermediate between the wild type and the Δpnp mutant strains complemented by wild type and single RBD mutant-PNPase; iii) remarkably, the zeocin survival curve of the strain expressing the enzymatically inactive allele pnp^{S438A} was superimposable to that of the non-complemented Δpnp mutant. These results suggest that PNPase enzymatic activity is absolutely required to complement zeocin sensitivity, whereas the RBDs appear to modulate and/or participate to some extent in the overall response to the drug.

RNase II overexpression is known to suppress some of the traits associated to the pleiotropic PNPase deficiency [2, 30]. As shown in Fig. 1b, ectopic expression of RNase II increased zeocin resistance of the Δpnp mutant more than PNPase complementation, thus indicating that

RNase II can compensate for PNPase deficiency. Somewhat surprisingly, a non-complemented Δrnb mutant was more resistant than the wild type strain. This could be explained by the fact that PNPase expression level is higher in RNase II-deficient mutants [31] and fits with the observation that ectopic overexpression of PNPase did not substantially alter the survival curve of the Δrnb strain.

Frequency of mutants induced by zeocin is decreased in E. coli Δpnp

In addition to DSBs, the radiomimetic zeocin may cause other types of damage [29] and the survival rate may be the results of different processes in which PNPase may directly or indirectly be implicated. Ideally, a DNA repair process may either restore the original genetic information or introduce mutations [18, 32–35] and inactivation of error free and error prone pathways may lead to increase and decrease of mutation rate, respectively. To specifically analyze whether PNPase participates in repairing DNA DSBs, we assessed the frequency of mutants induced by zeocin treatment. Exponentially growing cultures treated with 100 µg/ml zeocin for 30 min and the non-treated controls were plated on LD-agar plates without and with 100 µg/ml rifampicin to assay for viable cells and rifampicin resistant (Rif^R) mutants, respectively, as described in Materials and Methods. In these experiments survival of zeocin treated cells was from 30 to 40%.

The ratio of zeocin-induced mutants to the spontaneous mutants (fold induction) in the different strain and the ratio of the fold induction of each strain relative to the

Table 2 Efficiency of P1 transduction: complementation by different pnp mutant alleles

Plasmid[a]	pnp allele on plasmid	trpE		leuA	
		TF[b] (×10^7)	TF fraction[c]	TF[b] (×10^7)	TF fraction[c]
pAZ101	wt	2.53 (±0.09)	1.00	4.40 (±0.35)	1.00
pGZ119HE	none	0.75 (±0.20)	0.30	1.34 (±0.30)	0.30
pAZ1112	S438A	2.07 (±0.52)	0.82	2.70 (±0.46)	0.60
pAZ1113	ΔKH	0.55 (±0.15)	0.22	1.18 (±0.24)	0.26
pAZ1114	ΔS1	0.53 (±0.19)	0.21	1.19 (±0.25)	0.27
pAZ133	ΔKH-S1	0.06 (±0.02)	0.02	0.17 (±0.06)	0.04

[a]in C-5691 (Δpnp) strain
[b]transduction frequency (TF) is the ratio of transductants to adsorbed phage
[c]ratio to TF of wild type strain

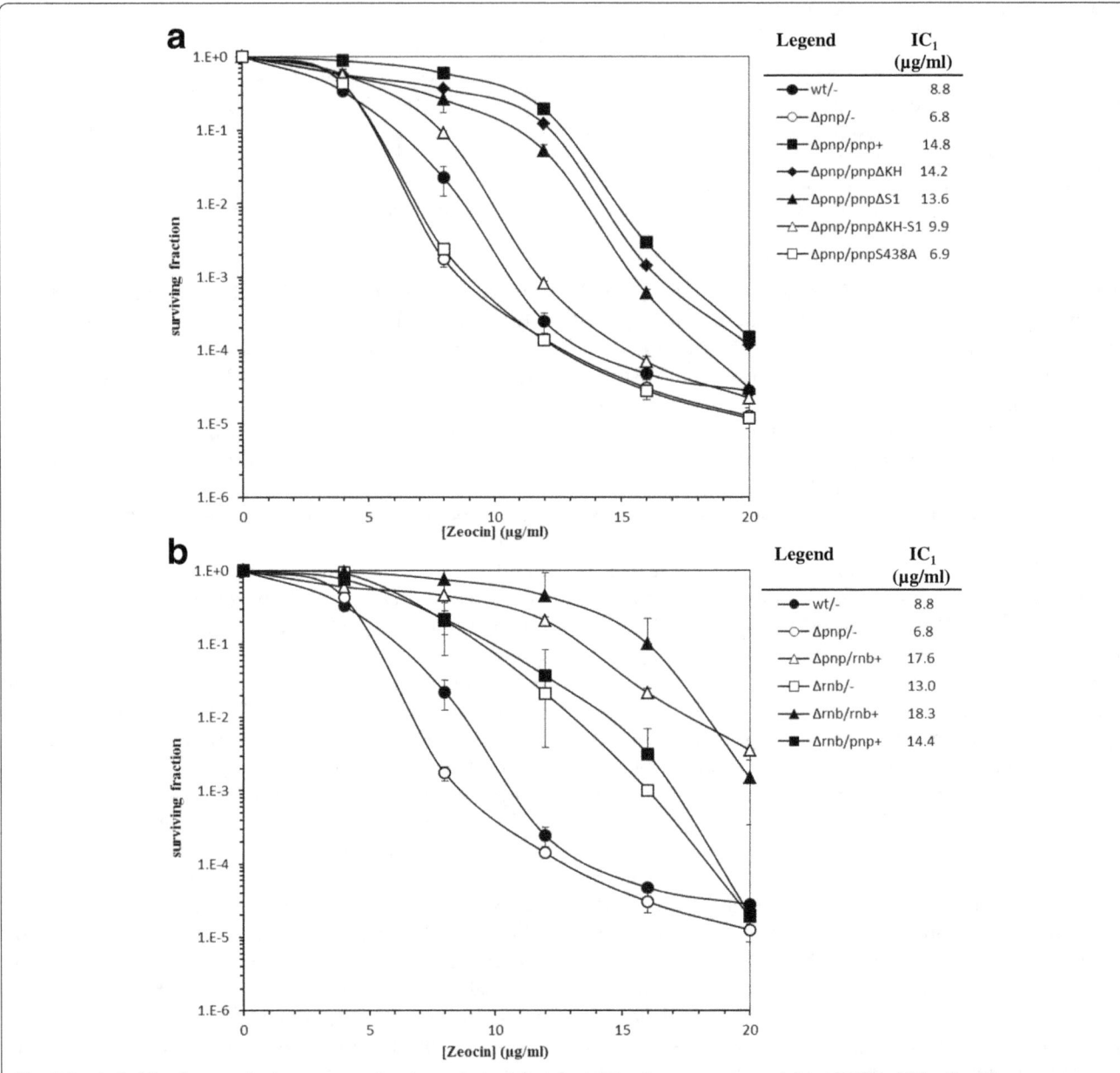

Fig. 1 Survival of *E. coli pnp* and *rnb* mutants to chronic zeocin treatment. Bacterial cultures grown overnight at 37 °C in LD broth with chloramphenicol were serially diluted and plated onto LD-agar with chloramphenicol and 0, 4, 8, 12, 16, 20 μg/ml zeocin; the plates were incubated at 37 °C for 2 days. The surviving fraction was calculated as the ratio of colony forming units (CFUs) at different zeocin concentration to the untreated control. The means of at least two independent experiments and the standard deviations (*error bars*) are shown. Zeocin inhibitory concentration giving 1% survival (IC_1) was extrapolated from the survival curves and reported as μg/ml on the *right column* in the panel legends. **a** Complementation of *E. coli pnp* deletion mutant with different *pnp* alleles. **b** Complementation of *E. coli pnp* and *rnb* deletion mutants. The *pnp* or *rnb* alleles harbored by the host chromosome and or by the plasmids are indicated by the *labels* before and after the *slash*, respectively. Bacterial strains: wt, C-1a; Δpnp, C-5691; Δrnb, C-5981. Plasmids: –, pGZ119HE; pnp+, pAZ101; rnb+, pAZ1115; pnpΔKH, pAZ1113; pnpΔS1, pAZ1114; pnpΔKH-S1, pAZ133; pnpS438A, pAZ1112

control (Δ*pnp* mutant ectopically complemented by the wild type allele) are reported in Table 3. A four-fold induction of Rif[R] mutant was observed in both the control (C-5691/pAZ101) and wild type (C-1a/pGZ119HE9 strains upon zeocin treatment, whereas this value was reduced to less than a half in the Δ*pnp* strain non-complemented or complemented with a PNPase defective in the catalytic activity. Rif[R] mutants induction was either comparable to or even higher than the control in Δ*pnp* strains complemented by PNPase lacking both or either KH and S1 RBDs. On the contrary, RNase II did not seem to affect zeocin-induced mutagenesis as a Δ*rnb* mutant, like the control strain, exhibited a fourfold increase of zeocin-induced Rif[R] mutants. Likewise, ectopic expression of

Table 3 Zeocin-induced mutagenesis in Δ*pnp* strains

Strain	Chrom. alleles	Plasmid allele	Fraction surviving (±SD)	Spontaneous mutants (×10⁹)	Zeocin-induced mutants (×10⁹)	Fold induction	Ratio to *pnp*⁺
C-1a/pGZ119HE	wild	none	0.40 (±0.014)	273.7 (±44.6)	1116.3 (±125)	4.08	0.95
C-5691/pAZ101	Δ*pnp*	*pnp*⁺	0.37 (±0.028)	77.4 (±19.6)	331.4 (±82.8)	4.28	1.00
C-5691/pGZ119HE	Δ*pnp*	none	0.38 (±0.015)	278.2 (±70.8)	530.7 (±98.0)	1.91	0.45
C-5691/pAZ1112	Δ*pnp*	*pnp*S438A	0.40 (±0.026)	123.9 (±55.2)	222.4 (±85.2)	1.80	0.42
C-5691/pAZ1113	Δ*pnp*	*pnp*-ΔKH	0.31 (±0.073)	15.8 (±11.7)	108.0 (±75.9)	6.81	1.59
C-5691/pAZ1114	Δ*pnp*	*pnp*-ΔS1	0.39 (±0.035)	16.7 (±7.8)	119.0 (±55.4)	7.14	1.67
C-5691/pAZ133	Δ*pnp*	*pnp*-ΔKHS1	0.33 (±0.035)	49.6 (±14.2)	208.4 (77.3)	4.20	0.98
C-5691/pAZ1115	Δ*pnp*	*rnb*⁺	0.41 (±0.023)	97.4 (±15.9)	189.3 (±16.2)	1.94	0.45
C-5981/pAZ1115	Δ*rnb*	*rnb*⁺	0.39 (±0.029)	48.6 (±23.8)	204.0 (±85.1)	4.20	0.98
C-5981/pGZ119HE	Δ*rnb*	none	0.36 (±0.027)	228.7 (±50.8)	992.5 (±222.8)	4.34	1.01
C-5981/pAZ101	Δ*rnb*	*pnp*⁺	0.38 (±0.019)	153.9 (±38.1)	638.4 (±159.4)	4.15	0.97

RNase II did not suppress the phenotype of the Δ*pnp* strain.

Overall these data suggest that PNPase participates with its catalytic activity in some zeocin-induced error prone repair pathways, whereas the RBDs are not required for such an activity. On the contrary, RNase II does not seem to differentially affect zeocin-induced error prone and error free DNA repair processes.

Catalytic and binding activities of KH and S1 mutant PNPase with RNA and DNA substrates

PNPase could indirectly participate in repair of zeocin-induced damage by regulating the expression of genes mechanistically implicated in this process at the level of mRNA stability. Alternatively, PNPase could directly be involved in DNA repair mechanisms. It was shown that the RBDs of PNPase from *Mycobacterium smegmatis* (*Ms*PNPase) differently affect the catalytic and binding activities of the enzyme on RNA or DNA substrates [15, 36]. We thus explored whether the biochemical properties of the KH-S1 truncated PNPase from *E. coli* (*Ec*PNPase) on the two substrates could provide support to either of the above hypotheses. We compared the ability of wild type and RBDs mutant PNPase to degrade RNA and DNA oligonucleotides and to use them as primers for template-independent ribo- and deoxyribonucleotide polymerization. RNA and DNA oligonucleotide-primed reactions were performed in the presence of Mg and Mn divalent cations, respectively.

As shown in Fig. 2a, the lack of either or both RBDs severely impaired degradation of the ribo-oligonucleotide, whereas degradation efficiency of the DNA oligonucleotide appeared to be only slightly reduced (Fig. 2b; compare the disappearance of the full length probe intensities in mutant vs wild type enzyme lanes, quantified in Fig. 2c and d). Figure 3a shows that deletion of either or both RBDs severely impaired the initiation efficiency of template-independent polymerization of ribonucleotides using an RNA primer and ribonucleotide diphosphates as substrates (see the rate of disappearance of the RNA primer in Fig. 3c) without seemingly affecting the reaction processivity, as indicated by the high molecular weight of the reaction products. On the contrary, using a DNA primer and deoxyribonucleotide diphosphates as substrates, the lack of the RBDs only slightly reduced the efficiency of polymerization initiation whereas it strongly impaired processivity/elongation rate, as evaluated by the rate of disappearance of the DNA oligonucleotide signal and by the average length of the reaction products, respectively (Fig. 3b and d). Noticeably, although deletion of either RBDs impaired to some extent DNA polymerization initiation efficiency (as indicated by the rate of disappearance of the primer), the KH domain appeared to play the major role in processivity/elongation.

As for the RNA and DNA binding activities of the PNPase mutants in the RBDs, we performed electrophoretic mobility shift assays (EMSA) on RNA and DNA in the presence of Mg²⁺ and Mn²⁺, respectively (Fig. 4). In the absence of both divalent cations, deletion of the S1 or both RBDs impaired RNA binding, whereas a milder effect was observed with the ΔKH mutant. Surprisingly, each divalent cation reduced binding to the RNA oligonucleotide of both wild type and mutant PNPases (Fig. 4a). In agreement with previous observation [37], wild type PNPase also bound with similar efficiency to the DNA oligonucleotide. However, divalent cations did not reduce PNPase binding to the DNA probe. The lack of the KH domain only mildly affected DNA binding that, on the contrary, was impaired by deletion of S1 or both RBDs. Noticeably, Mn²⁺ seemed to improve DNA binding to the ΔS1 and ΔKH-S1 mutant enzymes (Fig. 4b). Overall these data suggest that S1 domain contributes to ssDNA binding and to processivity of DNA synthesis without a marked effect on either polymerization or phosphorolysis efficiency; on the other end, as previously shown, lack of RBDs not only impairs RNA binding but also PNPase catalytic activity (see [2]).

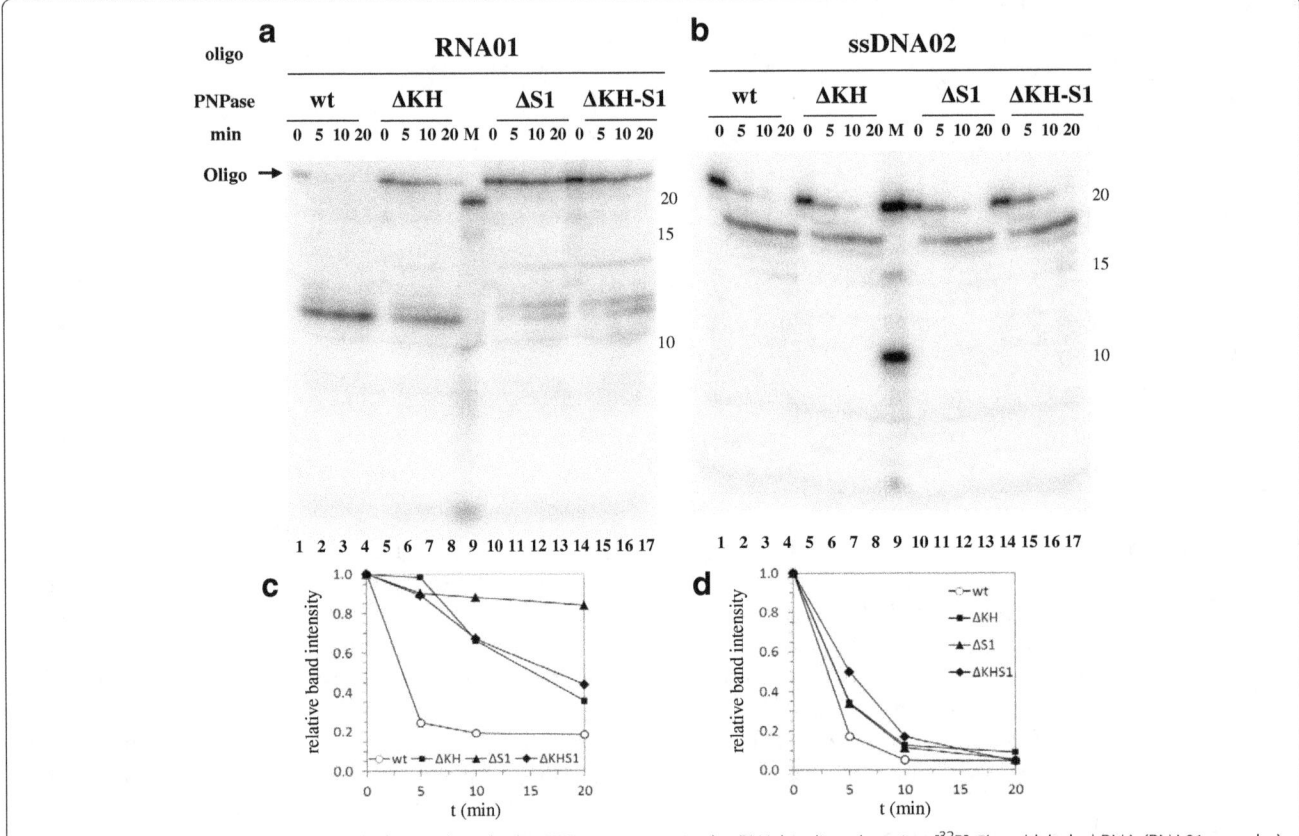

Fig. 2 Phosphorolysis of RNA and ssDNA oligonucleotides by PNPase mutants in the RNA binding domains. [^{32}P]-5'-end labeled RNA (RNA01; panel **a**) or ssDNA (DNA02; panel **b**) 20-mers, 4 nM each, were incubated at 26 °C with purified His-tagged PNPase or PNPase mutants in the RBDs (10 nM each) in buffer **a** with 10 μM Pi for the times indicated. The reaction products were fractionated by denaturing 6% PAGE. The oligonucleotide band intensities were evaluated by ImageQuant, normalized to the intensity of the 0 min sample, and plotted versus the time. Data from panels **a** and **b** are shown in panels **c** and **d**, respectively

Discussion

In this work we present evidence that different features of PNPase participate in HR upon P1 transduction and in error prone DNA repair of DSBs induced by zeocin in *E. coli*. Although PNPase is not essential for these processes, its contribution was readily and consistently detected. PNPase is implicated in mRNA and sRNA decay and processing, thus controlling the expression of several genes by various mechanisms [2]. A key question is whether PNPase may directly participate in recombination and repair pathways or indirectly controls these processes by regulating the expression of other genes directly involved in recombination and repair, and/or through metabolic pathways. This issue will be discussed below separately for recombination and repair.

PNPase structurally participates in HR protein complexes and modulates HR via its KH and S1 domains

We show that in *E. coli* mutants lacking PNPase P1 phage-mediated transduction frequency is reduced by up to five-six folds. Transductional recombination is thought to mainly depend on RecBCD recombination pathway,

whereas the RecF pathway seems to be responsible for less than 10% of the recombinants [26]. However, only a minor fraction of the transduced DNA undergoes recombination and gives rise to recombinant progeny. On the contrary, up to 90% of the total remains within the cytoplasm in a stable, supercoiled form, likely linked to specific phage encoded protein (s), and neither replicates nor is degraded for at least 5 h after infection (abortive transduction). Full transduction frequency may be increased at the expenses of abortive transduction by UV irradiating the donor strain or the phage lysate and by phage mutations [26]. We cannot rule out that PNPase regulates transduction efficiency by converting the abortive form of transduced DNA into the recombination proficient configuration. We favour, however, a hypothesis that implicates PNPase in the recombination pathway. In mutants expressing a PNPase devoid of catalytic activity but conserving the KH and S1 RBDs [24] transduction frequency was only marginally affected (Table 2). This suggests that PNPase is not implicated in transduction via processing of recombination intermediates or degradation of specific mRNAs coding for proteins implicated in this process. On the

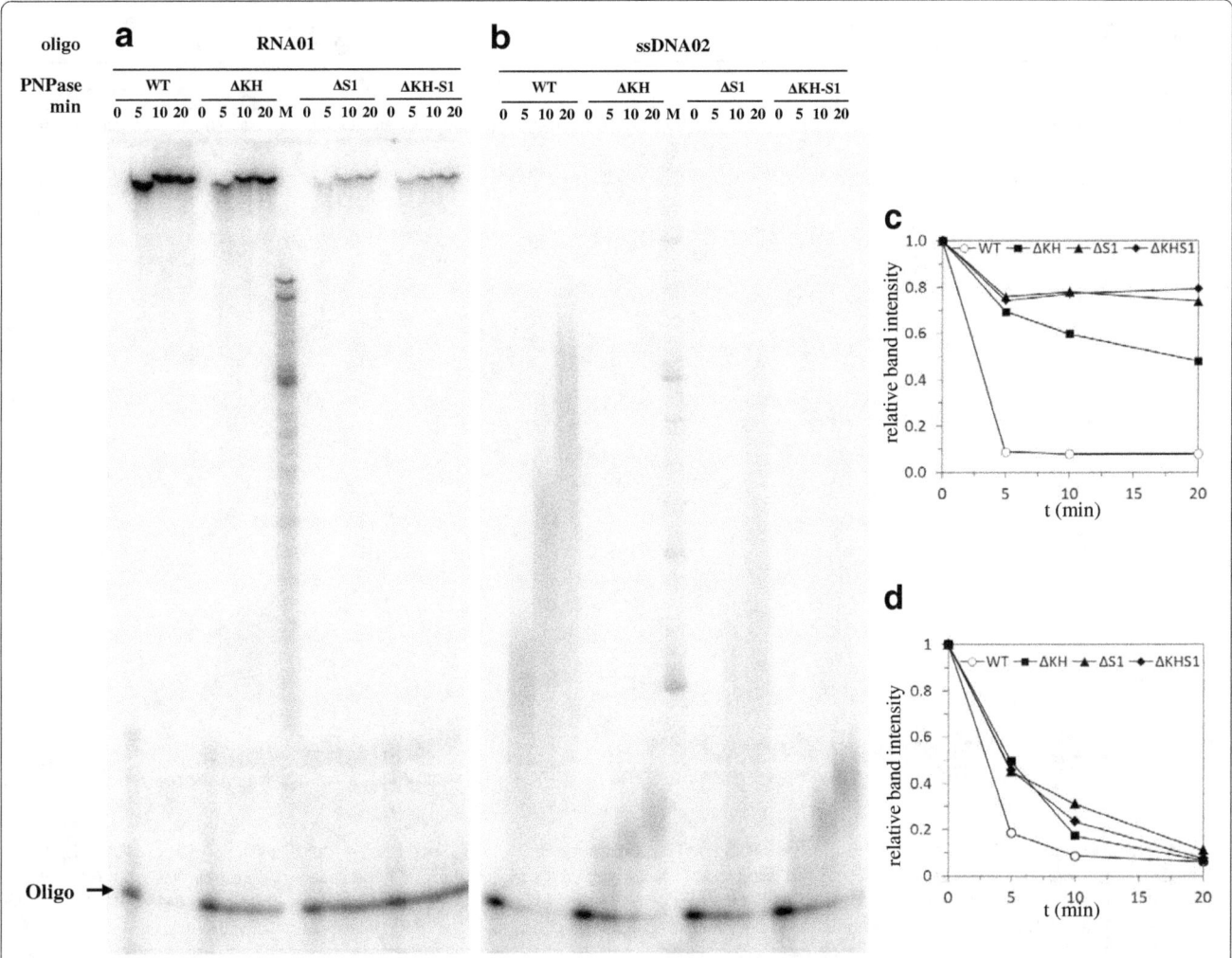

Fig. 3 Template-independent synthesis of ssDNA or RNA catalyzed by different purified His-tagged PNPases. [^{32}P]-5'-end labeled RNA (RNA01; panel **a**) or ssDNA (DNA02; panel **b**) 20-mers, 4 nM each, were incubated at 26 °C with purified His-tagged PNPase or PNPase mutants in the RBDs (10 nM each) in buffer **a** with 100 μM ADP (Panel **a**) or 100 μM dADP (for ssDNA Panel **b**) for the times indicated. The reaction products were fractionated by denaturing 6% PAGE. The oligonucleotide band intensities were evaluated by ImageQuant, normalized to the intensity of the 0 min sample, and plotted versus the time. Data from panels **a** and **b** are shown in panels **c** and **d**, respectively

other end, transduction frequency was reduced 3–5-folds in mutants in either RBD, thus suggesting that PNPase may play a regulatory role, possibly through interactions with RNA or DNA. Surprisingly, deletion of both KH and S1 domains dramatically impaired transduction frequency much more than the lack of PNPase itself. A working model to rationalize such results is that PNPase associates with other elements of the HR machinery through its structural core and facilitates some steps of the recombination process through its RBDs. The lack of both domains may not impair the association of the PNPase core with the component (s) of the recombination machinery but would poison the complex, thus inhibiting recombination.

Interestingly, it was observed that *E. coli* RecA protein filaments may contain RNA and a putative PNPase activity [38, 39]. These observations lend support to the

idea that in *E. coli* PNPase may interact with elements of the recombination machinery and regulate their function independently of its catalytic activity. In contrast with this hypothesis Rath et al. [16], based on the observation that the recovery of recombinants was 1–2% of the recipient number for both wild and *pnp* mutant strains, claim that conjugational recombination is unaffected by PNPase. Unfortunately these authors did not provide sufficient details, such as the donor to recipient ratio, to critically evaluate their results. Anyway, it should be noted that conjugation and generalized transduction, although largely dependent on the RecBCD pathway, differ in several features that might affect some critical steps. For example, whereas in transduction a linear dsDNA molecule is injected by P1, upon conjugation DNA is processively transferred into the recipient as a single stranded molecule

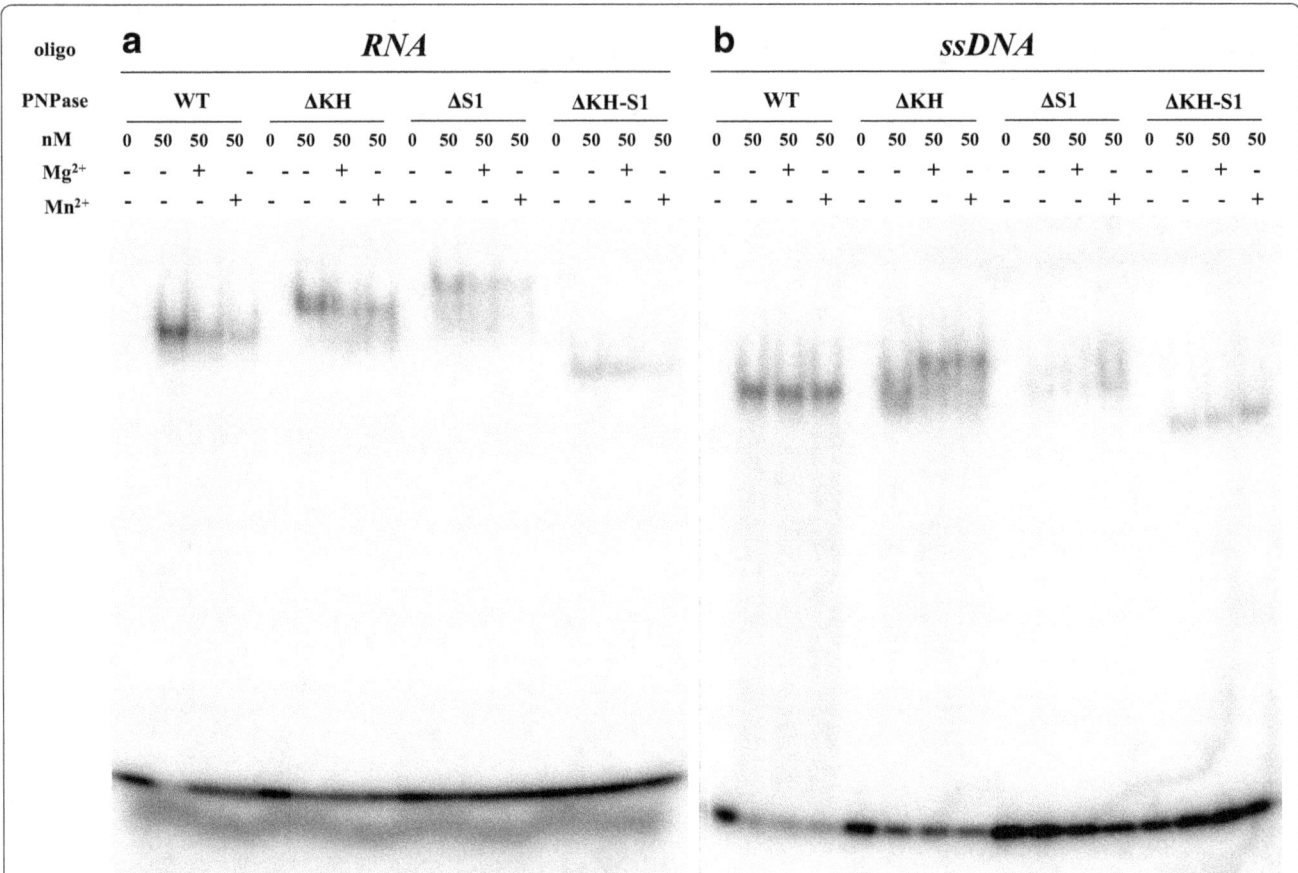

Fig. 4 Electrophoretic mobility shift assay of PNPase with RNA and DNA oligonucleotides. [^{32}P]-5'-end labeled RNA (RNA01; panel **a**) or ssDNA (DNA02; panel **b**) 20-mers were incubated 20 min at 21 °C with wild type or ΔKHS1 mutant His-tagged PNPase at the concentrations indicated on the *top* of the panels. 2 mM MnCl$_2$ or MgCl$_2$ were added to Binding Buffer as specified. PNPase-nucleic acid complexes were resolved by native 5% PAGE. After electrophoresis the gels were dried and the autoradiographic images analyzed by phosphorimaging and ImageQuant software

with its 5′-end covalently linked to the relaxase to be then replicated discontinuously [40]. Therefore a different involvement of PNPase in transduction- and conjugation-driven HR could be justified.

PNPase participates in error-prone DSBR with its catalytic activity

Recombination and DNA repair are intimately intertwined processes. In *E. coli* and several other bacteria one of the most relevant recombination pathways is the RecBCD/AddAB/AdnAB-dependent DSBR system, which can fix DNA breakages in an error free manner [41]. On the other side, template-independent non-homologous end joining (NHEJ) pathways can repair chromosome breaks at the cost of mutations [42]. Although evolutionarily conserved, the classical NHEJ pathway is missing in many bacteria such as *E. coli*, which lacks the gene encoding two signature factors of the NHEJ system, namely the Ku and ligase D proteins [42]. Nevertheless, an alternative template-independent pathway, termed alternative-end joining (A-EJ) appears to be operating in *E. coli* [43].

To test whether and how PNPase could participate not only in HR but also in DNA repair we assessed in different PNPase mutants the frequency of Rif[R] mutants induced by zeocin, a radiomimetic that causes DNA DSBs in vivo [29]. In our experimental conditions we observed that induction of Rif[R] mutants in strains lacking PNPase or expressing a catalytically inactive enzyme is about the half than in strains expressing the wild type protein or variants lacking either or both RBDs (Table 3). It thus appears that PNPase is implicated in a mutagenic DSB repair pathway and that, contrary to what we observed in HR, its catalytic activity is strictly required. Obviously these data do not rule out that PNPase may be also implicated in error free repair pathways through HR and that the overall effect could be the summation of both (or more) processes.

It should be noted that deletion of PNPase RBDs strongly impairs RNA degradation and template-independent polymerization efficiency, whereas it affects to a lesser extent these catalytic activities on DNA substrates (only polymerization processivity appears to be strongly impaired; Figs. 2 and 3). Since PNPase catalytic activity is required for induction of mutants, these observations fit the hypothesis that PNPase is directly implicated in DNA transactions in an error prone repair pathway rather than controlling expression of a DNA repair pathway at the level of mRNA stability. In addition, the lack of PNPase RBDs (in particular the KH domain) impairs in vitro polymerization (in particular polymerization processivity) with dNTPs as a substrate, whereas it seems to minimally affect ssDNA phosphorolysis (Figs. 2 and 3). It might thus be inferred that ssDNA degradation rather than synthesis is required for the mutagenic pathway. For example, PNPase might participate, together with RecBCD, in resection of single strand protruding ends and/or could remove "dirty" (not ligatable) ends thus contributing to generate substrates for the A-EJ pathway. We cannot rule out, however, that the residual template-independent polymerizing activity of PNPase lacking the RBDs may suffice to support the mutagenic repair pathway and that both (limited) resection and synthesis of ssDNA ends may be implicated in the process.

In *B. subtilis* PNPase copurifies with RecN, a key protein for the repair of DNA DSBs [44], and it is required for the formation of RecN-promoted discrete repair centers upon DSBs induction. In this context PNPase provides the RecN-associated $3' \rightarrow 5'$ ssDNA exonucleolytic activity. Moreover, PNPase catalytic activity on ssDNA is modulated in vitro by RecN, RecA and SSB [13, 14]. Thus PNPase from a distantly related bacterium seems to physically and functionally interact with recombination and DSB repair systems.

It has been shown [15, 36] that in *Ms*PNPase deletion of the RBDs impairs RNA phosphorylase and polymerase activities whereas it enhances the DNA polymerase and phosphorylase activity. Moreover, lack of the S1 RBD enhances divalent cation-dependent (catalytic) binding to ssDNA and DNA polymerase activity, thus suggesting that the S1 domain of *Ms*PNPase on the one hand helps capturing an RNA polynucleotide substrate for processive $3'$ end polymerization, on the other one provides a specificity filter that selects against a DNA polynucleotide substrate. *Ec*PNPase, however, seems to differ from *Ms*PNPase in that deletion of RBDs seems to impair, albeit mildly, PNPase catalytic activity on a DNA substrate, with a marked reduction of polymerization processivity. In Mycobacteria three distinct pathways are known to participate in DSB repair of DNA, namely HR, NHEJ and single-strand annealing [45]. It is possible that differences in biochemical properties related to DNA transactions of PNPase from these two distantly related bacteria may reflect different roles of this enzyme in the recombination and repair pathways.

Overall our data suggest that in *E. coli* PNPase may participate to recombination and repair pathways in different ways. PNPase may interact with HR pathways and modulate some steps through its RBDs, whereas its catalytic activity

does not appear to be implicated in this process. On the other hand, this enzyme may participate in an error-prone DSB repair process through its catalytic activity.

PNPase has been previously implicated also in mutagenesis caused by spontaneous misincorporation errors that occur during replication and are normally corrected by the mismatch repair (MMR) systems. Genetic interactions between *pnp* and the MMR system encoded by *mutS*, *mutL*, *mutH* and *uvrD* have been shown by both Jawali and coworkers [16] and by Miller and coworkers [17]. The former authors found that deletion of *pnp* partially suppresses the mutator phenotype of a *uvrD* deficient mutant and suggested that PNPase may regulate the expression of a hypothetical functional homologue of *uvrD*. The latter group showed that deletion of *pnp* suppresses the mutator phenotype of *mutS* or *mutL* mutants. The underlying mechanism suggested by these authors [17] is that phosphorolysis, by generating ribonucleotides diphosphates that serve as substrates for the salvage biosynthetic pathways of dNTPs, affects the nucleotide pool concentration and the rate of base misincorporation during replication [46, 47]. Thus PNPase might contribute in different ways to different mutagenic pathways.

Cytotoxicity of genotoxic agents in *pnp* mutants

Genotoxicity of toxic chemicals may account for only a quota of their cytotoxicity [48]; therefore, survival of cells exposed to genotoxic agents may also depend on mechanisms not implicated in DNA repair. We have shown that PNPase features required to complement a Δ*pnp* mutant for zeocin resistance are not completely superimposable to those required for zeocin mutagenesis or HR. Notably, zeocin sensitivity of *pnp* mutants, unlike zeocin-induced mutagenesis, is suppressed by overexpression of RNase II, thus suggesting that, at least in part, zeocin detoxification occurs via post-transcriptional control of gene expression and/or degradation of damaged RNA.

PNPase deficient mutants have been shown to be more sensitive to genotoxic agents in different species. *E. coli* mutants are more sensitive to hydrogen peroxide treatment which increases the levels of oxidized ribo- and deoxyriboguanosines; moreover, *E. coli* PNPase was suggested to contribute to cell survival to oxidative stress by acting as a specific scavenger of oxidized RNA [49]. *E. coli* *pnp* mutants also exhibit UV sensitivity. This phenotype is not epistatic to the *uvrABC* nucleotide excision repair (NER) system and to *recJ*, *recQ*, *recG*, thus suggesting that PNPase is implicated neither in NER nor in the single-stranded gap repair. However, UV sensitivity of *pnp* mutants is epistatic to *uvrD*, *recB* and *ruvA*, thus implicating PNPase in the recombinational repair process [16]. *B. subtilis* lacking PNPase is more sensitive to oxidative stress (chronic exposure to H_2O_2) but shows increased tolerance to other DNA damaging agents such as methyl methane sulfonate, 4-nitroquinoline-1-oxide or mitomycin C, as compared to wild type cells [13]. It thus appears that different repair and detoxification pathways may be differently activated or repressed by PNPase.

Conclusions

The scenario emerging from our results and those discussed hereby is that in *E. coli* various features of PNPase influence several different pathways implicated in DNA metabolism, such as homologous and non-homologous recombination, DNA repair, spontaneous and induced mutagenesis through a variety of mechanisms. RNA degradation may not only modulate expression of genes implicated in the above processes, but also impact the relative composition of nucleotide pool, thus affecting the rate of misincorporation at replication, and destroy RNA damaged by genotoxic agents, which may be toxic to the cell. In addition, as this enzyme can both degrade and synthesize ssDNA in a template-independent manner, it may directly participate in error-prone repair pathways that depend on these features. Finally, PNPase appears to non-enzymatically interact with components of HR machinery whereby modulating HR efficiency via its nucleic acid binding activity. It is remarkable that this non-essential gene has evolved so many subtle regulatory interactions with core cellular processes implicated in RNA and DNA metabolism.

Methods

Bacterial strains and media

Bacterial strains and plasmids are described in Tables 4 and 5, respectively. Recombinant plasmids used in complementation experiments were constructed in pGZ119HE [50], a derivative of the ColD high copy number plasmid (15–20 plasmids per chromosome in the exponential phase [51]). Unless otherwise stated, bacteria were grown at 37 °C in LD broth and LD agar plates [52]; auxotrophy screenings and selections were performed in M9-agar minimal medium [24]. Culture media were supplemented, as needed, with

0.2% arabinose, 0.2% glucose, 100 µg/ml ampicillin, 30 µg/ml chloramphenicol, 50 µg/ml kanamycin, 50 µg/ml streptomycin, 100 µg/ml rifampicin, 0.1 mM IPTG, and zeocin (InvivoGen) at the concentrations indicated.

Transduction

P1 HFT plate stocks were grown on the prototrophic donor strain C-1a from a single plaque as described by Miller [53]. The auxotrophic recipient strains were grown in LD broth with 5 mM $CaCl_2$ at 37 °C up to 0.5 OD_{600}, spun down (4500 rpm for 10 min at 4 °C), resuspended in 1/10 volume of MC buffer (10 mM $MgSO_4$; 5 mM $CaCl_2$), and incubated 30 min at 37 °C with aeration. Cells were infected with P1 HFT at a multiplicity of infection (m.o.i.) of 0.5 at 37 °C. To determine unadsorbed phage titer, 20 min upon infection a 0.02 ml sample was transferred into a tube with 2 ml of ice-cold M9-citrate buffer (10 mM Na-citrate in M9 medium) and a drop of chloroform, vigorously shaken, centrifuged 5 min at 13,000 rpm, and the supernatant assayed for plaque forming units. At the same time the infected cells were diluted 5-fold in M9-citrate buffer, washed twice in one volume and resuspended in 1/10 volume of the same buffer. The prototrophic transductants were then selected by plating on M9-glucose agar plates. Loss of the Kan^R marker associated to auxotrophy was assayed by replica plating of either all transductants obtained, if the total number was smaller than 45, or at least 45 individual transductant colonies per each transduction. Frequency of transduction was calculated as the ratio of auxotrophic transductants to the adsorbed phage.

Zeocin-induced mutagenesis

Bacterial cultures grown in LD at 37 °C with aeration up to $OD_{600} = 0.25$ were split in two aliquots and zeocin (100 µg/ml as indicated) was added to one culture. After 30 min of further incubation, both cultures were washed twice by centrifugation and resuspended in 400 µl of LD broth. 10 µl were used to assay the total viable counts on LD agar, whereas the remaining was plated on four LD-

Table 4 Bacterial strains

Strain	Parental	Relevant characters	Construction/Reference
BW25113	Prototype	*E. coli* K12	[54]
C-1a	Prototype	*E. coli* C, prototrophic	[55]
C-5691	C-1a	Δpnp751	[56]
C-5883	C-1a	ΔleuA::kan	by P1HFT * JW0073 transduction; this work
C-5884	C-5691	ΔleuA::kan Δpnp751	by P1HFT * JW0073 transduction; this work
C-5885	C-1a	ΔtrpE::kan	by P1HFT * JW1256 transduction; this work
C-5886	C-5691	ΔtrpE::kan Δpnp751	by P1HFT * JW1256 transduction; this work
C-5981	C-1a	Δrnb::kan	by P1HFT * JW1279 transduction; this work
JW0073	BW25113	ΔleuA::kan	[57]
JW1256	BW25113	ΔtrpE::kan	[57]

Table 5 Plasmids

Plasmid	Relevant characters	Reference
pAZ101	pGZ119HE derivative, harbours the pnp+ allele	[58]
pAZ1112	pAZ101 derivative; harbours the pnp-S438A allele encoding a catalytically inactive PNPase	[24]
pAZ1113	pAZ101 derivative; harbours the pnp-74 allele (ΔKH 603–615) under the control of pnp P2 promoter. Obtained by cloning the AgeI-BsiWI fragment of pnp-74 from pEJ04 in the large AgeI-BsiWI fragment of pAZ101.	this work
pAZ1114	pAZ101 derivative; harbours the pnp-78 allele (ΔS1 622–633) under the control of pnp-p2 promoter. Obtained by cloning the AgeI-BsiWI fragment of pnp-78 from pEJ08 in the large AgeI-BsiWI fragment of pAZ101.	this work
pAZ1115	pGZ119HE derivative; harbours the rnb+ allele under the control of its own rnb-p promoter. Obtained by cloning into pGZ119HE a BamHI-HindIII-digested PCR fragment amplified from MG1655 DNA with primers FG3063 (CG**GGATCC**TGCAAGGGCGAAAATG) and FG3086 (CCC**AAGCTT**CATGAAATTAACGGCGGC) encompassing chromosomal coordinates 1,349,209-1,344,800 (NCBI Sequence ID U00096.3).	this work
pAZ133	pAZ101 derivative, harbours the Δpnp-833 allele encoding Pnp-ΔKHS1	[23]
pEJ04	Harbours the pnp-74 allele under T5 promoter-lacO operator control	[22]
pEJ08	Harbours the pnp-78 allele under T5 promoter-lacO operator control	[22]
pGZ119HE	ColD, Cam^R	[50]

agar plates supplemented with 50 μg/ml rifampicin. Rif^R mutants were scored after 48 h incubation at 37 °C. For strains harbouring a plasmid, the growth media were supplemented with 30 μg/ml of chloramphenicol. Mutants frequency was assessed from three independent cultures.

EMSA and in vitro RNA degradation and polymerization assays

^{32}P-radiolabelled RNA and DNA probes for EMSA and in vitro degradation and polymerization assays were prepared as follows. 10 pmol of a 20-mer RNA oligonucleotide (RNA01: 5′-ACUGGACAAAUACUCCGAGG-3′ [24]; and a 20-mer ssDNA oligonucleotide (DNA02: 5′-ACTGGACAAATACTCCGACG-3′; were labelled at their 5′ end with [γ-^{32}P] ATP by T4 polynucleotide kinase (New England Biolabs) in 20 μl of polynucleotide kinase buffer provided by the manufacturer; after ethanol precipitation in the presence of 1 mg/ml glycogen to remove non-incorporated nucleotides, the samples were suspended in the same volume of RNase-free water. For EMSA, 50 fmol of either RNA01 or DNA02 probes were incubated for 20 min at 21 °C in Binding Buffer (50 mM Tris HCl pH 7.4, 50 mM NaCl, 0.5 mM DTT, 0.025% NP40 (Fluka), 10% glycerol) with increasing amounts of purified His-tagged PNPases in a final volume of 10 μl. The samples were fractionated by native 5% polyacrylamide gel electrophoresis (PAGE) at 4 °C. For in vitro template independent polymerization experiments, 80 fmol of radiolabelled RNA or ssDNA primers were incubated in buffer A (10 mM Tris–HCl, pH 7.5, 10 mM KCl, 2 mM MgCl$_2$ (for RNA) or 2 mM MnCl$_2$ (for ssDNA), 0.75 mM DTT, 2% PEG-6000) containing 100 μM ADP (for RNA) or 100 μM dADP (for ssDNA) and 10 nM PNPase at 26 °C in a final volume of 20 μl. For in vitro degradation experiments, 80 fmol of radiolabelled RNA or ssDNA primers were incubated in buffer A with 10 μM Pi

and 10 nM PNPase at 26 °C in a final volume of 20 μl. Three μl aliquots were withdrawn at different time points, diluted in 5 μl of RNA loading dye (2 mg/ml xylene cyanol and bromophenol blue, 10 mM EDTA in formamide) and fractionated by denaturing 6% PAGE. After electrophoresis the gels were dried and the autoradiography images analyzed by phosphorimaging and ImageQuant software.

Abbreviations

A-EJ: Alternative-end joining; CFU: Colony forming unit; dNDP: Deoxyribonucleotide; dNTP: Deoxyribonucleoside triphosphate; DSB: Double strand break; DSBR: Double strand break repair; dsDNA: Double strand DNA; DTT: Dithiothreitol; e.o.p.: Efficiency of plating; EcPNPase: PNPase from E. coli; EMSA: Electrophoretic mobility shift assays; HR: Homologous recombination; IC$_1$: Inhibitory concentration giving 1% survival; IPTG: Isopropyl β-D-1-thiogalactopyranoside; Kan$^{S/R}$: Kanamycin sensitive/resistant; m.o.i.: Multiplicity of infection; MMR: Mismatch repair; MsPNPase: PNPase from M. smegmatis; NDP: Nucleoside diphosphate; NER: Nucleotide excision repair; NHEJ: Non-homologous end joining; OD$_{600}$: Optical density at 600 nm wavelength; PAGE: Polyacrylamide gel electrophoresis; PAP: Polyadenylpolymerase; Pi: Inorganic phosphate; PNPase: Polynucleotide phosphorylase; RBD: RNA binding domain; Rif^R: Rifampicin resistant; rpm: Revolutions min^{-1}; sRNA: Small RNA; ssDNA: Single strand DNA; TF: Transduction frequency; Tris: Tris (hydroxymethyl) aminomethane; UV: Ultraviolet light; Δ: Deletion

Acknowledgements

We are grateful to JC Alonso for helpful discussions and reading of the manuscript.

Funding

This work was supported by Regione Lombardia-MIUR, project ID 30190679 (to GD). TC was supported by a Type A fellowship from Università degli Studi di Milano. The funding bodies had no role in the study design, data production, analysis and interpretation, and in manuscript writing process.

Authors' contributions

GD, TC and FB, conceived the project and designed the experiments. GD wrote the manuscript. TC and GS designed and performed the experiments. All authors read and approved the final manuscript.

Competing interests

The authors declare that they have no competing interests.

Author details

[1]Dipartimento di Bioscienze, Università degli Studi di Milano, via Celoria 26, Milan 20133, Italy. [2]Present address: Dipartimento di Biotecnologie mediche e medicina traslazionale, Università degli Studi di Milano, via F.lli Cervi 93, Segrate, MI 20090, Italy. [3]Present address: Eurofins BioPharma Product Testing Italy, Eurofins Biolab srl, via Bruno Buozzi, 2, Vimodrone 20090, Italy.

References

1. Grunberg-Manago M, Ortiz PJ, Ochoa S. Enzymatic synthesis of nucleic acidlike polynucleotides. Science. 1955;122:907–10.
2. Briani F, Carzaniga T, Dehò G. Regulation and functions of bacterial PNPase. Wiley Interdiscip Rev RNA. 2016;7:241–58.
3. Ochoa S. Enzymatic synthesis of ribonucleic acid. In: Nobel Lectures 1959 Stockholm. 1959. p. 146–64.
4. Hurwitz J. The discovery of RNA polymerase. J Biol Chem. 2005;280:42477–85.
5. Deutscher MP, Li Z. Exoribonucleases and their multiple roles in RNA metabolism. Prog Nucleic Acid Res Mol Biol. 2001;66:67–105.
6. Mohanty BK, Kushner SR. Polynucleotide phosphorylase functions both as a 3' right-arrow 5' exonuclease and a poly (A) polymerase in Escherichia coli. Proc Natl Acad Sci U S A. 2000;97:11966–71.
7. Slomovic S, Portnoy V, Yehudai-Resheff S, Bronshtein E, Schuster G. Polynucleotide phosphorylase and the archaeal exosome as poly (A)-polymerases. Biochim Biophys Acta. 2008;1779:247–55.
8. Mohanty BK, Kushner SR. The majority of Escherichia coli mRNAs undergo post-transcriptional modification in exponentially growing cells. Nucleic Acids Res. 2006;34:5695–704.
9. Kaufmann G, Littauer UZ. Deoxyadenosine diphosphate as substrate for polynucleotide phosphorylase from Escherichia coli. FEBS Lett. 1969;4:79–83.
10. Chou JY, Singer MF. Deoxyadenosine diphosphate as a substrate and inhibitor of polynucleotide phosphorylase of Micrococcus luteus.I. Deoxyadenosine diphosphate as a substrate for polymerization and the exchange reaction with inorganic ^{32}P. J Biol Chem. 1971;246:7486–96.
11. Gillam S, Waterman K, Doel M, Smith M. Enzymatic synthesis of deoxyribo-oligonucleotides of defined sequence. Deoxyribo-oligonucleotide synthesis. Nucleic Acids Res. 1974;1:1649–64.
12. Beljanski M. De novo synthesis of DNA-like molecules by polynucleotide phosphorylase in vitro. J Mol Evol. 1996;42:493–9.
13. Cardenas PP, Carrasco B, Sanchez H, Deikus G, Bechhofer DH, Alonso JC. Bacillus subtilis polynucleotide phosphorylase 3'-to-5' DNase activity is involved in DNA repair. Nucleic Acids Res. 2009;37:4157–69.
14. Cardenas PP, Carzaniga T, Zangrossi S, Briani F, Garcia-Tirado E, Dehò G, Alonso JC. Polynucleotide phosphorylase exonuclease and polymerase activities on single-stranded DNA ends are modulated by RecN, SsbA and RecA proteins. Nucleic Acids Res. 2011;39:9250–61.
15. Unciuleac M-C, Shuman S. Distinctive effects of domain deletions on the manganese-dependent DNA polymerase and DNA phosphorylase activities of Mycobacterium smegmatis polynucleotide phosphorylase. Biochemistry. 2013;52:2967–81.
16. Rath D, Mangoli SH, Pagedar AR, Jawali N. Involvement of pnp in survival of UV radiation in Escherichia coli K-12. Microbiology. 2012;158:1196–205.
17. Becket E, Tse L, Yung M, Cosico A, Miller JH. Polynucleotide phosphorylase plays an important role in the generation of spontaneous mutations in Escherichia coli. J Bacteriol. 2012;194:5613–20.
18. Haber JE. Genome Stability: DNA Repair and Recombination. New York: Garland Science; 2013.
19. Symmons MF, Jones GH, Luisi BF. A duplicated fold is the structural basis for polynucleotide phosphorylase catalytic activity, processivity, and regulation. Structure. 2000;8:1215–26.
20. Bermúdez-Cruz RM, Fernández-Ramírez F, Kameyama-Kawabe L, Montañez C. Conserved domains in polynucleotide phosphorylase among eubacteria. Biochimie. 2005;87:737–45.
21. Shi Z, Yang WZ, Lin-Chao S, Chak KF, Yuan HS. Crystal structure of Escherichia coli PNPase: central channel residues are involved in processive RNA degradation. RNA. 2008;14:2361–71.
22. Matus-Ortega ME, Regonesi ME, Piña-Escobedo A, Tortora P, Dehò G, García-Mena J. The KH and S1 domains of Escherichia coli polynucleotide phosphorylase are necessary for autoregulation and growth at low temperature. Biochim Biophys Acta. 2007;1769:194–203.
23. Briani F, Del Favero M, Capizzuto R, Consonni C, Zangrossi S, Greco C, De Gioia L, Tortora P, Dehò G. Genetic analysis of polynucleotide phosphorylase structure and functions. Biochimie. 2007;89:145–57.
24. Carzaniga T, Mazzantini E, Nardini M, Regonesi ME, Greco C, Briani F, De Gioia L, Dehò G, Tortora P. A conserved loop in polynucleotide phosphorylase (PNPase) essential for both RNA and ADP/phosphate binding. Biochimie. 2014;97:49–59.
25. Nurmohamed S, Vaidialingam B, Callaghan AJ, Luisi BF. Crystal structure of Escherichia coli polynucleotide phosphorylase core bound to RNase E, RNA and manganese: implications for catalytic mechanism and RNA degradosome assembly. J Mol Biol. 2009;389:17–33.
26. Masters M. Generalized transduction. In: Neidhardt FC, editor. Escherichia coli and Salmonella: cellular and molecular biology, vol. 2nd. 2nd ed. Washington D.C: ASM Press; 1996. p. 2421–41.
27. Smith GR. Conjugational recombination in E. coli: myths and mechanisms. Cell. 1991;64:19–27.
28. Haber JE. Evolution of models of homologous recombination. In: Genome Stability: DNA Repair and Recombination. New York: Garland Science; 2013. pp. 396.
29. Chankova SG, Dimova E, Dimitrova M, Bryant PE. Induction of DNA double-strand breaks by zeocin in Chlamydomonas reinhardtii and the role of increased DNA double-strand breaks rejoining in the formation of an adaptive response. Radiat Environ Biophys. 2007;46:409–16.
30. Awano N, Inouye M, Phadtare S. RNase activity of polynucleotide phosphorylase is critical at low temperature in Escherichia coli and is complemented by RNase II. J Bacteriol. 2008;190:5924–33.
31. Zilhão R, Cairrão F, Régnier P, Arraiano CM. PNPase modulates RNase II expression in Escherichia coli: implications for mRNA decay and cell metabolism. Mol Microbiol. 1996;20:1033–42.
32. Goodman MF. Error-prone repair DNA polymerases in prokaryotes and eukaryotes. Annu Rev Biochem. 2002;71:17–50.
33. Bridges BA. Error-prone DNA repair and translesion DNA synthesis. II: The inducible SOS hypothesis. DNA Repair. 2005;4:725–6. 739.
34. Rodgers K, McVey M. Error-prone repair of DNA double-strand breaks. J Cell Physiol. 2016;231:15–24.
35. Kuzminov A, Stahl FW. Overview of Homologous Recombination and Repair Machines. In: Higgins NP, editor. The Bacterial Chromosome. Washington, D.C.: ASM Press; 2005. pp. 349–367.
36. Unciuleac M-C, Shuman S. Discrimination of RNA from DNA by polynucleotide phosphorylase. Biochemistry. 2013;52:6702–11.
37. Bermúdez-Cruz RM, García-Mena J, Montañez C. Polynucleotide phosphorylase binds to ssRNA with same affinity as to ssDNA. Biochimie. 2002;84:321–8.
38. Register III JC, Griffith J. 10 nm RecA protein filaments formed in the presence of Mg^{2+} and ATPγS may contain RNA. Mol Gen Genet. 1985;199:415–20.
39. Roberts JW, Roberts CW, Craig NL, Phizicky EM. Activity of the Escherichia coli recA-gene product. Cold Spring Harb Symp Quant Biol. 1979;43(Pt 2):917–20.
40. Firth N, Ippen-Ihler K, Skurray RA. Structure and function of the F factor and mechanism of conjugation. In: Neidhardt FC, editor. Escherichia coli and Salmonella: cellular and molecular biology, vol. 2nd. 2nd ed. Washington D. C: ASM Press; 1996. p. 2377–422.
41. Wigley DB. Bacterial DNA repair: recent insights into the mechanism of RecBCD, AddAB and AdnAB. Nat Rev Microbiol. 2013;11:9–13.
42. Pitcher RS, Brissett NC, Doherty AJ. Nonhomologous end-joining in bacteria: a microbial perspective. Annu Rev Microbiol. 2007;61:259–82.
43. Chayot R, Montagne B, Mazel D, Ricchetti M. An end-joining repair mechanism in Escherichia coli. Proc Natl Acad Sci U S A. 2010;107:2141–6.
44. Alonso JC, Cardenas PP, Sanchez H, Hejna J, Suzuki Y, Takeyasu K. Early steps of double-strand break repair in Bacillus subtilis. DNA Repair. 2013;12:162–76.
45. Gupta R, Barkan D, Redelman-Sidi G, Shuman S, Glickman MS. Mycobacteria exploit three genetically distinct DNA double-strand break repair pathways. Mol Microbiol. 2011;79:316–30.
46. Danchin A. Comparison between the Escherichia coli and Bacillus subtilis genomes suggests that a major function of polynucleotide phosphorylase is to synthesize CDP. DNA Res. 1997;4:9–18.

47 Tse L, Kang TM, Yuan J, Mihora D, Becket E, Maslowska KH, Schaaper RM, Miller JH. Extreme dNTP pool changes and hypermutability in *dcd ndk* strains. Mutat Res. 2016;784–785:16–24.

48 Vock EH, Lutz WK, Hormes P, Hoffmann HD, Vamvakas S. Discrimination between genotoxicity and cytotoxicity in the induction of DNA double-strand breaks in cells treated with etoposide, melphalan, cisplatin, potassium cyanide, Triton X-100, and gamma-irradiation. Mutat Res. 1998;413:83–94.

49 Wu J, Jiang Z, Liu M, Gong X, Wu S, Burns CM, Li Z. Polynucleotide phosphorylase protects *Escherichia coli* against oxidative stress. Biochemistry. 2009;48:2012–20.

50 Lessl M, Balzer D, Lurz R, Waters VL, Guiney DG, Lanka E. Dissection of IncP conjugative plasmid transfer: definition of the transfer region Tra2 by mobilization of the Tra1 region in trans. J Bacteriol. 1992;174:2493–500.

51 Frey J, Timmis KN. ColD-derived cloning vectors that autoamplify in the stationary phase of bacterial growth. Gene. 1985;35:103–11.

52 Ghisotti D, Chiaramonte R, Forti F, Zangrossi S, Sironi G, Dehò G. Genetic analysis of the immunity region of phage-plasmid P4. Mol Microbiol. 1992;6:3405–13.

53 Miller JH. Experiments in Molecular Genetics, vol. 1. Cold Spring Harbor: Cold Spring Harbor Laboratory; 1972.

54 Datsenko KA, Wanner BL. One-step inactivation of chromosomal genes in *Escherichia coli* K-12 using PCR products. Proc Natl Acad Sci U S A. 2000;97:6640–5.

55 Sasaki I, Bertani G. Growth abnormalities in Hfr derivatives of *Escherichia coli* strain C. J Gen Microbiol. 1965;40:365–76.

56 Regonesi ME, Del Favero M, Basilico F, Briani F, Benazzi L, Tortora P, Mauri P, Dehò G. Analysis of the *Escherichia coli* RNA degradosome composition by a proteomic approach. Biochimie. 2006;88:151–61.

57 Baba T, Ara T, Hasegawa M, Takai Y, Okumura Y, Baba M, Datsenko KA, Tomita M, Wanner BL, Mori H. Construction of *Escherichia coli* K-12 in-frame, single-gene knockout mutants: the Keio collection. Mol Syst Biol. 2006;2:2006.0008.

58 Regonesi ME, Briani F, Ghetta A, Zangrossi S, Ghisotti D, Tortora P, Dehò G. A mutation in polynucleotide phosphorylase from *Escherichia coli* impairing RNA binding and degradosome stability. Nucleic Acids Res. 2004;32:1006–17.

6

Transcriptome profiling analysis reveals metabolic changes across various growth phases in *Bacillus pumilus* BA06

Lin-Li Han[1,2†], Huan-Huan Shao[1,2†], Yong-Cheng Liu[1,2], Gang Liu[1,2], Chao-Ying Xie[1,2], Xiao-Jie Cheng[1,2], Hai-Yan Wang[1,2], Xue-Mei Tan[1,2] and Hong Feng[1,2*] (iD)

Abstract

Background: *Bacillus pumilus* can secret abundant extracellular enzymes, and may be used as a potential host for the industrial production of enzymes. It is necessary to understand the metabolic processes during cellular growth. Here, an RNA-seq based transcriptome analysis was applied to examine *B. pumilus* BA06 across various growth stages to reveal metabolic changes under two conditions.

Results: Based on the gene expression levels, changes to metabolism pathways that were specific to various growth phases were enriched by KEGG analysis. Upon entry into the transition from the exponential growth phase, striking changes were revealed that included down-regulation of the tricarboxylic acid cycle, oxidative phosphorylation, flagellar assembly, and chemotaxis signaling. In contrast, the expression of stress-responding genes was induced when entering the transition phase, suggesting that the cell may suffer from stress during this growth stage. As expected, up-regulation of sporulation-related genes was continuous during the stationary growth phase, which was consistent with the observed sporulation. However, the expression pattern of the various extracellular proteases was different, suggesting that the regulatory mechanism may be distinct for various proteases. In addition, two protein secretion pathways were enriched with genes responsive to the observed protein secretion in *B. pumilus*. However, the expression of some genes that encode sporulation-related proteins and extracellular proteases was delayed by the addition of gelatin to the minimal medium.

Conclusions: The transcriptome data depict global alterations in the genome-wide transcriptome across the various growth phases, which will enable an understanding of the physiology and phenotype of *B. pumilus* through gene expression.

Keywords: *Bacillus pumilus*, Bacterial flagella, Chemotaxis, Gene expression, Gene regulation, Protease, RNA-seq, Sporulation, Stress response, Transcriptome, Tricarboxylic acid cycle

Background

Bacillus pumilus is an endospore-forming, gram-positive, rod-shaped bacterium. Due to its metabolic diversity and spore dispersal, *B. pumilus* is ubiquitous in various environments and commonly resistant to extreme environmental conditions [1, 2]. Similar to other *Bacillus* species, *B. pumilus* is able to secrete a large number of industrial enzymes, such as lipases [3], xylanases [4], and proteases [5–7]. Therefore, *B. pumilus* has attracted attention in biotechnology and was selected to engineer novel industrial production strain [8, 9]. In addition, some strains of *B. pumilus* were used to produce valuable small molecules [10, 11] and have served as biocontrol agents to manage plant diseases [12]. However, studies pertaining to the physiological and metabolic processes of *B. pumilus* are extremely limited in comparison with the other *Bacillus* species.

* Correspondence: hfeng@scu.edu.cn
†Equal contributors
[1]Key Laboratory of Bio-resources and Eco-environment, Ministry of Education, Sichuan Key Laboratory of Molecular Biology and Biotechnology, Sichuan University, Chengdu 610064, Sichuan, People's Republic of China
[2]College of Life Sciences, Sichuan University, Chengdu 610064, Sichuan, People's Republic of China

Along with the great advances in genomics, many strains of B. pumilus have been selected for genome sequencing. To date, more than 30 genomes of B. pumilus are available at NCBI. Based on genomic alignments, B. pumilus is closer to B. subtilis, B. licheniformis, and B. amyloliquefaciens [13], suggesting that similar metabolic or physiological processes may exist between B. pumilus and the model organism B. subtilis. Although the genome is considered the blueprint of life, much information regarding the physiological or metabolic processes is not directly accessible from the genome [14]. Therefore, various omics technologies such as transcriptomics, proteomics and metabolomics, have been employed as essential steps toward gaining insights into cell physiology from the genome. For instance, a combined omics-based approach has been applied to B. pumilus to try to understand cell physiology in response to oxidative stress and protein secretion [15, 16]. For this purpose, RNA-seq -based transcriptomics analysis is one of the most powerful tools that not only provides important insights into the functional elements of the genome, gene expression patterns and regulation [17], but also offers a simpler and more cost-effective approach [18]. Therefore, the RNA-seq method has been widely applied to many Bacillus species. For example, B. subtilis [19, 20], B. licheniformis [21], and B. thuringiensis [22] were examined by RNA-seq analysis.

B. pumilus BA06 has been isolated from proteinaceous soil and is demonstrated to be able to secrete extracellular proteases that exhibited great potential in leather processing [5, 23, 24]. However, the production of extracellular proteases is considered to occur at the stationary growth phase and is extensively regulated in many species of Bacillus [25, 26]. For example, nitrogen and carbon sources have great impacts on the production of extracellular proteases [27]. Thus, medium components have usually been optimized for fermentation of proteases [28]. In addition, the other physiological processes of secondary metabolite synthesis, protein secretion, and sporulation occur during the stationary growth phase

[29–31]. Based on studies in B. subtilis, a transition point occurs between the exponential growth and stationary growth phases [32]. Over the various growth phases, the expression of many genes may be turned off and another set of genes may be turned on. For example, more than 100 genes, whose expression was induced at the onset of the stationary growth phase, have been assigned to the SigB regulon [33, 34].

By taking advantage of the RNA-seq technology, a time-resolved transcriptomic analysis to cover the various growth stages of B. pumilus BA06 was performed. The results clearly indicated that changes in gene expression or metabolic pathways were occurring during the various growth phases; this observation will be helpful for us toward understanding the physiology and phenotype of B. pumilus.

Results

Cell growth, sporulation and extracellular protease activity

To gain insights into the temporal transcriptome changes of Bacillus pumilus BA06, two sets of cultures were established in 50 -ml of MM (minimal medium) and GM (MM plus 2% gelatin) in 250 ml flasks. Initially, the growth curve was monitored by measurement of OD_{600} at 6-h intervals (data not shown). Unexpectedly, the cell density suddenly declined after 12 h. To reflect the intrinsic cell growth, the numbers of vegetative cells and endospores in the same cultures were further calculated by plate -counting. Fig. 1a shows the growth curves of the total cell and endospore counts at various time points; it is evident that the total cell number arrived at its peak at 12 h and then declined at 24 h. The addition of gelatin to the medium did not change the pattern of cell growth except that the cell number was slightly higher. However, endospore formation started at approximately 24 h, and then increased greatly up to a peak at 60 h and 72 h in MM and GM, respectively. Therefore, the cell growth of B. pumilus BA06 could be obviously divided into two phases: the exponential

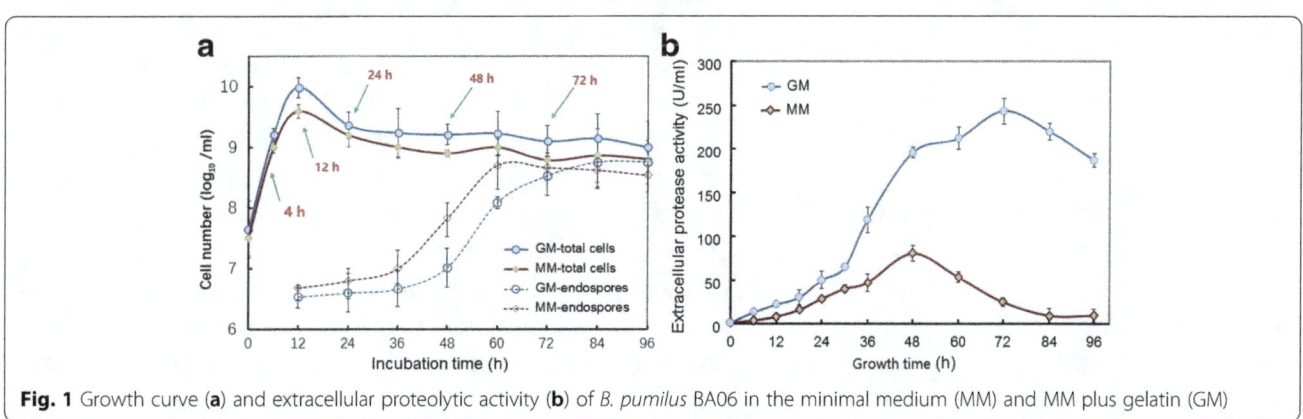

Fig. 1 Growth curve (**a**) and extracellular proteolytic activity (**b**) of B. pumilus BA06 in the minimal medium (MM) and MM plus gelatin (GM)

growth and stationary phases with a transition point at 12 h under this condition [32].

Meanwhile, the extracellular protease activity was also monitored during the growing period. Fig. 1b shows that the extracellular protease activity mainly occurred at the onset of the stationary phase with a peak at 48 h in MM, which was consistent with previous results [25]. Moreover, the addition of gelatin led to a higher peak at the delayed time (72 h) in extracellular protease activity (Fig. 1b), indicating that the nitrogen source (gelatin) affected the production of extracellular proteases.

RNA sequencing and identification of differentially expressed genes (DEGs)

Based on the growth curve, cell samples were collected at 5 time points (4 h, 12 h, 24 h, 48 h, and 72 h) across the exponential growth phase, transition point, and early and late stationary growth phases from three independent cultures of MM and GM; and total RNAs were subsequently isolated (Additional file 1). Three RNA samples for each time point were mixed equally, and used for Illumina sequencing. Finally, approximately 1.5 G clean data were generated for each time point. The summary of the sequence data is presented in Table 1. It was found that more than 98% of the clean reads from each sample could be mapped to the reference genome, indicating that the transcriptome data were sufficient for further analysis.

Based on the number of mapped reads against the genome with the Bowtie software, the expression level of each gene was calculated in terms of the FPKM value (A complete list of annotated genes with the FPKM value is presented in Additional file 2.). The DEGs were extracted using the edgeR software with p value <0.05 and $\log_2$(fold-change) > 1. Consequently, 1418 and 1499 DEGs were identified between the two different time points in MM and GM, respectively. The numbers of DEGs between the various time points during the growth course of *B. pumilus* BA06 were analyzed using a Venn diagram (Fig. 2a). Globally, two striking changes

of the DEG numbers between the two time points could be observed in the MM cultures, one for the transition point (12 h) and one for the stationary growth phase (48 h). There were 815 and 868 DEGs identified between 12 h/4 h and 48 h/ 24 h (Fig. 2a), respectively, indicating that the metabolic transition and endospore development were an intrinsic consequence of the gene expression change. However, the addition of gelatin to the MM did not change the gene expression pattern of *B. pumilus* BA06 significantly, except that more DEGs were also found between 72 h/48 h (Fig. 2b).

All the DEGs were categorized by the RAST system [35]. Table 2 shows the top 12 subsystems with the numbers of the involved DEGs. The most abundant subsystems of DEGs were related to metabolism of "Carbohydrates", followed by "Amino Acids and Derivatives", which suggested that the changes in primary metabolism were great between the various growth phases. In addition, the DEGs involved in "Mobility and Chemotaxis" and "Dormancy and Sporulation" were also great in number (see details in the following sections), indicating that the physiology or phenotype would be changed greatly. It was noticed that the change in "Iron Acquisition and Metabolism" would be great for this bacterium since the expression levels of many genes included in iron metabolism were altered between the different growth phases. For example, several operons responsive to iron compound uptake (peg.2296–2299) and siderophore biosynthesis (peg.34–39) were up-regulated upon entry into the transition point (12 h) (Additional file 2). In fact, the operon encoding siderophore biosynthesis is missing in model organism *B. subtilis* 168, suggesting that something in the physiological or metabolic processes is different among the various *Bacillus species*.

Changes in the metabolic pathways at the transition point of growth

A transition point is recognized as a growth phase where the cells cease exponential growth and enter the stationary

Table 1 Summary of RNA-seq and the reads mapped to the genome of *B. pumilus* BA06

Sample name	Left size (M)	Right size (M)	Number of reads	Mapped reads (%)	Paired reads (%)	GC (%)
MM-04	866.19	875.33	23,415,852	23,010,464 (98.27)	23,005,388 (98.27)	42.37
MM-12	847.89	854.26	24,131,710	24,009,599 (99.49)	24,009,599 (99.49)	42.49
MM-24	729.02	738.72	20,629,536	20,492,441 (99.34)	20,492,441 (99.34)	42.77
MM-48	755.99	761.63	20,882,864	20,732,600 (99.28)	20,732,600 (99.28)	42.27
MM-72	829.13	834.27	23,020,742	22,782,607 (98.97)	22,782,607 (98.97)	42.40
GM-04	770.57	777.42	21,326,144	21,223,188 (99.52)	21,223,188 (99.52)	42.33
GM-12	888.51	898.43	25,386,720	25,258,250 (99.49)	25,258,250 (99.49)	42.67
GM-24	792.57	799.42	9,023,902	8,970,697 (99.41)	8,970,697 (99.41)	43.21
GM-48	796.46	803.54	17,332,139	17,212,840 (99.31)	17,212,840 (99.31)	42.61
GM-72	749.20	753.69	20,854,598	20,625,390 (98.9)	20,625,390 (98.9)	42.22

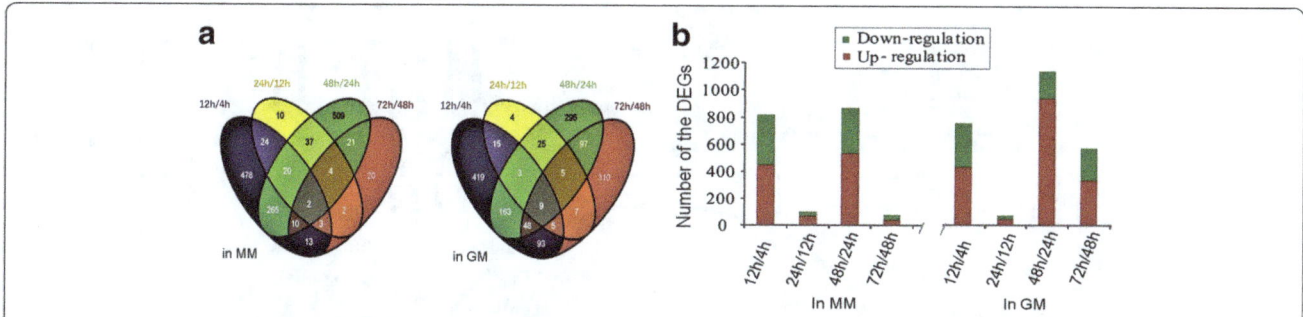

Fig. 2 Analysis of the differentially expressed genes of *B. pumilus* BA06 grown in MM and GM over the entire growth curve. **a** Venn diagram analysis and (**b**) numbers of DEGs between toe time points

growth phase in *Bacillus* [32]. Previous studies have revealed that the expression levels of many genes involved in various metabolic pathways could change between the various growth phases in *B. subtilis* [36, 37]. Similarly, a large number DEGs in *B. pumilus* BA06 were identified at the transition point (12 h) from exponential growth in both cultures of MM and GM (Fig. 2). By KEGG analysis, several metabolic pathways were significantly enriched. First, the tricarboxylic acid cycle (TCA) pathway was down-regulated, since the expression levels of almost all the genes involved in the TCA cycle were decreased significantly (Fig. 3). There was no large difference for the cells cultivated in MM and GM (Additional file 3). However, the metabolic pathway from acetyl-CoA to phosphoenol-pyruvate via oxaloacetate was up-regulated [38], which may provide sufficient substrate for the downstream pathway of glycolysis. It was noticed that the accompanying respiration metabolism was also down-regulated (Additional file 3).

Second, down-regulation of genes encoding flagellar structural proteins and genes involved in chemotaxis signaling was also observed (Fig. 4), indicating that the

swarming mobility of *B. pumilus* BA06 may decrease when entering the transition point. In *B. subtilis*, the sigma factor (*sigD*) and the *swrA* gene were responsive toward activating the expression of the flagellar operon [39]. Similarly, the homologs encoding sigD (peg.585) and swrC (peg.2650) were identified in *B. pumilus* BA06, which were demonstrated to be down-regulated at the transition point (Additional file 4). For swarming mobility, biosynthesis of biosurfactin is also necessary [40]. The expression of the *srf* operon, which encodes genes (peg.2905–2910) for the biosynthesis of biosurfactin in *B. pumilus*, was also down-regulated (Additional file 4). Altogether, the swarming mobility of BA06 may be seriously attenuated when the cells enter the transition point and thereafter.

Continuous up-regulation of sporulation-related genes

For *Bacillus* species, a critical characteristic is the formation of the endospore, a kind of dormant cells with high resistance to environmental stress. Endospore development starts at the onset of the stationary growth phase, which is subject to highly hierarchical regulation [41]. By

Table 2 Top 12 functional classification of the DEGs of *B. pumilus* BA06 during the entire growth curve by RAST analysis

Subsystems	in MM				in GM			
	12/04	24/12	48/24	72/48	12/04	24/12	48/24	72/48
Carbohydrates	98	2	45	5	79	1	37	22
Amino Acids and Derivatives	42	2	44	1	45	8	29	12
Motility and Chemotaxis	55	7	55	5	55	10	34	33
Cofactor/Vitamin/Prosthetic Group	28	4	12	0	28	1	15	23
Clustering-based Subsystems	25	0	28	0	27	1	42	17
Dormancy and Sporulation	20	0	59	0	16	0	34	31
Cell Wall and Capsule	16	1	17	2	14	0	10	16
Iron Acquisition and Metabolism	18	2	9	0	22	0	14	14
Stress Response	16	0	13	2	13	0	15	13
Nucleosides and Nucleotides	16	1	4	1	16	0	5	7
Protein Metabolism	11	1	14	1	10	0	17	8
Respiration	12	0	6	0	12	0	4	2

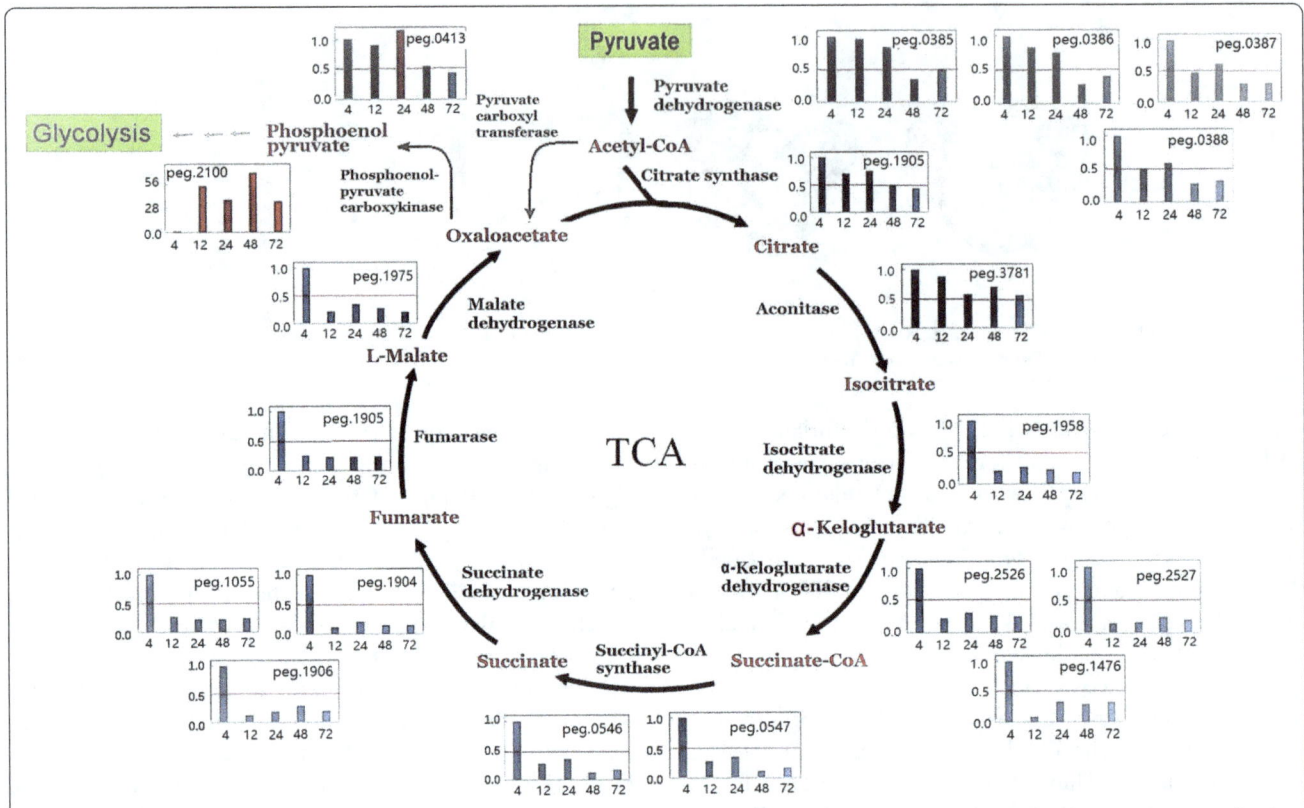

Fig. 3 Down-regulation of the tricarboxylic acid cycle of *B. pumilus* BA06 grown in MM. The gene expression level of each gene was normalized to 1.00 for the first -time point (4 h)

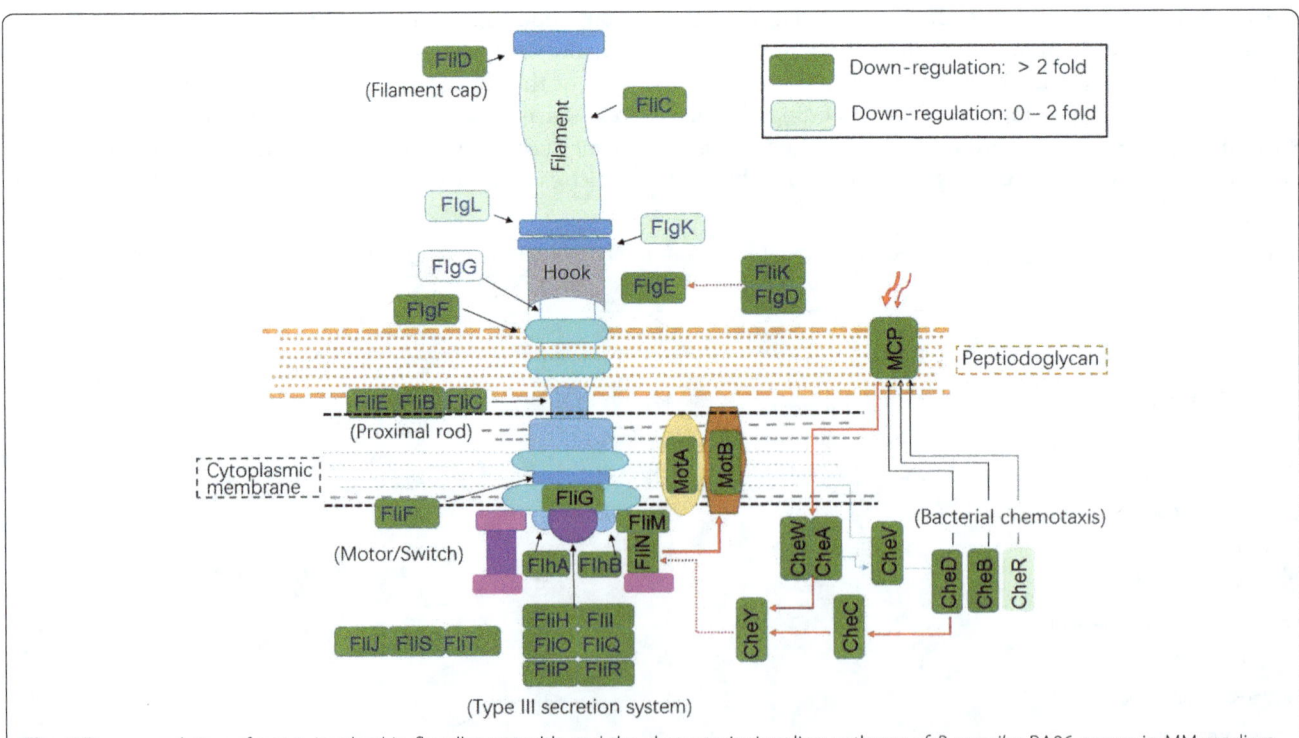

Fig. 4 Down-regulation of genes involved in flagellar assembly and the chemotaxis signaling pathway of *B. pumilus* BA06 grown in MM medium at the transition from the exponential growth phase

means of RAST annotation and manually searching, 123 genes were categorized into "Dormancy and Sporulation" (Additional file 5). A cluster analysis of all the sporulation-related genes is shown in Fig. 5, which was divided into three groups (G-1, G-2 and G-3). A great number of the genes in group G-1 displayed a sharp surge in expression at 12 h and then at 48 h in MM. However, the addition of gelatin to the MM led to an extra surge in expression at 72 h, which was consistent with the observation that endospore formation was delayed in GM in comparison with MM (Fig. 1b).

A master regulator protein, Spo0A, governs sporulation initiation through phosphorylation by a series of kinases and phosphotransferase [39]. All of these genes were identified in *B. pumilus* BA06 (Additional file 5) and were clustered into group G-2 (Fig. 5). The expression profile of these genes was generally different from the G-1 group. Many of these genes (G-2) were repressed at the later growth phases, perhaps indicating their leading role only in sporulation initiation.

During the procedure of endospore formation, four sigma factors (SigE, SigF, SigG, and SigK) play important roles in *B. subtilis* [41]. Their counterparts in *B. pumilus* BA06 were identified (group G-3), and all of their expression levels were continuously up-regulated across all growth phases (Fig. 5 and Additional file 5). In contrast, expression of *sigH* (peg.2382) fluctuated over the growth course.

Extracellular proteases and the protein secretion systems

Since *B. pumilus* produces large extracellular proteases that are of interest to the field of biotechnology, the expression of several extracellular proteases was examined. Table 3 showed that the genes for *aprE*, *aprX* and *wprA*, which encode extracellular proteases, were more highly expressed in the stationary growth phase, which was consistent with the activity assay (Fig. 1b). It was notable that the addition of gelatin led to higher expression level of *apr*E at 72 h. However, *apr*E was the major component in the extracellular proteolytic activity in terms of the transcription level. In contrast, the *epr* and *subE* genes were largely expressed at the exponential growth phase (4 h) or the transition phase (12–24 h). However, the expression of *vpr* fluctuated over the entire growth course.

Protein secretion is generally associated with the *Bacillus* species, which is a critical consideration for the development of the cell factory [42]. By KEGG analysis, two protein secretion pathways were enriched in *B. pumilus* BA06: the Sec-dependent pathway and Tat system. The expression pattern for these two pathways was similar in both MM and GM cultures (Fig. 6 and Additional file 6). However, the major components involved in the Sec-dependent pathway were expressed with two peaks at 4 h and 48 h. In contrast, the expression of the Tat system was increased in the transition phase (12–24 h). Furthermore, various signal peptidases displayed different expression patterns (Fig. 6). These results implied that various secretion systems may function at various growth phases.

Sigma factors and regulator proteins

Bacillus species usually employ different sigma factors to regulate various physiological processes. For example, SigB in *B. subtilis* mediates the stress response by regulating a large group of genes [43]. Therefore, the expression pattern of the sigma factors was also examined in *B. pumilus* BA06 (Fig. 7a and Additional file 7). The *sigB* gene (peg.3035) displayed a sharp surge in expression in the transition phase (12 h and 24 h), indicating that a stress response may occur at the transition point, which was similar to an observation in *B. subtilis* [44]. Although a SigB regulon in *B. pumilus* was not identified, we expect that a similar regulon may exist in this bacterium.

The other sigma factors exhibited different patterns of expression (Additional file 7). For example, the expression of *sigV* (peg.3701), *rpoN* (peg.1161) and *rpoE* (peg.399) was up-regulated mainly at the stationary growth phase (48 h). In contrast, the expression of *sigV* (peg.2600) and *sig70* (peg.775) was repressed upon entry into the transition phase and thereafter (Fig. 7a).

In *B. subtilis*, several regulatory proteins were recognized as transition-state regulators, such as Hpr (ScoC), AbrB, and SinR. All their homologs were encoded by the BA06 genome. Their expression pattern is shown in Fig. 7b

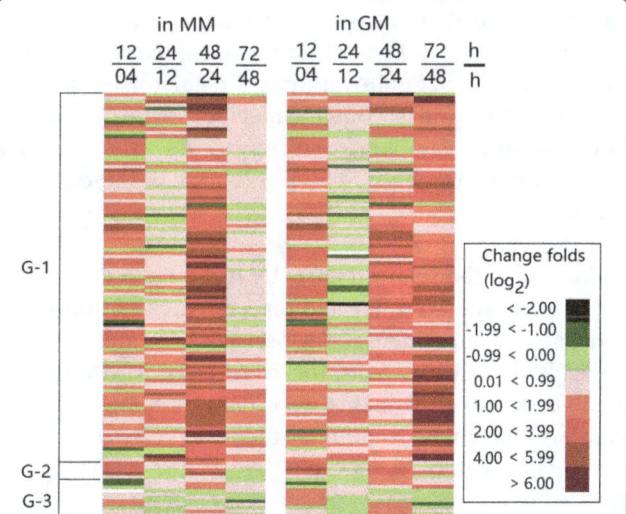

Fig. 5 Clustering analysis of the expression changes to genes involved in sporulation of *B. pumilus* BA06. G-1 represents the sporulation-related genes; G-2 displays the genes encoding the sporulation kinases and sporulation initiation phosphotransferases that are involved in sporulation initiation; G-3 shows five regulatory genes encoding sporulation sigma factors SigE, SigF, SigG, SigK, and SigH

Table 3 Relative expression level of the extracellular proteases of *B. pumilus* BA06 during the entirety of each growth phase

Gene ID	Gene	Protein	Medium	Expression level in FPKM value				
				4 h	12 h	24 h	48 h	72 h
peg.2284	*aprE*	serine alkaline protease (AprE)	MM	128	782	8665	18,789	22,894
			GM	71	575	2578	20,671	35,163
peg.658	*aprX*	alkaline serine protease (AprX)	MM	14	42	192	862	4011
			GM	11	40	62	70	4820
peg.1414	*subE*	subtilisin Carlsberg (SubE)	MM	528	437	467	66	37
			GM	376	688	597	220	38
peg.2809	*epr*	Alkaline serine proteinase (Epr)	MM	107	96	76	48	37
			GM	112	115	97	70	42
peg.3435	*vpr*	extracellular protease (Vpr)	MM	879	332	1106	344	703
			GM	747	561	862	1758	801
peg.2794	*wprA*	Wall-associated protease (WprA)	MM	52	185	258	217	316
			GM	97	203	275	198	221

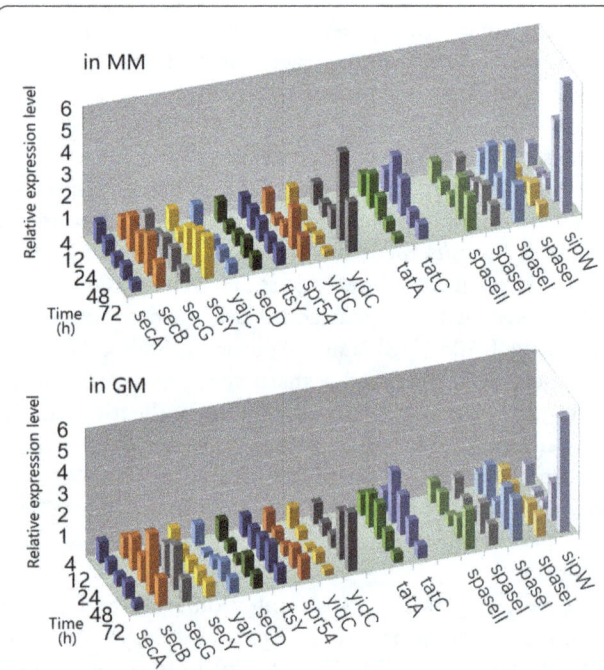

Fig. 6 Expression changes in genes involved in the protein secretion pathways of *B. pumilus* BA06 grown in MM and GM during the entire growth curve. The corresponding gene IDs are peg.3162 for *SecA*; peg.2383 for *SecB*; peg.1131 for *SecG*; peg.2418 for *SecY*; peg.1755 for *yajC*; peg.1750 for *SecD*; peg.533 for *ftsY*; peg.535 for *spr54*; peg.3727 for *yidC*; peg.1460 for *yidC*; peg.2558 for *tatA*; peg.2557 for *tatC*; peg.2146 for *Spase II*; peg.2304, peg.364 and peg.482 for *Spase I*, and peg.1525 for *sipW*. The gene expression level (FPKM) of each gene was normalized to 1.00 for the first -time point (4 h)

(Additional file 7). The gene for *hpr* was highly expressed at the exponential growth phase (4 h) and then declined. In contrast, expression of *SinR* and *pai2* was increased at 48 h and later, especially in GM. Another transition-state regulator, AbrB, was strongly repressed at 48 h, which is consistent with a similar observation in *B. subtilis* [36].

Global regulators play important roles in the process of gene transcription by binding to the promoter elements. In *Bacillus*, CodY and Spo0A have been recognized as global regulators. Cody, a GTP-binding protein, regulates more than one hundred genes that are typically repressed during rapid (exponential) growth and induced when cells experience nutrient deprivation [45]. Our data indicated that expression of *codY* was higher at the exponential growth phase and then declined. In contrast, the expression of Spo0A, a master regulator of sporulation initiation and secondary metabolism in *B. subtilis* [46], continuously increased up to 48 h. A similar expression pattern for *Spo0A* was also observed in *B. subtilis* [36].

Validation of the selected DEGs by real-time PCR

To confirm the accuracy and reproducibility of the transcriptome data, 8 genes were selected for qPCR validation. RNA samples from the cultures of MM and GM at different growth phases were used as templates. The data are shown in Additional file 8, indicating that the two sets of data between the RNA-seq and RT-PCR analyses were almost consistent.

Discussion

In this study, we determined the transcriptome profiles of *B. pumilus* across the various growth phases in both MM and GM cultures. Through DEG analysis and KEGG enrichment, transcriptional changes to genes that

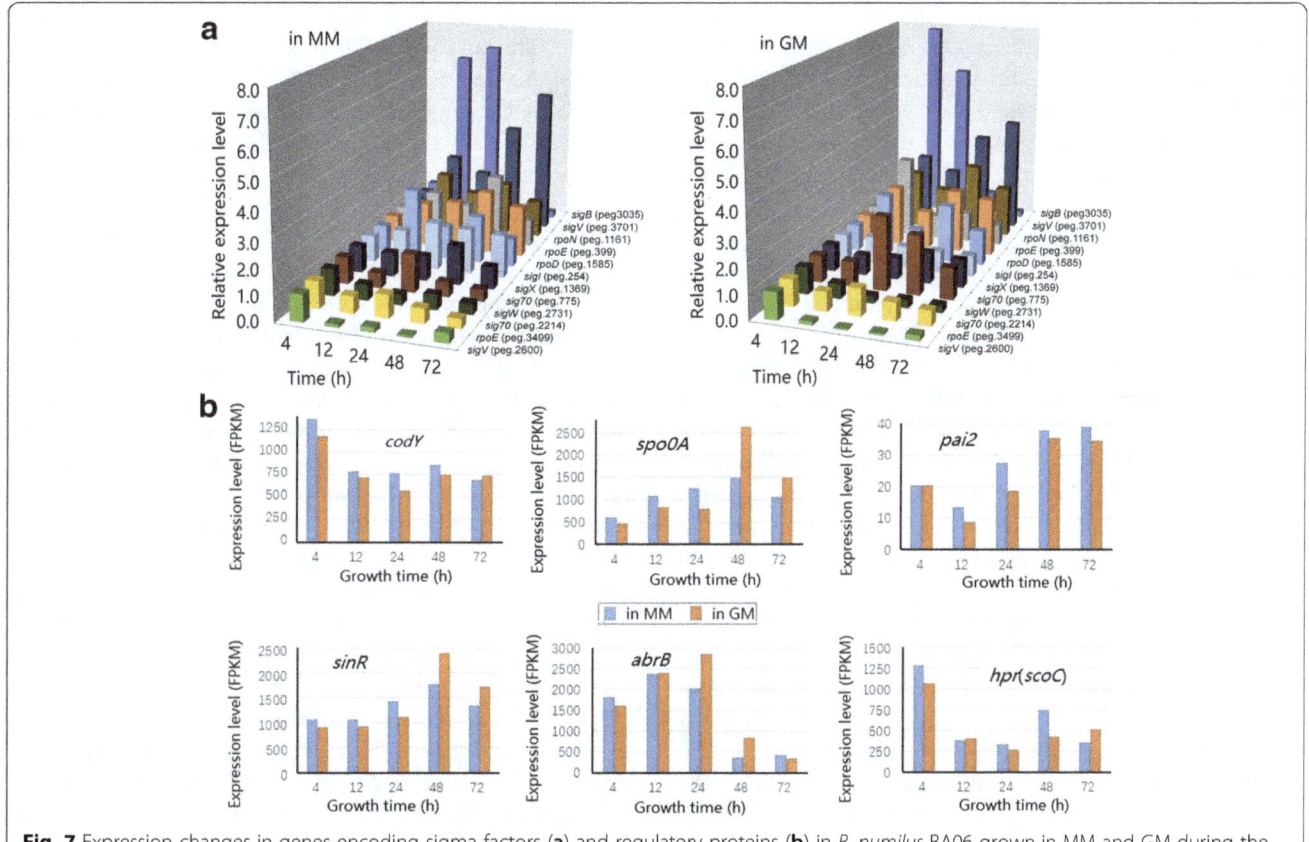

Fig. 7 Expression changes in genes encoding sigma factors (**a**) and regulatory proteins (**b**) in *B. pumilus* BA06 grown in MM and GM during the entire growth curve. The gene expression level (FPKM) of each gene was normalized to 1.00 for the first -time point (4 h)

are specific to certain metabolic pathways were unveiled at the various growth phases.

During the exponential growth phase, cellular growth is most active along with a quick depletion of nutrients. When the culture enters the transition phase, stress may occur at least through nutrient depletion [47]. At this transition point, a critical alteration in TCA metabolism was revealed in *B. pumilus* BA06. Our results showed that the TCA cycle was most active during the exponential growth phase (Fig. 3). Similarly, TCA -cycle enzymes were frequently identified from the *B. pumilus* cells during the exponential growth phase in the proteomic analysis [16]. The TCA cycle is a central metabolic pathway that not only unifies the carbohydrate, fat and protein metabolic processes, but also produces energy and reducing power. Therefore, the slowdown in the TCA cycle at the transition point and the later growth stage may be a signature of reduced cell activity. Previous studies showed that the expression levels of some genes involved in the TCA cycle decreased when *B. subtilis* was grown under anaerobic conditions [48]. In addition, glucose may regulate the TCA cycle by repressing several genes for the TCA [49]. In fact, at the transition phase and stationary growth phase, glucose in the culture is consumed. Therefore, regulation of the TCA

cycle may include other factors. However, secondary metabolism and sporulation will be triggered in the stationary growth phase, which requires the TCA metabolism to maintain carbohydrate flux and energy. Based on our transcriptome data, the carbohydrate flux could be recycled via another branch pathway, from acetyl-CoA to phosphoenol-pyruvate via oxaloacetate, since the genes encoding pyruvate carboxyl transferase (peg.413) and phosphoenolpyruvate carboxykinase (peg.2100) were not repressed but were greatly induced in expression upon entry into the transition phase and later growth phases (Fig. 3).

Another significant change at the transition point is down-regulation of the genes involved in flagellar assembly and the chemotaxis signaling pathway. The flagellum is an organ for mobility, which enables bacterial cells to colonize an ecological niche [50]. Mobility is also regulated by chemotaxis factors [39]. There are several chemoreceptors known as methyl-accepting chemotaxis proteins (MCPs) that are encoded by *B. pumilus*, most of which were revealed as being down-regulated after the exponential growth phase (Additional file 4). However, for other factors, such as surfactin, extracellular proteolytic activity was required to support mobility [51, 52]. The operon *srf* (peg.2905–2910) for biosurfactin biosynthesis was also down-regulated in *B. pumilus* BA06 at the same growth phase (Additional file 4). In fact, *B. pumilus* BA06 could produce surfactant at 24 h of culture, although the yield was lower [53]. Together, these data indicate the attenuation of the swarming mobility of *B. pumilus* upon entering the transition point and thereafter.

Sporulation is a complex cellular process and can be triggered by nutrient limitations [41]. For *B. pumilus* BA06, the formation of endospores was quick during the stationary growth phase (Fig. 1). Meanwhile, our transcriptome data indicated that almost all the genes involved in sporulation continued to be up-regulated after exponential growth up to the stationary growth phase (Fig. 5 and Additional file 5). The addition of gelatin to the MM led to a delay in both sporulation and expression of the sporulation-related genes. However, sporulation is subject to highly hierarchical regulation. In general, the transcriptional factor Spo0A governs the decision to initiate sporulation by phosphorylation via a series of kinases and phosphotransferases [54]. The expression level of *Spo0A* (peg.1484) increased slowly from 4 h to 48 h, which was consistent with a previous study in *B. subtilis* [36]. Since the phosphorylation state is a key factor for Spo0A to initiate sporulation [36], the transcription level observed here may not be relevant to the explanation of its function in sporulation initiation. Five kinases were identified in *B. pumilus*, of which the expression of *KinA* (peg.311, peg.312), *KinC* (peg.373), *KinD* (peg.276), and *KinE* (peg.265) was up-regulated. In contrast, *KinB* (peg.909) was down-regulated at the transition point (Additional file 5). In addition, the sporulation initiation phosphotransferase (Spo0F, peg.3334) was also down-regulated at 12 h and then up-regulated at 48 h and thereafter. Overall, the expression levels of the conserved sporulation kinases and phosphotransferases and Spo0A displayed a pattern that was different from the other sporulation-related genes, whose expression was continuously up-regulated until the later stationary growth phase (Fig. 5). The interruption may be ascribed to their role as only being responsive toward initiating sporulation. Recently, KinD in *B. subtilis* was shown to delay the onset of sporulation [55]. However, the expression level of *KinD* in *B. pumilus* was highest among the five sporulation kinases, which may directly phosphorylate Spo0A [41]. Therefore, KinD may play a more important role in sporulation initiation in *B. pumilus*.

During the sporulation process, the roles of several specific sigma factors SigE, SigF, SigG, and SigK were well documented in regulation of sporulation. These four factors were sequentially activated after sporulation initiation: sigF in the forespore, sigE in the mother cell, sigG in the forespore, and sigK in the mother cell [54]. Our transcriptome data indicated that all four sigma factors were expressed in a similar way with a peak at the later stationary growth phase (Fig. 5). However, another sigma factor, SigH (peg.2382), was found to participate in the regulation of sporulation in *Clostridium difficile* [56]. In *B. pumilus*, *sigH* was expressed in a manner that was different from the above four sigma factors. SigH may regulate the other genes involved in mobility and cell division. Therefore, the *sigH* gene may not be specific to sporulation in *B. pumilus*.

Extracellular protease production and protein secretion are processes of interest to the field of biotechnology. Previously, AprE, Epr, WprA, and Vpr were identified to be extracellular protease in *B. pumilus* SCU11 [57]. Our transcriptome data indicated that the expression patterns of various extracellular proteases were different in *B. pumilus* (Table 3). Three genes (*aprE*, *aprX*, and *wprA*) were expressed more highly during the stationary growth phase. Based on the expression levels, these proteases may contribute to the high extracellular proteolytic activity observed in the culture supernatant. In contrast, *epr* and *subE* were highly expressed in the exponential growth phase (Table 3), which could be explained by the fact that the expression of *epr* was controlled by SigD in *B. subtilis* [58]. However, these results suggested that different regulatory mechanisms and secretion pathways may be employed for the various proteases. In *B. subtilis*, the regulatory mechanism of the *aprE* expression has been extensively studied, indicating that at least two positive regulatory proteins (DegU and Spo0A) are involved in regulation of the *aprE* expression [59, 60]. The expression of *degU* and *spo0A* in *B. pumilus* was up-regulated at 48 h in GM (Additional file 1 and Fig. 7), which may contribute to the observed increase of *aprE* expression in the medium supplemented with gelatin.

Two protein secretion pathways were enriched by KEGG analysis in *B. pumilus*: The Sec-dependent pathway and Tat system, which were also identified by the previous proteomics analysis [16]. The Tat system responds by to secreting twin-arginine (RR/KRP) signal peptides, through which 44 proteins are predicated to be secreted in *B. subtilis* [61]. WprA has been proposed to be secreted through the Tat system in *B. pumilus* [62]. Regarding the expression pattern during the growth course, the Tat system (including TatA and TatC) may respond by secreting limited proteins and may play a major role in the transition phase. In contrast, the large number of genes involved in the Sec secretion pathway are down-regulated at the transition point, and then up-regulated at 48 h. In general, many proteins have been observed to be secreted extracellularly during the stationary growth phase in *B. subtilis* [62]. Therefore, the Sec secretion system may respond by secreting many proteins during the stationary growth phase. For example, AprE, Vpr, and aprX may be secreted via this system [63]. However, 65 proteins were detected from the late exponential growth cultures in *B. pumilus*; this number is much lower than the predicted protein species (513) [57]. Therefore, the Sec-dependent protein secretion system may be the major pathway, especially during the stationary growth phase in *B. pumilus*.

In *Bacillus*, alternative sigma factors are involved in the regulation of certain genes or specific metabolic processes. For example, SigB is generally recognized in response to stress. In *B. subtilis*, approximately 150 general stress-associated genes have been identified as the SigB regulon [64]. The expression of the *sigB* gene (peg.3035) is also up-regulated greatly at the transition phase in *B. pumilus*, which is similar with the observation in *B. subtilis* [37]. Although the sigB regulon has not been identified for *B. pumilus*, the *rsb* operon (peg.3029–3033) is induced upon entry into the transition point (Additional file 9). Other sigma factors such as SigW and SigX have been reported to respond to stress. For example, SigW in *B. subtilis* responds by regulating detoxification and the production of antimicrobial compounds [65]. However, the genes involved in the osmotic and oxidative responses, such as *opuAA* (peg. 2747, peg.3708), *opuAB* (peg.3707), *opuCB* (peg.1139), *opuCA* (peg.1140), *perR* (peg.820), and *cat* (peg.2205, peg.3703), have been shown to be up-regulated at the transition point in *B. pumilus* (Additional file 8: Table S8); some of these genes have also been revealed to be involved in the response to oxidative stress in another study using *B. pumilus* Jo2 and a microarray-based transcriptome analysis [15]. Therefore, *B. pumilus* may suffer from stress when entering the transition phase and subsequently activates a set of stress-related genes to be expressed in a manner that is similar to *B. subtilis*.

Conclusions

In conclusion, an RNA-seq-based transcriptome analysis was first applied to *B. pumilus* BA06 to monitor the transcriptional profile over the entire growth course in a defined MM or GM medium. Changes to the specific metabolic pathways in terms of gene expression were shown to relate to the transition from the exponential growth phase and the onset of the stationary growth phase. Upon entry into the transition point, one of the striking changes was down-regulation of central TCA metabolism and oxidative phosphorylation. Furthermore, the cellular mobility ability was also reduced because of down-regulation of genes involved in flagellar assembly and the chemotaxis signaling pathway. In contrast, many stress-responding genes including the SigB regulon, were induced in expression after entering the transition point, indicating that the cells may suffer from stress. During the

stationary growth phase, a significant change occurred in the genes involved in sporulation, and more than 100 sporulation-related genes were induced, which was consistent with the process of sporulation formation. However, the addition of gelatin to the MM medium did not cause a great impact on the transcriptome profile, except for the genes encoding sporulation-related proteins and extracellular proteases with delayed expression.

Methods
Bacterial strain and growth conditions

B. pumilus BA06 was routinely maintained on Luria-Bertani (LB, 10 g/l tryptone, 5 g/l yeast extract, 10 g/l NaCl, pH 7.5) agar plate. A single colony of *B. pumilus* BA06 was selected for transfer into 10 ml LB broth and incubated at 37 °C overnight with shaking at 140 rpm. Afterward, 500 µl of an overnight culture was transferred into 50 ml of minimal medium [MM, 1.0 g/l sodium citrate, 2.0 g/l $(NH_4)_2SO_4$, 14.0 g/l K_2HPO_4, 6.0 g/l KH_2PO_4, 0.2 g/l $MgSO_4$, 2.5 g/l yeast extract, 5.0 g/l D-glucose] and the gelatin-amended minimal medium (GM) (2.0 g/l gelatin) in 250-ml flasks. The cultures were incubated at 37 °C with shaking at 140 rpm for the indicated time points.

At various time points, the cell density was measured by reading the OD_{560} on a spectrometer. The total number of cells and endospores was also determined. To account for the total cell number, a 10-fold dilution of the fresh culture was achieved by serial dilution in sterile PBS buffer. Finally, 0.1 ml of the cell suspension was dispensed onto LB agar plates, and the colonies assigned as the total cells were counted after incubation at 37 °C. To count the number of endospores, an aliquot of fresh culture was sampled and incubated at 55 °C for 15 min to kill the vegetative cells; and the samples were diluted as above and dispensed onto LB agarose plates. The colonies formed on these plates were regarded as endospores. Meanwhile, extracellular protease activity was also assayed using casein as the substrate as described previously [24]. All experiments were performed in triplicate.

RNA isolation, library construction and Illumina sequencing

The cell samples were pelleted by centrifugation at 8000 rpm at the indicated time points (4, 12, 24, 48, and 72 h) from the *B. pumilus* BA06 cultures in MM and GM, and the cells were then suspended in TE buffer supplemented with 1.5 mg/ml lysozyme and incubated at 37 °C for 10 min, to which the TRIzol reagent (Invitrogen, Invitrogen, Carlsbad, CA) was added. The cell suspension was mixed extensively while using the gauge to disrupt the cells completely. Finally, total RNA was isolated following the instructions provided with the

TRIzol reagent. The genomic DNA was removed using the Genome DNA Eraser kit (Takara, Dalian, China).

Subsequently, the rRNA was removed using the Ribo-ZeroTM rRNA Removal kit (Epicentre Biotech, Madison, WI). The resulting mRNAs were fragmented and reverse transcribed using random hexamers as the primer. Second strand cDNA synthesis was performed using DNA Polymerase I and RNase H. The cDNA fragments were processed for end repair and ligated to paired end adaptors. Finally, the library was constructed and sequenced on an Illumina HiSeqTM2000 sequencing platform.

Mapping and identification of the differentially expressed genes

Clean data were obtained from the raw data by removing the sequences of the adapters and low-quality reads. The clean reads were aligned to the *B. pumilus* genome (GenBank accession number: AMDH00000000) [13] using Bowtie2 with default parameters, allowing up to one-base mismatches [66]. The aligned read files were processed by Cufflinks v2.2.1 [67]. The relative abundances of the transcripts were calculated as the fragments per kilobase of transcript per million fragments mapped (FPKM). Differentially expressed genes (DEGs) among the different samples were extracted by using edgeR in the Bioconductor package [68]. The DEGs were defined with an FDR (false discovery rate) ≤ 0.05 and $\log_2$fold-change ($\log_2$FC) ≥ 1.

Functional classification and KEGG analysis

Functional classification of DEGs was performed online by RAST (Rapid Annotation using Subsystem Technology, http://www.nmpdr.org/FIG/wiki/view.cgi/FIG/ RapidAnnotationServer). RAST is a fully automated service that is especially designed for annotating bacterial and archaeal genomes [35]. Once one gene is annotated, it can be classified into various subsystems.

Meanwhile, the DEGs were further assigned to KEGG (Kyoto Encyclopedia of Genes and Genomes) pathways on the KEGG Automatic Annotation Server (http://www.genome.jp/kegg).

Quantitative real-time PCR analysis

The expression levels of the selected eight genes (*degS*, *aprX*, *glnR*, *hpr*, *vpr*, *sinR*, *yqkD*, and *spo0A*) of *B. pumilus* BA06 growing under the same conditions, were validated by real-time RT-PCR analysis. The real-time RT-PCR was performed using an iCyclerMyiQ Real-Time PCR System (Bio-Rad, Hercules, CA). The PCR conditions were set up as 95 °C for 2 mins, followed by 40 cycles of 95 °C for 10 s, 65 °C for 15 s and 72 °C for 20 s. A melting curve analysis of the amplification products was performed at the end of each PCR run to ensure that unique products were amplified. The specific

primers used for the selected genes are listed in Additional file 10. The expression level was normalized to the internal control gene 16S rRNA, using the $2^{-\Delta\Delta Ct}$ method [69].

Additional files

Additional file 1: Electrophoresis analysis of RNA samples of *Bacillus pumilus* BA06.

Additional file 2: The gene annotation by RAST and expression levels (FPKM) across various growth stages of *Bacillus pumilus* BA06.

Additional file 3: List of the differentially expressed genes in the tricarboxylic acid cycle and oxidative phosphorylation by KEGG analysis.

Additional file 4: List of the differentially expressed genes involved in flagellar assembly and the chemotaxis signaling pathway enriched by KEGG analysis.

Additional file 5: List of the differentially expressed genes involved in dormancy and sporulation by RAST analysis.

Additional file 6: List of the differentially expressed genes assigned to protein secretion systems by KEGG analysis.

Additional file 7: List of the differentially expressed genes encoding regulatory proteins and the sigma factors.

Additional file 8: Comparison of expression levels of the selected eight genes between the transcriptome and qPCR methods.

Additional file 9: List of the differentially expressed genes assigned as the "Stress Response" subsystem by RAST analysis.

Additional file 10: Primers used for real-time PCR.

Acknowledgements

We thank the American Journal Experts (www.aje.com) for language edition for our manuscript.

Funding

This study was financially supported by the National Natural Science Fund of China (31171204). The funding sponsor had no role in the design of the study, collections, analysis, interpretation of data, writing of the manuscript, or decision to publish the results.

Authors' contributions

HF and HYW conceived and designed the experiments; LLH, HHS, GL, CYX and XJC performed the experiments; LLH, HHS, YCL and HF analyzed the data; HF and TXM contributed reagents and materials; HF, LLH and HHS wrote the manuscript. All authors read and approved the final manuscript.

Competing interests

The authors declare that they have no competing interest.

References

1. Liu Y, Lai Q, Dong C, Sun F, Wang L, Li G, et al. Phylogenetic diversity of the *Bacillus pumilus* group and the marine ecotype revealed by multilocus sequence analysis. PLoS One. 2013;8:e80097.

2. Connor N, Sikorski J, Rooney AP, Kopac S, Koeppel AF, Burger A, et al. Ecology of speciation in the genus *Bacillus*. Appl Environ Microbiol. 2010;76:1349–58.

3. Kim HK, Choi HJ, Kim MH, Sohn CB, Oh TK. Expression and characterization of Ca²⁺-independent lipase from *Bacillus pumilus* B26. Biochim Biophys Acta. 2002;1583:205–12.

4. Battan B, Sharma J, Dhiman SS, Kuhad RC. Enhanced production of cellulase-free thermostable xylanase by *Bacillus pumilus* ASH and its potential application in paper industry. Enzy Microbial Technol. 2007;41:733–9.

5. Huang Q, Peng Y, Li X, Wang H, Zhang Y. Purification and characterization of an extracellular alkaline serine protease with dehairing function from *Bacillus pumilus*. Curr Microbiol. 2003;46:169–73.

6. Kumar AG, Swarnalatha S, Gayathri S, Nagesh N, Sekaran G. Characterization of an alkaline active-thiol forming extracellular serine keratinase by the newly isolated *Bacillus pumilus*. J Appl Microbiol. 2011;104:411–9.

7. Jaouadi B, Ellouz-Chaaboum S, Rhimi M, Bejar S. Biochemical and molecular characterization of a detergent-stable serine alkaline protease from *Bacillus pumilus* CBS with high catalytic efficiency. Biochimie. 2008;90:1291–305.

8. Küppers T, Steffen V, Hellmuth H, O'Connell T, Bongaerts J, Maurer KH, et al. Developing a new production host from a blueprint: *Bacillus pumilus* as an industrial enzyme producer. Microbial Cell Fact. 2014;13:46.

9. Wemhoff S, Meinhardt F. Generation of biologically contained, readily transformable, and genetically manageable mutants of the biotechnologically important *Bacillus pumilus*. Appl Microbiol Biotechnol. 2013;97:7805–19.

10. Hua D, Ma C, Lin S, Song L, Deng Z, Maomy Z, et al. Biotransformation of isoeugenol to vanillin by a newly isolated *Bacillus pumilus* strain: identification of major metabolites. J Biotechnol. 2007;130:463–70.

11. Srivastava RK, Jaiswal R, Panda D, Wangikar PP. Megacell phenotype and its relation to metabolic alterations in transketolase deficient strain of *Bacillus pumilus*. Biotechnol Bioeng. 2009;102:1387–97.

12. *Bacillus pumilus* strain GB 34 (006493) Fact Sheet. https://www3.epa.gov/pesticides/chem_search/reg_actions/registration/fs_PC-006493_13-Mar-03.pdf.

13. Zhao CW, Wang HY, Zhang YZ, Feng H. Draft genome sequence of *Bacillus pumilus* BA06, a producer of alkaline serine protease with leather-dehairing function. J Bacteriol. 2012;194:6668–9.

14. Wiegand S, Dietrich S, Hertel R, Bongaerts J, Evers S, Volland S, et al. RNA-Seq of *Bacillus licheniformis*: active regulatory RNA features expressed within a productive fermentation. BMC Genomics. 2013;14:1.

15. Handtke S, Schroeter R, Jürgen B, Methling K, Schlüter R, Albercht D, et al. *Bacillus pumilus* reveals a remarkably high resistance to hydrogen peroxide provoked oxidative stress. PLoS One. 2014;9:e85625.

16. Handtke S, Volland S, Methling K, Albrecht D, Becher D, Nehls J, et al. Cell physiology of the biotechnical relevant bacterium *Bacillus pumilus*—an omics-based approach. J Biotechnol. 2014;192:204–14.

17. Sorek R, Cossart P. Prokaryotic transcriptomics: a new view on regulation, physiology and pathogenicity. Nat Rev Genet. 2010;11:9–16.

18. Nagalakshmi U, Waern K, Snyder M. RNA-Seq: a method for comprehensive transcriptome analysis. Curr Protocol Mol Biol. 2010;4:4.11.11–14.11.13.

19. Nicolas P, Mäder U, Dervyn E, Rochat T, Leduc A, Pigeonneau N, et al. Condition-dependent transcriptome reveals high-level regulatory architecture in *Bacillus subtilis*. Science. 2012;335:1103–6.

20. Brinsmade SR, Alexander EL, Livny J, Stettner AI, Segrè D, Rhee KY, et al. Hierarchical expression of genes controlled by the *Bacillus subtilis* global regulatory protein CodY. Proc Natl Acad Sci U S A. 2014;111:8227–32.

21. Guo J, Cheng G, Gou XY, Xing F, Li S, Han YC, et al. Comprehensive transcriptome and improved genome annotation of *Bacillus licheniformis* WX-02. FEBS Lett. 2015;589:2372–81.

22. Bassi D, Colla F, Gazzola S, Puglisi E, Delledonne M, Cocconcelli PS. Transcriptome analysis of *Bacillus thuringiensis* spore life, germination and cell outgrowth in a vegetable-based food model. Food Microbiol. 2016;55:73–85.

23. Wang HY, Liu DM, Liu Y, Cheng CF, Ma QY, Huang Q, et al. Screening and mutagenesis of a novel *Bacillus pumilus* strain producing alkaline protease for dehairing. Lett Appl Microbiol. 2006;44:1–6.

24. Wan MY, Wang HY, Zhang YZ, Feng H. Substrate specificity and thermostability of the dehairing alkaline protease from *Bacillus pumilus*. Appl Biochem Biotechnol. 2009;159:394–403.

25. Liu RF, Huang CL, Feng H. Salt stress represses production of extracellular proteases in *Bacillus pumilus*. Genet Mol Res. 2015;14:4339–948.

26. Rao MB, Tanksale AM, Ghatge MS, Deshpande V. Molecular and biotechnological aspects of microbial proteases. Microbiol Mol Biol Rev. 1998;62:597–635.

27. Gupta R, Beg QK, Lorenz P. Bacterial alkaline proteases: molecular approaches and industrial applications. Appl Microbiol Biotechnol. 2002;59:15–32.

28. Fakhfakh-Zouari N, Haddar A, Hmidet N, Frikha F, Nasri M. Application of statistical experimental design for optimization of keratinases production by *Bacillus pumilus* A1 grown on chicken feather and some biochemical properties. Process Biochem. 2010;45:617–26.

29. Schallmey M, Singh A, Ward OP. Developments in the use of *Bacillus* species for industrial production. Can J Microbiol. 2003;50:1–17.

30. Manabe K, Kageyama Y, Morimoto T, Shimizu E, Takahashi H, Kanaya S, et al. Improved production of secreted heterologous enzyme in *Bacillus subtilis* strain MGB874 via modification of glutamate metabolism and growth conditions. Microbial Cell Fact. 2013;12:18.

31. Setlow P. Spores of *Bacillus subtilis*: their resistance to and killing by radiation, heat and chemicals. J Appl Microbiol. 2006;101:514–25.

32. Strauch MA, Hoch JA. Transition-state regulators: sentinels of *Bacillus subtilis* post-exponential gene expression. Mol Microbiol. 1993;7:337–42.

33. Brigulla M, Hoffmann T, Krisp A, Völker A, Bremer E, Völker U. Chill induction of the SigB-dependent general stress response in *Bacillus subtilis* and its contribution to low-temperature adaptation. J Bacteriol. 2003;185:4305–14.

34. Akbar S, Gaidenko TA, Kang CM, O'Reilly M, Devine KM, Price CW. New family of regulators in the environmental signaling pathway which activates the general stress transcription factor sigma B of *Bacillus subtilis*. J Bacteriol. 2001;183:1329–38.

35. Aziz RK, Bartels D, Best AA, DeJongh M, Disz T, Edwards RA, et al. The RAST server: rapid annotations using subsystems technology. BMC Genomics. 2008;9:75.

36. Blom E-J, Ridder AN, Lulko AT, Roerdink JB, Kuipers OP. Time-resolved transcriptomics and bioinformatic analyses reveal intrinsic stress responses during batch culture of *Bacillus subtilis*. PLoS One. 2011;6:e27160.

37. Yang CK, Tai PC, Lu CD. Time-related transcriptome analysis of *B. subtilis* 168 during growth with glucose. Curr Microbiol. 2014;68:12–20.

38. Zamboni N, Maaheimo H, Szyperski T, Hohmann HP, Sauer U. The phosphoenolpyruvate carboxykinase also catalyzes C3 carboxylation at the interface of glycolysis and the TCA cycle of *Bacillus subtilis*. Metabolic Eng. 2004;6:277–84.

39. Mukherjee S, Kearns DB. The structure and regulation of flagella in *Bacillus subtilis*. Ann Rev Genet. 2014;48:319–40.

40. Kearns DB, Losick R. Swarming mobility in undomesticated *Bacillus subtilis*. Mol Microbiol. 2003;49:581–90.

41. Higgins D, Dworkin J. Recent progress in *Bacillus subtilis* sporulation. FEMS Microbiol Rev. 2011;36:131–48.

42. Liu L, Liu Y, Shin H, Chen RR, Wang NS, Li J, et al. Developing *Bacillus* spp. as a cell factory for production of microbial enzymes and industrially important biochemicals in the context of systems and synthetic biology. Appl Microbiol Biotechnol. 2013;97:6113–27.

43. Voelker U, Voelker A, Maul B, Hecker M, Dufour A, Haldenwang WG. Separate mechanisms activate sigma B of *Bacillus subtilis* in response to environmental and metabolic stresses. J Bacteriol. 1995;177:3771–80.

44. Zhang S, Haldenwang WG. Contributions of ATP, GTP, and redox state to nutritional stress activation of the *Bacillus subtilis* Sigma B transcription factor. J Bacteriol. 2005;187:7554–60.

45. Sonenshein AL. CodY, a global regulator of stationary phase and virulence in gram-positive bacteria. Curr Opin Microbiol. 2005;8:203–7.

46. Molle V, Fujita M, Jensen ST, Eichenberger P, González-Pastor JE, Liu JS, et al. The Spo0A regulon of *Bacillus subtilis*. Mol Microbiol. 2003;50:1683–701.

47. Sung H-M, Yasbin RE. Adaptive, or stationary-phase, mutagenesis, a component of bacterial differentiation in *Bacillus subtilis*. J Bacteriol. 2002;184:5641–53.

48. Nakano MM, Zuber P, Sonenshein AL. Anaerobic regulation of *Bacillus subtilis* Krebs cycle genes. J Bacteriol. 1998;180:3304–11.

49. Sonenshein AL. Control of key metabolic intersections in *Bacillus subtilis*. Nat Rev Microbiol. 2007;5:917–27.

50. Harshey RM, Matsuyama T. Dimorphic transition in *Escherichia coli* and *Samonella typhimurium*: surface-induced differentiation into hyperflagte swarmer cells. Proc Natl Acad Sci U S A. 1994;91:8631–5.

51. Connelly MB, Young GM, Sloma A. Extracellular proteolytic activity plays a central role in swarming mobility in *Bacillus subtilis*. J Bacteriol. 2004;186:4159–67.

52. Kerrns DB, Losick R. Swarming mobility in undomesticated *Bacillus subtilis*. Mol Microbiol. 2003;49:581–90.

53. Ji Y, Feng H. Optimal of fermentation medium for producing surfactin by *Bacillus pumilus* BA06. J Sichuan University (Nat Sci ed). 2016;53:925–30.

54. Fimlaid KA, Shen A. Diverse mechanisms regulate sporulation sigma factor activity in the Firmicutes. Curr Opin Microbiol. 2015;24:88–95.

55. Aguilar C, Vlanakis H, Guzman A, Losick R, Kolter R. KinD is a checkpoint protein linking spore formation to extracellular-matrix production in Bacillus subtilis biofilms. mBio. 2010;1:e00035–10.

56. Saujet L, Monot M, Dupuy B, Soutourina O, Martin-Verstraete I. The key sigma factor of transition phase, sigH, control sporulation, metab0lism, and virulence factor expression in *Clostridium difficile*. J Bacteriol. 2011;193:3186–96.

57. Wang C, Yu S, Song T, He T, Shao H, Wang H. Extracellular proteome profiling of *Bacillus pumilus* SCU11 producing alkaline protease for dehairing. J Microbiol Biotechnol. 2016;26:1993–2005.

58. Dixit M, Murudkar CS, Rao KK. *epr* is transcribed from σ^D promoter and is involved in swarming of *Bacillus subtilis*. J Bacteriol. 2002;184:596–9.

59. Yasumura A, Abe S, Tanaka T. Involvement of nitrogen regulation in *Bacillus subtilis degU* expression. J Bacteriol. 2008;190:5162–71.

60. Kodama T, Endo K, Ara K, Ozaki K, Kakeshita H, Yamane K, et al. Effect of Bacillus Subtilis spo0A mutation on cell wall lytic enzymes and extracellular proteases, and prevention of cell lysis. J Biosci Bioeng. 2007;103:13–21.

61. Jongbloed JDH, Antelmann H, Hecker M, Nijland R, Pries F, Koski P, et al. Selective contribution of the twin-arginine translocation pathway to protein secretion in *Bacillus subtilis*. J Biol Chem. 2002;277:44068–78.

62. Tjalsma H, Antlmann H, Jongbloed JDH, Braun PG, Darmon E, Dorenbos R, et al. Proteomics of protein secretion by *Bacillus subtilis*: separating the "secrets" of the secrtome. Microbiol Mol Biol Rev. 2004;68:207–33.

63. Antelmann H, Tjalsma H, Voigt B, Ohlmeier S, Bron S, van Dijl JM, et al. A proteomic view on genome-based signal peptide predictions. Genome Res. 2001;11:14984–502.

64. Nannapaneni P, Hertwig F, Depke M, Hecker M, Mäder U, Völker U, et al. Defining the structure of the general stress regulon of *Bacillus subtilis* using targeted microarray analysis and random forest classification. Microbiol. 2012;158:696–707.

65. Turner MS, Helmann JD. Mutations in multidrug efflux homologs, sugar isomerases, and antimicrobial biosynthesis genes differentially elevate activity of the sigma X and sigma W factors in *Bacillus subtilis*. J Bacteriol. 2000;182:5202–10.

66. Langmead B, Salzberg SL. Fast gapped-read alignment with bowtie 2. Nat Method. 2012;9:357–9.

67. Trapnell C, Roberts A, Goff L, Pertea G, Kim D, Kelley DR, et al. Differential gene and transcript expression analysis of RNA-seq experiments with TopHat and cufflinks. Nat Protocol. 2012;7:562–78.

68. Robinson MD, McCarthy DJ, Smyth GK. edgeR: a bioconductor package for differential expression analysis of digital gene expression data. Bioinformatics. 2010;26:139–40.

69. Livak KJ, Schmittgen TD. Analysis of relative gene expression data using real-time quantitative PCR and the $2^{-\Delta\Delta CT}$ method. Method. 2001;25:402–8.

7

Do biofilm communities respond to the chemical signatures of fracking? A test involving streams in North-central Arkansas

Wilson H. Johnson[1], Marlis R. Douglas[1], Jeffrey A. Lewis[1], Tara N. Stuecker[1], Franck G. Carbonero[2], Bradley J. Austin[1], Michelle A. Evans-White[1], Sally A. Entrekin[3] and Michael E. Douglas[1*]

Abstract

Background: Unconventional natural gas (UNG) extraction (fracking) is ongoing in 29 North American shale basins (20 states), with ~6000 wells found within the Fayetteville shale (north-central Arkansas). If the chemical signature of fracking is detectable in streams, it can be employed to bookmark potential impacts. We evaluated benthic biofilm community composition as a proxy for stream chemistry so as to segregate anthropogenic signatures in eight Arkansas River catchments. In doing so, we tested the hypothesis that fracking characteristics in study streams are statistically distinguishable from those produced by agriculture or urbanization.

Results: Four tributary catchments had UNG-wells significantly more dense and near to our sampling sites and were grouped as 'potentially-impacted catchment zones' (PICZ). Four others were characterized by significantly larger forested area with greater slope and elevation but reduced pasture, and were classified as 'minimally-impacted' (MICZ). Overall, 46 bacterial phyla/141 classes were identified, with 24 phyla (52%) and 54 classes (38%) across all samples. PICZ-sites were ecologically more variable than MICZ-sites, with significantly greater nutrient levels (total nitrogen, total phosphorous), and elevated Cyanobacteria as bioindicators that tracked these conditions. PICZ-sites also exhibited elevated conductance (a correlate of increased ion concentration) and depressed salt-intolerant Spartobacteria, suggesting the presence of brine as a fracking effect. Biofilm communities at PICZ-sites were significantly less variable than those at MICZ-sites.

Conclusions: Study streams differed by Group according to morphology, land use, and water chemistry but not in biofilm community structure. Those at PICZ-sites covaried according to anthropogenic impact, and were qualitatively similar to communities found at sites disturbed by fracking. The hypothesis that fracking signatures in study streams are distinguishable from those produced by other anthropogenic effects was statistically rejected. Instead, alterations in biofilm community composition, as induced by fracking, may be less specific than initially predicted, and thus more easily confounded by agriculture and urbanization effects (among others). Study streams must be carefully categorized with regard to the magnitude and extent of anthropogenic impacts. They must also be segregated with statistical confidence (as herein) before fracking impacts are monitored.

Keywords: 16S ribosomal RNA, Anthropogenic impacts, Bioindicators, Fayetteville shale, Groundwater, Microbiome

* Correspondence: med1@uark.edu
[1]Department of Biological Sciences, University of Arkansas, Fayetteville, AR, USA
Full list of author information is available at the end of the article

Background

Unconventional natural gas (UNG) extraction has been promoted as a potential fuel source in North America, as well as a bridge to a cleaner energy economy [1]. It is now ongoing in over 30 states, particularly those containing appropriate geologic 'plays,' i.e., geographic areas that contain fine-grained sedimentary rock with an appropriate clay-to-silt particle size. In North America, these include: Bakken (ND), Barnett (TX), Haynesville (LA), Fayetteville (AR), Antrim (MI), Woodford (OK), Green River (WY), Denver (CO), Marcellus and Utica (PA, OH, WV) [2] (Fig. 1a). Shale gas is termed 'unconventional' in that it is trapped in strata with low porosity and permeability and requires additional extraction processes beyond those normally employed in more traditional petroleum exploitations.

UNG extraction is initiated by drilling downward then horizontally into shale strata, followed by injection of 8000–50,000 m^3 of pressurized local groundwater to fracture shale and release trapped hydrocarbons, a process termed 'fracking' [3]. The injected water contains numerous chemical additives [4, 5] as well as 'proppants' (i.e., sand/silica) that lodge into fractures, allowing oil and gas to flow outward as fluid pressure subsides. Of the injected water, less than half is quickly

returned to the surface (i.e., as flowback), whereas the majority (i.e., produced water) lingers underground and is slowly mobilized as gas is removed [5].

The fracking process can generate numerous environmental impacts [6], the majority of which stem from poor well integrity, improper wastewater disposal, and surface spills [3, 7], with the latter either anthropogenic or environmental (i.e., due to rainwater and/or storm flooding). Of serious concern are those that transport toxic chemicals into surface and ground water [8], with contamination directly correlated to the proximity of the drill site [9]. Impacts are most often gauged by monitoring 'indicator species' i.e., organisms whose presence, absence, or abundance can reflect a specific environmental condition [10], particularly in the context of adaptive stream management.

Biofilm communities in streams (*sensu lato*) are composed of sessile organisms on substrata [11] and thus have an intimate contact with, and long-term exposure to flowing waters. They provide a matrix within which fundamental ecosystem processes occur [12] and, as such, are functionally employed as bioindicators. For example, the Cyanobacterial component of biofilm can contribute >80% of the primary production in a system [13], whereas other biofilm components such as heterotrophic bacteria employ complex metabolic pathways

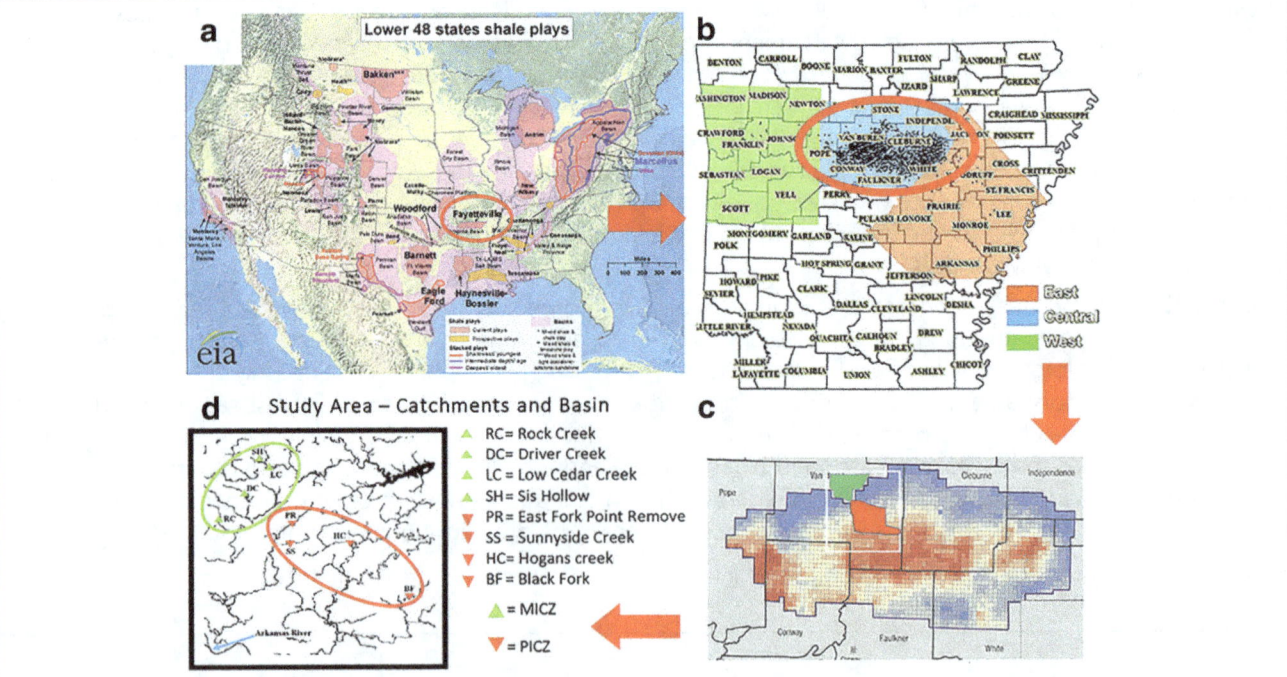

Fig. 1 a Map depicting shale plays located in the United States, with the Fayetteville Shale circled in *red* [44]; (**b**) Map of Arkansas counties showing the topographic location of the Fayetteville shale, with eastern (*red*), central (*blue*), and western (*green*) sections highlighted. The study region is *circled in red*, with closed *black circles* designating the locations of unconventional natural gas (UNG) well sites; **c** Close-up of the northern Arkansas counties within which the Fayetteville shale is distributed. The region in *red* is designated as the 'potentially impacted catchment zone' (=PICZ), a region with high UNG well density, whereas the region in *green* indicates the 'minimally impacted catchment zone' (MICZ). **d** Map depicting the locations of the eight study sites, with inverted *red triangles* designating sites grouped as PICZ, and *green triangles* depicting sites grouped as MICZ. The *blue arrow*, lower left, indicates the location of the Arkansas River

that can quickly remediate harmful substances [14]. The composition of biofilm is radically transformed by alterations in stream conditions [15], with deterioration directly impacting the aquatic food base, such that ramifications are quickly translated into higher trophic levels [16]. Although biofilm communities play a major role in the dynamics of stream ecosystems, they have been traditionally difficult to monitor, due largely to a time-consuming process of optical identification coupled with an inability to initiate and/or sustain laboratory cultures for identification [17].

Molecular advances have now largely ameliorated these issues by facilitating identification and quantification of bacterial constituents in the biofilm community. From this, a much broader perspective on stream metabolism can be developed, in that numerous concurrent samples can be rapidly, simultaneously, and accurately characterized. For example, microbial traits are not only conserved in a phylogenetic context but also linked across clades through biochemical and genetic complexities. Important ecological traits such as pH- and salinity preferences are not only characteristic in a phylogenetic sense but also drive stream metabolism and fulfill ecosystem services [18].

Genomic approaches that characterize microbial communities are also utilized to interpret their dynamics. Here, the 16S ribosomal RNA region has been the molecular marker of choice, as it contains both conserved and hyper-variable regions that are well suited for phylogenetic analyses. Furthermore, the advent of high-throughput DNA sequencing technologies has improved accuracy and reduced costs [17], making community characterization an attractive procedure with which to gauge ecosystem health.

A molecular genetic approach was utilized in the current study to assay biofilm communities of selected streams within a 932-km² region of Fayetteville shale located in the Boston Mountains of northwest Arkansas (Fig. 1b, c). The topography of this region is a limestone-based karst, with numerous emergent ground and spring-fed streams. Previous studies have assessed the potential impacts of fracking in these streams by focusing on either stream metabolism [19] or the presence/absence of aquatic insects as bioindicator species [20].

The objectives of this study were to characterize and compare the biofilm communities at sampling sites a priori characterized by fracking impacts. These sites were first evaluated across a series of abiotic and anthropogenic factors, then compared and contrasted using univariate and multivariate statistical approaches. Our results could then be evaluated against biofilm communities recorded within other shale play studies, as well as those utilizing non-microbial indicators within the Fayetteville shale [19, 20]. In addition to assaying for potential effects of fracking on stream biofilm communities, other potential anthropogenic

effects that drive biofilm communities such as agriculture, silviculture, urbanization, etc., were also considered so as to guide the adaptive management of regional streams. This, in turn, provides broader insights into the manner by which the functioning of stream ecosystem can vary locally and regionally with regard to anthropogenic land manipulations, and nationally with regard to fossil fuel extraction. It also allowed us the opportunity to test if potential fracking effects could be parsed from those engendered by other anthropogenic activities.

Results

Biofilm collection

To understand the potential effects of fracking on stream microbial communities, we collected biofilm at eight stream sites. We grouped our sampling sites using two parameters that denoted their proximity to UNG wells. These were 'inverse flow length' (IFL) and 'well density.' Four sampling locations quite distant from UNG wells were allocated as MICZ-sites (i.e., 'minimally-impacted catchment zone;' = Group 1), whereas four that were significantly proximal to UNG wells were defined PICZ-sites (i.e., 'potentially-impacted catchment zone;' = Group 2). For easier reference, MICZ-sites and PICZ-sites are listed with an affiliated letter (i.e., A-D) that designates sampling locations in each Group (Table 1).

For each stream site, we collected two biofilm samples (one from the downstream and one from the upstream

Table 1 Study sites (Sites) characterized by unconventional natural gas (UNG) activities within 1 km² catchment radius

Sites	Density	IFL	Group
A = Rock creek	0.12	0.18	1
B = Driver creek	0.00	0.00	1
C = Cedar creek	0.04	0.00	1
D = Sis hollow	0.00	0.00	1
A = East fork	2.32	2.35	2
B = Sunnyside creek	3.64	0.31	2
C = Hogans creek	1.77	1.7	2
D = Black fork	0.69	1.3	2
F-value	11.30	10.17	
Probability	0.015[a]	0.019[a]	

Sites are geographically depicted in Fig. 1; Density is the number of unconventional natural gas (UNG) wells within a km² of each site; Inverse Flow Length (IFL) represents the length of flow from each well to the stream channel, corrected for slope, and calculated for wells upstream of each sampling location using the flow length tool in ArcGIS [19]. The inverse of each flow length was summed across all wells for each catchment area such that wells more proximal had a higher value and thus a greater potential effect; Group is based on threshold values of > =0.25 wells/km² and IFL >0.05, with Group 1 indicating presence within a 'minimally impacted catchment zone' (=MICZ), whereas Group 2 are within a 'potentially impacted catchment zone' (PICZ) with greater density of, and proximimty to, UNG wells; F-value is the F-statistic recorded in a 1-way analysis of variance (ANOVA) by Group [i.e., = MICZ (1) versus PICZ (2)] as derived in R [41]. Probability represents the statistical significance of each F-value as determined by Bonferroni adjusted alpha = 0.025, with significance indicated by an[a]

boundaries of the pool), extracted DNA, and used Illumina sequencing to evaluate a 16S rDNA molecular marker that delimits representative biofilm communities. The biofilm samples (2/site; $N = 16$) averaged 153 mg/sample (wet weight), with significantly greater amounts from downstream sections of pools as compared to those upstream (average lower = 174.4 mg, average upper = 131.1 mg; F = 7.09, $P < 0.011$, one-way ANOVA [21]). DNA concentration per sample averaged 39.8 ng/μl and did not differ by site or pool location (results not shown).

Univariate analyses of site, hydrology, land use, and stream chemistry

We characterized ten variables at each site so as to determine whether our designated Groups differed with regard to environmental or anthropogenic factors that could, in turn, affect microbial communities. Four stream morphology variables (i.e., 'elevation,' 'stream order,' '%-slope,' and 'watershed area') were non-significant by Group at the Bonferroni-adjusted P-value (results not shown). With regard to land use characteristics, MICZ-sites reflected significantly greater '%-forested area' and significantly less '%-pasture' (Table 2). The two Groups did not differ significantly with regard to '%-urban area' at the Bonferroni-adjusted probability. In the stream chemistry analyses, PICZ-sites showed significantly greater mean values for 'total nitrogen' and 'total phosphorus' than did MICZ-sites (Table 3), suggesting more nutrient-rich catchments. Elevated values for 'stream conductivity'

Table 2 Land use characterization for Location (sampling sites) and Group (sites grouped in Table 1)

Location	Forest	Pasture	Urban	Group
A = Rock creek	1.22	0.04	0.01	1
B = Driver creek	1.29	0.02	0.01	1
C = Cedar creek	1.10	0.09	0.01	1
D = Sis hollow	0.94	0.14	0.01	1
A = East fork	0.69	0.24	0.02	2
B = Sunnyside creek	0.51	0.41	0.01	2
C = Hogans creek	0.82	0.23	0.03	2
D = Black fork	0.40	0.52	0.02	2
F-value	19.45	13.66	6.00	
Probability	0.005[a]	0.010[a]	0.050	

Sites are geographically depicted in Fig. 1; Allocation of sites to Group is provided in Table 1; Sites labeled as Group 1 are within a 'minimally impacted catchment zone' (=MICZ), whereas sites labeled as Group 2 are within a 'potentially impacted catchment zone' (PICZ) that contains a significantly greater density of unconventional natural gas (UNG) wells; Forest, Pasture, and Urban represent arcsin transformed values originally recorded as percentage within a 1 km² radius of the catchment area; F-value is the F-statistic recorded in a 1-way analysis of variance (ANOVA) by Group [i.e., = MICZ (1) versus PICZ (2)] as derived in R [41]; Probability represents statistical significance of each F-value determined by Bonferroni adjusted alpha = 0.017, with significance indicated by an[a]

Table 3 Water chemistry for each Site (sampling location) and Group (sites group in Table 1)

Site	Tot-N	Tot-Ph	Conductivity	Group
A = Rock creek	0.21	0.014	0.014	1
B = Driver creek	0.07	0.012	0.012	1
C = Cedar creek	0.07	0.01	0.01	1
D = Sis hollow	0.07	0.01	0.01	1
A = East fork	0.3	0.032	0.032	2
B = Sunnyside creek	0.72	0.032	0.032	2
C = Hogans creek	0.86	0.016	0.016	2
D = Black fork	0.41	0.038	0.038	2
F-value	11.93	13.99	6.49	
Probability	0.014[a]	0.010[a]	0.043	

Sites are geographically depicted in Fig. 1; Allocation of sites to Group is provided in Table 1; Total nitrogen (Tot-N), total Phosphorus (Tot-Ph), and Conductivity values were originally recorded as μg/L (Tot-N and Tot-Ph) and millisieverts/cm (Conductivity) but have been log10-transformed; F-value is the F-statistic recorded in a 1-way analysis of variance (ANOVA) by Group [i.e., = MICZ (1) versus PICZ (2)] as derived in R [41]. Probability represents the statistical significance of each F-value as determined by Bonferroni adjusted alpha = 0.017, with significance indicated by an[a]

did not differ by Group at the adjusted Bonferroni-probability level.

Microbial community composition

We performed Illumina sequencing of a 16S rDNA marker as a means of identifying and quantifying microbial biofilm communities at each site. De-replication (i.e., merging of identical reads) condensed the data by 89% [i.e., from 761,914 reads into a unique set of 83,441 OTUs (operational taxonomic units)]. Elimination of singletons (i.e., OTUs that occurred but once) further reduced the total to 48,802 (a 41.5% reduction). The removal of chimeric sequences (i.e., hybrid sequences consisting of multiple OTUs) eliminated an additional 3753 (7.7%). A comparison of sequences against a reference database excluded an additional 50 (0.1%), and alignment with the core set database [22] removed an additional 345, yielding 6965 unique OTUs as a final total.

We generated rarefaction curves that estimated alpha-diversity for each site to determine whether depth of sampling and sequencing were sufficient to adequately capture microbial community diversity. These curves approached horizontal asymptotes when plotted against number of sequence reads, suggesting sufficient sequencing depth (Fig. 2). A total of 46 phyla were represented, with 24 of these found across all samples. Average per sample = 36 (range = 32–39), with several phyla dominating across all samples: Cyanobacteria (37.4%); Proteobacteria (31.7%); Bacteroidetes (7.6%); Planctomycetes (5.3%); and Actinobacteria (4%) [21].

A total of 141 microbial classes were also represented, with 54 found across all samples. Those with average

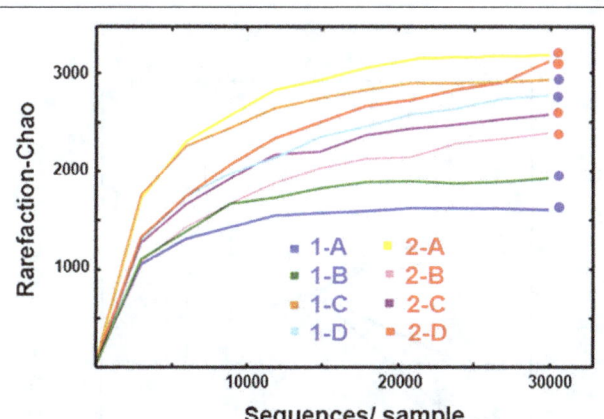

Fig. 2 Graph depicting the number of 16S ribosomal DNA sequences generated for each of the eight study sites located in the Fayetteville shale of north-central Arkansas (X-axis) plotted according to their rarefaction scores (Chao statistic, Y-axis) as generated by the program QIIME [40]. Color of the rarefaction curve indicates study site, dots at terminus reflects 'potentially impacted catchment zones' (=PICZ) in *red*, or 'minimally impacted catchment zone' (=MICZ) in *blue*. PICZ-sites have significantly greater density of unconventional natural gas (UNG) well sites

abundance >2% ($N = 20$) are presented in Fig. 3. Of these, Alphaproteobacteria was the most dominant, averaging 18.9% across samples, with Betaproteobacteria averaging 8.4%. A total of 310 genera were subsequently identified, with 116 (37%) identified across all sites and 297 (95.8%) found at ≥ 4 sites [21].

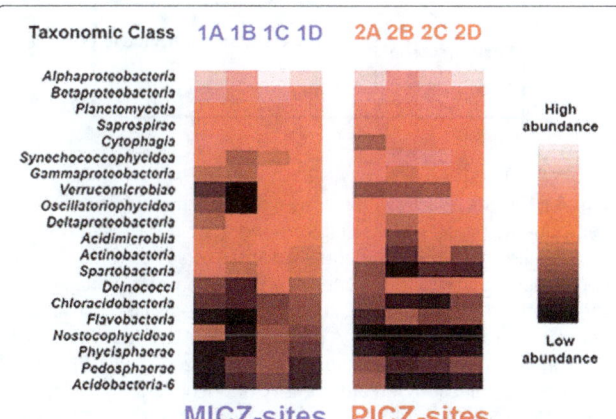

Fig. 3 Heat map reflecting abundance of the 20-most abundant bacterial classes across the eight study sites located in the Fayetteville shale of north-central Arkansas. Columns represent study sites (X-axis) and rows are bacterial classes. The heat map was generated by the program QIIME [40] with intensities of colors (=heat) reflecting abundances as depicted by the scale to the right of the map. Study sites within 'minimally impacted catchment zones' (MICZ) are on the left (1-A through 1-D), whereas sites within 'potentially impacted catchment zones' (=PICZ) are on the right (2-A through 2-D). PICZ-sites have significantly greater density of unconventional natural gas (UNG) well sites

Univariate analyses of biofilm communities

We performed univariate analyses to determine whether microbial community diversity and/or individual membership varied across MICZ and PICZ sites. Values for Shannon entropy, evenness, and number of OTUs/site did not differ significantly by Group [21], suggesting in turn that differences between groups did not broadly affect microbial diversity. The top five most abundant and the bottom three least abundant bacterial classes (Fig. 3) did not differ significantly when compared by Group [21]. We did observe differences between Groups for four other bacterial classes: the 6th (Synechoccophycideae: $F_{(1,6)} = 8.24$, $P < 0.028$); the 9th (Oscillatoriophycideae: $F_{(1,6)} = 9.36$, $P < 0.022$); 13th (Spartobacteria $F_{(1,6)} = 6.36$, $p < 0.045$); and 17th (Nostocophycideae: $F_{(1,6)} = 14.23$, $p < 0.009$), with only the latter significant at an adjusted Bonferroni-value (Table 4). Synechoccophycideae and Oscillatoriophycideae are Cyanobacteria (=primary producers), and each was more prevalent at PICZ-sites, whereas Spartobacteria and Nostocophycideae were most prevalent at MICZ-sites (Fig. 3).

The class Synechoccophycideae was represented by six genera, listed in descending abundance as: *Arthronema*, *Acaryochloris*, *Leptolyngbya*, *Pseudanabaena*, *Paulinella*, and *Synechococus*. In turn, seven genera composed the class Oscillatoriophycideae: *Microcystis*, *Chroococcus*, *Cyanobacterium*, *Chroococciddoipsis*, *Phoridium*, and *Planktothrix*. *Microcystis* was particularly elevated at PICZ-site 2-D (at 5.54%) [21]. Of the five Spartobacteria genera, two were identified as *Xiphinematobacter* and *Chthoniobacter* (family Chthoniobacteraceae), while the remaining three were not identified to genus. The

Table 4 Four dominant microbial classes found at study sites (Site) and analyzed by Group

Site	Sparto	Synecho	Oscillato	Nostoc	Group
A = Rock creek	0.0261	0.0162	0.0050	0.0105	1
B = Driver creek	0.0075	0.0063	0.0001	0.0026	1
C = Cedar creek	0.0158	0.0076	0.0086	0.0089	1
D = Sis hollow	0.0078	0.0163	0.0105	0.0073	1
A = East fork	0.0051	0.0269	0.0125	0.0001	2
B = Sunnyside creek	0.0031	0.0636	0.0766	0.0011	2
C = Hogans creek	0.0011	0.0549	0.0638	0.0017	2
D = Black fork	0.0029	0.0223	0.0443	0.0003	2
F-value	6.36	8.24	9.36	14.13	
Probability	0.045	0.028	0.022	0.009[a]	

Sites are geographically depicted in Fig. 1; Allocation of sites to Group is provided in Table 1; Sparto = Bacterial class Spartobacteria, Synecho = Synechococcophycideae, Oscillato = Oscillatoriophycideae, and Nostoc = Nostocophycideae, with values representing arcsin-transformed percentages of abundance (Fig. 3); *F*-value is the F-statistic recorded in a 1-way analysis of variance (ANOVA) by Group (i.e., = MICZ versus PICZ) as derived in R [41]. Probability represents the statistical significance of each *F*-value at Bonferroni adjusted alpha = 0.017, with significance indicated by an[a]

implications with regard to the abundances of these microbial classes and genera between Groups are discussed below.

Multivariate comparisons among group

In Fig. 4a, a bi-plot depicts relationships within and between Groups based upon the first two principle components (PCs) of the stream morphology, anthropogenic land use, and water chemistry variables. Sites are identified according to Group (number) and Site (letter) with MICZ-sites in blue (1-A through 1-D), and PICZ-sites in red (2-A through 2-D), respectively (per Table 1). PC-1 accommodated 60% of the variation in the data, and PC-2 absorbed an additional 17% (77% total). MICZ-sites clustered to the positive (right) side of PC-1 with congruent loadings for 'slope,' 'elevation,' and '%-forest.' Separation on PC-2 was more prominent for MICZ-sites, largely due to the negative values that associated sites 1-B and 1-C with '%-forest' and 'slope.' On the positive side of PC-2, MICZ-sites 1-A and 1-D were and allied with 'elevation.'

PICZ-sites grouped instead to the far left of the PC-1 axis, quite distinct from MICZ-sites. They still separated into quite distinct pairs, with sites 2-A and 2-C on the negative side of this axis and consistent with vectors depicting 'watershed' size and '%-urban.' PICZ-sites 2-B and 2-D fell more distant on the positive side of the PC-2 axis, and in alliance with vectors depicting 'conductance,' 'total nitrogen', total phosphorus,' and '%-pasture.' The acute angles of these four vectors reflected their close correlation. In this regard, PICZ-site 2-D was more strongly affected than 2-B. Scores on PC-1 differed significantly by Group ($P < 0.003$; results not shown), whereas those for PC-2 did not.

In Fig. 4b, a second biplot depicted relationships within and between Groups, but in relation to the composition of their bacterial communities, with MICZ-sites in blue and PICZ-sites in red (as above). PC-1 accommodated 58% of the variation in the data, and PC-2 absorbed an additional 25% (83% total). Of the 20 bacterial classes evaluated, 16 clustered quite closely with one another and were represented by an ellipse in the plot. Four bacterial classes clearly separated from the ellipse, with arrows designating the magnitude and direction of their trajectories. PICZ-sites 2-B, 2-C, and 2-D aligned with vectors depicting classes Synechoccophycideae and Oscillatoriophycideae, whereas site 2-A grouped within the ellipse. MICZ-site 1-C was well separated and in conjunction with the class Spartobacteria, whereas class Planctomycetia separated but little from the ellipse. MICZ-site 1-D fell at the edge of the ellipse, but sites 1-B and 1-A were more distant, with 1-B particularly so.

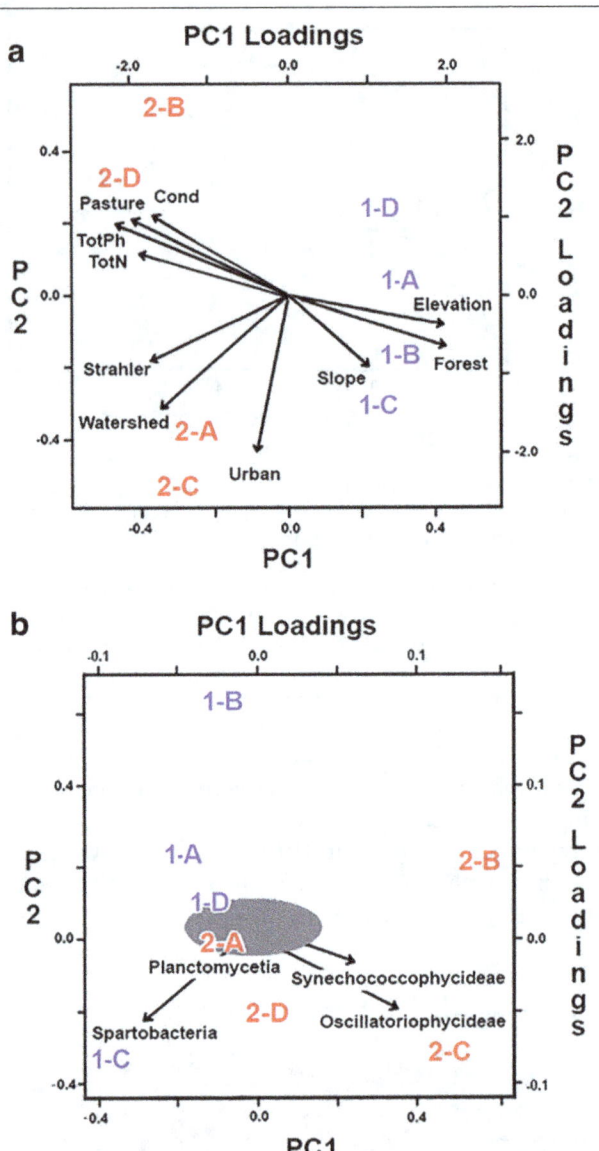

Fig. 4 a Results of a biplot analysis where the first two principal components depict relationships among the eight sites in the Fayetteville shale of north-central Arkansas versus principal component loadings for a suite of ten environmental variables in three defined categories (i.e., stream morphology, land use, and water chemistry) using library "prcomp" in R [41]. Sites in *red* text are within a 'potentially impacted catchment zone' (=PICZ) that signifies a significantly greater density of unconventional natural gas (UNG) well sites, whereas those in *blue* text are found within a 'minimally impacted catchment zone' (MICZ). Variables in the biplot are represented as vectors, and the angle at their origin reflects pairwise correlations (i.e., the more acute the angle, the greater the correlation). TotPh = Total Phosphorus, TotN = Total Nitrogen, Cond = Conductance, Strahler = Stream order. **b** Results of a biplot analysis in which the first two principal components reflect relationships among the eight sites in the Fayetteville shale of north-central Arkansas versus principal component loadings for the 20-most abundant bacterial classes, where densities are represented as arcsin-transformed percentages

Linear iscriminant analysis Effect Size (LEfSe) analyses corroborated much of the above by delineating 20 OTUs with an LDA score > 2.7. Subclasses Oscillatoriophycideae and Synechococcophycideae (Cyanobacteria), and Roseiflexales (a filamentous bacteria often found with Synechococcophycideae and deemed tolerant of eutrophication and/or poor water quality) were abundant at PICZ-sites. Bioindicators of healthy streams [i.e., families Rhodocyclaceae (Proteobacteria), Stigonematales (Cyanobacteria), and Rivulariaceae (Cyanobacteria)] were abundant at MICZ-sites, as were 'negative bioindicators' ($N = 5$; primarily Spartobacteria) whose abundances covary negatively with particular impacts such as elevated pH or salt concentrations.

Discussion

During the past decade, shale resources have been heavily developed in the United States, an industry that will steadily increase over the next several years [10]. The majority of environmental impacts that stem from these activities parallel those recorded for traditional petroleum-extraction, and as such can be predictably monitored [23]. Others are instead UNG-specific, such as poor well integrity and accidental wastewater release, and are compounded by the geographic distribution of shale plays across the continent [24] (Fig. 1a). Environmental risks associated with UNG are hence more difficult to predict and to track, in that sufficient data regarding their breadth and depth have yet to accumulate. This, in turn, delays the designation of appropriate environmental policies that would otherwise provide for their regulation [25, 26].

Research activities that evaluate these impacts are ongoing in the Fayetteville Shale of northwest Arkansas [19, 20] (Fig. 1b, c), and have now been expanded so as to encompass biofilm communities as biological indicators of study catchments (this study). The composition of microbial communities reflects sensitivity and exposure of these catchments to anthropogenic activities [14], such as urbanization, deforestation, agricultural development, habitat fragmentation, and others [27], including UNG-extraction. It is of interest to potentially parse these situations according to the manner by which they drive stream microbial diversity. Similarly, ecosystem processes are also driven by hydrology, stream gradient, stream order, and stream chemistry (among others), and these also modulate the composition of biofilm communities [15]. Given this, we first tested (and rejected) the hypothesis that environmental variability was similar among our minimally impacted (MICZ) versus potentially impacted (PICZ) study sites.

Ecological variation among study sites

Instead, we found significant differences among several test variables, as evaluated by Group. For example, MICZ-sites reflected catchments with significantly greater '%-forested' area, but significantly less '%-pasture' (Table 2). Of interest is the fact that several other variables showed elevated but not significantly different values, as gauged by the Bonferroni-corrected probability value for multiple comparisons. We comment on this situation below.

Significant environmental differences between the two groups were also noted when multivariate analyses incorporated the ten variables across the three categories. Sites separated along PC-1, with strong positive (MICZ) and negative (PICZ) loadings manifested according to stream morphology, anthropogenic land use, and water chemistry (Fig. 4a). There was also considerably more variance among PICZ-sites, with paired catchments (i.e., 2-A/2-C and 2-D/ 2-B) well separated on PC-2. Somewhat surprisingly, these sites also showed a strong and concerted response to those variables not deemed significant in the univariate analyses. MICZ-sites displayed much less variability, yet were similarly separated on PC-2 according to a composite of variables that were significant (i.e., '%-forest') and non-significant (i.e., 'slope' and 'elevation').

The univariate statistics provided differentiation by Group according to individual variables evaluated singly whereas the multivariate analyses provided broader patterns much more interpretable at the ecosystem level, yet not apparent from the separate univariate analyses. This was due largely to the reduced degrees of freedom in the univariate analyses, as constrained by small sample sizes parsed between groups and gauged with Bonferroni-adjusted probabilities. Although our multivariate analyses did not provide statistical probabilities within a hypothesis-testing framework, they more easily depicted the disparity within- and among-Groups, as promoted by watershed, land use, and water chemistry.

The variability in biofilm communities among study sites

Having established the environmental context for our sampling sites by Group, we could then contrast their biofilm communities (Fig. 4b). Groups again separated in multivariate space, albeit less distinctively and with a greater spread among PICZ-sites along PC-1. PICZ-sites also exhibited less variation than did MICZ-sites along PC-2. Clearly, microbial composition varies both among- and within-sites, but with different microbial taxa driving this result in each Group.

For example, Spartobacteria (non-significant in the 1-way ANOVA) clearly associated with MICZ-site 1-C, and differentiated it from all others, whereas site 1-B (associated with 1-C in Fig. 4a) was arrayed quite distantly from other MICZ-sites on PC-2. Furthermore, one site from each Group (i.e., 1-D/ 2-A) fell close to the origin of the PC-axes, suggesting a low overall diversity in their microbial composition (results not shown).

These differences indicate a disparity between biofilm community composition and environmental variability among sites, a result that parallels those from other studies. For instance, microbial assemblages will often group according to historical (i.e., phylogenetic) events [28], but also in response to more contemporary biochemical conditions associated with streams [29]. In this study, numerous factors (per Tables 1, 2 and 3) obviously impact the relationship between biofilm communities and their environment. This result confounds any 'cause-and-effect' scenarios for the observed patterns. However, we draw two strong conclusions from the multivariate analyses: PIZC-sites display greater ecological variability within the environmental matrix (Fig. 4a), yet are much less variable when embedded within the biofilm community matrix (Fig. 4b).

An alternate approach would be to contrast our results with those from studies in other shale plays that employed biofilm communities as bioindicators of fracking. Yet many of the latter are tangential to the present study, in that they examined biofilm communities in either flowback [30] or produced waters [30, 31]. Few evaluated stream catchments into which groundwater from fracked sites would eventually percolate, as herein. One study that did so evaluated headwater streams in the Marcellus shale (PA) (Fig. 1a), and found significantly lower species richness and evenness values at sites impacted by fracking [24]. Several of these sites also contained an abundance of bacterial OTUs that correlated positively with decreasing pH, suggesting more acidic stream environments. The diversity of carbon sources available in a stream promotes the functioning of its biofilm communities, and as diversity decreases, so does the community [32]. In this sense, a reduction in carbon sources would be an ecological explanation for the observed reduction in species richness at these sites, although this was not stated as such.

The richness and evenness of species within biofilm communities

In our study, the species richness, evenness, and number of OTUs in biofilm communities were not significantly different when compared between MICZ- and PICZ-sites. Yet such comparisons often mask the interactions among OTUs within these communities. For example, a decrease in abundance of some taxa can also stimulate growth in others normally more rare, a situation that would promote rather than depress evenness [29, 32]. Those streams with lowest values for evenness in each of our Groups [i.e. 1-B and 2-D; 21] may indeed reflect this consideration. For example, 1-B is a headwater stream (stream order = 1) with the greatest '%-forest' in the study (=96%) both of these environmental aspects would promote deposition of leaf litter into the stream that, in turn, must be decomposed. This similarly constrains the biofilm community.

The two least diverse streams in an ecological sense (i.e., 1-A and 1-B) also had low numbers of OTUs [21], again suggesting the potential for a reduction in available nutrients [19]. In a similar vein, PICZ-site 2-D had the highest value for '%-pasture' in the study, and was also associated with elevated levels of available phosphates and nitrates (Fig. 4a), both of which can promote a few dominant species. This was represented at PICZ-sites by the elevated abundances of two Cyanobacterial classes (i.e., Synechococcophycideae and Oscillatoriophycideae). Cyanobacteria are primary producers that seemingly track the significantly elevated levels of nitrogen and phosphorus found in these streams.

A second but related limitation with regard to species richness and evenness is the strong competition among bacteria and hyphomycetes (stream fungi), as promoted by the reduction in dissolved organic matter (DOM) [27]. Dissolved nitrogen primarily exists as nitrates within ground and surface waters, and must be transformed by microbes before entering into and moving through the ecosystem. This, in turn, could promote bacterial OTUs more strongly competitive, at the expense of those less competitive, a situation that would also constrain biofilm community diversity. In addition, and as a second consideration, elevated nitrogen levels are often associated with UNG well sites [19].

Additionally, the removal of pollutants can, paradoxically, also reduce microbial diversity and evenness [33], suggesting (as above) that external sources of carbon can promote the development of OTUs normally more rare. These caveats, in turn, provide numerous potential corollaries to explain the low values for evenness at sites, particularly when carbon sources have become more limited due to fracking [24, 32].

Biofilm communities as bioindicators

The function of many bacterial lineages is not well understood at the ecosystem level, despite their abundances in soil and aquatic systems, and this in turn makes it more difficult to ascertain their status as potential bioindicators. Despite this, general functions are indeed assignable to some clades. Many Synechococcophycideae, for example, employ unique metabolic pathways that allow them to persist in highly acidic environments such as volcanic seeps. The Oscillatoriophycideae is an equally diverse clade that can also serve a bioindicator for organic pollutants. For example, Microcystis (a genus of Oscillatoriophycideae) is abundant at PICZ-sites, and its presence may point to the presence of elevated polycyclic aromatic hydrocarbons (PAHs) that in turn promote its growth [34].

In addition, the genomes of aquatic Spartobacteria encode for a diversity of glycoside hydrolases that are employed in the degradation of complex carbohydrates [35]. This physiological aspect also explains its common co-occurrence with Cyanobacteria, in that the former metabolizes the complex carbohydrates produced by the latter [12]. Spartobacteria should thus positively correlate with Cyanobacteria at PICZ-sites, but was instead found to be significantly reduced. Brine contamination is a well-known fracking by-product, and it continues to be pulled upwards from deeper strata long after drilling has subsided [36]. In addition, PICZ-sites also reflected greater conductance in their water chemistry. Spartobacteria has a pronounced intolerance for salt, and these environmental conditions at PICZ-sites would impede its expected proliferation.

In an attempt to gain a more comprehensive perspective, we can also contrast results from this study with those from earlier studies at the same sites. For example, UNG development had a definite impact on stream macroinvertebrate communities, with short-lived generalists being more abundant at those sites [20]. Yet these effects were difficult to parse across specific taxa, or to specifically associate with the benthic habitat found at PICZ-site.

A second study [19] found increased primary production and eutrophication at sites impacted by UNG activities, and this was interpreted as a potential response to the enhanced levels of nitrogen these sites displayed. Our data support these conclusions in that two classes of Cyanobacteria were clearly more abundant at impacted sites, suggesting the presence of an environment that is beneficial for primary production. Our statistical analyses also verified significant levels of 'total nitrogen' and 'total phosphorus' at these sites, as well as heightened conductance.

Overall, the observed differences between MICZ-sites and PICZ-sites may reflect the accessibility of sites chosen for UNG well construction, and as such, may add an additional consideration for the design controlled studies to gauge the effects of fracking (see Discussion).

Conclusions

Biofilm communities have complex roles in freshwater stream metabolism, and consequently drive numerous critical processes: Primary production [12], biogeochemical cycling [17], nitrogen cycles [28], and the remediation of deleterious carbon sources [14], among many. Microbial communities are also extraordinarily diverse, composed of numerous rare OTUs, and display a rapid response to changes in temperature, pH, and stream metabolism [16]. This also provoke taxonomic turnover in stream biofilm communities as an ecosystem-scale response [29]. Given this, stream biofilm communities can be employed to only to gauge ecosystem health [28], but also its potential impacts on humankind [17]. Unfortunately, the breadth and depth of biofilm communities are also confounding factors that can limit diagnostic and taxonomic projections, particularly with regard to bioremediation.

Region specific issues also predominate [4]. For example, biofilm communities are quite sensitive to changes in land use [15]. This is important in that both the Fayetteville and Barnett shale catchments display pre-existing anthropogenic disturbances [26] that can easily confound more focused analyses regarding the impacts of UNG-activity. In addition, habitat and water chemistry data collected prior to the onset of fracking are necessary baselines from which potential impacts on both freshwater streams and their biofilm communities can be assessed. These data were lacking herein, and similarly lacking in other studies that employed biofilm communities as a means to adjudicate fracking activities [24]. As a result, the statistical analyses employed to contrast these sites were similarly limited.

Unfortunately, necessary data are often unavailable at the national level, and a mandate for their collection has not as yet been established in state or federal management plans. This, in turn, cripples the development of conservation measures that may promote the sustainability of stream ecosystems. Resource managers require these data so as to guide local development projects, and to reduce possible environmental effects particularly in light of the interactive effects produced by multiple stressors in a warming climate [12]. The evaluation of anthropogenic impacts, whether fracking or otherwise, also depends upon rigorous statistical analyses conducted in a comparative manner (as herein). This, too, is often lacking with regard to those projects that attempt to recognize and define biodiversity elements, or conserve and restore habitats.

Our data mirror similar conditions found in other systems with long-term disturbance, such as elevated conductance/lack of Spartobacteria, and elevated nitrogen/elevated Cyanobacteria, and these, in turn, suggest potential impacts from UNG wells. Our data are also confounded by pre-existing conditions such as development of pasture and the extent of urbanization, as well as naturally occurring aspects such as stream order that likewise influence the constituents of biofilm communities, and biodiversity in general. These limitations argue for an a priori selection of pre- versus post-impact study sites, in that a variety of anthropogenic endeavors can drive biofilm communities in concurrent directions and it is difficult if not impossible to separate these effects a posteriori. The complexities of anthropogenic/environmental interactions also necessitate the development of a rigorous statistical framework, one within which variability can be tested among- and between-groups. This study provides a set of guidelines with regard to study design that can avoid the former, while establishing a strong statistical framework for the latter.

Methods
Sampling sites and environmental data for catchments
Eight sites from an ongoing stream ecology project [19, 20] (Fig. 1b, c, d) were assigned to 'Group' using two

parameters that relate to UNG-well activity: 'well density' and 'inverse flow length' (IFL). 'Well density' is defined as the number of UNG well sites within a 1-km^2 radius (=catchment area), whereas 'IFL' represents the length of flow from each well site to the stream channel, corrected for slope, and calculated for wells upstream of each sampling location via the flow length tool in the 'Spatial Analyst Toolkit' of ArcGIS [19]. The inverse of each flow length was summed across all well sites for each catchment area, such that wells more proximal had a higher value that corresponded to a greater potential effect. Sites with an IFL < 0.25 and Well Density (no./km2) < 0.5 were scored as '1,' and designated as a 'minimally-impacted catchment zone' (MICZ), whereas those with an IFL ≥ 0.25 and a Well Density ≥ 0.5 were scored as '2' and grouped as a 'potentially-impacted catchment zone' (PICZ).

We characterized ten variables distributed across three categories at each site so as to ascertain if designated Groups differed with regard to environmental or anthropogenic factors that could, in turn, affect microbial communities. The first category related to stream morphology and employed four variables (i.e., 'elevation,' 'stream order,' '%-slope,' and 'watershed area'). The second utilized three variables that summarized anthropogenic land use (i.e., '%-forest,' '%-pasture,' and '%-urban'). The third recorded three water-chemistry parameters (i.e., 'total nitrogen,' 'total phosphorus,' and 'conductivity') deemed important in gauging relationships between stream metabolism and bacterial communities [19]. Abundance of nutrients was measured as μg/L, whereas dissolved salt/ions was in microSiemens (uS)/cm, with higher values signaling an elevated presence of ions.

Biofilm collection, DNA extraction, and Illumina sequencing

At each site, a pool was identified peripheral to the greatest stream flow and a biofilm-covered rock was then selected at downstream (lower) and upstream (upper) boundaries and scrubbed with a sterile Nasco Whirl-Pak Speci-Sponge™. Sponges were immediately re-sealed in the sterile Whirl-Pak and placed onto dry ice for transport to the lab where they were stored at −80°C until processed. For DNA extraction, 20 ml of phosphate buffered saline solution (PBS; 137 mM NaCl, 2.7 mM KCl, 4.3 mM Na_2HPO_4, 1.47 mM KH_2PO_4, pH 7.4) was added to each sample, and the sponge squeezed manually for 5 min to suspend biofilm. Suspensions were transferred to individual centrifuge tubes and pelleted by centrifugation (8000 g for 20 min), with biofilm quantified via wet weight (mg). Standard laboratory protocols were used for all procedures to prevent sample contamination.

DNA from pelleted biofilm was extracted for all 16 samples (2 per site) using a MOBIO commercial kit (PowerBiofilm® DNA Isolation Kit) following manufacturer's instructions. DNA was quantified (ng/ul) using a Qubit 2.0 Fluorometer (Invitrogen®). Extractions were subjected to PCR using primers that amplified the hypervariable V4 region of the 16S structural subunit rRNA gene [37]. Multiplexed 16S metagenomic libraries were constructed using standard Illumina protocols, and were sequenced on an Illumina MiSeq platform. Raw Illumina reads were de-multiplexed (MiSeq Reporter software™) and downloaded from the Illumina BaseSpace® cloud.

Bioinformatics

Sequences were trimmed to 251 bp and quality filtered at an expected error of <1% using USEARCH v8.0 [38]. A pipeline developed by the Brazilian Microbiome Project [39] was employed to correct any Illumina formatting issues for subsequent analyses in QIIME v1.7 [40]. OTUs were selected with the UCLUST method (as implemented in QIIME) and taxonomy assigned using the Greengenes 16S rRNA gene database [22], with subsequent conversion into an OTU table (QIIME).

Univariate analyses

Prior to analyses, nine variables were transformed: Percentages ($N = 4$) were arcsin transformed to radians; areas ($N = 1$) reduced to square root; and quantitative variables ($N = 5$) transformed to log_{10}. 'Stream order' was evaluated as recorded. Each category was test by Group using a 1-way analysis of variance in R [41], with statistical significance assigned according to Bonferroni-corrected probabilities.

Shannon entropy was computed in QIIME to gauge the number of unique bacterial taxa in each community (i.e., richness) and the evenness of their distributions, with results compared by Group using a 1-way ANOVA in R. Species richness (with repeated subsampling) was then plotted by site as rarefaction curves, so as to estimate whether sampling at each site was of sufficient depth to accurately characterize biofilm communities. Analyses were carried out with the default number of Monte-Carlo permutations ($N = 999$) at a p-value of 0.05. UniFrac analyses (in QIIME) were used to derive beta (or between sample) diversity estimates using both unweighted data (i.e., OTU presence/absence) and weighted (by relative abundance) [42]. To identify potential bioindicators, a heat map was generated in QIIME using the 20-most abundant taxonomic classes of bacteria. Potential bioindicators were then identified and compared by Group using a 1-Way ANOVA in R with Bonferroni-corrected probabilities.

Multivariate analyses

A principal components analysis (PCA) was performed using a matrix of correlations among sites based on the ten variables across the three categories (i.e., stream morphology, anthropogenic land use, and stream chemistry) using library "prcomp" in R [41]. The first two principal components depicted relationships among the eight sites (i.e., PC-scores) and were contrasted against principal component loadings for the variables. Both scores and loading were visualized in a single plot (hence the term, 'biplot'), so as to promote the interpretation of the component axes in relation to the variables. Those in the biplot were represented as vectors, and the angle at their origin(s) reflects pairwise correlations (i.e., the more acute the angle, the greater the correlation). We then compared the first six principal components by Group in R, using a 1-way ANOVA with Bonferroni-corrected probabilities.

A principal component analysis was also used to contrast densities of the 20-most abundant bacterial classes among study sites (using library "prcomp" in R [41]), with densities represented as arcsin-transformed percentages. The first two principal components depicted relationships among the eight sites (i.e., PC-scores) and were contrasted against principal component loadings for the 20-most abundant classes.

The biomarker discovery algorithm LEfSe (Linear discriminant analysis Effect Size) was used to designate potential bioindicators among biofilm communities [43]. The program employs a linear discriminant analysis (LDA) with effect size estimated by linking output to the level-6 (Kingdom to Genus) taxonomic summary in QIIME. Parameters employed were: an alpha value of 0.05 for the Kruskal-Wallis (KW) test, an LDA score threshold of >2.7, and a pairwise Group-comparison. Initially, LEfSe conducts the KW rank-sum test as a means of detecting OTUs that differed significantly in abundances between Groups. Biological significance was then investigated with the (unpaired) Wilcoxon rank-sum test. Finally, LDA was then employed to evaluate each OTU with an effect size > 2.7, and with biological indicator gauged via habitat and metabolism.

Acknowledgments
This research was completed by WHJ in partial fulfillment of the M.S. degree in Biological Sciences at University of Arkansas/Fayetteville. We thank The Nature Conservancy (TNC; Fayetteville) for supplying GIS-data, and V.S. Pylro (Microbiology Department, Universidade Federal de Viçosa, Minas Gerais, Brazil) for advice and assistance in biofilm community analyses. The Huttenhower Lab (C. Huttenhower, Harvard University; https://huttenhower.sph.harvard.edu) provided access to its Galaxy server and the LEfSe software. The Arkansas High Performance Computing Center (AHPCC), with support from multiple National Science Foundation (NSF) grants and the Arkansas Economic Development Commission, facilitated the analyses of data.

Funding
This research was supported by endowments to the University of Arkansas (MRD: Bruker Professorship in Life Sciences; MED: 21st Century Chair in Global Change Biology), and the Arkansas Biosciences Institute (Arkansas Settlement Proceeds Act of 2000). Ecological data were collected under Arkansas State Wildlife Grants T31-03 and T33-01 (MAEW and SAE) from the U.S. Fish and Wildlife Service through an agreement with the Arkansas Game and Fish Commission.

Authors' contributions
WHJ and BJA collected samples; MRD, JAL, and TNS extracted DNA and amplified 16S rRNA genes; FGC performed MiSeq sequencing; BJA, MAEW, and SAE provided ecological data; WJH and MED completed data analyses; WHJ, MRD, JAL, TNS, and MED drafted the manuscript. MAEW and SAE initiated the original project to investigate fracking effects in Arkansas streams. All authors approved the final manuscript.

Competing interests
The authors declare that they have no competing interests.

Author details
[1]Department of Biological Sciences, University of Arkansas, Fayetteville, AR, USA. [2]Department of Food Sciences, University of Arkansas, Fayetteville, AR, USA. [3]Department of Biology, University of Central Arkansas, Conway, AR 72035, USA.

References
1. Gold R. The Boom: How fracking ignited the American energy revolution and changed the world. New York City: Simon and Schuster Publishers; 2014 (http://www.ebooksdownloads.xyz/search/the-boom).
2. Lampe DJ, Stolz JF. Current perspectives on unconventional shale gas extraction in the Appalachian Basin. J Environ Sci Heal A. 2015;50:434–46. doi:10.1080/10934529.2015.992653.
3. Burton Jr GA, Basu N, Ellis BR, Kapo KE, Entrekin S, Nadelhoffer K. Hydraulic "fracking:" Are surface water impacts an ecological concern? Environ Toxicol Chem. 2014;33:1679–89. doi:10.1002/etc.2619.
4. Vidic RD, Brantley SL, Vandenbossche JM, Yoxtheimer D, Abad JD. Impact of shale gas development on regional water quality. Science. 2013;340: 1235009. doi:10.1126/science.1235009.
5. Ferrer I, Thurman EM. Analysis of hydraulic fracturing additives by LC/Q-TOF-MS. Anal Bioanal Chem. 2015;407:6417–28. doi:10.1007/s00216-015-8780-5.
6. Allred BW, Smith WK, Twidwell D, Haggerty JH, Running SW, Naugle DE, Fuhlendorf SD. Ecosystem services lost to oil and gas in North America. Science. 2015;348:401–2. doi:10.1126/science.aaa4785.
7. Rahm BG, Vedachalam S, Bertoia LR, Mehta D, Vanka VS, Riha SJ. Shale gas operator violations in the Marcellus and what they tell us about water resource risks. Energ Policy. 2015;82:1–11. doi:10.1016/j.enpol.2015.02.033.
8. Vengosh A, Jackson RB, Warner N, Darrah TH, Kondash A. A critical review of the risks to water resources from unconventional shale gas development and hydraulic fracturing in the United States. Environ Sci Technol. 2014;48: 8334–48. org/10.1021/es405118y.
9. Jackson RB, Vengosh A, Carey JW, Davies RJ, Darrah TH, O'Sullivan F, Pétron G. The environmental costs and benefits of fracking. Annu Rev Environ Resour. 2014;39:327–62. doi:10.1146/annurev-environ-031113-144051.
10. Foissner W, Berger H. A user-friendly guide to the ciliates (Protozoa, Ciliophora) commonly used by hydrobiologists as bioindicators in rivers, lakes, and waste waters, with notes on their ecology. Freshw Biol. 1996;35: 375–482.
11. Geesey GG, Mutch R, Costerton JW, Green RB. Sessile bacteria: an important component of the microbial population in small mountain streams. Limnol Oceanogr. 1978;23:1214–23.

12. Battin TJ, Besemer K, Bengtsson MM, Romani AM, Packmann AI. The ecology and biogeochemistry of stream biofilms. Nat Rev Microbiol. 2016; 14:251–63. doi:10.1038/nrmicro.2016.15.

13. Lyon DR, Ziegler SE. Carbon cycling within epilithic biofilm communities across a nutrient gradient of headwater streams. Limnol Oceanogr. 2009;54: 439–49. doi:10.4319/lo.2009.54.2.0439.

14. Gadd GM. Metals, minerals and microbes: geomicrobiology and bioremediation. Microbiology. 2010;156:609–43. doi:10.1099/mic.0.037143-0.

15. Gibbons SM, Jones E, Bearquiver A, Blackwolf F, Roundstone W, Scott N, Hooker J, Madsen R, Coleman ML, Gilbert JA. Human and environmental impacts on river sediment microbial communities. PLoS One. 2014;9, e97435. doi:10.1371/journal.pone.0097435.

16. Findlay S. Stream microbial ecology. J N Am Benth Soc. 2010;29:170–81. doi:10.1899/09-023.1.

17. Zarraonaindia I, Smith DP, Gilbert JA. Beyond the genome: community-level analysis of the microbial world. Biol Philos. 2013;28:261–82. doi:10.1007/ s10539-012-9357-8.

18. Martiny JBH, Jones SE, Lennon JT, Martiny AC. Microbiomes in light of traits: a phylogenetic perspective. Science. 2015;350:aac9323. doi:10.1126/science.aac9323.

19. Austin BJ, Hardgrave N, Inlander E, Gallipeau C, Entrekin SA, Evans-White MA. Stream primary producers relate positively to watershed natural gas measures in north-central Arkansas streams. Sci Tot Environ. 2015;529:54–64. doi:10.1016/j.scitotenv.2015.05.030.

20. Johnson E, Austin BJ, Inlander E, Gallipeau C, Evans-White MA, Entrekin SA. Stream macroinvertebrate communities across a gradient of natural gas development in the Fayetteville Shale. Sci Tot Environ. 2015;530–531:323–32. doi:10.1016/j.scitotenv.2015.05.027.

21. Johnson WH. Stream Microbial Communities as Potential Indicators of River and Landscape Disturbance in North-Central Arkansas, Unpubl. Master's Thesis. Fayetteville: University of Arkansas; 2016.

22. DeSantis TZ, Hugenholtz P, Larsen N, Rojas M, Brodie EL, Keller K, Huber T, Dalevi D, Hu P, Andersen GL. Greengenes, a chimera-checked 16S rRNA gene database and workbench compatible with ARB. Appl Environ Microb. 2006;72:5069–72. doi:10.1128/AEM.03006-05.

23. Brittingham MC, Maloney KO, Farag AM, Harper DD, Bowen ZH. Ecological risks of shale oil and gas development to wildlife, aquatic resources and their habitats. Environ Sci Technol. 2014;48:11034–47. doi:10.1021/es5020482.

24. Trexler R, Solomon C, Brislawn CJ, Wright JR, Rosenberger A, McClure EE, Grube AM, Peterson MP, Keddache M, Mason OU, Hazen TC, Grant CJ, Lamendella R. Assessing impacts of unconventional natural gas extraction on microbial communities in headwater stream ecosystems in northwestern Pennsylvania. Front Microbiol. 2014;5:1–13. doi:10.3389/fmicb.2014.00522.

25. Entrekin SA, Evans-White M, Johnson B, Hagenbuch E. Rapid expansion of natural gas development poses a threat to surface waters. Front Ecol Environ. 2011;9:503–11. doi:10.1890/110053.

26. Entrekin SA, Maloney KO, Kapo KE, Walters AW, Evans-White MA, Klemow KM. Stream vulnerability to widespread and emergent stressors: A focus on unconventional oil and gas. PLoS One. 2015;10(9):e0137416. doi:10.1371/ journal.pone.0137416.

27. Zeglin L. Stream microbial diversity responds to environmental changes: review and synthesis of existing research. Front Microbiol. 2015;6:454. doi:10. 3389/fmicb.2015.00454.

28. Martiny JBH, Bohannan BJM, Brown JH, Colwell RK, Fuhrman JA, Green JL, Horner-Devine MC, Kane M, Krumins JA, Kuske CR, Morin PJ, Naeem S, Øvreås L, Reysenbach A-L, Smith VH, Staley JT. Microbial biogeography: putting microorganisms on the map. Nat Rev Microbiol. 2006;4:102–12. doi:10.1038/nrmicro1341.

29. Portillo MC, Anderson SP, Fierer N. Temporal variability in the diversity and composition of stream bacterioplankton communities. Environ Microbiol. 2012;14:2417–28. doi:10.1111/j.1462-2920.2012.02785.x.

30. Cluff MA, Hartsock A, MacRae JD, Carter K, Mouser PJ. Temporal changes in microbial ecology and geochemistry in produced water from hydraulically fractured Marcellus Shale gas wells. Environ Sci Technol. 2014;48:6508–17. doi:10.1021/es501173p.

31. Wuchter C, Banning E, Mincer TJ, Drenzek NJ, Coolen MJL. Microbial diversity and methanogenic activity of Antrim Shale formation waters from recently fractured wells. Front Microbiol. 2013;4:367. doi:10.3389/fmicb.2013.00367.

32. Peter H, Ylla I, Gudasz C, Romaní AM, Sabater S, Tranvik LJ. Multifunctionality and diversity in bacterial biofilms. PLoS One. 2011;6(8), e23225. doi:10.1371/journal. pone.0023225.

33. Pholchan MK, Baptista JC, Davenport RJ, Sloan WT, Curtis TP. Microbial community assembly, theory and rare functions. Front Microbiol. 2013;4:68. doi:10.3389/fmicb.2013.00068.

34. Zhua X, Kong H, Gao Y, Wu M, Kong F. Low concentrations of polycyclic aromatic hydrocarbons promote the growth of *Microcystis aeruginosa*. J Hazard Mater. 2012;237–238:371–5. doi:10.1016/j.jhazmat.2012.08.029.

35. Herlemann DPR, Lundin D, Labrenz M, Jürgens K, Zheng Z, Aspeborg H, Andersson AF. Metagenomic de novo assembly of an aquatic representative of the Verrucomicrobial Class Spartobacteria. Mbio. 2013;4(3):e00569–12. doi:10.1128/mBio.00569-12.

36. Myers T. Potential contaminant pathways from hydraulically fractured shale to aquifers. Groundwater. 2012;50:872–82. doi:10.1111/j.1745-6584.2012.00933.x.

37. Klindworth A, Pruesse E, Schweer T, Peplies J, Quast C, Horn M, Glöckner FO. Evaluation of general 16S ribosomal RNA gene PCR primers for classical and next-generation sequencing-based diversity studies. Nucl Acids Res. 2012;41: 1–11. doi:10.1093/nar/gks808.

38. Edgar RC. USEARCH: Ultra-fast sequence analysis. 2016. http://www.drive5. com/usearch/. Accessed 26 May 2016.

39. Pylro VS, Roesch LFW, Morais DK, Clark IM, Hirsch PR, Tótola MR. Data analysis for 16S microbial profiling from different benchtop sequencing platforms. J Microbiol Methods. 2014;107:30–7. doi:10.1016/j.mimet.2014.08.018.

40. Caporaso JG, Kuczynski J, Stombaugh J, Bittinger K, Bushman FD, Costello EK, Fierer N, Gonzalez Peña A, Goodrich JK, Gordon JI, Huttley GA, Kelley ST, Knights D, Koenig JE, Ley RE, Lozupone CA, McDonald D, Muegge BD, Pirrung M, Reeder J, Sevinsky JR, Turnbaugh PJ, Walters WA, Widmann J, Yatsunenko T, Zaneveld J, Knight R. QIIME allows analysis of high-throughput community sequencing data. Nat Methods. 2010;7:335–6. doi: 10.1038/nmeth.F.303.

41. R Core Team. R: A language and environment for statistical computing, R-version 3.2.2. Vienna: R Foundation for Statistical Computing; 2012. http://www.R-project.org/.

42. Lozupone C, Knight R. UniFrac: a new phylogenetic method for comparing microbial communities. Appl Environ Microb. 2005;71:8228–35. doi:10.1128/ AEM.71.12.8228-8235.2005.

43. Segata N, Izard J, Waldron L, Gevers D, Miropolsky L, Garrett WS. Metagenomic biomarker discovery and explanation. Genome Biol. 2011;12:R60. doi:10.1186/ gb-2011-12-6-r60.

44. U.S. Energy Information Administration. Shale gas and oil plays, North America, 2011. http://www.eia.gov/maps/maps.htm#shaleplay

8

Illumina MiSeq sequencing analysis of fungal diversity in stored dates

Ismail M. Al-Bulushi[1], Muna S. Bani-Uraba[1], Nejib S. Guizani[1], Mohammed K. Al-Khusaibi[1] and Abdullah M. Al-Sadi[2*]

Abstract

Background: Date palm has been a major fruit tree in the Middle East over thousands of years, especially in the Arabian Peninsula. Dates are consumed fresh (*Rutab*) or after partial drying and storage (*Tamar*) during off-season. The aim of the study was to provide in-depth analysis of fungal communities associated with the skin (outer part) and mesocarp (inner fleshy part) of stored dates (*Tamar*) of two cultivars (*Khenizi* and *Burny*) through the use of Illumina MiSeq sequencing.

Results: The study revealed the dominance of *Ascomycota* (94%) in both cultivars, followed by *Chytridiomycota* (4%) and *Zygomycota* (2%). Among the classes recovered, *Eurotiomycetes*, *Dothideomycetes*, *Saccharomycetes* and *Sordariomycetes* were the most dominant. A total of 54 fungal species were detected, with species belonging to *Penicillium*, *Alternaria*, *Cladosporium* and *Aspergillus* comprising more than 60% of the fungal reads. Some potentially mycotoxin-producing fungi were detected in stored dates, including *Aspergillus flavus*, *A. versicolor* and *Penicillium citrinum*, but their relative abundance was very limited (<0.5%). PerMANOVA analysis revealed the presence of insignificant differences in fungal communities between date parts or date cultivars, indicating that fungal species associated with the skin may also be detected in the mesocarp. It also indicates the possible contamination of dates from different cultivars with similar fungal species, even though if they are obtained from different areas.

Conclusion: The analysis shows the presence of different fungal species in dates. This appears to be the first study to report 25 new fungal species in Oman and 28 new fungal species from date fruits. The study discusses the sources of fungi on dates and the presence of potentially mycotoxin producing fungi on date skin and mesocarp.

Keywords: *Phoenix dactylifera* L, Population structure, Fungal diversity, Fungal pathogens, Date palm

Background

Dates palm (*Phoenix dactylifera* L.) is one of the oldest and most important fruit trees in the Middle East [1, 2]. The total worldwide production of dates is around 7.2 million tons, with approximately 5.1 million tons produced by countries in the Middle East [3]. The top 10 producers of dates are Egypt, Iran, Saudi Arabia, Algeria, Iraq, Pakistan, Oman, UAE, Tunisia and Libya. Besides being an important source of vitamins, minerals and other beneficial nutrients, date fruits were the main sources of calories for people living in this part of the world. There are hundreds of date palm cultivars grown in the Middle East, varying in their types from one

country to the other. In Oman, there are over 200 different date palm cultivars. *Khalas*, *Khenizi*, *Naghal*, *Burny*, *Um Al-Sella*, *Shahla*, *Mabsali* and *Fardh* are some of the common cultivars in Oman, occupying more than 50% of the area devoted for date palm production [4, 5].

Date fruits are usually harvested and either consumed directly or dried, packed and consumed at a later stage. The fresh and directly consumed dates are referred to as '*Rutab*', while the dried and stored dates are referred to as '*Tamar*'. The traditional way of drying dates involves exposing them to direct sun for a certain period of time (few days to weeks). This is followed by packing and storing dates for several months until they are consumed. Since most date palm production in the Middle East is usually within the period from May to October, most people rely on the consumption of fresh dates (*Rutab*) after harvesting. The duration of consumption of fresh dates is variable, as it depends on the cultivars

* Correspondence: alsadi@squ.edu.om; http://www.researchgate.
net/profile/Abdullah_Al-Sadi
[2]Department of Crop Sciences, College of Agricultural and Marine Sciences, Sultan Qaboos University, P.O. Box-34, Al-Khod 123, Oman
Full list of author information is available at the end of the article

which are grown on a specific location. Some cultivars mature early (e.g. by April to May), while other mature late, sometimes up to October and November. However, after this period, people start consuming the stored dates (*Tamar*) until the next cycle of date's harvest and production. Some low quality dates are fed to animals because they either come from low quality cultivars or their quality is affected during harvest or storage.

Previous studies reported on the potential contamination of date fruits with some fungal species, including *Aspergillus flavus*, *A. niger*, *Penicillium chrysogenum* and many others [6–10]. These studies raised concerns from the potential contamination of dates with certain mycotoxin-producing fungal species. However, all the previous studies were limited in either being focused on certain fungal types or being dependent on only culture-based approaches for fungal detection [7, 9]. Thus, the amount of information available on the fungal species associated with date fruits is still very limited. This imposes a barrier towards predicting sources of fungal communities and the presence of potentially mycotoxin producing species.

The detection of fungal species in plant material, including date fruits, depended largely on the use of serial dilution or different baiting techniques [7, 8, 10]. However, with the development in molecular techniques, several DNA-based approaches were developed which enabled the detection of several fungal species that are either difficult to grow on synthetic media, or those which are slow growing and usually outgrown by fast growing species. These include the use of pyrosequencing or MiSeq sequencing which made the detection and identification of fungal and bacterial species easy, not only from plant and food material but also from environmental samples such as water and soil [1, 11–15].

The main objective of this study was to characterize the main fungal species associated with dates at the *Tamar* stage. Specific objectives include: (1) to investigate the common fungal species in dates using MiSeq sequencing; and (2) to investigate whether different date parts or date cultivars could differ in their fungal community structure. Understanding fungal diversity in date fruits can help establish a database of the common fungi in these fruits and predict the date fruit parts which are more vulnerable for fungal contamination. It will also help find out the presence of potentially mycotoxin-producing fungi in date fruits.

Methods
Collection of samples
The experiment focused on two common date cultivars: *Burny* and *Khenizi*. *Burny* and *Khenizi* cultivars were grown in Oman in two separate fields, in Ibra and Samail, respectively. Date samples were harvested and immediately exposed to direct sun for approx. 2 weeks. Drying was on the surface of a mat made from dry date leaves. The drying place did not follow any standard hygienic procedures as dates were exposed to natural air without sterilization, which is a usual practice in several places in the Arabian Peninsula. Three different date samples (500 g each) were collected at the *Tamar* stage from each cultivar after partial drying under the sun. The date samples were healthy without any visual symptoms of any disease. The samples were stored in sterile polyethylene plastic bags at 25–30 °C for 3 months prior to analysis. The water activity was measured for each sample using a water activity meter (Ro-tronic Hygrolab, Switzerland). Water activity was measured at the beginning of the storage time and 3 months later (at the microbial analysis time). Three individual date fruits were selected from each cultivar. The skin and the mesocarp of each fruit were separated using sterile forceps and scalpel.

DNA extraction
DNA was extracted from three skin samples and three mesocarp samples of each date cultivar using the CTAB method with slight modifications [16]. The skin and mesocarp of each sample were ground separately using liquid nitrogen. Then, 0.1 g of date tissue was mixed with 500 μl of pre-warmed 2x CTAB buffer (2% CTAB, 100 mM Tris pH 8.0, 20 mM EDTA pH 8.0, 1.4 M NaCl, 1% PVP-40, 0.2% ß-mercapto-ethanol) and incubated at 65 °C for 30 min. Then 750 μl of phenol: chloroform: isoamyl alcohol (25:24:1) was added to the mixture, vortexed and centrifuged at 10,000 RCF for 15 min. Precooled isopropanol was added to the supernatant and incubated at −40 °C for two hr. Then, the mixture was centrifuged at 10,000 RCF for 5 min and the pellet was washed using 70% ethanol. The DNA pellet was resuspended in 100 μl sterile distilled water and was stored at −60 °C.

Illumina MiSeq
Illumina MiSeq was carried out for the six samples from each date cultivar. Amplification of samples was carried out in a two-step process, with the first step to amplify genomic regions of interest and the second step to add sequencing adaptors and sample-specific indices to samples. Construction of the forward primer was done using the Illumina i5 sequencing primer (TCGTCGGCAGCGTCA-GATGTGTATAAGAGACAG) and the ITS1F primer (CTTGGTCATTTAGAGGAAGTAA) [17]. The reverse primer was constructed with the Illumina i7 sequencing primer (GTCTCGTGGGCTCGGAGATGTGTATAAGA-GACAG) and the ITS2aR primer (GCTGCGTTCTT CATCGATGC) [1, 18]. The first PCR was conducted in 25 μl reaction mixture consisting of 1 μl of template

DNA, 1 µl of each 5 µM primer and Qiagen HotStar Taq master mix (Qiagen Inc, Valencia, California). The reaction conditions were as follows: an initial denaturation step of 95°C for 5 min, then 25 cycles of denaturation at 94°C for 30 sec, annealing at 54°C for 40 sec, and extension at 72°C for 1 min. The final extension was performed at 72°C for 10 min.

Products from the first stage amplification were subjected to a second PCR. Primers for the second PCR were designed based on the Illumina Nextera PCR primers as follows: Forward - AATGATACGGCGACC ACCGAGATCTACAC[i5index]TCGTCGGCAGCGTC and Reverse - CAAGCAGAAGACGGCATACGAGA-T[i7index]GTCTCGTGGGCTCGG. The second stage amplification was run the same as the first stage except for 10 cycles.

Amplification products were visualized and then pooled equimolar. Size selection of each pool was done in two rounds followed by quantification using the Quibit 2.0 fluorometer (Life Technologies). Then it was loaded on an Illumina MiSeq (Illumina, Inc. San Diego, California) 2x300 flow cell at 10pM [19].

BioInformatic analysis

All sequencing reads were run through Research and Testing Laboratory's (RTL, Lubbock, TX, USA) standard microbial analysis pipeline. The data analysis pipeline consisted of the denoising and chimera detection stage and the microbial diversity analysis stage. In the first stage, denoising was carried out to remove short sequences, singleton sequences, and noisy reads using the USEARCH [20] and UPARSE [21] algorithms. Then, chimera detection was used to remove chimeric sequences using the UCHIME chimera detection software in *de novo* mode [22]. Finally, the remaining sequences were then corrected per-base to help remove errors in sequencing.

During the diversity analysis stage, all samples were assembled into OTU clusters at 97% identity using the UPARSE [21] algorithm and then globally aligned using the USEARCH [20] global algorithm against a database of high quality ITS fungal gene sequences from GenBank, compiled by RTL, to determine taxonomic classifications. After OTU selection was performed, a phylogenetic tree was constructed in Newick format from a multiple sequence alignment of the OTUs done in MUSCLE [23, 24] and generated in FastTree [25]. Then fungi were classified at the appropriate taxonomic levels using trimmed taxa which takes confidence values into account at each taxonomic level. Individual analysis was carried out for the percentage of sequences assigned to each fungal phylogenetic level for each pooled sample in order to provide the relative abundance for individual samples. The data were filtered at

97% similarity threshold. The mean number of raw reads was 33272, 44517, 40643, 54067 before filtering and 26543, 42272, 37194, 51628 after filtering for *Burny* (mesocarp), *Burny* (skin), *Khenizi* (mesocarp) and *Khenizi* (skin), respectively.

The data were analyzed using the R software [26]. This included the generation of a rarefaction curve plot of the number of OTUs versus the number of sequences, and estimating Richness and Shannon Diversity indices as explained by Kazeerroni and Al-Sadi [1]. Fungal diversity was also estimated using Bray-Curtis similarities followed by analyzing differences in fungal diversity between groups of samples using 'Permutational Multivariate Analysis of Variance Using Distance Matrices' function ADONIS [27–29].

Statistical analysis

Differences among samples in the mean value of water activity were analyzed using Tukey's Studentized range test (SAS, SAS Institute Inc., USA).

Results

Water activity

The water activity of the date samples significantly decreased from 0.65 to 0.60 for *Burny* and 0.62 to 0.59 *Khenizi* from the first day of storage to 3 months after that (at the day of microbial analysis) ($P < 0.05$).

Fungal diversity estimates

Analysis showed the presence of variable levels of fungal diversity in the two date cultivars (*Burny* and *Khenizi*) and in the skin and mesocarp of date fruits (Fig. 1). No significant differences were observed in Chao Richness estimates between the mesocarp and skin of date fruits and also between the two cultivars (Fig. 2; $P = 0.0684$), which was due to the slightly high intra-sample diversity within the *Burny*-skin and *Khenizi*-mesocarp treatments. Similarly, no significant differences were observed in Shannon diversity between the fruit cultivars or fruit parts (Fig. 3; $P = 0.7739$).

Dominant fungal groups

Ascomycota was the most dominant phylum in the skin and mesocarp of the two date cultivars. It accounted for 81 to over 99% of the fungal reads in the samples. *Basidiomycota* was present in the skin of both cultivars and in the mesocarp of *Khenizi*. *Chytridiomycota* accounted for 16% of the fungal populations in the mesocarp of *Khenizi* (Fig. 4).

Eurotiomycetes was the most dominant fungal class in the samples, followed by *Dothideomycetes, Saccharomycetes* and *Sordariomycetes* (Fig. 4). *Eurotiomycetes, Dothideomycetes* and *Sordariomycetes* were detected in all four samples, while the remaining classes were

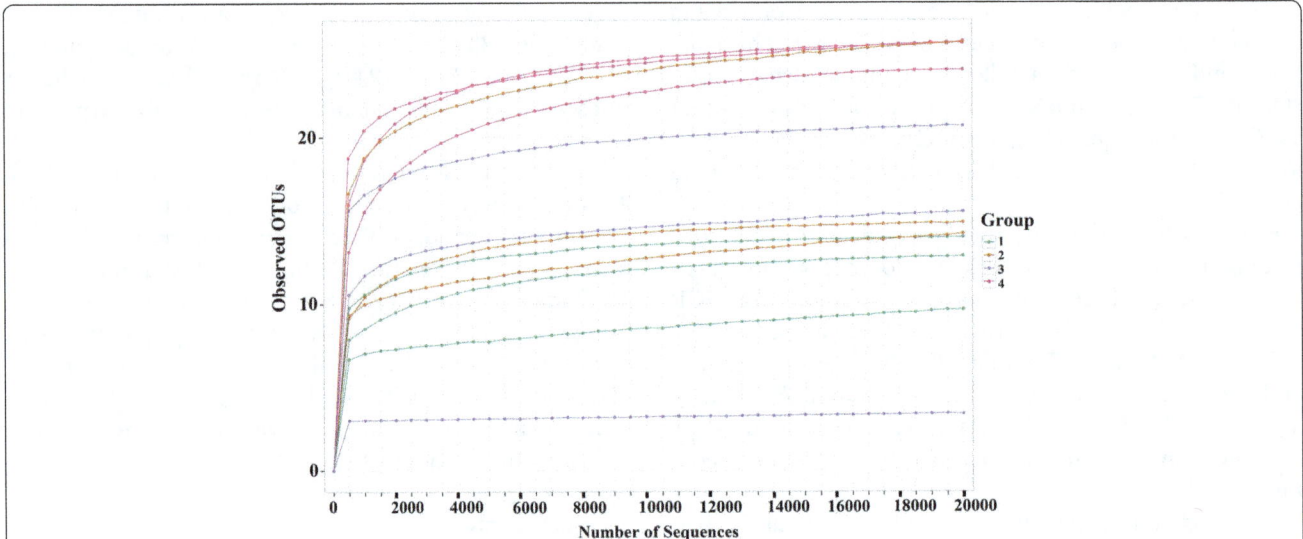

Fig. 1 Rarefaction plot of species richness, subsampling from 0 to 20,000 reads in increments of 500 reads. Groups 1, 2, 3 and 4 represent *Burny* (mesocarp), *Burny* (skin), *Khenizi* (mesocarp) and *Khenizi* (skin), respectively

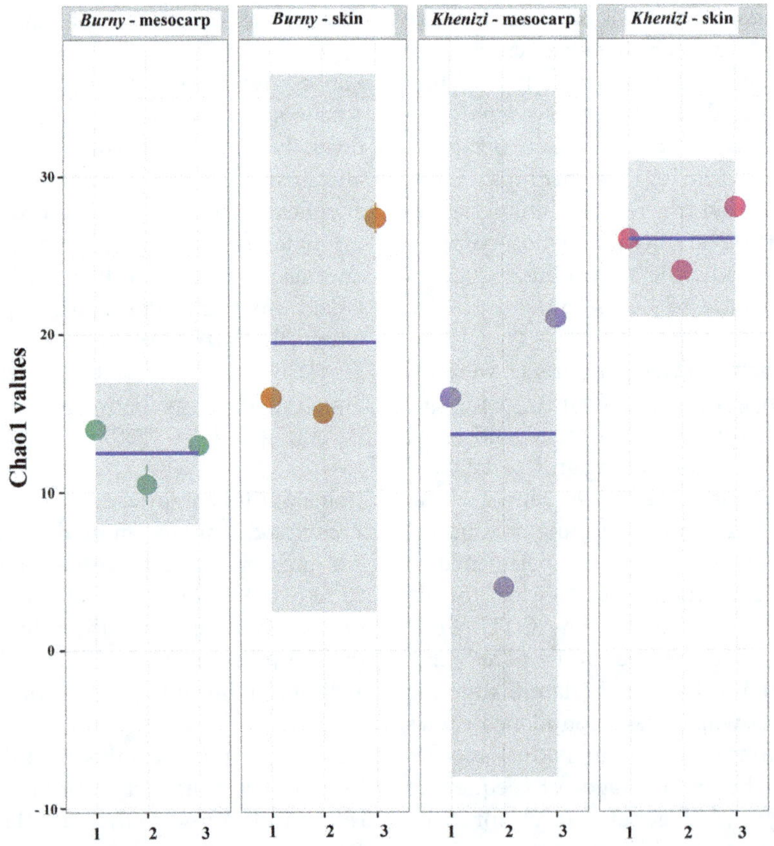

Fig. 2 Chao1 richness estimates for the four date samples. The mean value (*line*) and confidence interval (*shaded*) in each group also are illustrated

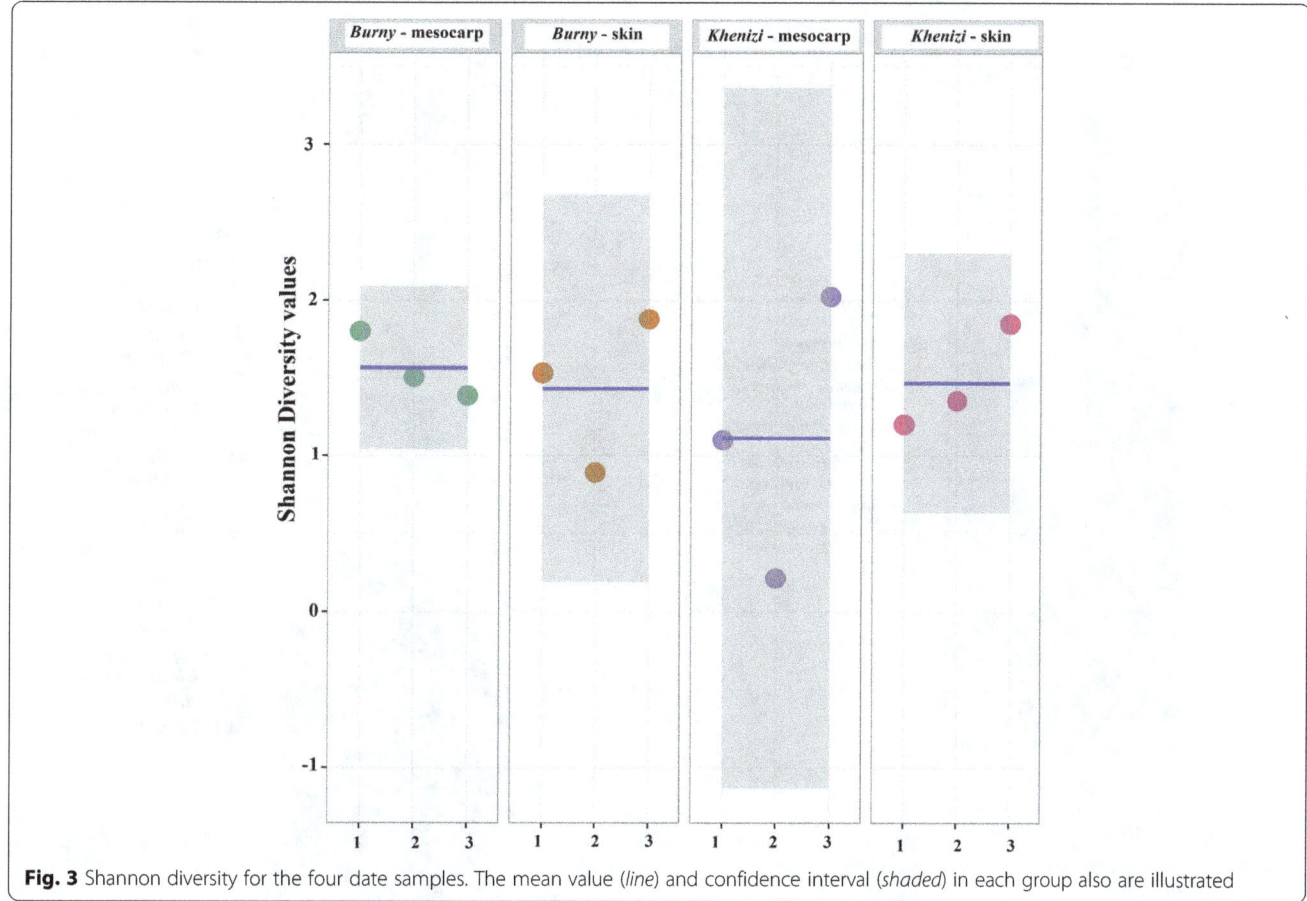

Fig. 3 Shannon diversity for the four date samples. The mean value (*line*) and confidence interval (*shaded*) in each group also are illustrated

detected in some of the samples. Tremellomycetes was detected in the skin and mesocarp of *Khenizi* but not in *Burny*.

Analysis of fungal species in the date samples revealed the presence of 54 different fungal species. Eleven of the fungal taxa could not be resolved to the genus or species level, nine were only resolved to the genus level while 34 were identified to the species level (Fig. 5; Table 1). *Penicillium*, *Alternaria*, *Cladosporium* and *Aspergillus* species were the most common in most samples. *Penicillium griseofulvum* was the most common fungal species in all samples, making up 13 to 42% of the total fungal reads. This was followed by *Alternaria* sp., *Aspergillus tubingensis*, *Fusarium* sp. and *Cladosporium cladosporioides*.

Twelve fungal species were detected from the skin and mesocarp of *Burny*, 17 were detected in skin but not the mesocarp and two were detected in mesocarp but not in skin. In *Khenizi*, 17 fungal species occurred in both the skin and mesocarp tissues, 19 occurred only in the skin and 11 occurred only in the mesocarp (Table 1). *Trichoderma asperellum*, *Aspergillus versicolor* and *Pichia* sp were detected only in the mesocarp and skin of *Khenizi* but not in *Burny*. *Aspergillus flavus* and *Zygosaccharomyces rouxii* were detected only in the skin of *Khenizi*

and *Burny* cultivars (Table 1). Some fungal species were detected for the first time in date fruits or in Oman (Table 1).

Analysis of community composition across samples

PerMANOVA analysis based on Bray-Curtis distances indicated the presence of insignificant differences in the fungal community structure between the mesocarp and skin of *Burny* ($R^2 = 0.346$, $P = 0.150$) and *Khenizi* ($R^2 = 0.310$, $P = 0.150$) cultivars. Also, no significant differences were observed in the fungal community structure between the *Burny* and *Khenizi* cultivars (Table 2).

Discussion

Ascomycota was the most common phylum in the skin and mesocarp of dates. *Ascomycota* is a very common fungal phylum, previously reported to dominate fungal groups in plant tissues and different soil types and fertilizers [1, 7, 11, 13]. Previous studies on date fruits using culture-based techniques also revealed that *Ascomycota* is the dominant phylum in date fruits [6, 8]. Eurotiomycetes was the most dominant class in date fruits, mainly because it contains two of the most dominant genera in date fruits: *Penicillium* and *Aspergillus*.

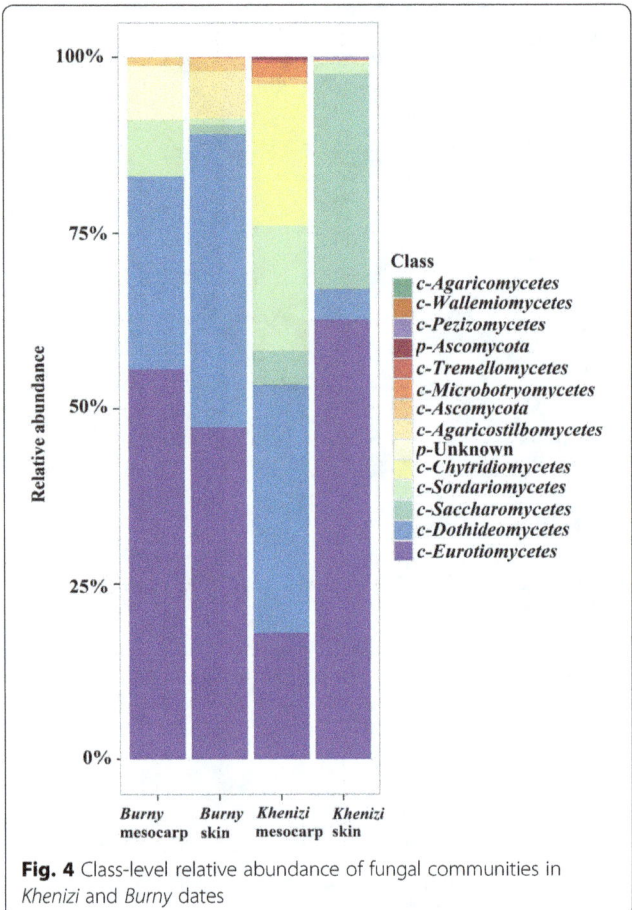

Fig. 4 Class-level relative abundance of fungal communities in *Khenizi* and *Burny* dates

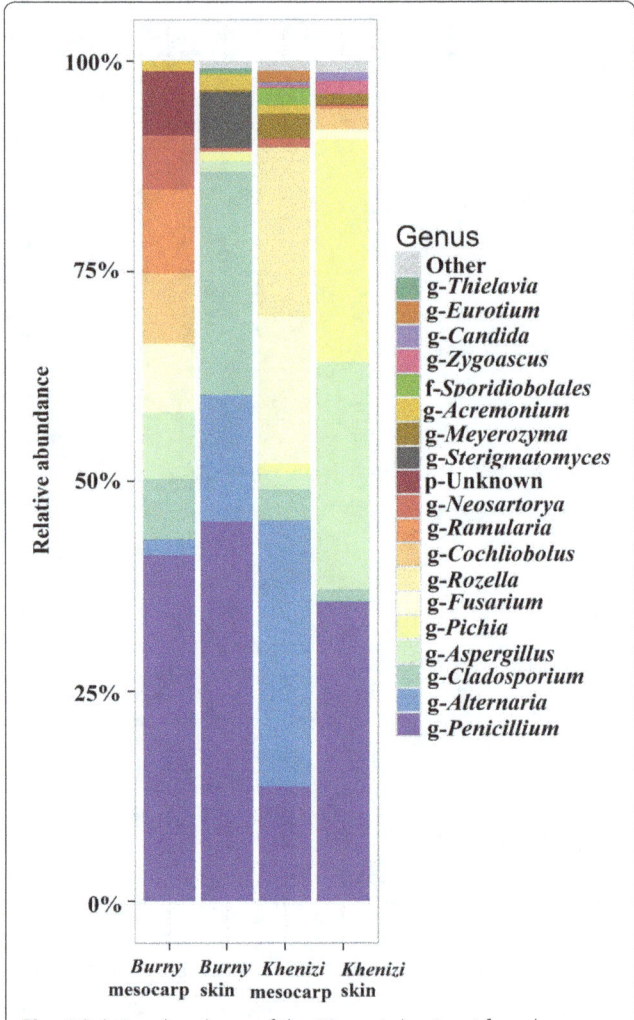

Fig. 5 Relative abundance of the 19 most dominant fungal genera in *Khenizi* and *Burny* dates

Penicillium, Alternaria, Aspergillus and *Cladosporium* were the most dominant fungal genera in date fruits, comprising more than 60% of the genera observed in date fruits. These fungi, especially *Penicillium* and *Aspergillus*, are very common airborne fungi that produce thousands of spores and they are common on date fruits. Previous studies reported the association of *Alternaria* spp., *Aspergillus* spp., *Cladosporium* spp, *Dreschlera spicifera, Eurotium amstelodami, E. chevalieri, Fusarium* spp., *Mucor racemosus, Myrothecium verrucaria, Penicillium* spp., *Rhizopus stolonifer, Ulocladium atrum* and others with date fruits [6–10, 30–32]. In the current study, 28 fungal species appear to be reported for the first time on date fruits, of which 12 were found on skin and mesocarp, 15 were only on skin and one was only in the mesocarp. This indicates that date fruit, especially the outer skin, is exposed to several fungal species.

The majority of the detected fungal taxa in date fruits are either spoilage fungi (e.g. *Alternaria* spp.) or saprophytes (e.g. *Trichoderma asperellum*). Although date palm is known to be affected by several fungal diseases including bayoud disease (*Fusarium oxysporum* f.sp. *albedenis*) and black scorch (*Ceratocystis radicicola*) [4, 33], the causal agents of these diseases

were not detected in date fruits. Three potentially mycotoxin producing fungi were detected on date fruits, namely *Aspergillus flavus, A. versicolor* and *Penicillium citrinum*. Although several reports indicated that these are potential mycotoxin-producing fungi in several crops and food types [30, 34–38], our findings showed that they were found to make up less than 0.5% of the total fungal reads in date fruits. In addition, *A. flavus* was only detected in the skin of both cultivars, not in the fleshy part, which may impose less risk on humans. However, more studies should be done in the future to examine the potential presence of mycotoxin and mycotoxin-producing fungi in dates at different stages of maturation and from different cultivars. In addition, it is unclear whether many of the several fungal species detected in this study could impose a potential risk to humans after consuming contaminated dates or they have a possible role in chemical changes in stored date fruits. Future

Table 1 Frequency of occurrence of different fungal isolates in the skin and mesocarp of *Burny* and *Khenizi* date cultivars

Species	Burny		Khenizi	
	Mesocarp	Skin	Mesocarp	Skin
Penicillium griseofulvum*	42.132	40.873	12.511	26.625
Alternaria sp	1.698	20.16	40.983	0.168
Aspergillus tubingensis*	7.344	1.105	1.299	37.294
Fusarium sp	8.056	0.077	14.909	2.366
Cladosporium cladosporioides	3.07	16.768	2.421	1.823
Pichia kudriavzevii*	0	1.44	0.99	16.474
Rozella sp	0	0.004	16.612	0
Cochliobolus sp	10.004	0.002	0	5.143
Unknown 1	3.184	7.221	0.635	0.504
Ramularia eucalypti*	8.891	0	0	0.688
Neosartorya pseudofischeri*	5.848	0.572	0.953	0.129
Unknown 2	7.004	0	0	0.048
Sterigmatomyces elviae*	0	5.651	0	0
Meyerozyma guilliermondii*	0	0.181	2.587	2.227
Acremonium implicatum*	1.432	1.58	0.761	0
Cladosporium perangustum*	1.296	1.385	0.209	0.471
Zygoascus meyerae*	0	0.002	0.319	2.243
Unknown 3	0	0.134	1.786	0
Candida tropicalis*	0	0	0	1.195
Eurotium amstelodami	0	0	1.174	0
Pichia sp	0	0	0.014	1.001
Cephaliophora tropica*	0	0	0	0.723
Unknown 4	0	0.513	0	0
Exophiala oligosperma*	0	0.432	0.049	0
Penicillium pinophilum*	0	0.438	0	0
Cladosporium sphaerospermum*	0	0.428	0	0
Alternaria alternata	0.001	0.4	0.001	0.003
Aspergillus versicolor	0	0	0.32	0.065
Unknown 5	0	0	0.376	0
Unknown 6	0	0	0.356	0
Hannaella sinensis*	0	0	0.335	0
Aspergillus flavus	0	0.266	0	0.057
Unknown 7	0	0	0	0.221
Nigrospora sp	0	0	0	0.192
Myrothecium inundatum*	0	0.177	0	0
Acremonium sp	0	0	0.136	0
Unknown 8	0	0	0.118	0
Trichoderma asperellum*	0	0	0.109	0.008
Penicillium citrinum*	0.041	0.003	0.032	0.008
Cladosporium sp	0	0.076	0	0
Kodamaea ohmeri*	0	0	0	0.073
Unknown 9	0	0	0	0.068
Zygosaccharomyces rouxii*	0	0.053	0	0.005

Table 1 Frequency of occurrence of different fungal isolates in the skin and mesocarp of *Burny* and *Khenizi* date cultivars (*Continued*)

*Rhodosporidium kratochvilovae**	0	0	0	0.052
*Symbiotaphrina kochii**	0	0	0	0.05
Unknown 10	0	0.046	0	0.003
*Cryptococcus albidus**	0	0	0	0.036
*Candida pimensis**	0	0	0	0.015
Phoma sp	0	0.01	0	0
*Melanocarpus albomyces**	0	0	0	0.008
*Rhodotorula mucilaginosa**	0	0	0	0.008
*Wallemia sebi**	0	0	0	0.006
Penicillium corylophilum	0	0.004	0	0
Unknown 11	0	0	0.002	0

Species in bold are reported in this study for the first time in Oman, while species with (*) symbol are reported for the first time on date fruits. Unknown fungi could not be resolved to the species level. Full data are available through this link http://rtlgenomics.com/ (Project ID: Al-Sadi 4317 Fungal)

experiments on these fungi could reveal some of their risks or benefits.

Although several fungal species were detected in dates at the *Tamar* stage, no spoilage was observed in any of the date fruits which were subject to analysis. As opposite to dates at the *Rutab* stage which usually spoil quickly because of the high water activity, spoilage of *Tamar* is not common mainly because of the reduced water activity. Findings from this study revealed that water activity in the stored dates decreased for *Burny* and *Khenizi* dates from 0.64 to 0.62 at the storage time to 0.61 and 0.59, respectively 3 months later. Previous studies reported that many of the food spoilage fungi usually grow at water activity ranges from 0.7 to 0.94 [39].

PerMANOVA analysis indicated that fungal communities in the skin of dates are not significantly different from the communities in the mesocarp for both cultivars. This may suggest that fungal species contaminating the outer part of dates' fruit (skin) may have the ability to grow into the mesocarp. In our study, 39% and 35% of the fungal species contaminating the skin were also detected in the mesocarp of *Burny* and *Khenizi* cultivars, respectively. Contamination of the dates' skin and mesocarp with the same fungal species could have occurred while dates were on trees or immediately after harvest. This is because drying of dates can reduce water activity to levels that may not favor fungal growth [6–8, 39].

This may impose a problem to consumers, as even if they remove the skin of dates, they may not get rid of all fungi because many of the fungi are in the fleshy part, the mesocarp. It is therefore important to find out the stage at which contamination occurs to help reduce fungal contamination in dates.

Analysis indicated that 23 unique fungal species were observed in *Khenizi* but not in *Burny*, while 7 unique fungal species were observed in *Burny* but not in *Khenizi*. Also, *Penicillium griseofulvum* was found to make up 41–42% of the species in *Burny* compared to 13–27% of the species in *Khenizi*. However, PerMANOVA analyses did not reveal any significant differences in fungal diversity between the two date cultivars (*P* >0.05). Although the dates from the two cultivars were obtained from two different areas, there appears to be no effect of location or cultivars on the fungal community structure of date fruits.

The presence of different fungal species in date fruits as shown by the analyses of alpha diversity (Shannon index, richness estimates) and beta diversity (perMANOVA analysis of Bray-Curtis similarities) raises questions concerning the sources of these fungi. The low level of water activity in dates may lower the chance for dates to be infected at the drying/storage stage. However, the ripening stage of dates is the stage at which contamination by fungi may occur [8, 31]. Since our study did not evaluate this stage, a future study on the

Table 2 Effect of date parts and date cultivars on fungal diversity revealed using PerMANOVA analysis

Parameter	Treatment	F model	R^2	P adjusted
Date part	Skin X mesocarp (Burny cultivar)	2.113499	0.34571	0.150
	Skin X mesocarp (Khenizi cultivar)	1.799876	0.31033	0.150
Cultivar	Burny X Khenizi (Mesocarp)	1.621053	0.28839	0.150
	Burny X Khenizi (Skin)	2.899156	0.420219	0.150

possible contamination of dates at different stages of maturation and storage may reveal the stage at which contamination is at high. This may help reduce the chance of date contamination with fungi.

Conclusion

Alpha-based analyses of fungal diversity in date palm fruits at the *Tamar* stage indicated the presence of different fungal species. The study appears to be the first report of 25 fungal species in Oman and 28 fungal species on date fruits, with some species being potential producers of mycotoxins. Beta analysis of fungal communities showed that they are not related to specific date cultivars or date part (skin and mesocarp), indicating the possible contamination of date cultivars and date parts with the same species of fungi. Future studies should address the source of these fungi in date fruits. They should also address fungal contamination in dates at different stages of maturation/drying and the role of fungi in date spoilage, especially at the *Rutab* stage. In addition, attention should be given to evaluating the effect of date processing on reducing contamination of dates with harmful fungi.

Abbreviations

CTAB: Cetyltrimethylammonium bromide; ITS: Internal transcribed spacer; OUT: Operational taxonomic unit

Acknowledgments

Thanks are due to Mr. Issa Al-Mahmooli and Mr. Waleed Al Busaidi for help in DNA extraction and to Jeremy Wilkinson for help in data analysis.

Funding

Authors would like to acknowledge Sultan Qaboos University, The Research Council and Oman Animal and Plant Genetic Resources Center for financial support of the study through the project EG/AGR/CROP/16/01.

Author's contributions

IA, MB, NG, MA and AMA planned the study; MB and AMA carried out the work; IA, MB and AMA analyzed data; IA, MB, NG, MA and AMA wrote the manuscript; IA, MB, NG, MA and AMA revised and approved the final version of the paper.

Competing interests

The authors declare that the research was conducted in the absence of any commercial or financial relationships that could be construed as a potential conflict of interest. The authors declare that they have no competing interests.

Author details

[1]Department of Food Science and Nutrition, College of Agricultural and Marine Sciences, Sultan Qaboos University, P.O. Box-34, Al-Khod 123, Oman. [2]Department of Crop Sciences, College of Agricultural and Marine Sciences, Sultan Qaboos University, P.O. Box-34, Al-Khod 123, Oman.

References

1. Kazeerooni EA, Al-Sadi AM. 454-pyrosequencing reveals variable fungal diversity across farming systems. Front Plant Sci. 2016;7:314.
2. Chao CT, Krueger RR. The date palm (Phoenix dactylifera L.): Overview of biology, uses, and cultivation. HortSci. 2007;42(5):1077–82.
3. FAOSTAT. [http://www.fao.org/faostat/en/#data/QC/visualize]
4. Al-Sadi AM, Al-Jabri AH, Al-Mazroui SS, Al-Mahmooli IH. Characterization and pathogenicity of fungi and oomycetes associated with root diseases of date palms in Oman. Crop Protect. 2012;37:1–6.
5. Al-Yahyai R, Khan MM. Date palm status and perspective in oman. In: Date Palm Genetic Resources and Utilization: Volume 2: Asia and Europe. 2015. p. 207–40.
6. Gherbawy YA, Elhariry HM, Bahobial AAS. Mycobiota and mycotoxins (aflatoxins and ochratoxin) associated with some Saudi date palm fruits. Foodborne Pathog Dis. 2012;9(6):561–7.
7. Al-Sheikh H. Date-palm fruit spoilage and seed-borne fungi of Saudi Arabia. Res J Microbiol. 2009;4(5):208–13.
8. Shenasi M, Aidoo KE, Candlish AAG. Microflora of date fruits and production of aflatoxins at various stages of maturation. Int J Food Microbiol. 2002; 79(1–2):113–9.
9. Atia MMM. Efficiency of physical treatments and essential oils in controlling fungi associated with some stored date palm fruits. Aust J Basic Appl Sci. 2011;5(6):1572–80.
10. Jogee SP, Ingle AP, Gupta IR, Bonde SR, Rai MK. Detection and management of mycotoxigenic fungi in nuts and dry fruits. Acta Hortic. 2012;963:69–77.
11. Al-Mazroui SS, Al-Sadi AM. 454 pyrosequencing and direct plating reveal high fungal diversity and dominance by saprophytic species in organic compost. Int J Agric Biol. 2016;18(1):98–102.
12. Al-Sadi AM, Al-Mazroui SS, Phillips AJL. Evaluation of culture-based techniques and 454 pyrosequencing for the analysis of fungal diversity in potting media and organic fertilizers. J Appl Microbiol. 2015;119(2):500–9.
13. Abed RMM, Al-Sadi AM, Al-Shehi M, Al-Hinai S, Robinson MD. Diversity of free-living and lichenized fungal communities in biological soil crusts of the Sultanate of Oman and their role in improving soil properties. Soil Biol Biochem. 2013;57:695–705.
14. Huang X, Liu L, Wen T, Zhu R, Zhang J, Cai Z. Illumina MiSeq investigations on the changes of microbial community in the Fusarium oxysporum f.sp. cubense infected soil during and after reductive soil disinfestation. Microbiol Res. 2015;181:33–42.
15. Qi X, Liu B, Song Q, Zou B, Bu Y, Wu H, Ding L, Zhou G. Assessing fungal population in soil planted with Cry1Ac and CPTI transgenic cotton and its conventional parental line using 18S and ITS rDNA sequences over four seasons. Front Plant Sci. 2016;7:1023.
16. Doyle J, Doyle JL. Isolation of plant DNA from fresh tissue. Focus. 1990;12: 13–5.
17. Gardes M, Bruns T. ITS primers with enhanced specificity for basidiomycetes – application to the identification of mycorrhizae and rusts. Mol Ecol. 1993; 2:113–8.
18. White TJ, Bruns T, Lee S, Taylor J. Amplification and direct sequencing of fungal ribosomal RNA genes for phylogenetics. In: Innis MA, Gelfand DH, Sninsky JJ, White TJ, editors. PCR protocols: A Guide to Methods and Applications. New York: Academic; 1990. p. 315–22.
19. Ruff S, Kuhfuss H, Wegener G, Lott C, Ramette A, Wiedling J, Knittel K, Weber M. Methane seep in shallow-water permeable sediment harbors high diversity of anaerobic methanotrophic communities, Elba, Italy. Front Microbiol. 2016;7:374.
20. Edgar RC. Search and clustering orders of magnitude faster than BLAST. Bioinformatics. 2010;26:2460–1.
21. Edgar RC. UPARSE: highly accurate OTU sequences from microbial amplicon reads. Nat Methods. 2013;10:996–8.
22. Edgar RC, Haas BJ, Clemente JC, Quince C, Knight R. UCHIME improves sensitivity and speed of chimera detection. Oxford Journal of Bioinformatics. 2011;27:2194–200.
23. Edgar RC. MUSCLE: multiple sequence alignment with high accuracy and high throughput. Nucleic Acids Res. 2004;32:1792–7.
24. Edgar RC. MUSCLE: a multiple sequence alignment method with reduced time and space complexity. BMC Bioinformatics. 2004;5:1.

25. Price M, Dehal P, Arkin A. FastTree 2-approximately maximum-likelihood trees for large alignments. Plos One. 2010;5, e9490.

26. Team RDC. R: A Language and Environment for Statistical Computing. Vienna: R Foundation for Statistical Computing; 2011.

27. Oksanen J, Blanchet FG, Kindt R, Legendre P, O'Hara RB, Simpson GL, Solymos P, Henry M, Stevens H, Wagner H. VEGAN: Community Ecology Package. 2011, R package version:1.17-18.

28. Hartmann M, Frey B, Mayer J, Mäder P, Widmer F. Distinct soil microbial diversity under long-term organic and conventional farming. ISME J. 2015; 9(5):1177–94.

29. Moll J, Hoppe B, König S, Wubet T, Buscot F, Krüger D. Spatial distribution of fungal communities in an arable soil. Plos One. 2016;11, e0148130.

30. Abbas HK, Shier WT, Plasencia J, Weaver MA, Bellaloui N, Kotowicz JK, Butler AM, Accinelli C, de la Torre-Hernandez ME, Zablotowicz RM. Mycotoxin contamination in corn smut (Ustilago maydis) galls in the field and in the commercial food products. Food Control. 2017;71:57–63.

31. Shenasi M, Candlish AAG, Aidoo KE. The production of aflatoxins in fresh date fruits and under simulated storage conditions. J Sci Food Agric. 2002; 82(8):848–53.

32. Aidoo KE, Tester RF, Morrison JE, MacFarlane D. The composition and microbial quality of pre-packed dates purchased in Greater Glasgow. Int J Food Sci Technol. 1996;31(5):433–8.

33. Siala R, Chobba IB, Vallaeys T, Triki MA, Jrad M, Cheffi M, Ayedi I, Elleuch A, Nemsi A, Cerqueira F, et al. Analysis of the cultivable endophytic bacterial diversity in the date palm (Phoenix dactylifera L.) and evaluation of its antagonistic potential against pathogenic Fusarium species that cause date palm bayound disease. Journal of Applied and Environmental Microbiology. 2016;4(5):93–104.

34. Kumar M, Dwivedi P, Sharma AK, Sankar M, Patil RD, Singh ND. Apoptosis and lipid peroxidation in ochratoxin A- and citrinin-induced nephrotoxicity in rabbits. Toxicol Ind Health. 2014;30(1):90–8.

35. Kocić-Tanackov SD, Dimić GR, Lević JT, Pejin DJ, Pejin JD, Jajić IM. Occurrence of potentially toxigenic mould species in fresh salads of different kinds of ready-for-use vegetables. Acta Periodica Technologica. 2010;41:33–45.

36. Engelhart S, Loock A, Skutlarek D, Sagunski H, Lommel A, Färber H, Exner M. Occurrence of toxigenic Aspergillus versicolor isolates and sterigmatocystin in carpet dust from damp indoor environments. Appl Environ Microbiol. 2002;68(8):3886–90.

37. Mills JT. Mycotoxins and toxigenic fungi on cereal grains in western Canada. Can J Physiol Pharmacol. 1990;68(7):982–6.

38. Yogendrarajah P, Vermeulen A, Jacxsens L, Mavromichali E, De Saeger S, De Meulenaer B, Devlieghere F. Mycotoxin production and predictive modelling kinetics on the growth of Aspergillus flavus and Aspergillus parasiticus isolates in whole black peppercorns (Piper nigrum L). Int J Food Microbiol. 2016;228:44–57.

39. Jay J, Loessner M, Golden D. Modern Food Microbiology. 7th ed. New York: Springer Science and Business Media; 2005.

9

Disrupting ROS-protection mechanism allows hydrogen peroxide to accumulate and oxidize Sb(III) to Sb(V) in *Pseudomonas stutzeri* TS44

Dan Wang[1], Fengqiu Zhu[1], Qian Wang[1], Christopher Rensing[2], Peng Yu[1], Jing Gong[1] and Gejiao Wang[1*]

Abstract

Background: Microbial antimonite [Sb(III)] oxidation converts toxic Sb(III) into less toxic antimonate [Sb(V)] and plays an important role in the biogeochemical Sb cycle. Currently, little is known about the mechanisms underlying bacterial Sb(III) resistance and oxidation.

Results: In this study, Tn5 transposon mutagenesis was conducted in the Sb(III)-oxidizing strain *Pseudomonas stutzeri* TS44 to isolate the genes responsible for Sb(III) resistance and oxidation. An insertion mutation into *gshA*, encoding a glutamate cysteine ligase involved in glutathione biosynthesis, generated a strain called *P. stutzeri* TS44-gshA$_{540}$. This mutant strain was complemented with a plasmid carrying *gshA* to generate strain *P. stutzeri* TS44-gshA-C. The transcription of *gshA*, the two superoxide dismutase (SOD)-encoding genes *sodB* and *sodC* as well as the catalase-encoding gene *katE* was monitored because *gshA*-encoded glutamate cysteine ligase is responsible for the biosynthesis of glutathione (GSH) and involved in the cellular stress defense system as are superoxide dismutase and catalase responsible for the conversion of ROS. In addition, the cellular content of total ROS and in particular H_2O_2 was analyzed. Compared to the wild type *P. stutzeri* TS44 and TS44-gshA-C, the mutant *P. stutzeri* TS44-gshA$_{540}$ had a lower GSH content and exhibited an increased content of total ROS and H_2O_2 and increased the Sb(III) oxidation rate. Furthermore, the transcription of *sodB*, *sodC* and *katE* was induced by Sb(III). A positive linear correlation was found between the Sb(III) oxidation rate and the H_2O_2 content ($R^2 = 0.97$), indicating that the accumulated H_2O_2 is correlated to the increased Sb(III) oxidation rate.

Conclusions: Based on the results, we propose that a disruption of the pathway involved in ROS-protection allowed H_2O_2 to accumulate. In addition to the previously reported enzyme mediated Sb(III) oxidation, the mechanism of bacterial oxidation of Sb(III) to Sb(V) includes a non-enzymatic mediated step using H_2O_2 as the oxidant.

Keywords: *Pseudomonas stutzeri*, Sb(III) oxidation, H_2O_2, Transposon mutagenesis, *gshA*, Reactive oxygen species (ROS)

* Correspondence: gejiao@mail.hzau.edu.cn
[1]State Key Laboratory of Agricultural Microbiology, College of Life Science and Technology, Huazhong Agricultural University, Wuhan, People's Republic of China
Full list of author information is available at the end of the article

Background

Antimony (Sb) is a metal belonging to group V in the periodic table. It is present in aquatic systems and soil, with stibnite (Sb_2S_3) representing the most common mineral form [1]. Antimonials have been used to treat leishmaniasis for over 60 years [2, 3]. However, the U.S. Environmental Protection Agency recognized Sb as a priority pollutant [4], and the WHO has proposed 5 μg/L to be the highest acceptable Sb concentration in potable water [5]. Although Sb pollution has gained increased attention in recent decades [6–8], the exact mechanism of Sb toxicity in both mammals and microorganism remains unclear [9, 10].

A comparison of two common inorganic forms of Sb revealed that antimonite [Sb(III)] was more toxic than antimonate [Sb(V)] [7]. The abiotic Sb(III) oxidation in the natural environment is extremely slow [11]. In contrast, Sb(III)-oxidizing bacteria can oxidize Sb(III) at a relatively high rate. Thus, Sb(III) oxidation was proposed to be a microbial detoxification process that could be useful for environmental Sb bioremediation [12–14]. A few Sb(III)-oxidizing bacteria had been reported in earlier literature [15, 16]. Recently, dozens of Sb(III)-oxidizing bacteria have been isolated and identified both by our group and by others [12, 13, 17, 18]. The arsenite oxidase AioAB in *Agrobacterium tumefaciens* was also found to be able to oxidize Sb(III) [19]. In addition, the oxidoreductase AnoA was shown to be responsible for bacterial Sb(III) oxidation [20]. However, the disruption of both of these genes did not result in a complete loss of Sb oxidation, indicating the existence of other mechanisms responsible for bacterial Sb(III) oxidation.

In general, aerobic respiration is inevitably accompanied by the production of reactive oxygen species (ROS). ROS include superoxide ($O_2^{\cdot-}$), hydroxyl ($OH^{\cdot}$), hydroperoxyl ($HO_2^{\cdot}$), and peroxyl ($RO_2^{\cdot}$) radicals and other oxidizing agents, such as hydrogen peroxide (H_2O_2) [21]. Among the various ROS, $O_2^{\cdot-}$ is the primary radical that results from the univalent reduction of molecular oxygen (O_2) via the bacterial respiratory chain. Subsequently, $O_2^{\cdot-}$ can be converted into H_2O_2 and O_2 by superoxide dismutase (SOD). Both $O_2^{\cdot-}$ and H_2O_2 can generate highly reactive hydroxyl radicals via the Haber-Weiss reaction or the Fenton reaction [22]. ROS can damage different types of macromolecules; thus, bacteria have evolved defense mechanisms against ROS and induce specific genes in response to oxidative stress [23]. Glutathione (GSH) is the central substrate involved in cellular protection against ROS and their toxic products in eukaryotic cells [24]. In bacteria GSH plays multiple role in protection against environmental stresses such as osmotic shock and acidity, as well as against toxins and oxidative stress. Indeed, GSH represents the primary low molecular weight thiol in many Gram-negative bacteria [25]. Two enzymes (glutamate cysteine ligase, the rate-limiting enzyme, and GSH synthetase) catalyze the *de novo* synthesis of GSH. These two enzymes are encoded by distinct genes (*gshA* and *gshB*, respectively) in most bacteria [26].

Previously, the aerobic, arsenite-oxidizing bacterium *Pseudomonas stutzeri* TS44 was isolated from arsenic-contaminated soil [27] and the genome sequence was published (Accession No. AJXE00000000, [28]). In this study, we showed that strain TS44 is able to oxidize Sb(III) to Sb(V). Therefore, in an effort to determine the molecular mechanism underlying bacterial Sb(III) oxidation, more than 3000 Tn5 transposon insertion mutants were generated and screened. We isolated a mutant of *gshA* (TS44-gshA$_{540}$) that showed an increase in Sb(III) oxidation rate. The Sb(III) oxidation rates of the wild type, mutant and complemented strain, the gene transcription of *gshA*, *sod* and *katE*, as well as the cellular contents of ROS, GSH and H_2O_2 were analyzed since *gshA* is responsible for the biosynthesis of GSH and therefore involved in the cellular stress defense system.

Methods

Bacterial strains, plasmids, primers and culture conditions

The strains, plasmids and primers used in this study are listed in Table 1 and Additional file 1: Table S1. The *Pseudomonas stutzeri* strains were grown in a chemically defined medium (CDM) [29] containing 0.114 mmol/L phosphate (0.07 mmol/L $K_2HPO_4 \cdot 3H_2O$ and 0.044 mmol/L KH_2PO_4) and shaken under aerobic conditions at 28 °C. The *Escherichia coli* strains were cultured at 37 °C in Luria-Bertani medium [30]. Rifampin (Rif, 50 mg/mL), kanamycin (Kan, 50 mg/mL), tetracycline (Tet, 5 mg/mL) or chloromycetin (Cm, 50 mg/mL) were added when needed.

Isolation of a Sb(III) sensitive transposon mutant and complementation of the mutant strain

Plasmid pRL27-Kan was transferred into *P. stutzeri* TS44 by conjugation from *E. coli* strain S17-1 carrying Tn5 as described previously [31]. To obtain Sb(III) sensitive mutants, the colonies of transformants from the mating plates were spread onto LB agar plates containing 50 μg/mL Kan and 50 μg/mL Rif, in the presence or absence of 0.4 mmol/L $C_8H_4K_2O_{12}Sb_2S_3(H_2O)$ [as Sb(III)]. The mutants did not grow in the presence of the indicated concentration of Sb(III). The insertion sequence of the mutant was determined using a plasmid rescue strategy as described by Zheng et al. [31]. The sequences were compared to the draft genome sequence of *P. stutzeri* TS44 using BLAST [32] and compared to the protein sequence database at GenBank using the BlastX algorithm [33]. Finally, the Sb-sensitive Tn5-insertion mutant TS44-gshA$_{540}$ was isolated.

Table 1 The strains and plasmids used in this study

Strain/plasmid	Relevant properties or derivation	Source or reference
Pseudomonas stutzeri		
TS44	Wild-type, As(III)-oxidizing phenotype	[27]
TS44-gshA$_{540}$	Kan^r; *gshA* mutant by Tn5 random transposon mutagenesis	This study
TS44-gshA-C	Kan^r Cm^r; *gshA* complemented strain	This study
TS44 (P*gshA*)	Kan^r; TS44 with pLSP-P*gshA*	This study
TS44 (P*sodB*)	Kan^r; TS44 with pLSP-P*sodB*	This study
TS44 (P*sodC*)	Kan^r; TS44 with pLSP-P*sodC*	This study
TS44-gshA$_{540}$ (P*gshA*)	Kan^r; TS44-gshA$_{540}$ with pLSP-P*gshA*	This study
TS44-gshA$_{540}$ (P*sodB*)	Kan^r; TS44-gshA$_{540}$ with pLSP-P*sodB*	This study
TS44-gshA$_{540}$ (P*sodC*)	Kan^r; TS44-gshA$_{540}$ with pLSP-P*sodC*	This study
Escherichia coli		
DH5α	supE44 lacU169(j80lacZM15) hRDR17 recA1 endA1 gyrA96 thi-1 relA1	[35]
S17-1	Pro$^-$ Mob$^+$; conjugation donor	[36]
pir116	mcrA, Δ(mrr-hsdRMS-mcrBC) recA1 R6Kγ lacZΔM15 λpir	Epicenter, Madison, WI
Plasmids		
pRL27-Kan	Kan^r Transposon vector, *ori*R6K	[37]
pCT-Zori	broad host vector, pUC ori, Cm^r	[34]
pCT-Zori-*gshA*	gshA complementation vector	This study
pLSP-kt2lacZ	Kan^r*oriV*; *lacZ* fusion vector used for *lacZ* fusion constructs	T. R. McDermott, MSU
pLSP-P*gshA*	pLSP-kt2lacZ containing *gshA* promoter region	This study
pLSP-P*sodB*	pLSP-kt2lacZ containing *sodB* promoter region	This study
pLSP-P*sodC*	pLSP-kt2lacZ containing *sodC* promoter region	This study

The construction of the *gshA* complementary strain TS44-gshA-C was accomplished using the high-copy broad host range plasmid pCT-Zori containing a Kan resistance determinant ([34], Table 1). Briefly, a 2023 bp DNA fragment containing the complete *gshA* coding region along with 156 bp upstream and 289 bp downstream sequence was PCR-amplified using the gshA-F/gshA-R primers (Additional file 1: Table S1) and subsequently cloned into *Hin*dIII + *Bam*HI-digested pCT-Zori. The resulting plasmid pCT-Zori-*gshA* was transformed into *E. coli* S17-1 and conjugated into the mutant TS44-gshA$_{540}$, yielding the complementary strain TS44-gshA-C. The integrity of the mutant and the complementary strain was confirmed by PCR amplification and subsequent DNA sequencing.

Growth and Sb(III) oxidation tests

Overnight cultures of *P. stutzeri* strains TS44, TS44-gshA$_{540}$ and TS44-gshA-C (OD$_{600}$ = 1.0) were inoculated into 100 mL of CDM medium with or without 0.2 mmol/L Sb(III) and incubated at 28 °C for 48 h with shaking at 170 rpm. Culture samples were collected to determine the OD$_{600}$ value using a UV spectrophotometer (DU800, Beckman, USA) at the designated time points. Sb(III)/Sb(V) concentrations were monitored using hydride-generation atomic fluorescence spectroscopy combining HPLC (HPLC-HG-AFS, Beijing Titan Instruments Co., Ltd., China) according to the method described by Li et al. [13]. The measured data were analyzed with single factor analysis of variance (one way ANOVA) method in Excel program.

GshA activity and determination of GSH content

To detect GshA activity and GSH content, *P. stutzeri* strains TS44, TS44-gshA$_{540}$ and TS44-gshA-C were cultured under aerobic condition with shaking in 100 mL of liquid CDM medium at 28 °C. Sb(III) was added at a final concentration of 0.2 mmol/L when the OD$_{600}$ reached 0.3. Cultures without the addition of Sb(III) were used as controls. One mililiter of cells was centrifuged at 13,400 × g at 4 °C after 30 min of cultivation. The pellet was washed twice using phosphate-buffered saline (PBS, pH 7.0), then resuspended in 1 mL of PBS and sonicated on ice to dissolve the cell membranes.

The total protein content of the sonicate was measured by the Coomassie brilliant blue G-250-staining method [38]. Bovine serum albumin was used as the standard.

The GshA activity and GSH content were determined using 2, 3-naphthalenedicarboxyaldehyde (NDA) (Aladdin Industrial Co., Shanghai, China) as described previously [39]. To measure GshA activity, an equal volume of cell lysate was mixed with the GshA reaction buffer and substrate solution (50 μL each). Fifty microliter of reaction terminator (1 mol/L Na_2CO_3) was added after 45 min of incubation at 28 °C. The mixture was incubated on ice for 20 min and centrifuged at $8000 \times g$ for 5 min. Then, 20 μL of the supernatants were transferred to wells of black Microlon 96-well plates (Greiner). The samples were mixed with 180 μL of NDA derivatization solution (50 mmol/L Tris–HCl, pH = 10, 0.5 N NaOH, and 10 mmol/L NDA in Me_2SO, v/v/v 1.4/0.2/0.2). Fluorescence was measured (472 ex/528 em) with an EnVision® Multimode Plate Reader (Perkin Elmer) after 60 min of incubation at 37 °C and converted to the γ-glutamylcysteine concentration using appropriate calibration standards. The procedure for measuring GSH content was almost the same as the procedure to detect GshA activity, except that the reaction terminator was added immediately when the equal volume of cell lysate was mixed with the GshA reaction buffer and substrate solution (50 μL each).

Determination of the ROS content

The cellular ROS content was tested with the fluorescent probe 2, 7-dichlorofluorescin diacetate (DCFH-DA) (Sigma Chemical Co., St. Louis, MO, USA). The cells were cultivated and collected for processing as described above. Cells from 0.5 mL cultures were washed two times with PBS buffer and resuspended in 0.5 mL of PBS buffer. Then, a 0.2 mL cell suspension was mixed with 5 μL of 0.1 mmol/L DCFH-DA and incubated at 37 °C for 30 min to develop the fluorescent product DCF. The cells were harvested by centrifugation for 3 min at $13,400 \times g$ and washed two times using PBS buffer (pH 7.0) to remove background fluorescence. The fluorescence was measured (488 ex/535 em) on the EnVision® Multimode Plate Reader (Perkin Elmer) [40]. The ROS contents were normalized to the total protein content as described above.

Quantification of gene expression using a *lacZ* reporter gene fusion and qRT-PCR

Quantitative reverse transcription PCR (qRT-PCR) was employed to test the transcription of *gshA*, two superoxide dismutase (SOD)-encoding genes (*sodB* and *sodC*) and the catalase *katE*. The strains were inoculated into 100 mL of CDM medium. 0.2 mmol/L of Sb(III) was added (or not) when the OD_{600} reached 0.3. The cells were harvested after 30 min. Total RNA was extracted using the TRIzol®

Reagent (Invitrogen) and treated with DNaseI following the manufacturer's instructions. The synthesis of cDNA from 300 ng of total RNA was performed using the RevertAid First Strand cDNA Synthesis Kit (Thermo) [41]. The resulting cDNA was used as a template for qRT-PCR with the SYBR-Green® qPCR Master Mix (Takara). Primers RT-gshA-F/RT-gshA-R, RT-sodB-F/RT-sodB-R, RT-sodC-F/RT-sodC-R and RT-katE-F/RT-katE-R were used to test the expression of *gshA*, *sodB*, *sodC* and *katE*, respectively. The RT-PCRs were performed using the AB ViiA 7 RT-PCR system (Life Technologies) following the manufacturer's recommended protocol. The annealing temperature for *gshA* and *sodB* was 55 °C, while for *sodC* and *katE* it was 48 °C.

For the *lacZ* reporter fusion analysis, plasmid pLSP-kt2lacZ was used to construct the *lacZ* fusions. DNA fragments containing the predicted promoter regions of *gshA*, *sodB* and *sodC* were amplified by PCR using the primers PgshA-F/PgshA-R, PsodB-F/PsodB-R and PsodC-F/PsodC-R, respectively. The DNA fragments were digested using *Eco*RI and *Bam*HI and ligated into the double-digested pLSP-kt2lacz. The resulting plasmids pLSP-PgshA, pLSP-PsodB, and pLSP-PsodC were separately transferred into strains TS44 and TS44-gshA$_{540}$ by conjugation employing *E. coli* S17-1. The resulting strains containing the above constructs were inoculated into 100 mL of CDM medium and cultivated with shaking at 28 °C. 0.2 mmol/L of Sb(III) was added when the OD_{600} reached 0.3. The samples were collected and the β-galactosidase activity was tested according to the method described by Miller [42] after 30 min of incubation.

Determination of the linear correlation between H_2O_2 and Sb(III) oxidation rate

To test the H_2O_2 content, strains TS44, TS44-gshA540 and TS44-gshA-C were incubated as described above. To eliminate the H_2O_2 content difference caused by the different amount of cell collection, Sb(III) was added when the OD_{600} reached 0.3, and then 2 mL of cells were harvested after a 30 min incubation with Sb(III). The cells were resuspended in 1 mL of K_3PO_4 (pH 7.8) after washing twice with 50 mmol/L K_3PO_4 (pH 7.8) and sonicated on ice. Then, the sonicated cell lysates were centrifuged at $13,400 \times g$ at 4 °C, and 0.1 mL of the supernatant was transferred to wells of black Microlon 96-well plates (Greiner). The samples were mixed with 50 μL of amplex red (AR) (Sigma Chemical Co., St. Louis, MO, USA) and 50 μL of horseradish peroxidase (HRP) (F. Hoffmann-La Roche Ltd, Shanghai, China), then incubated at 37 °C for 15 min. Fluorescence was measured (530 ex/587 em) as described above using the EnVision® Multimode Plate Reader (Perkin Elmer) and converted to the H_2O_2 concentration using appropriate calibration standards [43].

To detect the dynamic variation of the H_2O_2 and Sb(III) contents, strains TS44, TS44-gshA$_{540}$ and TS44-gshA-C were inoculated into 100 mL of CDM medium supplemented with 0.2 mmol/L Sb(III). The cells were harvested and sonicated as described above at the designated time points. The Sb(III) contents were monitored using HPLC-HG-AFS as described above, while the H_2O_2 contents were determined by the AR/HRP method as described above. A boiled CDM culture of strain TS44-gshA$_{540}$ was used as a control. To test the effect of H_2O_2 on bacterial growth, strain TS44 was inoculated onto CDM plates containing different concentrations of H_2O_2 and cultivated at 28 °C for 48 h.

In vitro oxidation of Sb(III) by H_2O_2 was performed using an un-inoculated liquid CDM medium containing 0.2 mmol/L of Sb(III) and the media with the addition of 0, 0.02, 0.05, 0.1, 0.3, 0.4 and 0.5 mmol/L of H_2O_2 and reacted within 10 min.

Results

A Sb(III)-sensitive mutant could be generated by transposon mutagenesis

More than 3000 transposon insertions were isolated and screened for loss of Sb(III) resistance with one mutant displaying lower Sb(III) resistance. This strain, TS44-gshA$_{540}$, was selected for further characterization (data not shown). The Tn5 transposon had inserted into *gshA* encoding a putative glutamate-cysteine ligase at nucleotide 540 (Additional file 2: Figure S1A). The *gshA* encoded glutamate-cysteine ligase is the rate-limiting enzyme in *de novo* GSH biosynthesis and therefore is involved in conferring resistance to ROS and their toxic products [44]. Additionally, a strain complementing the *gshA* insertion was constructed and named TS44-gshA-C. Diagnostic PCRs were used to confirm the transposon mutation and the complementation (Additional file 2: Figure S1B-C).

An insertion in *gshA* led to an increased Sb(III) oxidation rate

Cells of *P. stutzeri* TS44, TS44-gshA$_{540}$ and TS44-gshA-C were incubated in CDM medium lacking Sb(III) supplementation after thorough washing. The growth of strain TS44-gshA$_{540}$ was slightly slower compared to the wild type strain (Fig. 1a). The growth of strain TS44-gshA$_{540}$ was further delayed by supplementation with 0.2 mmol/L Sb(III), causing strain TS44-gshA$_{540}$ to need an extra 24 h to reach the lag phase (Fig. 1b). The *gshA* complementing strain TS44-gshA-C showed no difference in growth compared to the wild type strain regardless of whether Sb(III) was added or not. Although the growth of TS44-gshA$_{540}$ was delayed, strain TS44-gshA$_{540}$ oxidized 88% Sb(III) to Sb(V) in 48 h (Fig. 2b). In contrast, strains TS44 and TS44-gshA-C both oxidized only 48% Sb(III) to Sb(V) in 48 h (Fig. 2a and c). The Sb(III)-oxidation rate of TS44-gshA$_{540}$ significantly increased by 83% in the monitored 48 h time frame ($p < 0.01$) (Fig. 2).

An insertion in *gshA* eliminated GshA activity and influenced cellular contents of GSH and ROS

We investigated both GshA activity and the GSH and ROS content to elucidate how the presence of GshA affected Sb(III) oxidation. The GshA activity in strains TS44 and TS44-gshA-C was slightly increased following the addition of Sb(III); however, GshA activity was undetectable in the insertional mutant strain TS44-gshA$_{540}$ (Fig. 3a), indicating that the transposon insertion functionally disrupted *gshA* in strain TS44. Consistent with this result, the GSH content was decreased by approximately 66% in the mutant compared to strains TS44 and TS44-gshA-C regardless of whether Sb(III) was provided (Fig. 3b).

As a next step, cellular ROS content was determined because GSH is the central substrate for cellular protection against ROS and their toxic products [24]. The cellular ROS content in strain TS44-gshA$_{540}$ was increased by approximately three-fold compared to strains TS44 and TS44-gshA-C regardless of whether Sb(III) was

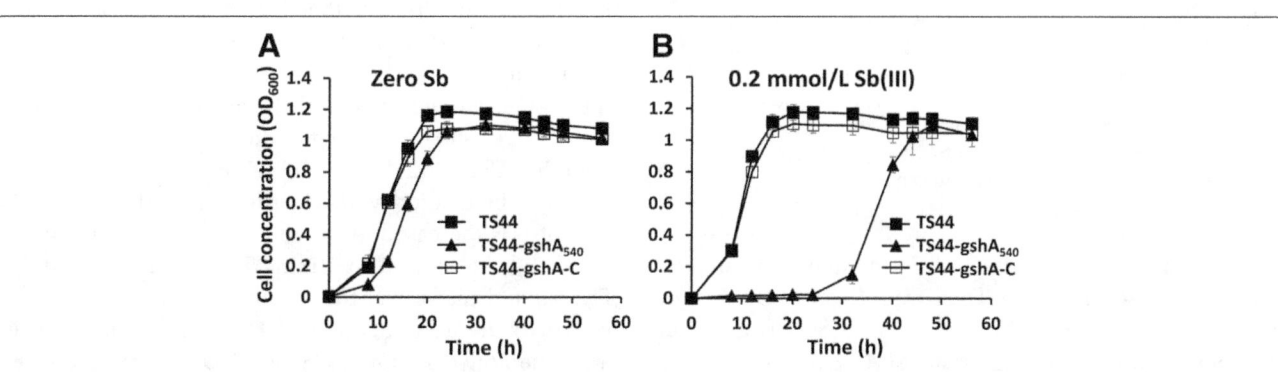

Fig. 1 Growth curves of *P. stutzeri* strains TS44, TS44-gshA$_{540}$ and TS44-gshA-C without (**a**) or with (**b**) the addition of 0.2 mmol/L Sb(III). Error bars correspond to the standard deviations of the means from three independent experiments

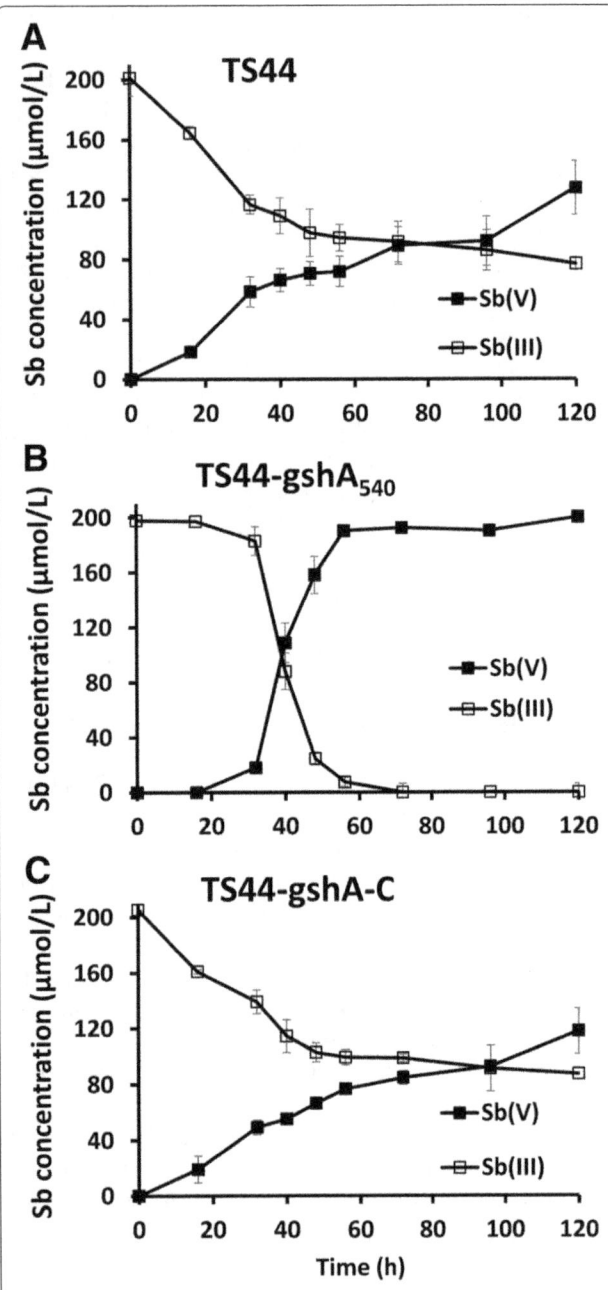

Fig. 2 Sb(III) oxidation in *P. stutzeri* strains TS44, TS44-gshA$_{540}$ and TS44-gshA-C with the addition of 0.2 mmol/L Sb(III). The amounts of Sb(III) and Sb(V) in culture fluids were calculated based on culture volume to normalize for the total amount of antimony added. Sb(III) and Sb(V) concentrations in the culture fluids were measured using HPLC-HG-AFS. Error bars correspond to the standard deviations of the means from three independent experiments

quickly in 30 min, and decreased significantly in the strain TS44-gshA$_{540}$ (Fig. 3d). In such a short time, the oxidation rate of Sb(III) did not significantly change. The Sb(III) oxidation mainly occurred in the 36–48 h time frame (Fig. 2).

Sb(III) affected transcription of *gshA, sodB, sodC* and *katE* differently

The expression of *gshA, sodB, sodC* and *katE* was monitored, because GshA, SodB, SodC and KatE are all involved in changes of GSH, ROS or H$_2$O$_2$. There are two distinct genes (*sodB* and *sodC*) encoding superoxide dismutase (SOD) on the genome of *P. stutzeri* TS44 that encode for Fe SOD and Cu-Zn SOD, respectively. These two enzymes shared over 83% and 58% amino acid sequence similarity with the respective enzymes of other species of *Pseudomonas*, respectively, and can convert O$_2^{\cdot-}$ to H$_2$O$_2$ and O$_2$. In the *lacZ* reporter fusion analysis, we only transformed the *lacZ* fusions designated pLSP-P*gshA*, pLSP-P*sodB* and pLSP-P*sodC* into strains TS44 and TS44-gshA$_{540}$ because the *gshA* complementary strain TS44-gshA-C carried a plasmid that might be incompatible with the *lacZ* fusion.

The transcription levels of *gshA, sodB, sodC* and *katE* were tested by qRT-PCR in all three strains. Sb(III) had no effect on the transcription of *gshA* in strains TS44, TS44-gshA$_{540}$ or TS44-gshA-C (Fig. 4a and d). The *gshA::lacZ* fusion only showed a low level of background expression levels in the *gshA* insertional mutation strain TS44-gshA$_{540}$, (Fig. 4a), which indicated a very low promoter activity of *gshA*, while qRT-PCR analysis showed no transcription of *gshA* in strain TS44-gshA$_{540}$ (Fig. 5d). Moreover, *lacZ* fusion and qRT-PCR both showed that the *sodB* gene was strongly induced by Sb(III) in all three strains (Fig. 4b and e). Sb(III) only induced transcription of *sodC* and *katE* in the mutant strain TS44-gshA$_{540}$, but not in strains TS44 and TS44-gshA-C (Fig. 4c, f and g).

Disruption of GSH synthesis allowed accumulation of H$_2$O$_2$

In order to examine whether the ROS-protection system is involved in H$_2$O$_2$ production, we tested the H$_2$O$_2$ content in the presence and absence of Sb(III) in all strains. The H$_2$O$_2$ content was slightly higher in *gshA* mutant strain TS44-gshA$_{540}$ without addition of Sb(III) compared to strains TS44 and TS44-gshA-C (Fig. 5a). Notably, after a 30 min incubation with 0.2 mmol/L Sb(III), the H$_2$O$_2$ content was increased in the mutant strain TS44-gshA$_{540}$, but slightly decreased in strains TS44 and TS44-gshA-C (Fig. 5a), indicating that the disruption of GSH synthesis is related to the accumulation of H$_2$O$_2$. In addition, it appears the conversion of ROS by SOD was more efficient than the H$_2$O$_2$ conversion by KatE and thus would both *sod* and *katE* were induced by Sb(III).

provided or not. The cellular ROS content was significantly decreased in strain TS44-gshA$_{540}$ after 30 min of incubation with Sb(III) compared to no Sb(III) addition (Fig. 3b-c), despite the fact that *gshA* was mutated and the GSH content was obviously decreased (Fig. 3a-b). After addition of Sb(III), the ROS content changed

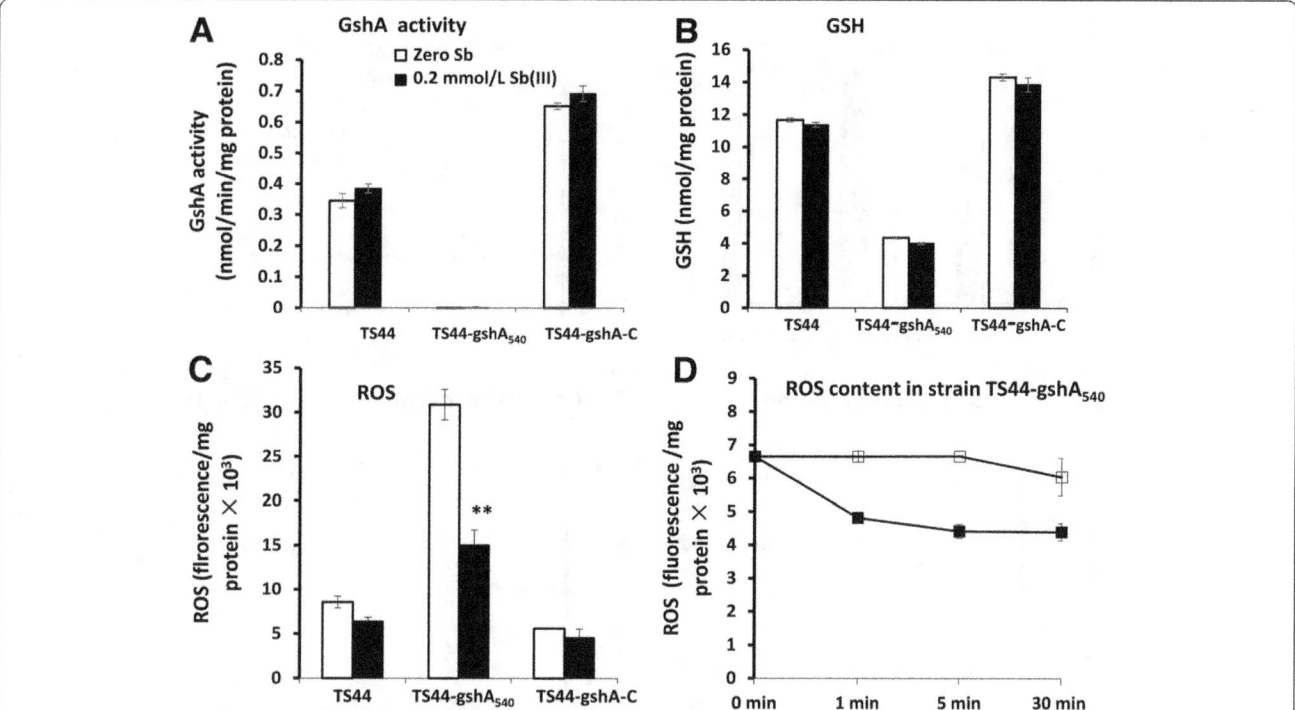

Fig. 3 *gshA* insertion affected GSH content, GshA activity and ROS content. **a**, GshA activity, (**b**), GSH content, (**c**), ROS content and (**d**) ROS content in strain TS44-gshA$_{540}$. Data symbols shown in panels (**a**), (**b**) and (**c**) are the same. Data are expressed as the mean ± SD, $N = 3$. **Indicates a significant difference from the control ($p < 0.01$)

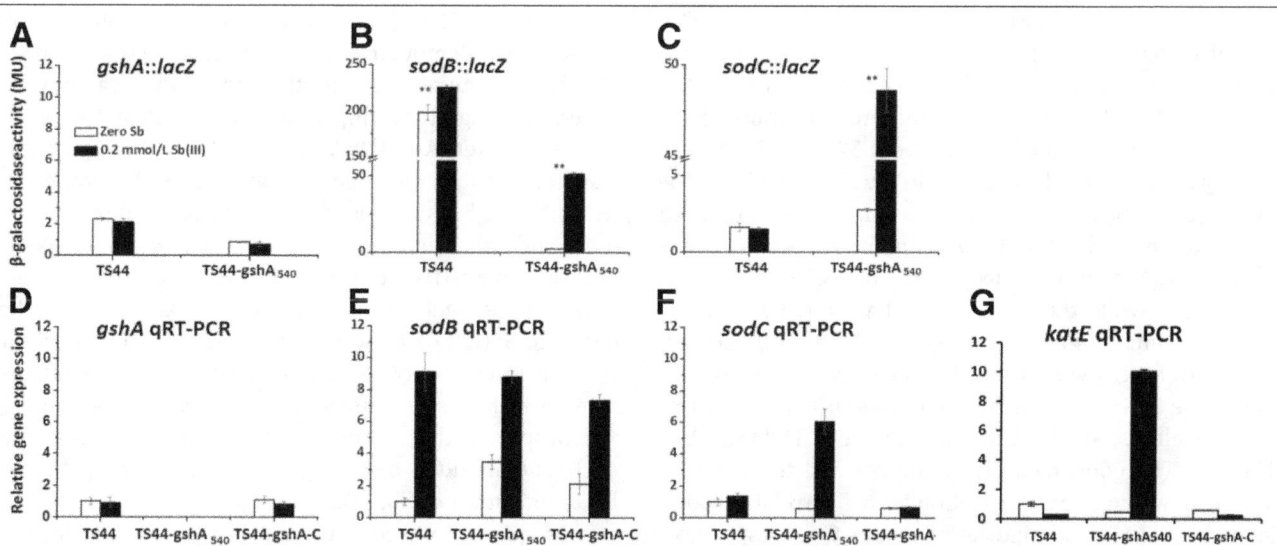

Fig. 4 The effect of Sb(III) to the transcription of *gshA*, *sodB*, *sodC* and *katE*. The transcription of *gshA*, *sodB* and *sodC* was tested by a *lacZ* reporter fusion and qRT-PCR, while the *katE* transcription was only tested by qRT-PCR. For *lacZ* reporter fusion (**a**, **b** and **c**), data are expressed as the mean ± SD, $N = 3$. **Indicates a significant difference from the control ($p < 0.01$). For qRT-PCR, error bars correspond to the standard deviations of the means from three biological replicates. Gene expression was normalized to the 16S rRNA gene. The results are presented as the mean gene expression normalized to mRNA levels in Sb(III)-free CDM. Data symbols shown in all panels are the same

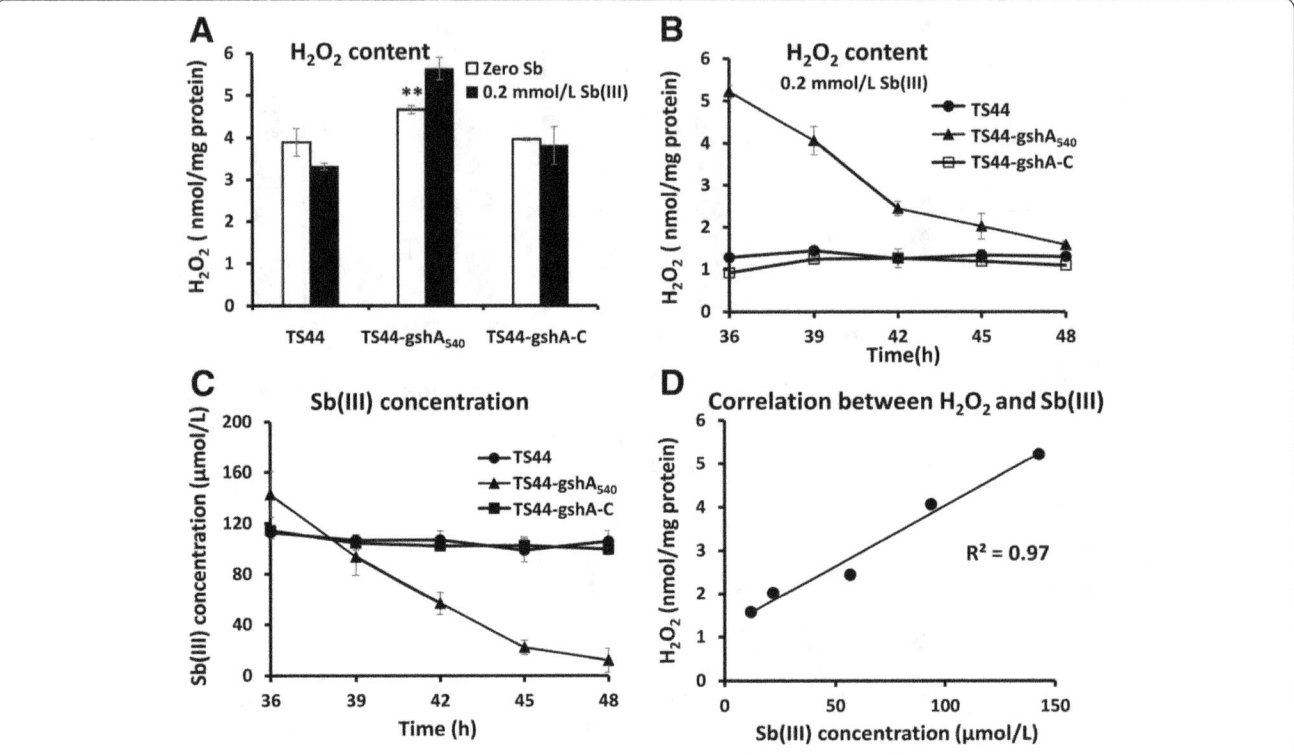

Fig. 5 The H_2O_2 content is correlated with bacterial Sb(III) oxidation. The *P. stutzeri* strains were cultured as described above. **a** The H_2O_2 concentration was tested after 30 min of incubation with Sb(III). **Indicates a significant difference from the control ($p < 0.01$). **b** H_2O_2 content and (**c**) Sb(III) concentration in strains TS44, TS44-gshA_540 and TS44-gshA-C from 36 to 48 h of incubation in cultures supplemented with 0.2 mmol/L Sb(III). Data are expressed as the mean ± SD, $N = 3$. **d** Correlation between H_2O_2 and Sb(III) concentrations in strain TS44-gshA_540

H_2O_2 is responsible for Sb(III) oxidation

H_2O_2 was shown to be an efficient Sb(III) oxidant in vitro (Additional file 3: Figure S2) as the un-inoculated control showed no Sb(III) oxidation (Additional file 3: Figure S2). A control with dead cells after boiling the culture also showed no oxidation of Sb(III) (data not shown). As the efficiency of Sb(III) oxidation in the mutant strain TS44-gshA_540 was shown to be highest from 32 to 48 h (Fig. 2), the H_2O_2 and Sb(III) content in strains TS44, TS44-gshA_540 and TS44-gshA-C were also measured from 36 to 48 h. Interestingly, the H_2O_2 content in strain TS44-gshA_540 was significantly decreased from 5.2 to 1.1 nmol/mg protein, while the H_2O_2 content was stable at a low level in strains TS44 and TS44-gshA-C (Fig. 5b). Correspondingly, Sb(III) was oxidized to Sb(V) concomitant with a decrease in H_2O_2 content, while the Sb(III) concentration was almost stable in strains TS44 and TS44-gshA-C (Fig. 5c). The consumed H_2O_2 content and the oxidized Sb(III) showed a linear correlation with a correlation coefficient of 0.97 (Fig. 5d), indicating that H_2O_2 is responsible for Sb(III) oxidation. In addition, we could show that growth was inhibited with increasing concentration of H_2O_2 (Additional file 4: Figure S3), thus Sb(III) oxidation may contribute to the bacterial detoxification of H_2O_2 and Sb(III).

Discussion

At present, several Sb(III)-oxidizing strains have been reported [13, 14, 18]; some of these strains were also able to oxidize As(III) [13, 15, 19]. The arsenite oxidase AioAB was demonstrated to be capable of oxidizing Sb(III), but the AioAB kinetic rate of the reaction was orders of magnitude higher for As(III) than for Sb(III) [19]. Moreover, AnoA belonging to the SDR superfamily was reported to be able to catalyze Sb(III) oxidation using $NADP^+$ as a cofactor [20]. However, disruption of *aioA* and *anoA* in *Agrobacterium tumefaciens* caused a decrease in Sb(III) oxidation of only about 25% [19] and 27% [20], respectively, indicating the existence of other bacterial Sb(III) oxidation mechanisms. After discovering that the presence of *gshA* affected the Sb(III) oxidation rate by transposon mutagenesis in strain TS44-gshA_540, we proposed that some abiotic cellular components, such as GSH, ROS or H_2O_2, may have a role in the bacterial oxidation of Sb(III).

Subsequently, we conducted a comprehensive analysis. First, we found that a disruption of *gshA* caused a decrease in the cellular GSH amount; Second, it is conceivable that the *gshA* mutant strain resulted in an increase in cellular ROS content compared to the wild type; Third, we could also show a linear correlation between the decrease of H_2O_2 content and the increase in Sb(III) oxidation rate, indicating that Sb(III) oxidation consumed H_2O_2 and acts as a detoxification mechanism to counter this cellular stressor. In addition, it appeared that Sb(III) directly caused a disruption of the ROS-protection system in strain TS44-gshA$_{540}$ and allowed the accumulation of H_2O_2 instead of affecting GSH levels, since neither the activity of GshA nor the GSH content were influenced by Sb(III). Previously, it was reported that Sb(III) may consume residual GSH (forming a stable Sb(GS)$_3$ complex) in red blood cells [45]. However, we did not observe significant changes in GSH content when Sb(III) was added to wild type strain TS44 and the gshA complemented strain TS44-gshA-C, indicating little of this complex was formed and Sb(III) was mainly oxidized to Sb(V) by H_2O_2. We therefore propose a new model for Sb(III) oxidation in P. stutzeri TS44. i) the addition of Sb(III) would trigger the ROS-protective system by inducing the transcription of sodB, sodC and katE, with SodB and SodC catalyzing the conversion of ROS to H_2O_2, and KatE responsible for the depletion of excessive H_2O_2; ii) the increased cellular H_2O_2 content enhanced the Sb(III) oxidation rate; and iii) the addition of Sb(III) played a selection role on the characterization of the gshA insertion (Fig. 6); iv) in addition, the accumulated H_2O_2 was partially consumed by KatE (data not shown), since katE was induced by Sb(III) (Fig. 4 g), and similar result was described recently with transcription of the peroxidase-encoding gene katA being induced by both Sb(III) and H_2O_2 [46].

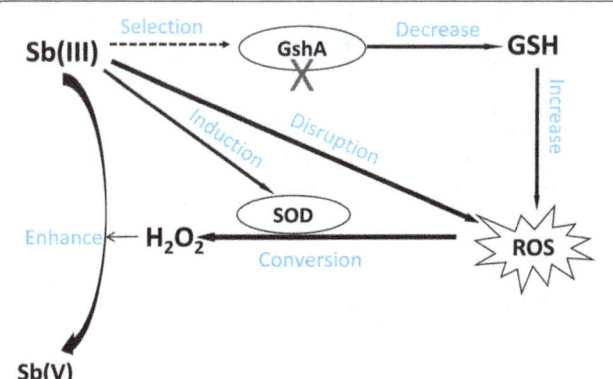

Fig. 6 The proposed model for Sb(III) bacterial oxidation in P. steutzeri TS44. In this study, i) the addition of Sb(III) would trigger the ROS-protective system by inducing the transcription of sodB, sodC and katE, with SodB and SodC catalyzing the conversion of ROS to H_2O_2, while KatE responsible for the degradation of excessive H_2O_2 (data not shown); ii) the increased cellular H_2O_2 content enhanced the Sb(III) oxidation rate; and iii) the addition of Sb(III) played a selection role on the characterization of gshA insertion; iv) in addition, the accumulated H_2O_2 is partially cosumed by the upregulated catalase KatE (data not shown)

Several chemical substances (i.e., amorphous iron, manganese oxyhydroxides and H_2O_2) have been reported to be capable of mediating Sb(III) oxidation in vitro [47–49]. OH$^{\bullet}$ was the oxidant in acidic solutions and H_2O_2 was the main oxidant in neutral and alkaline solutions involved in Sb(III) oxidation, with a pH range of 8.1 to 11.7 [48, 49]. Sb(III) was reported to cause cellular oxidative stress in lymphoid tumoral cells [50]. H_2O_2 is a substantial component of cellular oxidative stress [51] and can inhibit cellular growth [52]. In this study, the pH of the cultures changed from the initial 7.0 to approximately 8.0 following exposure to Sb(III), indicating that H_2O_2 may correlate with an increase in the Sb(III) oxidation rate. In addition, the in vitro experiment provided direct evidence that H_2O_2 can oxidize Sb(III) to Sb(V) (Additional file 4: Figure S3) and a correlation between the concentrations of Sb(III) and H_2O_2 was found in vivo (Fig. 5d). As an moderatedly reactive intermediate, $O_2^{\bullet-}$ may also have an effect on Sb(III) oxidation. However, in the presence of high level of SOD, the steady-state of $O_2^{\bullet-}$ would be no more than 0.1 nmol/L and thus the effect of $O_2^{\bullet-}$ is negligible [53]. Thus, we conclude that H_2O_2 oxidized Sb(III) to Sb(V).

Conclusions

This study proposed a novel mechanism for microbial antimonite oxidation involving changes in cellular content of components related to oxidative stress (GSH, ROS and H_2O_2). Although H_2O_2 was reported to be capable of oxidizing Sb(III) chemically as early as 2005 [11], this is the first study to demonstrate that H_2O_2 is responsible for Sb(III)-oxidation in a microbe. The results showed that a disruption of bacterial GSH-dependent ROS-protection mechanism allowed H_2O_2 to build up and thus promote the oxidation of Sb(III) to Sb(V). In addition, the oxidation of Sb(III) to Sb(V) is a detoxification process against the cellular stressor H_2O_2. We do not exclude the possibility of an additional enzymatic process responsible for Sb(III) oxidation in strain TS44, since genes encoding a putative arsenite oxidase AioBA were found on its genome [28] which may function as a Sb(III) oxidase. Our data and other findings [19, 20, 46] could show that microbial antimonite oxidation contains both biotic and abiotic components.

Additional files

Additional file 1: Table S1. Primers used in this study.

Additional file 2: Figure S1. The gene cluster containing gshA in strain TS44 and Diagnostic PCR confirming the transposon mutation to create mutant strain TS44-gshA$_{540}$ and complementation to create TS44-gshA540-C. (**A**), The transposon insertion site of the gshA mutant is shown by the vertical arrow. (**B**), PCR used primers R6K-F/R6K-R and TnpA-F/TnpA-R (Additional file 4: Table S1) using genomic DNA of strain TS44-gshA$_{540}$ as template. Lane 1 is the amplification of R6K fragment, while line 2 represents the TnpA fragment. (**C**), PCR used primers gshA-F/gshA-F (Additional file 4: Table S1). Lane 1, strain TS44, lane 2, *gshA* gene insertional inactivation strain TS44-gshA$_{540}$ and lane 3, the complemented strain TS44-gshA540-C. M, the molecular weight marker (DL 2000 plus). Amplicon identities were confirmed by DNA sequencing.

Additional file 3: Figure S2. In vitro oxidation of Sb(III) by H_2O_2. (**A**) Sb(III) was added into uninoculated liquid CDM medium to a final concentration of 0.2 mmol/L. (**B**) CDM containing 0.2 mmol/L of Sb(III) and with the addition of 0, 0.02, 0.05, 0.1, 0.3, 0.4 and 0.5 mmol/L of H_2O_2, respectively.

Additional file 4: Figure S3. The tolerance of *P. stutzeri* strains TS44, TS44-gshA$_{540}$ and TS44-gshA-C to H_2O_2. The strains were grown in liquid CDM medium until reaching an OD$_{600}$ of 1.0 and then serially diluted 10-fold. The 10^0, 10^{-1}, 10^{-3}, and 10^{-5} dilutions were spotted onto agar plates with different H_2O_2 concentrations.

Acknowledgments
Not applicable.

Funding
The present study was supported by the National Natural Science Foundation of China (31170106) to GW. The funders had no role in study design, data collection and analysis, decision to publish, or preparation of the manuscript.

Authors' contributions
DW performed the experiments and wrote the draft of the manuscript, FZ designed and performed the experiments and helped to draft the manuscript. QW and CR revised the draft of the manuscript. PY and JG performed the experiments. GW designed the study and revised the manuscript. All authors read and approved the final manuscript.

Competing interests
The authors declare that they have no competing interests.

Author details
[1]State Key Laboratory of Agricultural Microbiology, College of Life Science and Technology, Huazhong Agricultural University, Wuhan, People's Republic of China. [2]College of Resources and Environment, Fujian Agriculture and Forestry University, Fuzhou, People's Republic of China.

References
1. Ehrlich HL. Newman DK. Geomicrobiology: CRC press; 2008.
2. Goyeneche-Patino DA, Valderrama L, Walker J, Saravia NG. Antimony resistance and trypanothione in experimentally selected and clinical strains of *Leishmania panamensis*. Antimicrob Agents Ch. 2008;52(12):4503–6.
3. Liarte DB, Murta SM. Selection and phenotype characterization of potassium antimony tartrate-resistant populations of four New World *Leishmania* species. Parasitol Res. 2010;107(1):205–12.
4. Callahan MA. Water-related environmental fate of 129 priority pollutants. Office of Water Planning and Standards. Office of Water and Waste Management, US Environmental Protection Agency; Washington. 1979.
5. WHO. Antimony in Drinking-water-Background. Document for Development of WHO Guidelines for Drinking Water Quality. Geneva. 2003.
6. Filella M, Belzile N, Chen Y-W. Antimony in the environment: a review focused on natural waters: I. Occurrence. Earth-Sci Rev. 2002;57(1):125–76.
7. Filella M, Belzile N, Chen Y-W. 2002b. Antimony in the environment: a review focused on natural waters: II. Relevant solution chemistry. Earth-Sci Rev. 2002;59(1):265–85.
8. Wilson SC, Lockwood PV, Ashley PM, Tighe M. The chemistry and behaviour of antimony in the soil environment with comparisons to arsenic: a critical review. Environ Pollut. 2010;158(5):1169–81.
9. Gebel T. Arsenic and antimony: comparative approach on mechanistic toxicology. Chem-bio interact. 1997;107(3):131–44.
10. De Boeck M, Kirsch-Volders M, Lison D. Cobalt and antimony: genotoxicity and carcinogenicity. Mutat Res-fund Mol M. 2003;533(1):135–52.
11. Leuz A-K, Johnson CA. Oxidation of Sb III to Sb V by O$_2$ and H$_2$O$_2$ in aqueous solutions. Geochim Cosmochim Ac. 2005;69(5):1165–72.
12. Smichowski P. Antimony in the environment as a global pollutant: a review on analytical methodologies for its determination in atmospheric aerosols. Talanta. 2008;75(1):2–14.
13. Li J, Wang Q, Zhang SZ, Qin D, Wang GJ. Phylogenetic and genome analyses of antimony-oxidizing bacteria isolated from antimony mined soil. Int Biodet Biodegr. 2013;76:76–80.
14. Shi Z, Cao Z, Qin D, Zhu W, Wang Q, Li M, et al. Correlation models between environmental factors and bacterial resistance to antimony and copper. PLoS One. 2013;8(10), e78533.
15. Lehr CR, Kashyap DR, McDermott TR. New insights into microbial oxidation of antimony and arsenic. Appl Environ Microb. 2007;73(7):2386–9.
16. Lialikova N. *Stibiobacter senarmontii*–a new microorganism oxidizing antimony. Mikrobiologiia. 1974;43(6):941.
17. Hamamura N, Fukushima K, Itai T. Identification of antimony- and arsenic-oxidizing bacteria associated with antimony mine tailing. Microbes Environ/JSME. 2013;28(2):257–63.
18. Nguyen VK, Lee J-U. Antimony-oxidizing bacteria isolated from antimony-contaminated sediment–a phylogenetic study. Geomicrobiol J. 2015;32(1):50–8.
19. Wang Q, Warelow TP, Kang Y-S, Romano C, Osborne TH, Lehr CR, et al. Arsenite oxidase also functions as an antimonite oxidase. Appl Environ Microb. 2015: AEM. 02981–14
20. Li JX, Wang Q, Li MS, Yang BR, Shi MM, Guo W, et al. Proteomics and genetics for identification of a bacterial antimonite oxidase in *Agrobacterium tumefaciens*. Environ Sci Technol. 2015. doi:10.1021/es506318b.
21. Halliwell B, Gutteridge JM. Free radicals in biology and medicine. Clarendon: Oxford University Press; 1999.
22. Bayr H. Reactive oxygen species. Crit Care Med. 2005;33(12):S498–501.
23. Cabiscol E, Tamarit J, Ros J. Oxidative stress in bacteria and protein damage by reactive oxygen species. Int Microbiol. 2010;3(1):3–8.
24. Hayes JD, McLellan Ll. Glutathione and glutathione-dependent enzymes represent a co-ordinately regulated defence against oxidative stress. Free Radical Res. 1999;31(4):273–300.
25. Masip L, Veeravalli K, Georgiou G. The many faces of glutathione in bacteria. Antioxid Redox Sign. 2006;8(5–6):753–62.
26. Janowiak BE, Griffith OW. Glutathione Synthesis in *Streptococcus agalactiae* one protein accounts for γ-glutamylcysteine synthetase and glutathione synthetase activities. J Biol Chem. 2005;280(12):11829–39.
27. Cai L, Rensing C, Li X, Wang G. Novel gene clusters involved in arsenite oxidation and resistance in two arsenite oxidizers: *Achromobacter* sp. SY8 and *Pseudomonas* sp. TS44. Appl Microbiol Biot. 2009;83(4):715–25.
28. Li XY, Gong J, Hu Y, Cai L, Johnstone L, Grass G, et al. Genome sequence of the moderately halotolerant, arsenite-oxidizing bacterium *Pseudomonas stutzeri* TS44. J Bacteriol. 2012;194(16):4473–4.
29. Weeger W, Lievremont D, Perret M, Lagarde F, Hubert J-C, Leroy M, et al. Oxidation of arsenite to arsenate by a bacterium isolated from an aquatic environment. Biometals. 1999;12(2):141–9.
30. Sambrook J, Fritsch E, Maniatis T. Molecular Cloning: A Laboratory Manual. 2nd ed. New York: Cold Spring Harbor Laboratory; 1989.
31. Zheng S, Su J, Wang L, Yao R, Wang D, Deng Y, et al. Selenite reduction by the obligate aerobic bacterium *Comamonas testosteroni* S44 isolated from a metal-contaminated soil. BMC Microbiol. 2014;14(1):204. doi:10.1186/s12866-014-0204-8.
32. Overbeek R, Olson R, Pusch GD, Olsen GJ, Davis JJ, Disz T, et al. The SEED and the Rapid Annotation of microbial genomes using Subsystems Technology RAST. Nucleic Acids Res. 2014;42(D1):D206–14.
33. Altschul SF, Gish W, Miller W, Myers EW, Lipman DJ. Basic local alignment search tool. J Mol Biol. 1990;215:403–10.

34. Chen F, Cao Y, Wei S, Li Y, Li X, Wang Q, Wang G. Regulation of Arsenite Oxidation by the Phosphate Two-Component System PhoBR in *Halomonas* sp. HAL1. Front Microbiol. 2015;6:923.

35. Hanahan D. Studies on transformation of *Escherichia coli* with plasmids. J Mol Biol. 1983;166(4):557–80.

36. Simon R, Priefer U, Pühler A. A broad host range mobilization system for in vivo genetic engineering: transposon mutagenesis in gram negative bacteria. Nat Biotechnol. 1983;1(9):784–91.

37. Larsen RA, Wilson MM, Guss AM, Metcalf WW. Genetic analysis of pigment biosynthesis in *Xanthobacter autotrophicus* Py2 using a new, highly efficient transposon mutagenesis system that is functional in a wide variety of bacteria. Arch Microbiol. 2002;178(3):193–201.

38. Bradford MM. A rapid and sensitive method for the quantitation of microgram quantities of protein utilizing the principle of protein-dye binding. Anal Biochem. 1976;72(1):248–54.

39. White CC, Viernes H, Krejsa CM, Botta D, Kavanagh TJ. Fluorescence-based microtiter plate assay for glutamate–cysteine ligase activity. Anal Biochem. 2003;318(2):175–80.

40. Guo FF, Yang W, Jiang W, Geng S, Peng T, Li JL. Magnetosomes eliminate intracellular reactive oxygen species in *Magnetospirillum gryphiswaldense* MSR-1. Environ Microbiol. 2012;14(7):1722–9.

41. Wang Q, Lei Y, Xu X, Wang G, Chen L-L. Theoretical prediction and experimental verification of protein-coding genes in plant pathogen genome *Agrobacterium tumefaciens* strain C58. PLoS One. 2012;7(9), e43176. doi:10.1371/journal.pone.0043176.

42. Miller JH. Assay of β-galactosidase. In: Experiments in Molecular Genetics. New York: Cold Spring Harbor Laboratory Press; 1972. p. 352–55.

43. Seaver LC, Imlay JA. Alkyl hydroperoxide reductase is the primary scavenger of endogenous hydrogen peroxide in *Escherichia coli*. J Bacteriol. 2001; 183(24):7173–81.

44. Murata K, Kimura A. Cloning of a gene responsible for the biosynthesis of glutathione in *Escherichia coli* B. Appl Environ Microb. 1982;44(6):1444–8.

45. Sun H, Yan SC, Cheng WS. Interaction of antimony tartrate with the tripeptide glutathione. Eur J Biochem. 2000;267(17):5450–7.

46. Li J, Wang Q, Oremland R S, Kulp T R, Rensing C, Wang G. Microbial antimony biogeochemistry-enzymes, regulation and related metabolic pathways. Appl Environ Microb. 2016: AEM. 01375–16.

47. Belzile N, Chen Y-W, Wang Z. Oxidation of antimony III by amorphous iron and manganese oxyhydroxides. Chem Geol. 2001;174(4):379–87.

48. Leuz A-K, Hug SJ, Wehrli B, Johnson CA. Iron-mediated oxidation of antimony III by oxygen and hydrogen peroxide compared to arsenic III oxidation. Environ Sci Technol. 2006;40(8):2565–71.

49. Kong L, Hu X, He M. Mechanisms of Sb III oxidation by pyrite-induced hydroxyl radicals and hydrogen peroxide. Environ Sci Technol. 2015;49:3499–505.

50. Lecureur V, Le Thiec A, Le Meur A, Amiot L, Drenou B, Bernard M, et al. Potassium antimonyl tartrate induces caspase-and reactive oxygen species-dependent apoptosis in lymphoid tumoral cells. Brit J haematol. 2002;119(3):608–15.

51. Imlay JA, Chin SM, Linn S. Toxic DNA damage by hydrogen peroxide through the Fenton reaction in vivo and in vitro. Science. 1988;240(4852):640–2.

52. McLeod JW, Gordon J. Production of hydrogen peroxide by bacteria. Biochem J. 1922;16(4):499.

53. Imlay JA, Fridovich I. Assay of metabolic superoxide production in *Escherichia coli*. J Biol Chem. 1991;266(11):6957–65.

10

Transposon insertion sequencing reveals T4SS as the major genetic trait for conjugation transfer of multi-drug resistance pEIB202 from *Edwardsiella*

Yang Liu[1], Yanan Gao[1], Xiaohong Liu[1,2,3], Qin Liu[1,2,3], Yuanxing Zhang[1,2,3], Qiyao Wang[1,2,3] and Jingfan Xiao[1,2,3*]

Abstract

Background: Conjugation is a major type of horizontal transmission of genes that involves transfer of a plasmid into a recipient using specific conjugation machinery, which results in an extended spectrum of bacterial antibiotics resistance. However, there is inadequate knowledge about the regulator and mechanisms that control the conjugation processes, especially in an aquaculture environment where a cocktail of antibiotics may be present. Here, we investigated these with pEIB202, a typical multi-drug resistant IncP plasmid encoding tetracycline, streptomycin, sulfonamide and chloramphenicol resistance in fish pathogen *Edwardsiella piscicida* strain EIB202.

Results: We used transposon insertion sequencing (TIS) to identify genes that are responsible for conjugation transfer of pEIB202. All ten of the plasmid-borne type IV secretion system (T4SS) genes and a putative lipoprotein p007 were identified to play an important role in pEIB202 horizontal transfer. Antibiotics appear to modulate conjugation frequencies by repressing T4SS gene expression. In addition, we identified *topA* gene, which encodes topoisomerase I, as an inhibitor of pEIB202 transfer. Furthermore, the RNA-seq analysis of the response regulator EsrB encoded on the chromosome also revealed its essential role in facilitating the conjugation by upregulating the T4SS genes.

Conclusions: Collectively, our screens unraveled the genetic basis of the conjugation transfer of pEIB202 and the influence of horizontally acquired EsrB on this process. Our results will improve the understanding of the mechanism of plasmid conjugation processes that facilitate dissemination of antibiotic resistance especially in aquaculture industries.

Keywords: Plasmid conjugation, *Edwardsiella Piscicida*, Tis, T4SS, RNA-seq, EsrB

Background

Antibiotic resistance (ABR) raises global issues regarding to multidrug-resistant and overuse of antibiotics [1]. It has been established that the spread of ABR and virulence is often mediated by mobile genetic elements including insertion sequence, transposons, integrons, bacteriophage, genomic island (such as pathogenicity island), plasmids and combinations of these elements [2–4]. Plasmids and other mobile DNA elements can be

more easily transferred than chromosomal DNA between genera, phyla and even major domains by a mechanism known as conjugation [5–7]. Direct evidence has established that tetracycline resistance-encoding plasmids (R plasmid) disseminate between different *Aeromonas* species and *Escherichia coli* in various environments [8]. Horizontal gene transfer (HGT) is also important for genome plasticity. It has been reported that approximately 3/4 of genes in the bacterial genomes have been acquired by HGT [9], which is essential for bacteria to adapt to various growth niches. Though there are several reports about the genetic screen for bacterial plasmid conjugation, the knowledge about the detailed

* Correspondence: jfxiao@ecust.edu.cn
[1]State Key Laboratory of Bioreactor Engineering, East China University of Science and Technology, Shanghai 200237, China
[2]Shanghai Engineering Research Center of Maricultured Animal Vaccines, Shanghai, China
Full list of author information is available at the end of the article

processes and genetic basis that control the transfer of the plasmid to another host cell is still lacking.

Bacteria utilize secretion systems to transport numerous substrates across cellular membranes, mediating their virulence and survival. The type IV secretion system T4SS is unique and present in many pathogens to mediate both genetic exchange and the delivery of effector proteins to target eukaryotic cells [10]. T4SS in *Agrobacterium tumefaciens* delivers oncogenic nucleoprotein particles into plant cells, resulting in the development of crown-gall tumors [11]. *A. tumefaciens* T4SS is composed of 12 proteins, VirB1 ~ 11 and VirD4, while VirB proteins can be grouped into three classes: the putative channel components (VirB6 – VirB10); the energy components (the nucleoside triphosphatases VirB4 and VirB11); and the pilus-associated components (VirB2, and possibly VirB3 and VirB5) [12]. The machines assembled from VirB homologs are proposed as type IVA (T4AS) and are widespread to mediate the conjugative transfer of plasmids and thus promote dissemination of multiple-antibiotic resistance [13].

Edwardsiella piscicida is a bacterial pathogen causing edwardsiellosis in over 20 piscine species such as flatfish, eel and tilapia, resulting in huge economic losses in worldwide aquaculture industries [14]. *E. piscicida* is a facultative intracellular pathogen and develops the ability to resist killing by professional phagocytes and colonize and replicate in macrophages [15–17]. In our previous study, the genome of a typical highly virulent *E. piscicida* strain EIB202 was published [18]. The horizontally acquired two-component system EsrA-EsrB has been established to be essential for its pathogenesis mediated by type III and VI secretion systems (T3SS and T6SS) [18–20]. A plasmid (pEIB202) of 43,703 bp was identified from the assembled sequence [18]. Six genes in the sequence of pEIB202 were identified to be involved with ABR, providing genetic properties for multi-drug resistance in EIB202, including *tetA* and *tetR* for tetracycline, *strA* and *strB* for streptomycin, *sulII* for sulfonamide, and *catA3* for chloramphenicol resistance. Genetic analysis suggested that the chloramphenicol resistance might be recently acquired by the plasmid [18]. In addition, this plasmid encoding an incomplete set of T4AS proteins (VirB2, -B4, -B5, -B6, -B8, -B9, -B10, -B11, -D2, and -D4) [18]. It is unknown whether these T4AS proteins involved in the acquisition of the plasmid by the bacterium.

In this study, we identified that the self-transmissible pEIB202 is not involved in virulence and colonization of *E. piscicida* at least in the zebrafish model we used in this study. We used transposon insertion sequencing (TIS) technology to investigate genetic basis of pEIB202 on conjugation transfer of pEIB202, and identified all of the T4SS related proteins on pEIB202 and putative lipoproteins as horizontal transfer enhancer, and the TopA as the related inhibitor. Intriguingly, response regulator EsrB encoded in the chromosome was found to facilitate the conjugation through inducing the expression of genes associated with antibiotics resistance and T4SS. Our data unraveled the genetic basis of the conjugation transfer of pEIB202 and the influence of horizontally acquired EsrB on the plasmid transfer efficiency.

Methods

Bacterial strains, plasmids and culture conditions. Bacterial strains and plasmids used in this work were described in Table 1, respectively. *E. piscicida* strains were grown at 30 °C in tryptic soy broth (TSB) or Luria broth (LB). *E. coli* strains were cultured in LB at 37 °C. DH5α *λpir* were used for plasmid harvest, and SM10 *λpir* was used plasmid conjugation. Antibiotics were added to the following final concentrations: gentamicin (Gm, 25 μg/ml), polymyxin B (Col, 10 μg/ml), kanamycin (Km, 50 μg/ml), and carbenicillin (Carb, 100 μg/ml), chloramphenicol (Cm, 34 μg/ml), tetracycline (Tet, 12.5 μg/ml) and streptomycin (Str, 100 μg/ml). For the tests of the effect of antibiotics on conjugation frequency, the following sublethal concentrations (1/4 of minimum inhibition concentration (MIC)) of antibiotics were used: Tet, 0.312 μg/ml; Cm, 0.437 μg/ml; Str: 0.575 μg/ml.

Quantitative real-time PCR (qRT-PCR). RNA samples were extracted by using the RNA isolation kit (Tiangen). RNase-free DNase I (Promega) was used to remove DNA contamination in the RNA sample. One microgram RNA was used as a template for first strand cDNA synthesis with the PrimeScript reverse transcriptase (TaKaRa). qRT-PCR analysis was performed in a total volume of 20 μl, containing 1 μl of diluted cDNA, 1 μl of each primer (10 mM stock) (Table 2), 10 μl of FastStart Universal SYBR Green Master (Roche) and 7 μl deionized water. qRT-PCR was performed with the 7500 RealTime PCR System (Applied Biosystems) under the following conditions: 95 °C for 10 min; 40 cycles at 95 °C for 15 s, and 60 °C for 1 min. Melting curve analysis of amplification products was performed at the end of each PCR to confirm that only one PCR product had been amplified. Relative quantification was performed using the comparative CT ($2^{-\Delta\Delta CT}$) method [21], with the housekeeping gene *gyrB* as an internal control.

Bacterial conjugation and transfer efficiency. Equal amounts (1 ml) of secondary inocula of recipient and donor cells were washed twice with fresh LB to remove residual antibiotics, then mixed and resuspended in 400 μl LB. 35 μl of mixture was dropped on 0.45 μm millipore filter membranes (Sartorius) placed on LB agar plate and incubated at 30 °C for 8 h. For tests of the influence of antibiotics on conjugation, the filter

Table 1 Strains and plasmids used in this study

Strain or plasmid	Characteristics[a]	Reference
E. piscicida		
EIB202	Wild type (CCTC No. M 208068) carrying pEIB202, Col[r], Cm[r], Str[r]	[33]
EIB202-*lacZ*[+]	EIB202, insert a *lacZ* in *glmS*, Col[r], Str[r]	Lab collection
EIB202 ΔP	EIB202, pEIB202 cured, Col[r]	[41]
Δ*esrB*	EIB202, in-frame deletion of *esrB*, Col[r], Cm[r], Str[r]	[20]
Δ*topA*	EIB202, in-frame deletion of *topA*, Col[r], Cm[r], Str[r]	This study
Δ*p007*	EIB202, in-frame deletion of *p007*, Col[r], Cm[r], Str[r]	This study
Δ*virB9*	EIB202, in-frame deletion of *virB9*, Col[r], Cm[r], Str[r]	This study
esrB[+]	EIB202, Δ*esrB*, containing pAK*gfp1*::*flag-esrB*, Col[r], Carb[r]	Lab collection
topA[+]	EIB202, Δ*topA*, containing pAK*gfp1*::*topA*, Col[r], Carb[r]	This study
p007[+]	EIB202, Δ*p007*, containing pAK*gfp1*::*p007*, Col[r], Carb[r]	This study
virB9[+]	EIB202, Δ*virB9*, containing pAK*gfp1*::*virB9*, Col[r], Carb[r]	This study
ΔP::pNQ705K	EIB202 ΔP with pNQ705K inserted in downstream of *glms*, Col[r], Km[r]	This study
E. coli		
DH5α λpir	Δ (*lacZYA-argF*) U169 (Φ80 *LacZ* ΔM15), *pir* dependent *R6K*.	Lab collection
SM10 λpir	*thi thr leu tonA lacY supE recA::RP4–2-Tc::Mu*, *pir* dependent *R6K*, Km[r]	Lab collection
Vibrio harveyi		
BB170	*Vibrio harveyi*, reporter strain, Carb[r]	Lab collection
Vibrio alginolyticus		
ΔT6SS	*Vibrio alginolyticus*, T6SS genes were deleted, Carb[r]	Lab collection
Plasmid		
pMPR	Mariner transposon plasmid, *pir* dependent R6K, Gm[r], Carb[r]	Lab collection
pAKgfp1	pBBRMCS4 with *gfpmut3a*	Lab collection
pNQ705	Suicide plasmid, *pir* dependent R6K, Cm[r]	[53]
pNQ705K	pNQ705 derivative with Km fragment inserted in *SalI* site, Cm[r], Km[r]	This study
pNQ705K-*glmS*	pNQ705K derivative with *glmS* fragment inserted in *XbaI* site, Cm[r], Km[r]	This study
pDM4	Suicide plasmid, *pir* dependent, R6K, *sacBR*, Cm[r]	[54]
pDMK	pDM4 derivative with Km fragment inserted in *SalI* site, Km[r], Cm[r]	[25]
pD43B	pDMK derivative with *virB10* and *gyrB* fragment inserted in *XbaI* site, Km[r], Cm[r]	This study
p34S-Km	Cloning vector, Km[r]	[55]

[a]Col[r], polymyxin B resistance; Cm[r], chloroamphenicol resistance; Km[r], kanamycin resistance; Carb[r], carbenicillin resistance; Gm[r], gentamycin resistance

membranes were put on LB agar plates containing sublethal concentrations of indicated antibiotics (1/4 of MIC). Bacteria washed out from the membranes were serially diluted with PBS and plated to determine the bacterial loads. The conjugants were differentiated from other strains based on their antibiotics resistance. The ratios of conjugants counts to donor strains were used to determine the transfer efficiency.

Construction of ΔP::pNQ705K. The kanamycin open reading frame fragment was PCR-amplified from p34S-Km plasmid using the primer pairs KmCS-F (*SalI*)/R (*SalI*) (Table 2), digested with *SalI* and ligated with linearized pNQ705. The resulting plasmid pNQ705K was digested with *XbaI*. The *glms* fragment was amplified from EIB202 genomic DNA using the primer pairs

Glms-F/R and cloned into linearized pNQ705K by Gibson assembly [22]. The resulting plasmid pNQ705Km-*glms* were transformed into EIB202 ΔP by conjugation and the targeted EIB202 ΔP::pNQ705K was validated using the primer pair Glms-yz-F/KmCS-R (*SalI*).

Transposon insertion sequencing (TIS) and data analysis. We followed the protocol described previously to create transposon insertion libraries and perform TIS [23]. In brief, overnight cultures of SM10 λpir/pMPR (the donor of transposon, Tn) and EIB202/pEIB202 were mixed and incubated for 8 h. Exconjugant cells with the Gm[r] Tn inserted into the chromosome or plasmid of EIB202 carrying pEIB202 were recovered on a total of 30 plates as Col[r], Gm[r], and Str[r] colonies (~6000 cfu/plate) that were

Table 2 Primers used in this study

Oligonucleotides	Sequence
Glms-F	CCCCCCCGAGCTCAGGTTACCCGGATCTATGGAAATCGGCGTAGCGTCAACCAAG
Glms-R	CCCTCGAGTACGCGTCACTAGTGGGGCCCTTCGCGCTTTTATTCTACGGTAACCG
Glms-yz-F	AGAGATTGGCTACTTGGGATCGTTG
KmCS-F (SalI)	ACGCGTCGACATTGTGAGCGGATAACAATTTGTGG
KmCS-R (SalI)	ACGCGTCGACTAGATCCGGGTAACCTGAGCT
evpP-F	TCATCGCACATACAGAATAAACGCC
evpP-R	CCGTAACATTTCTTACAACACTGCG
virB9-P1	CCCCCCCGAGCTCAGGTTACCCGGATCTATGGCCACTGGTTTGTTGTAGGGCCAT
virB9-P2	CAAATTGCCATGGGCTGATGGCTGAGAACAGAGACGATCTGGA
virB9-P3	TGTTCTCAGCCATCAGCCCATGGCAATTTGTTGGACCGTT
virB9-P4	GAGTACGCGTCACTAGTGGGGCCCTTCTAGCGCGAGGGCTATCAGTGGGAAACCC
virB9-out-F	GCGCCCAGGCCGTCCGCTCGTTCAG
virB9-out-R	CCGCGTCGATAACAACACTGGCGTG
virB9-in-F	TTGACTCCCTCTAATTACTCGCTCA
virB9-in-R	GTCGGAAATCACATTTTCATCAAGC
p007-P1	CCCCCCCGAGCTCAGGTTACCCGGATCTATTGGCGCGGGTCGGTATATGCGGCAT
p007-P2	AGGAGAGTTCGAGTGGCTTGGTCATCCGAGGAATGGAGGC
p007-P3	CTCGGATGACCAAGCCACTCGAACTCTCCTTGATCAGTGT
p007-P4	GAGTACGCGTCACTAGTGGGGCCCTTCTAGTTTTTGAAAGCTGGCTAGGCATGGT
p007-out-F	CTGCGCTCCCCTGCCCTTTTCACCT
p007-out-R	TGCGCTTTCTCTCGTTGTGGCGTTC
topA-P1	CCCCCCCGAGCTCAGGTTACCCGGATCTATGGTCCAGGACCCAATCCACCCCTTC
topA-P2	TAGGGAGGACCAATGTGAGCTTGGCGGCAATCAAAGTTGT
topA-P3	TTGCCGCCAAGCTCACATTGGTCCTCCCTACCGTCAACCA
topA-P4	GAGTACGCGTCACTAGTGGGGCCCTTCTAGACCTCCAGTCGGTCGAGTTGAGCAA
topA-out-F	GGCCGCGATAATCAGGTCAACGATG
topA-out-R	CACCTGTGCGGCCCTGTCCGGGGCT
topA-in-F	ACCAAAAATCGTAACCCTTCTTGCC
topA-in-R	AGGGTGGGACGTGCAGGCCAGTGTC
pDMK-PF	AAAGCTCTCATCAACCGTGGC
pDMK-PR	TGCTCCAGTGGCTTCTGTTTC
topA-C-F	GATCCTCTAGATTTAAGAAGGAGATATACAATGAATCTAGTTATTGTTGA
topA-C-R	GATCCCCCGGGCTGCAGGAATTCGATATCATCAAATCTTTGGCTTGCCAC

Table 2 Primers used in this study (Continued)

p007-C-F	GATCCTCTAGATTTAAGAAGGAGATATACAGTGAACCTGAAAACACTAAG
p007-C-R	GATCCCCCGGGCTGCAGGAATTCGATATCATCAGTGATGGCCTCCATTCC
virB9-C-F	GATCCTCTAGATTTAAGAAGGAGATATACAATGATACGAGCAAAATCACT
virB9-C-R	GATCCCCCGGGCTGCAGGAATTCGATATCATCATTGACTCCCTCTAATTA
virB9-qRT-F	TCATGTTCGTGGTCGCATCA
virB9-qRT-R	CCATTTTGGCTTCTCCACGC
virB4-qRT-F	TGGGGGCCGTTTTGAGATAC
virB4-qRT-R	GCGGCAGCTTCAATAACCAG
avtA-qRT-F	TTACTCCGCAATCACCCGTC
avtA-qRT-R	CAGCTCACCGCATAGGGAG
P021-qRT-F	AAAGCCCCAAACCGTAAAGC
P021-qRT-R	ATGCGGGAATGGGTCAGTTT
gyrB-qRT-F	CCGATGATGGTACGGGTCTG
gyrB-qRT-R	GCTTTTCAGACAGGGCGTTC
p003-qRT-F	GCCGAAGCGTTCCCAAAAAT
p003-qRT-R	CCTGTGGAATCGCATCGAGA
pD43B-p003-F	GAGCTCAGGTTACCCGCATGCAAGATCTATATGATGGCTGAGAACAGAGACGATC
pD43B-p003-R	ATACGTATTTGACATTCAATCGGCTTTGAGGTCATATACC
pD43B-gyrB-F	CTCAAAGCCGATTGAATGTCAAATACGTATGACTCCTCAA
pD43B-gyrB-R	CCCTCGAGTACGCGTCACTAGTGGGGCCCTTTAAAAGTCCAGATTGGACGCTTTA

resuspended and collected as the input library. Total plasmid DNA was extracted from a 5 ml aliquot (~1/5) of the input library. Another aliquot of the input library (with the OD_{600} adjusted to match that of an overnight culture recipient) was used as the donor in a second round of conjugation with the new recipient EIB202 Δp::pNQ705K (Km[r]). Exconjugants were selected as Km[r], Gm[r], and Str[r] colonies; this scheme selects for Tn insertion in pEIB202 (not the chromosome) that are still capable of transmission. About 8×10^5 Km[r], Gm[r], and Str[r] colonies were collected and designated as the output library (n = 3). For each library, plasmid DNA (pDNA) from 5 ml bacteria was extracted via the TIANprep Mini Plasmid kit (Tiangen), diluted in 100 μl MiniQ water to a final concentration of 50 ng/μl in the 0.5 ml tube and sonicated by Bioruptor machine (Diagenode) with 30 s ON/90 s OFF for 12 cycles. The sheared DNA length was in the range of 200–600 bp as shown by electrophoresis in a 2% agarose gel. DNA libraries were constructed using the VAHTS Turbo DNA library preparation kit (Vazyme) and sequenced on an Illumina MiSeq platform (Illumina). The sequencing reads were mapped to *E. piscicida* pEIB202 by Bowtie [24]. After mapping, the reads per TA site were tallied and assigned to annotated genes or intergenic regions using previously described scripts [23].

Construction of deletion mutant. In-frame deletion mutants were generated by the *sacB*-based allelic exchange as previously described [25]. Upstream and downstream fragments were generated by PCR with primer pairs P1/P2 and P3/P4 listed in Table 2, respectively. The resulting fragments were cloned into *sacB* suicide vector pDMK using Gibson Assembly and transformed into DH5α *λpir*, clones were validated with the primer pair pDMK-PF/PR. After sequencing, the resulting plasmids were transformed into SM10 *λpir*, and then mated into EIB202 by conjugation. Double crossover processes were selected sequentially on TSA medium containing Col and Km and then on TSA with 12% (*w/v*) sucrose to complete homologous recombination. The targeted mutants were confirmed by PCR using the primer pairs in-F/R, out-F/R and *evpP*-F/R, and sequencing of the deleted region.

Determination of minimum inhibitory concentration (MIC). Inocula of WT and Δ*esrB* strains were seeded into LB culture with different amount of antibiotics in 96-well plates and incubated at 30 °C for 24 h.

Determination of pEIB202 copy number. As previous described [26], genomic DNA was extracted from log phase cultures with the gDNA purification kit (Tiangen) and subjected to qPCR in 7500 RealTime PCR System (Applied Biosystems) with FastStart Universal SYBR Green Master (Roche). *virB10* and *gyrB* loci were used for plasmid and chromosome quantification, respectively. A standard plasmid (pD43B) that contains chromosomal (*gyrB*) and plasmid DNA (*virB10*) fragments for qPCR analyses for measurement of pEIB202 copy number was constructed by first amplifying *virB10* and *gyrB* target regions and then inserting them into pDMK by Gibson Assembly. The primers used for these experiments are listed in Table 2.

Total RNA extraction and mRNA enrichment. The WT strain and the Δ*esrB* mutant were cultured in LB and DMEM, respectively, at 30 °C without shaking for 24 h. RNA samples were extracted using an RNA isolation kit (Tiangen). DNase I (Promega) and Ribo-Zero-rRNA removal kits for Gram-negative bacteria (Epicentre) were used to remove DNA and rRNA following the manufacturer's instructions. Prior to reverse transcription, regular PCR was routinely performed using the isolated RNA sample as a template to confirm that there was no DNA contamination. Samples used for RNA-seq were validated using an Agilent 2100 Bioanalyzer (Agilent Technologies), and the final concentration was measured using a Qubit 2.0 Fluorometer (Thermo Fisher).

RNA-seq transcriptome generation and data analysis. First-strand cDNA synthesis from rRNA-depleted samples was carried out using TruSeq RNA sample Prep (Illumina). The cDNA was purified using the RNA Clean and Concentrator-25 kit (Zymo Research). Following second-strand synthesis, the reactions were cleaned up with AMPure XP beads followed by end repair, adenylation of 3′ ends and ligation of adapters. The reaction products were cleaned with AMPure XP beads and treated with uracil-N-glycosylase using the AmpErase kit (Applied Biosystems). Finally, PCR (10 cycles) was used to amplify the library and to enrich the fragments that were ligated to the sequencing adapters. Libraries were sequenced on the HiSeq 2000 platform to yield 100-base-pair end-reads. Adapter sequences and low-quality bases (PHRED quality scores ≤5) were trimmed by the Trimmomatic package using the default parameters, and truncated reads smaller than 35 bp were discarded [27]. Then, the BWA program [28] was used to align the remaining reads to the reference sequences of *E. piscicida* EIB202 (Chr, CP001135.1; plasmid, CP001136.1). The number of reads mapped to each gene was determined by Picard tools (http://broadinstitute.github.io/picard/faq.html) and normalized to the reads per kilobase of genic region per million mapped reads (RPKM) to obtain the relative level of expression. An analysis of variance was performed on the average expression of the three biological replicates to identify genes that showed differential expression under the tested condition (adjusted $P < 0.05$ and 1 fold change). The differential expression analysis was performed using the DEGseq package [29]. The fold changes of genes of interest were validated using qRT-PCR. The RNA-seq datasets have been deposited in the NCBI GenBank under accession number SRP077869.

Zebrafish maintance, LD_{50} and competitive indices analysis. The zebrafish maintenance and challenge were performed as previously described [20]. Healthy zebrafish (*Dario rerio*) of 0.25 ± 0.05 g were raised in reverse osmosis purified water in a flow-through auto-controlled system at 25 °C and acclimatized at least 10 days. The strains of WT, ΔP, and Δ*esrB* were harvested and serially diluted with PBS, respectively. Zebrafish were anesthetized with tricaine methanesulfonate (MS-222, Sigma) at a concentration of 80 mg/L and intramuscularly (i.m.) injected with the strains at doses ranging from 5×10 to 5×10^5 CFU/fish. Thirty fish were injected with each dilution and divided randomly into three groups. Fish injected with PBS served as negative controls. The mortalities were recorded over a period of 7 days after infection. The LD_{50} values were calculated via the method described by Fernández et al. [30]. Competitive indices were determined between each strain. The indicated strains were equally mixed and serially diluted with PBS into

1×10^3 CFU/fish. At 72 h post injection, 7 fish were homogenized respectively and plated to determine the bacterial loads. WT (*lacZ*) was differentiated from WT using LBA containing X-Gal (blue and white colonies), and ΔP was differentiated from WT based on Str resistance.

Results

pEIB202 is not required for virulence and colonization of *E. piscicida*. Previous whole genome sequencing indicated that *E. piscicida* EIB202 (previously *E. tarda* EIB202) harbors a multi-drug resistance plasmid pEIB202 [18]. To characterize the roles of pEIB202 on virulence of *E. piscicida*, healthy zebrafish were intramuscularly (i.m.) infected with serial dilutions of bacterial suspensions of WT, ΔP, and $\Delta esrB$, respectively. The results indicated that LD_{50} value of EIB202 was 1.5×10^2 CFU/fish during 7 days of observation (Table 3). $\Delta esrB$ showed significant virulence attenuation with an $LD_{50} \sim 1000$-fold higher than that of WT, as previously reported [19, 20]. The LD_{50} value of ΔP showed no obvious difference with that of WT. We further analyzed the effect of pEIB202 curing on the colonization of *E. piscicida*. The WT carrying *lacZ*, WT (*lacZ*), was equally mixed with WT and inoculated into the turbot fish. Similarly, ΔP was also used to compete against WT in vivo in fish. After 72 h competition in zebrafish, WT (*lacZ*), WT, and ΔP showed similar colonization capacity (Fig. 1). Collectively, these results indicated that pEIB202 appears to be not required for virulence and in vivo colonization of *E. piscicida*.

pEIB202 is self-transmissible under various conditions. Our previous investigation indicated that pEIB202 can be transferred in LB medium [18]. To further analyze the transfer capacity of the plasmid, we generated the strain ΔP::pNQ705K, a pEIB202-cured strain which carries Km resistance gene inserted in a neutral site of the chromosome (Table 1). The conjugation of pEIB202 (carrying Cm resistance) from WT (donor, Str^r/Col^r) to ΔP::pNQ705K (recipient, Km^r/Col^r) was performed under various conditions and then the conjugants thus could be differentiated from donor and recipient by Km and Str resistance. The transfer efficiency of pEIB202 between WT and the pEIB202-cured strain ΔP::pNQ705K is ~0.1 conjugant/donor after 4 h of conjugation (Fig. 2a). The conjugation efficiency significantly increased to 0.3 conjugant/donor at the time of 8 h incubation (Fig. 2a).

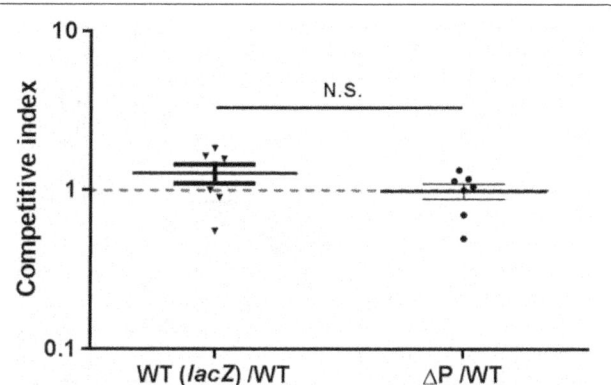

Fig. 1 Competitive indices of WT (lacZ)/WT and ΔP/WT in zebrafish at 72 h post-infection. The pEIB202-cured strain (ΔP) (Col^r) was differentiated from WT strains (Str^r/Col^r) based on Str and Col resistance. WT (*lacZ*) was differed from WT by blue or white colonies on LB plates supplemented with X-gal and Col. Results were presented as mean ± SD (*n* = 7). N.S., not significant, based on ANOVA followed by Bonferroni's multiple-comparison posttest

The plasmid pEIB202 can transfer through conjugation at the temperature ranging from 16 °C to 37 °C, though 30 °C is the optimum temperature for conjugation (Fig. 2b). We also investigated the conjugation condition with different antibiotics at sublethal concentrations (1/4 of MIC). The plasmid pEIB202 encodes genes against four antibiotics namely chloramphenicol, streptomycin, tetracycline, and sulfonamide [18]. The subinhibitory antibiotic concentrations didn't cause apparently reduced viability of recipient cells (data not shown). Notably, we found that sublethal of Str, Tet, and Cm significantly inhibited the pEIB202 conjugation transfer, as compared to the standard conjugation operation in laboratory without antibiotics (Fig. 2c). The investigation of conjugation in various media indicated that pEIB202 could transfer in 2216E, M9, and LB with similar efficiency, LBS significantly promotes the conjugation transfer, suggesting that 3% sodium is essential for conjugation processes (Fig. 2d). We also found that pEIB202 could be transferred from EIB202 to enterobacteria (*Escherichia coli*), and also other pathogenic bacteria (*Vibrio alginolyticus* and *V. harveyi*) in marine environments. Otherwise, pEIB202 is also able to transfer back to *E. piscicida* (Fig. 2e). Altogether, these data demonstrated that pEIB202 is able to transfer under various conditions.

Transposon insertion sequencing (TIS) reveals T4SS genes essential for pEIB202 transfer. We used TIS to screen genes in pEIB202 required for conjugation transfer. Suicide plasmid pMPR (Table 1) carrying Himar I mariner transposon was used for high-density mutagenesis of pEIB202. Our rationale is that if a plasmid gene made a positive contribution to (stimulated) conjugation transfer of pEIB202 to the recipient ΔP::pNQ705K, then

Table 3 Virulence towards zebrafish (LD_{50}) and MIC of each strain

Strains	LD_{50} (CFU/fish)	MIC (µg/ml)			
		Cm	Str	Col	Tet
WT	1.5×10^2	170	>6000	>600	250
ΔP	2.0×10^2	1.75	2.3	>600	1.25
$\Delta esrB$	1.4×10^5	136	4000	>600	190

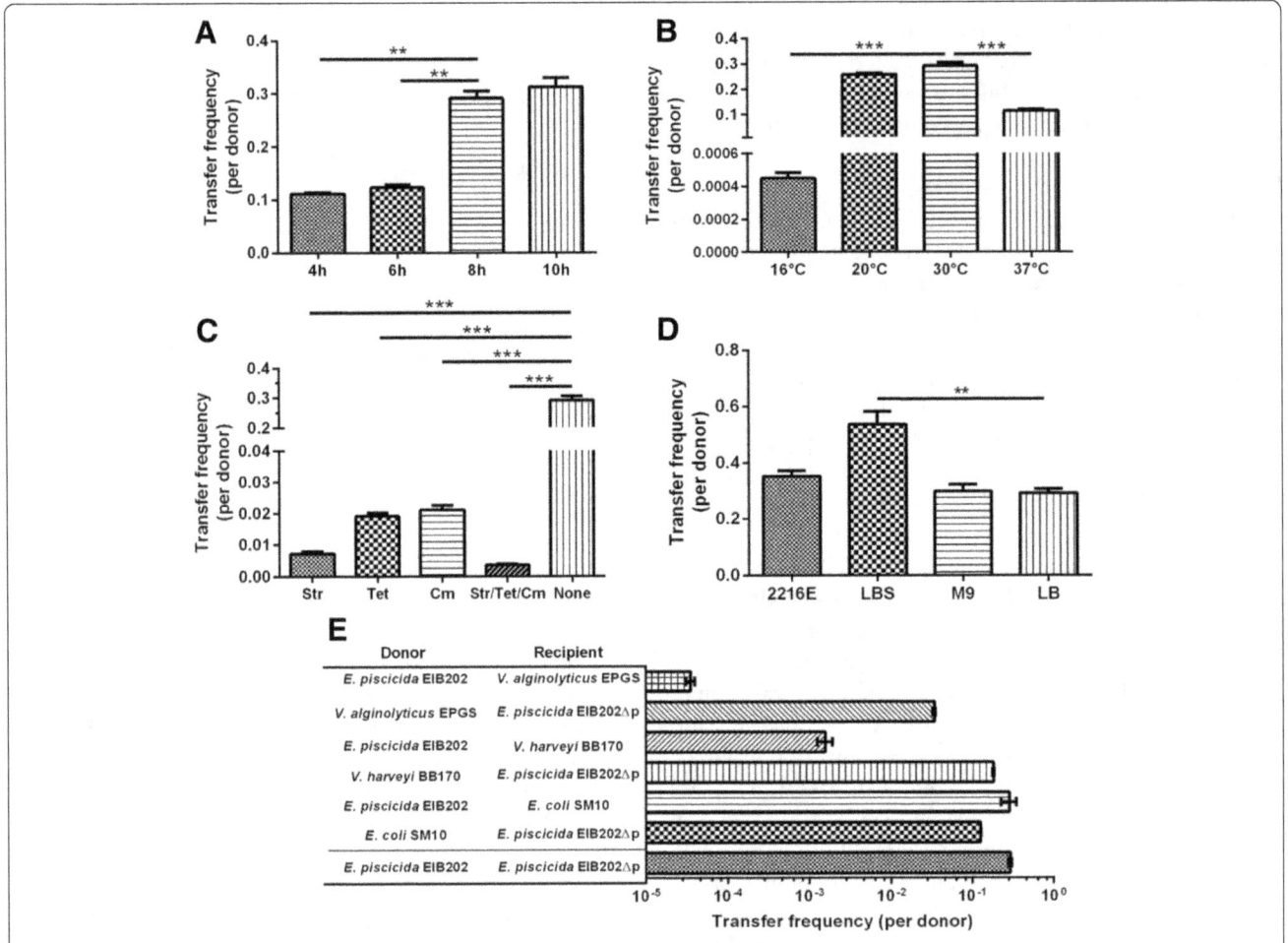

Fig. 2 Transfer frequency of pEIB202 under different conditions. The WT strain (donor, Str^r/Col^r) carrying pEIB202 plasmid was conjugated with the EIB202 plasmid-cured strain with Km resistant gene inserted in a neutral site of the chromosome (EIB202 ΔP::pNQ705K) (recipient, Km^r/Col^r) and the conjugants were selected on Str/Km/Col LB agar plates. **a** Strains were conjugated for different time under 30 °C. **b** Strains were conjugated under different temperatures for 8 h. **c** and **d** Strains were conjugated at different antibiotics at their sublethal concentrations (1/4 of MIC) and culture conditions under 30 °C for 8 h. **e** The transfer frequency between *E. piscicida* and other bacterial species. Results were presented as mean ± SD (*n* = 3). * *P* < 0.01, ** *P* < 0.001, ***P* < 0.0001, student's *t*-test

insertion mutations in that gene would be under-presented in a library after selection for tranconjugants. Conversely, if a gene made a negative contribution to (inhibited) conjugation, then mutations in that gene would be over-presented. We set out to construct a highly saturated Tn-insertion library in WT harboring pEIB202. The set of pEIB202 insertions were subsequently referred to as the input library and used to donate mutated pEIB202 to another recipient ΔP::pNQ705K and the resistant markers on the plasmid and transposon were used for selection of exconjugants. The set of pEIB202 insertion mutants in the ΔP::pNQ705K exconjugants is referred to below as the output library. We compared the distribution of transposon insertion sites in the input and output libraries and identify pEIB202 genes affecting plasmid transfer with the

previously described analysis pipeline [23, 31]. Insertions in genes in pEIB202 that are underrepresented or absent in the output library (relative to their abundance in the input library) likely correspond to loci that facilitate pEIB202 transfer, and over-represented in the output library likely negatively affect the plasmid transfer.

The input library included approximately 100,000 reads that mapped to pEIB202 and covered 88.1% of the plasmid TA sites, suggesting that the library is sufficiently saturated to identify genes essential for pEIB202 transfer. In the input library, the Tn insertions were fairly evenly distributed around pEIB202. The three output libraries contained 109,678 reads (covered 62.2% of TA), 110,322 reads (covered 62.4% of TA) and, 97,595 reads (62.1% of TA), respectively. Using the HMM-based analytic pipeline [23], we could profile the gene loci that possibly affect the replication, maintenance, conjugation transfer, or selection processes of pEIB202 in *E. piscicida*. Some genes, including the *tra*, *rep*, and antibiotics resistance genes, showed significantly increased or decreased output reads but with lower output/input ratios (< 4-fold change), suggesting that they might contribute to the maintenance and replication or the selection process after conjugation (data not shown). Their exact roles could not be assigned for the above-mentioned complex processes and for the possibility of redundancy for each of the gene from the chromosome or plasmid encoded genes. We identified 20 genes with output/input ratios higher than 4-fold which might significantly affect the conjugation transfer of pEIB202 (Table 4). Notably, all of the ten T4SS VirB proteins were identified as genes essential for pEIB202 transfer with a significantly ($P < 0.001$) decreased (by 2–3 log) output. The gene *topA*, encoding a DNA topoisomerase I, was the only candidate that inhibiting pEIB202 transfer (Table 4) (Fig. 3a). Insertions in the TA loci in this gene caused 4.30-fold increase ($P < 0.001$) in the output reads. All the regions related to the above-described 20 genes shared averaged input reads, suggesting that these genes were not associated with the maintenance or replication, but conjugation transfer in *E. piscicida*.

In order to further verify the TIS identified genes as the regulators or modulators controlling pEIB202 transfer, we constructed in-frame deletion mutants and corresponding complement strains for *topA*, *p007*, and *virB9* and determined the transfer efficiency of the mutated pEIB202. The mutant with a deletion in *virB9*, encoding the channel component protein essential for T4SS function [32], showed significantly reduced transfer frequency (Fig. 3b), suggesting that, as expected, the proper function of T4SS is required for pEIB202 horizontal transfer. Putative lipoprotein p007 was also proved to be important for pEIB202 transfer (Fig. 3b). The plasmid without gene *topA* has a significantly higher transfer frequency in comparison to that of the WT (Fig. 3b). The transfer frequencies of their complement strains were recovered to the WT level (Fig. 3b).

It was intriguingly to find that *p007* and *topA* are involved in the conjugation transfer. We further asked whether deletion of these genes in pEIB202 would affect the expression of *virB* genes, which subsequently modulates the T4SS function and activities to influence plasmid conjugation processes. We carried out qRT-PCR to test the expression of *virB9* and *virB4*, encoding two core proteins for T4SS function, in WT, $\Delta p007$ and $\Delta topA$ strains as well as their complement strains. The

Table 4 Genes identified by TIS in pEIB202 that affects maintenance or conjugation of pEIB202

Loci	Gene	Product	Output/input [a]	*P*-value
ETAE_p001	*virD4*	T4SS component VirD4	0.015	0.000548801
ETAE_p002	*virB11*	T4SS component VirB11	0.0058	0.000531746
ETAE_p003	*virB10*	T4SS component VirB10	0.00099	0.000531259
ETAE_p004		hypothetical protein	0.0022	0.000517762
ETAE_p005	*virB9*	T4SS component VirB9	0.0039	0.000520783
ETAE_p006	*virB8*	T4SS component VirB8	0.0014	0.000530141
ETAE_p007		putative lipoprotein	0.0034	0.000493853
ETAE_p008		hypothetical protein	0.10	0.0003746
ETAE_p009	*virB6*	T4SS component VirB6	0.011	0.000544597
ETAE_p010		putative lipoprotein	0.18	1.61153E-05
ETAE_p011	*virB5*	T4SS component VirB5	0.0027	0.000535705
ETAE_p012		hypothetical protein	0.026	0.0003982
ETAE_p013	*virB4*	T4SS component VirB4	0.0086	0.000541165
ETAE_p014	*virB2*	T4SS component VirB2	0.019	0.000514194
ETAE_p015		hypothetical protein	0.073	9.96994E-06
ETAE_p016	*topA*	DNA topoisomerase	4.30	0.000340276
ETAE_p026		transcriptional repressor protein	0.020	6.77947E-06
ETAE_p047	*mobC*	putative mobilisation protein	0.13	0.000639706
ETAE_p048	*virD2*	VirD2 component	0.028	0.000545018
ETAE_p053	*traC*	DNA primase TraC4	0.15	2.38855E-05

[a]The data indicate that the read number of a specific gene in output is significantly under-represented or over-represented as compared to that in input with a cutoff of less than 0.25 or higher than 4-fold and *p* < 0.001

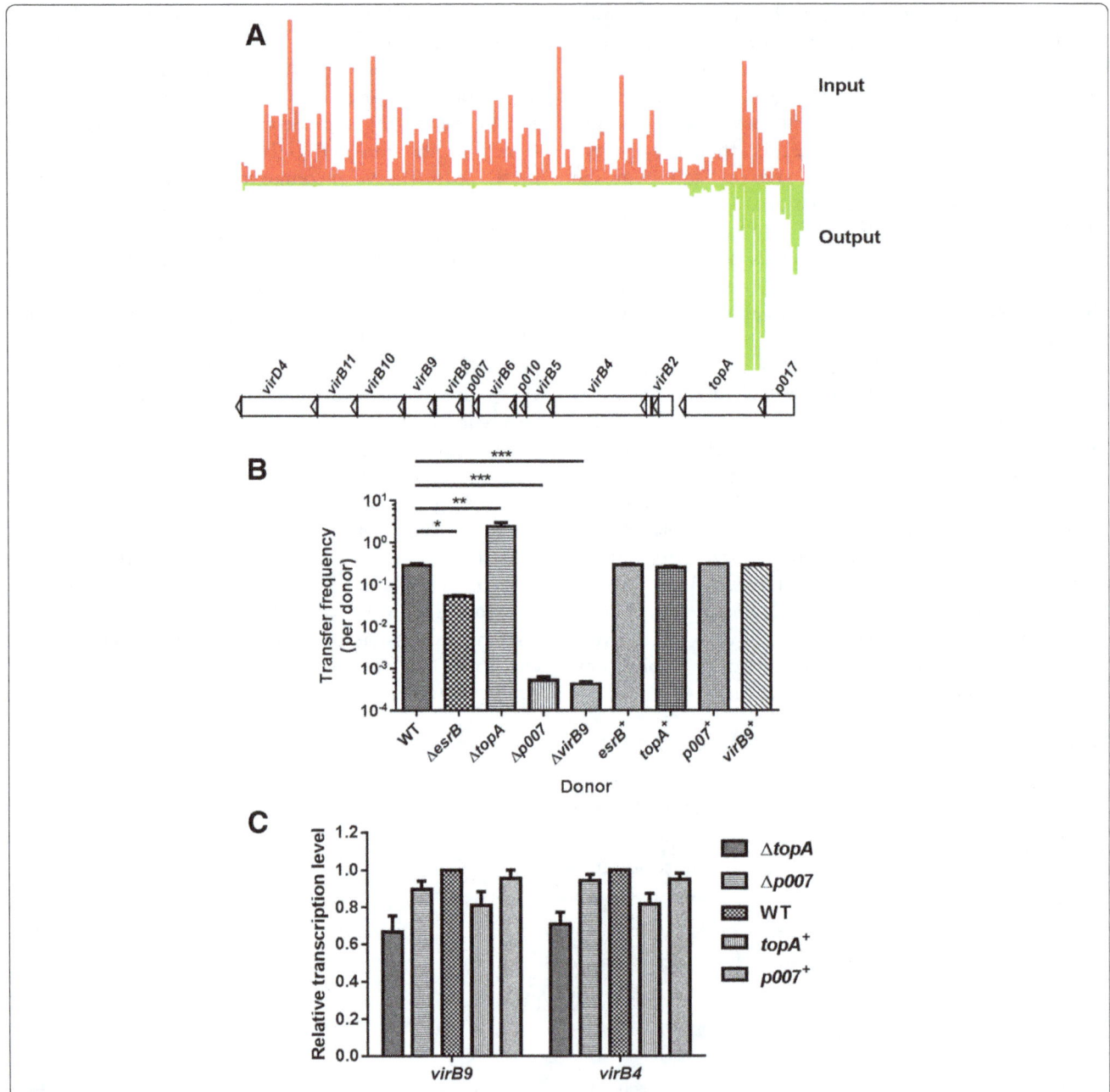

Fig. 3 TIS identification and verification of genes associated with pEIB202 transfer. **a** Artemis screenshot of abundance of reads in *topA*, *p007* and T4SS genes in input (*red*) and output (*green*) samples. The height of the *red* and *green* bars correlates with the number of reads. **b** The transfer frequency of pEIB202 from each strain to ΔP::pNQ705K. Strains were conjugated at 30 °C for 8 h. **c** qRT-PCR analysis of the expression of T4SS genes in each strain grown in LB. * $P < 0.01$, ** $P < 0.001$, ***$P < 0.0001$, student's *t*-test

expression levels of *virB* genes were somewhat decreased when *topA* was absent; while that of Δ*p007* were similar to that of the WT (Fig. 3c). In addition, the relative transcription levels of *virB9* and *virB4* were not changed much when *topA* or *p007* genes were back complemented into their respective deletion mutants, suggesting that the alterations of conjugation capacities in Δ*p007* and Δ*topA* might not be associated with the modulation of T4SS expression; Taken together, these data

demonstrated that *virB* genes, *p007* and *topA* are essential for the pEIB202 transfer.

Antibiotics modulate T4SS gene expression in pEIB202. We were intrigued that how antibiotics could modulate the pEIB202 conjugation frequency in *E. piscicida* as shown in Fig. 2c. We focused on the influences of antibiotics on plasmid conjugation as a whole process but did not consider the influence of antibiotics on the recipients' activities. We first asked whether the copy number of pEIB202 would be significantly changed with the addition of sublethal antibiotics (1/4 of MIC) Str, Tet, and Cm. The data indicated that the copy number of pEIB202 was 1.2–1.5 per cell and appeared not to be significantly affected by the presence of the sublethal antibiotics Str, Tet, and Cm during conjugation (Fig. 4a). In the TIS data, the 6 antibiotics resistance genes shared the averaged input reads and showed no variations in the output reads (Table 4), suggesting that the carrying of antibiotics resistance genes might not affect the pEIB202 transmission. qRT-PCR was further performed to analyze the expression of *virB9* and *virB4* in WT grown with or without sublethal concentrations of antibiotics Str, Tet, and Cm. The transcriptional level of T4SS genes (*virB4* and *virB9*) was significantly reduced due to the addition of antibiotics as compared to that without antibiotics and the expression of other unrelated genes on chromosome or the plasmid (Fig. 4b), suggesting an inhibitory effect of these antibiotics on the expression T4SS in pEIB202 at sublethal concentrations.

EsrB activates the expression of T4SS and antibiotic resistance genes. The response regulator EsrB encoded in the horizontal acquired T3SS gene cluster is essential for the bacterial virulence (Table 3) [20]. Intriguingly, our results also showed that absence of *esrB* in the chromosome significantly decreased the conjugation frequency in *E. piscicida* (Fig. 3b). The copy number of

pEIB202 was kept no change when *esrB* was absent as compared to the WT (Fig. 4a). In order to interrogate this, we performed RNA-seq analysis with EIB202 and Δ*esrB* RNA isolated from LB (n = 3) to compare their differential gene expression. RNA-Seq analysis was performed in triplicate and the reproducibility of the biological repeats was extremely high (a mean R^2–0.99, Fig. 5a). A total of 33 (62%) plasmid genes were up-regulated by at least 2-fold in WT, when WT and Δ*esrB* transcriptoms were compared (Fig. 5b). These data suggested that EsrB also serves as an activator of plasmid gene expression in *E. piscicida*. Notably, among all the 6 antibiotic resistance genes and 10 T4SS genes in pEIB202 nearly all antibiotics resistance (except *strA*) and T4SS genes (except *virB2* and *virB5*) were markedly activated by EsrB (Fig. 5c). We further compared the MIC of Cm, Str, Col, and Tet for WT and Δ*esrB*. The data indicated that, as compared to that of WT, Δ*esrB* showed significantly decreased resistance towards Cm, Str, and Tet, but remain similar Col resistance (Table 3), which is encoded on the chromosome. These data were well supporting the finding that EsrB activates the antibiotics gene expression in the plasmid. Collectively, our data demonstrated that EsrB is essential for the conjugation transfer of pEIB202, probably through regulating the expression of T4SS in LB.

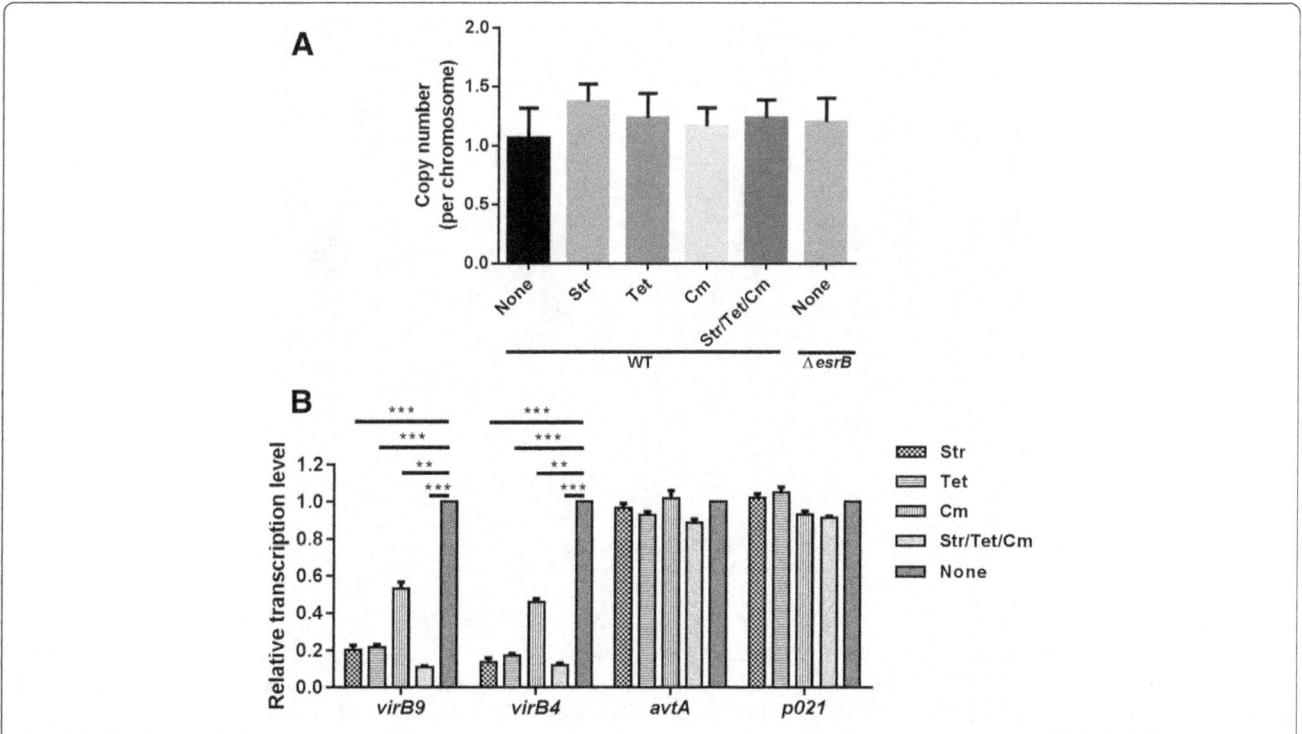

Fig. 4 Antibiotics inhibit the expression of T4SS but does not affect the copy number of pEIB202. **a** Copy number of pEIB202 in WT grown with or without antibiotics at their sublethal concentrations or in Δ*esrB* measured by qPCR. **b** qRT-PCR analysis of the expression of T4SS genes in WT grown with or without antibiotics at their sublethal concentrations (1/4 of MIC). * $P < 0.01$, ** $P < 0.001$, ***$P < 0.0001$, student's t-test

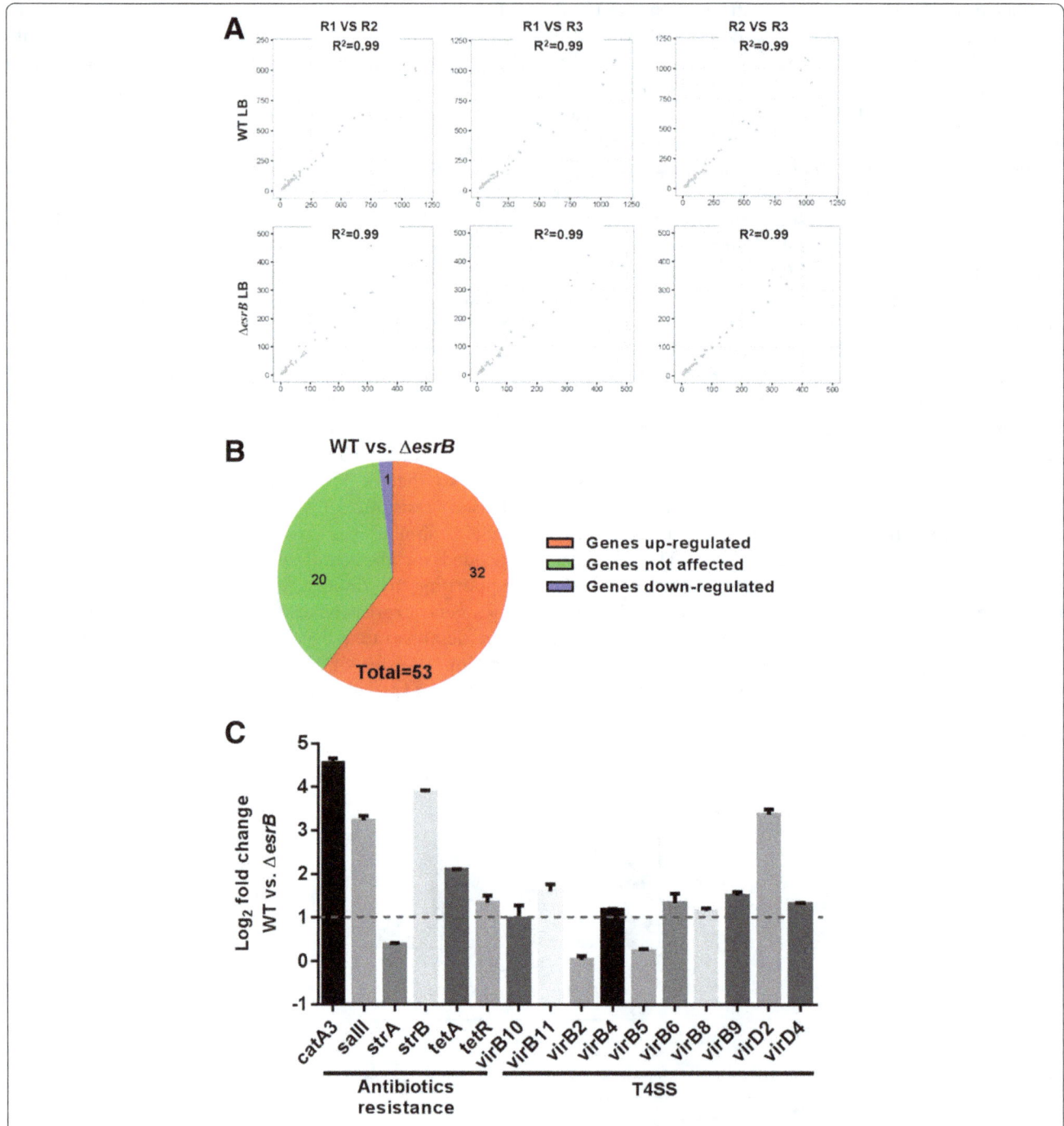

Fig. 5 RNA-seq analysis of differential plasmid gene expression of WT and ΔesrB cultured in LB (*n* = 3) based on the normalized transcript levels.
a Scatter plots of RPKM values per ORF in pEIB202 in biological replicates R1, R2, and R3. **b** Numbers of differentially transcribed plasmid genes (fold change >2). **c** Relative expression of antibiotics resistance and T4SS genes of pEIB202 that are differentially expressed in WT versus ΔesrB

Discussion

Bacterial antibiotics resistance has increasingly become a globally recognized concern. The capacities to exchange antibiotics resistance determinants conferred by horizontal gene transfer (HGT) have greatly intensified the issue. As a leading fish pathogen, *E. piscicida* harbors an array of antibiotics-resistance determinants in both of its chromosome and plasmids and well prepares for the antibiotics cocktail that might be present in the aquaculture environments [18]. Furthermore, *E. piscicida* is an intracellular pathogen residing various cell types including phagocytes and epithelial cells, which may help the bacterium to escape the antibiotics attack and persist in the ecosystems. The way antibiotics affect the conjugation process and interact with the genetic basis of conjugation is still confusing. Here we used the multi-drug resistance plasmid pEIB202 from a highly virulent *E. piscicida* strain EIB202 [33] to address this and our findings indicated that antibiotics or their mixture modulate the conjugation efficiencies. The T4SS genes in pEIB202 are required for the conjugation process. Antibiotics appear to modulate conjugation frequencies by repressing T4SS gene expression. Furthermore, the response regulator EsrB encoded on the chromosome is also essential to facilitate the conjugation by up-regulating the T4SS genes. Investigation of the gene transfer mediated by the plasmid conjugation in the growth conditions mimic of aquaculture farms will improve the understanding of requirements for plasmid conjugation processes and help evaluation of antibiotics treatment in the industries.

Genetic analysis of pEIB202 indicated that it belongs to IncP plasmid and is capable of replication and stable inheritance in a wide variety of gram-negative bacteria [18]. As compared to other reported plasmids carried by *Edwardsiella* bacterium, such as pCK41 (70 kb) from *E. piscicida* [34], p080813–1 (127 kb) from *E. anguillarum* [35], and pEI1/2 (3.9 kb and 4.8 kb) and similar plasmids from *E. ictaluri* [36, 37], pEIB202 (43 kb) found in *E. piscicida* strain EIB202 is medium sized. In some *Edwardsiella* strains, there are no plasmids. While some of the plasmids, i.e. pEI1/2 encoding T3SS genes, are highly associated to the bacterial pathogenesis [34, 36], many of the plasmids including pEIB202 in the bacterium bear multi-drug resistances and seems to mainly contribute to their survival in the antibiotics selection, while not the pathogenesis (Fig. 1 and Table 3). It is also notable that chloramphenicol resistance plasmids from *E. piscicida* were increasingly reported in recent years in the aquaculture farms [18, 35, 38, 39]. Given that the presence of *catA3* gene and the neighboring complex transposon ISSf1 containing IS4 family transposase, as well as the extremely low G + C content of this region (37.4%), as compared to average G + C content of the plasmid (57.3%) or of the genome (59.7%), the chloramphenicol resistance might emerge from a recent acquisition event through HGT in pEIB202.

To make the situations severe, pEIB202 carries a set of T4SS which is believed and now proved by our experiment here to potently facilitate the conjugation and dissemination of the antibiotics-resistance genes in the environment. This should not only make the *E. piscicida* EIB202 and other strains recalcitrant in aquaculture farms, but may also facilitate the HGT of pathogenic islands and their evolution for the emerging of other highly pathogenic strains like *E. anguillarum* [35, 40]. Indeed, pEIB202 can transfer through conjugation at various conditions with high frequencies between different bacterial species (Fig. 2). With this in mind, we

developed our recent licensed live attenuated vaccine WED from EIB202 by curing of pEIB202 for biosafety concerns [41]. The plasmid pEIB202 encodes an incomplete set of components of T4SS namely VirB2, –B4, –B5, –B6, –B8, –B9, –B10, –B11, –D2, and-D4 without –B1, –B3, –B7 [18]. Structural and biochemical analysis of T4SS indicated that the inner membrane protein VirB3 is essential for assembly of the T4SS translocon and the interaction of VirB3 with VirB7 and VirB8 is crucial for the stabilization [32, 42]. We identified all these rest of 10 proteins can promote the transfer of pEIB202 (Fig. 3a, Table 4). These data suggest that the chromosome- encoded proteins might be involved in rescuing the functions of VirB3 or other two VirB proteins when they are missing. The conjugation frequency of pEIB202 decreased significantly in ΔvirB4 and ΔvirB9 (Fig. 3b). All these results provided evidence that this incomplete T4SS is functional and plays an important role in the horizontal transfer of pEIB202. Whether and how the protein substrates could be transferred by the T4SS in pEIB202 should be further illuminated in the future.

Besides established T4SS genes, other genes were also revealed to be essential for pEIB202 conjugation transfer. The genes encoding lipoproteins p007, p010 and two hypothetical proteins and neighboring the T4SS genes were identified to be essential for pEIB202 transfer (Fig. 3b, Table 4). LpqM, a mycobacterial lipoprotein-metalloproteinase, was also required for conjugal DNA transfer, while the requirement of LpqM for conjugation is specific to donor strain but not recipient [43]. Lipoproteins constitute a significant proportion of the cell wall, and may play certain roles in the assembly or function of T4SS and further plasmid transfer. It is anticipated that *topA* influences conjugation processes in this study. The gene *topA* encodes topoisomerase I which is an enzyme essential for relaxation of DNA during a number of critical cellular process [44]. During the process of the plasmid horizontal transfer, topoisomerase I works as an inhibitor, as it breaks and reseals phosphodiester bonds to relax the supercoiled DNA. The role of *p026* encoding a transcriptional repressor in the conjugation transfer remains to be investigated (Table 4). Among all the established *mob/tra* proteins [45, 46], only *p047* for MobC and *p053* for TraC were observed to be essential for conjugation transfer, though with less effect as compared to T4SS genes (Table 4). How T4SS facilitate the *mob/tra* mediated conjugation processes and their interaction should be further investigated. Besides the DNA regions encoding specific proteins, some intergenic regions seem also to be required for plasmid maintenance or conjugation transfer. For example, there was no insertion in the regions ETAE_p015–ETAE_p016 and ETAE_p033–ETAE_p034 in the input libraries, suggesting their essentiality for the plasmid maintenance or replication in *E. pisicicida* (data not shown). While for the region ETAE_p004– ETAE_ p005 and ETAE_p007– ETAE_ p008, the no insertion may be resulted from the polar effect for the T4SS gene expression or they are the DNA regions for origin of transfer (*oriT*) per se (data not shown). In addition to the above-mentioned plasmid genes or DNA regions, EsrB was intriguingly identified to control the plasmid conjugation transfer (Fig. 3b). EsrA-EsrB is the two-component system essential for intracellular survival in *E. piscicida* [47], and plays a key role in regulating ~400 genes' expression (personal communication, Wang QY). However, no plasmid gene controlled by EsrA-EsrB was reported. Here, we showed that over half of the pEIB202 genes were induced by EsrB, which facilitate the conjugation in part by upregulating the T4SS genes (Fig. 5). It should be noted that although EsrB expression is significantly higher in the nutrient defined DMEM medium than in LB, but high level of EsrB expression could also be detected in cells grown in LB [48]. We only tested the modulation of pEIB202 genes' expression by EsrB when the cells are growing in LB, the same condition we used for conjugation experiment.

It has been generally assumed that antibiotics promote conjugation [49], while antibiotics of sublethal concentrations can increase the conjugation rate by either activating the excision of transferrable genes from the host chromosome or by inducing the expression of conjugation machinery [50–52]. There were also conflicting reports regarding whether or not antibiotics promote conjugation. Recent study identified that antibiotics of sublethal concentrations do not significantly increase the conjugation efficiency, but dictate antibiotic-mediated selection, which can both promote and suppress conjugation dynamics [1]. Here, we found that Str, Tet and Cm and their mixture at their sublethal concentrations inhibit the conjugation transfer probably mainly due to the repression of T4SS genes (Fig. 2c and 4b), although the influence of the antibiotics on the recipients' activities should be further considered in the future.

Conclusion
Our data demonstrated that T4SS is essential for the conjugation transfer of multiple-antibiotic resistant pEIB202 and suggested that antibiotics could affect conjugation processes through modulating T4SS. The results also indicated that the genetic elements in chromosome, especially the horizontally acquired response regulator EsrB could also affect the plasmid transfer through modulating the expression of T4SS. Our results will improve the understanding of the mechanism of plasmid conjugation processes that facilitate dissemination of antibiotic resistance especially in aquaculture industries.

Abbreviations
ABR: Antibiotic resistance; Carb: Carbenicillin; Cm: Chloramphenicol; Col: Polymyxin B; Gm: Gentamicin; HGT: Horizontal gene transfer; i.m: Intramuscularly; Km: Kanamycin; LB: Luria broth; MIC: Minimum inhibition concentration; pDNA: Plasmid DNA; qRT-PCR: Quantitative real-time PCR; R plasmid: Resistance-encoding plasmid; RPKM: Reads per kilobase of genic region per million mapped reads; Str: Streptomycin; T3SS: Type III secretion systems; T4SS: Type IV secretion system; T6SS: Type VI secretion systems; Tet: Tetracycline; TIS: Transposon insertion sequencing; TSB: Tryptic soy broth

Acknowledgements
Not applicable.

Funding
This work was supported by National Natural Science Foundation of China under grant number 31430090 (ZY), 31,602,200 (LX) and 31,672,696 (XJ); the Ministry of Agriculture of China under grant number CARS-50 and nyhyzx-201,303,047 (WQ); and the Shanghai Pujiang Program under grant number 16PJD018 (WQ).

Authors' contributions
LY, WQ and XJ conceived and designed the experiments. LY and GY performed the experiments. LY, LX, WQ, LQ and XJ analyzed the data. LY, LX, WQ, ZY and XJ wrote the paper. All authors read and approved the final manuscript.

Competing interests
The authors declare that they have no competing interests.

Author details
[1]State Key Laboratory of Bioreactor Engineering, East China University of Science and Technology, Shanghai 200237, China. [2]Shanghai Engineering Research Center of Maricultured Animal Vaccines, Shanghai, China. [3]Shanghai Collaborative Innovation Center for Biomanufacturing, 130 Meilong Road, Shanghai 200237, China.

References
1. Lopatkin AJ, Huang S, Smith RP, Srimani JK, Sysoeva TA, Bewick S, Karig DK, You L. Antibiotics as a selective driver for conjugation dynamics. Nat Microbiol. 2016;1(6):16044.
2. Sorensen SJ, Bailey M, Hansen LH, Kroer N, Wuertz S. Studying plasmid horizontal transfer in situ: a critical review. Nat Rev Microbiol. 2005;3(9):700–10.
3. Maiden MC. Horizontal genetic exchange, evolution, and spread of antibiotic resistance in bacteria. Clin Infect Dis. 1998;27(Suppl 1):S12–20.
4. Juhas M. Horizontal gene transfer in human pathogens. Crit Rev Microbiol. 2015;41(1):101–8.
5. Turner SL, Bailey MJ, Lilley AK, Thomas CM. Ecological and molecular maintenance strategies of mobile genetic elements. FEMS Microbiol Ecol. 2002;42(2):177–85.
6. van Elsas JD, Bailey MJ. The ecology of transfer of mobile genetic elements. FEMS Microbiol Ecol. 2002;42(2):187–97.
7. Bruto M, James A, Petton B, Labreuche Y, Chenivesse S, Alunno-Bruscia M, Polz MF, Le Roux F. *Vibrio crassostreae*, a benign oyster colonizer turned into a pathogen after plasmid acquisition. ISME J. 2016; doi:10.1038/ismej.2016.162.
8. Rhodes G, Huys G, Swings J, McGann P, Hiney M, Smith P, Pickup RW. Distribution of oxytetracycline resistance plasmids between aeromonads in hospital and aquaculture environments: implication of Tn1721 in dissemination of the tetracycline resistance determinant tet a. Appl Environ Microbiol. 2000;66(9):3883–90.

9. Popa O, Dagan T. Trends and barriers to lateral gene transfer in prokaryotes. Curr Opin Microbiol. 2011;14(5):615–23.

10. Fronzes R, Christie PJ, Waksman G. The structural biology of type IV secretion systems. Nat Rev Microbiol. 2009;7(10):703–14.

11. Cascales E, Christie PJ. The versatile bacterial type IV secretion systems. Nat Rev Microbiol. 2003;1(2):137–49.

12. Backert S, Meyer TF. Type IV secretion systems and their effectors in bacterial pathogenesis. Curr Opin Microbiol. 2006;9(2):207–17.

13. Alvarez-Martinez CE, Christie PJ. Biological diversity of prokaryotic type IV secretion systems. Microbiol Mol Biol Rev. 2009;73(4):775–808.

14. Park SB, Aoki T, Jung TS. Pathogenesis of and strategies for preventing *Edwardsiella tarda* infection in fish. Vet Res. 2012;43(1):67.

15. Tan YP, Zheng J, Tung SL, Rosenshine I, Leung KY. Role of type III secretion in *Edwardsiella tarda* virulence. Microbiology. 2005;151(Pt 7):2301–13.

16. Okuda J, Kiriyama M, Suzaki E, Kataoka K, Nishibuchi M, Nakai T. Characterization of proteins secreted from a type III secretion system of *Edwardsiella tarda* and their roles in macrophage infection. Dis Aquat Org. 2009;84(2):115–21.

17. Hou M, Chen R, Yang D, Nunez G, Wang Z, Wang Q, Zhang Y, Liu Q. Identification and functional characterization of EseH, a new effector of the type III secretion system of *Edwardsiella piscicida*. Cell Microbiol. 2017;19(1):e12638.

18. Wang Q, Yang M, Xiao J, Wu H, Wang X, Lv Y, Xu L, Zheng H, Wang S, Zhao G, et al. Genome sequence of the versatile fish pathogen *Edwardsiella tarda* provides insights into its adaptation to broad host ranges and intracellular niches. PLoS One. 2009;4(10):e7646.

19. Yang M, Lv Y, Xiao J, Wu H, Zheng H, Liu Q, Zhang Y, Wang Q. *Edwardsiella* comparative phylogenomics reveal the new intra/inter-species taxonomic relationships, virulence evolution and niche adaptation mechanisms. PLoS One. 2012;7(5):e36987.

20. Lv Y, Xiao J, Liu Q, Wu H, Zhang Y, Wang Q. Systematic mutation analysis of two-component signal transduction systems reveals EsrA-EsrB and PhoP-PhoQ as the major virulence regulators in *Edwardsiella tarda*. Vet Microbiol. 2012;157(1–2):190–9.

21. Livak KJ, Schmittgen TD. Analysis of relative gene expression data using real-time quantitative PCR and the 2(-Delta Delta C (T)) method. Methods. 2001;25(4):402–8.

22. Gibson DG, Young L, Chuang RY, Venter JC, Hutchison CR, Smith HO. Enzymatic assembly of DNA molecules up to several hundred kilobases. Nat Methods. 2009;6(5):343–5.

23. Chao MC, Pritchard JR, Zhang YJ, Rubin EJ, Livny J, Davis BM, Waldor MK. High-resolution definition of the *Vibrio cholerae* essential gene set with hidden Markov model-based analyses of transposon-insertion sequencing data. Nucleic Acids Res. 2013;41(19):9033–48.

24. Langmead B, Trapnell C, Pop M, Salzberg SL. Ultrafast and memory-efficient alignment of short DNA sequences to the human genome. Genome Biol. 2009;10(3):R25.

25. Xiao J, Wang Q, Liu Q, Xu L, Wang X, Wu H, Zhang Y. Characterization of *Edwardsiella tarda* rpoS. effect on serum resistance, chondroitinase activity, biofilm formation, and autoinducer synthetases expression. Appl Microbiol Biotechnol. 2009;83(1):151–60.

26. Yamaichi Y, Chao MC, Sasabe J, Clark L, Davis BM, Yamamoto N, Mori H, Kurokawa K, Waldor MK. High-resolution genetic analysis of the requirements for horizontal transmission of the ESBL plasmid from *Escherichia coli* O104:H4. Nucleic Acids Res. 2015;43(1):348–60.

27. Bolger AM, Lohse M, Usadel B. Trimmomatic. a flexible trimmer for Illumina sequence data. Bioinformatics. 2014;30(15):2114–20.

28. Li H, Durbin R. Fast and accurate short read alignment with burrows-wheeler transform. Bioinformatics. 2009;25(14):1754–60.

29. Anders S, Huber W. Differential expression analysis for sequence count data. Genome Biol. 2010;11(10):R106.

30. Fernandez AI, Perez MJ, Rodriguez LA, Nieto TP. Surface phenotypic characteristics and virulence of Spanish isolates of *Aeromonas salmonicida* after passage through fish. Appl Environ Microbiol. 1995;61(5):2010–2.

31. Pritchard JR, Chao MC, Abel S, Davis BM, Baranowski C, Zhang YJ, Rubin EJ, Waldor MK. ARTIST: high-resolution genome-wide assessment of fitness using transposon-insertion sequencing. PLoS Genet. 2014;10(11):e1004782.

32. Low HH, Gubellini F, Rivera-Calzada A, Braun N, Connery S, Dujeancourt A, Lu F, Redzej A, Fronzes R, Orlova EV, et al. Structure of a type IV secretion system. Nature. 2014;508(7497):550–3.

33. Xiao J, Wang Q, Liu Q, Wang X, Liu H, Zhang Y. Isolation and identification of fish pathogen *Edwardsiella tarda* from mariculture in China. Aquac Res. 2008;40(1):13–7.

34. Yu JE, Cho MY, Kim JW, Kang HY. Large antibiotic-resistance plasmid of *Edwardsiella tarda* contributes to virulence in fish. Microb Pathog. 2012;52(5):259–66.

35. Shao S, Lai Q, Liu Q, Wu H, Xiao J, Shao Z, Wang Q, Zhang Y. Phylogenomics characterization of a highly virulent *Edwardsiella* strain ET080813(T) encoding two distinct T3SS and three T6SS gene clusters: propose a novel species as *Edwardsiella anguillarum* sp. nov. Syst Appl Microbiol. 2015;38(1):36–47.

36. Zhao LJ, Lu JF, Nie P, Li AH, Xiong BX, Xie HX. Roles of plasmid-encoded proteins, EseH, EseI and EscD in invasion, replication and virulence of *Edwardsiella ictaluri*. Vet Microbiol. 2013;166(1–2):233–41.

37. Fernandez DH, Pittman-Cooley L, Thune RL. Sequencing and analysis of the *Edwardsiella ictaluri* plasmids. Plasmid. 2001;45(1):52–6.

38. Welch TJ, Evenhuis J, White DG, McDermott PF, Harbottle H, Miller RA, Griffin M, Wise D. IncA/C plasmid-mediated florfenicol resistance in the catfish pathogen *Edwardsiella ictaluri*. Antimicrob Agents Chemother. 2009;53:845–6.

39. Sun K, Wang H, Zhang M, Xiao Z, Sun L. Genetic mechanisms of multi-antimicrobial resistance in a pathogenic *Edwardsiella tarda* strain. Aquaculture. 2009;289(1):134–9.

40. Nakamura Y, Takano T, Yasuike M, Sakai T, Matsuyama T, Sano M. Comparative genomics reveals that a fish pathogenic bacterium *Edwardsiella tarda* has acquired the locus of enterocyte effacement (LEE) through horizontal gene transfer. BMC Genomics. 2013;14:642.

41. Xiao J, Chen T, Liu B, Yang W, Wang Q, Qu J, Zhang Y. *Edwardsiella tarda* mutant disrupted in type III secretion system and chorismic acid synthesis and cured of a plasmid as a live attenuated vaccine in turbot. Fish Shellfish Immunol. 2013;35(3):632–41.

42. den Hartigh AB, Rolan HG, de Jong MF, Tsolis RM. VirB3 to VirB6 and VirB8 to VirB11, but not VirB7, are essential for mediating persistence of *Brucella* in the reticuloendothelial system. J Bacteriol. 2008;190(13):4427–36.

43. Nguyen KT, Piastro K, Derbyshire KM. LpqM, a mycobacterial lipoprotein-metalloproteinase, is required for conjugal DNA transfer in *Mycobacterium smegmatis*. J Bacteriol. 2009;191(8):2721–7.

44. Wang JC. Cellular roles of DNA topoisomerases: a molecular perspective. Nat Rev Mol Cell Biol. 2002;3(6):430–40.

45. Meyer R. Identification of the mob genes of plasmid pSC101 and characterization of a hybrid pSC101-R1162 system for conjugal mobilization. J Bacteriol. 2000;182(17):4875–81.

46. Farrand SK, Hwang I, Cook DM. The tra region of the nopaline-type Ti plasmid is a chimera with elements from the transfer systems of RSF1010, RP4, and F. J Bacteriol. 1996;178(14):4233–47.

47. Chakraborty S, Sivaraman J, Leung KY, Mok YK. Two-component PhoB-PhoR regulatory system and ferric uptake regulator sense phosphate and iron to control virulence genes in type III and VI secretion systems of *Edwardsiella tarda*. J Biol Chem. 2011;286(45):39417–30.

48. Yin K, Wang Q, Xiao J, Zhang Y. Comparative proteomic analysis unravels a role for EsrB in the regulation of reactive oxygen species stress responses in *Edwardsiella piscicida*. FEMS Microbiol Lett. 2017;364(1):fnw269.

49. Andersson DI, Hughes D. Microbiological effects of sublethal levels of antibiotics. Nat Rev Microbiol. 2014;12(7):465–78.

50. Stevens AM, Shoemaker NB, Li LY, Salyers AA. Tetracycline regulation of genes on Bacteroides conjugative transposons. J Bacteriol. 1993;175(19):6134–41.

51. Whittle G, Shoemaker NB, Salyers AA. Characterization of genes involved in modulation of conjugal transfer of the Bacteroides conjugative transposon CTnDOT. J Bacteriol. 2002;184(14):3839–47.

52. Beaber JW, Hochhut B, Waldor MK. SOS response promotes horizontal dissemination of antibiotic resistance genes. Nature. 2004;427(6969):72–4.

53. Milton DL, Norqvist A, Wolf-Watz H. Cloning of a metalloprotease gene involved in the virulence mechanism of *Vibrio anguillarum*. J Bacteriol. 1992;174(22):7235–44.

54. Wang SY, Lauritz J, Jass J, Milton DL. A ToxR homolog from *Vibrio anguillarum* serotype O1 regulates its own production, bile resistance, and biofilm formation. J Bacteriol. 2002;184(6):1630–9.

55. Dennis JJ, Zylstra GJ. Plasposons: modular self-cloning minitransposon derivatives for rapid genetic analysis of gram-negative bacterial genomes. Appl Environ Microbiol. 1998;64(7):2710–5.

11

Biogeographical distribution analysis of hydrocarbon degrading and biosurfactant producing genes suggests that near-equatorial biomes have higher abundance of genes with potential for bioremediation

Jorge S. Oliveira[1,2,3*†], Wydemberg J. Araújo[1,3†], Ricardo M. Figueiredo[2], Rita C. B. Silva-Portela[1], Alaine de Brito Guerra[1], Sinara Carla da Silva Araújo[1], Carolina Minnicelli[1], Aline Cardoso Carlos[1], Ana Tereza Ribeiro de Vasconcelos[3], Ana Teresa Freitas[2] and Lucymara F. Agnez-Lima[1]

Abstract

Background: Bacterial and Archaeal communities have a complex, symbiotic role in crude oil bioremediation. Their biosurfactants and degradation enzymes have been in the spotlight, mainly due to the awareness of ecosystem pollution caused by crude oil accidents and their use. Initially, the scientific community studied the role of individual microbial species by characterizing and optimizing their biosurfactant and oil degradation genes, studying their individual distribution. However, with the advances in genomics, in particular with the use of New-Generation-Sequencing and Metagenomics, it is now possible to have a macro view of the complex pathways related to the symbiotic degradation of hydrocarbons and surfactant production. It is now possible, although more challenging, to obtain the DNA information of an entire microbial community before automatically characterizing it. By characterizing and understanding the interconnected role of microorganisms and the role of degradation and biosurfactant genes in an ecosystem, it becomes possible to develop new biotechnological approaches for bioremediation use. This paper analyzes 46 different metagenome samples, spanning 20 biomes from different geographies obtained from different research projects.

Results: A metagenomics bioinformatics pipeline, focused on the biodegradation and biosurfactant-production pathways, genes and organisms, was applied. Our main results show that: (1) surfactation and degradation are correlated events, and therefore should be studied together; (2) terrestrial biomes present more degradation genes, especially cyclic compounds, and less surfactation genes, when compared to water biomes; and (3) latitude has a significant influence on the diversity of genes involved in biodegradation and biosurfactant production. This suggests that microbiomes found near the equator are richer in genes that have a role in these processes and thus have a higher biotechnological potential.

(Continued on next page)

* Correspondence: oliveira.jorge.88@gmail.com
†Equal contributors
[1]Laboratório de Biologia Molecular e Genômica, Departamento de Biologia Celular e Genética, Centro de Biociências, Universidade Federal do Rio Grande do Norte, Natal, RN, Brazil
[2]INESC-ID/IST Instituto de Engenharia de Sistemas e Computadores/Instituto Superior Técnico, Universidade de Lisboa, Rua Alves Redol, 9, 1000-029 Lisbon, Portugal
Full list of author information is available at the end of the article

(Continued from previous page)

Conclusion: In this work we have focused on the biogeographical distribution of hydrocarbon degrading and biosurfactant producing genes. Our principle results can be seen as an important step forward in the application of bioremediation techniques, by considering the biostimulation, optimization or manipulation of a starting microbial consortia from the areas with higher degradation and biosurfactant producing genetic diversity.

Keywords: Hydrocarbon degradation, Biosurfactants, Environmental microbiology, Metagenomics, Metagenomics bioinformatics pipeline, Geographical ecology, Microbiome data analysis

Background

Studies evaluating the biogeographical influence in the diversity and/or abundance of alkane degradation and biosurfactant production genes may guide the creation of new industrial and biotechnological processes. These include bioremediation and biostimulation strategies that are important for preservation and environment planning [1, 2]. Although biogeographical studies of hydrocarbon degradation genes predominate in the literature [2], there is a relative lack of knowledge about the distribution of bacteria producing biosurfactants in the environment [3].

The synergic effects of biosurfactants on solubility, sorption and biodegradation of hydrophobic organic contaminants are known as they play an important role during biodegradation processes [4]. Biosurfactants can be synthesized by a myriad of microorganisms, which is influenced by the composition of the medium and environmental conditions [4]. However, because most studies of geographic distribution of bacteria oil-degrading genes in environments rely on the analysis of biomes that have been contaminated or enriched with crude oil, the understanding of the origin, abundance and natural role of degradation and surfactant genes on an ecosystem [3, 5, 6] has been hampered.

International microbial surveys [7–10] are good examples of large-scale coordinated efforts to explore soil and water taxonomic and functional diversity. In general, the generated datasets are available in public repositories like Sequence Read Archive (SRA). These datasets, combined with the appropriate computational pipelines, can reveal correlations between ecology and geography, based on taxonomic and functional characteristics of the biomes.

Metagenomic analysis software packages, like MG-RAST [11], MEGAN [12] and KRAKEN [13] include solutions for taxonomic, functional and comparative analyses. With these tools, metagenomic datasets are combined with global databases, which with the constantly growing size of these datasets, produces large and complex outputs that usually take several days to be analyzed. Other tools like MetAmos [14] work in a modular manner, allowing workflow customization and promise to reduce assembly errors and computational cost. However, its flexibility and modular construction makes the computational installation process time and space consuming.

Moreover, we have reached a state where the massive size of available data does not allow the use of classic brute-force bioinformatics approaches. It is thus clear that the use of domain specific studies and databases is essential to focus on a specific research scope and reduce the computational effort. In functional databases like KEGG, there are examples such as the ontology of degradation genes grouped with the beta-oxidation in the lipid metabolism pathway, or the synthesis of biosurfactants together with antibiotics in the nonribosomal peptide synthesis pathway, that make research on degradation, or surfactants individually much more difficult. To overcome this limitation, domain-specific databases, like BioSurfDB [15], reorganize the functional ontologies, thus allowing the focus, on biosurfactants and biodegradation. This domain-specific database also combines a set of tailored tools to enable efficient specific metagenomic analysis. The main goal of this tool is to support the identification of patterns of taxonomic and functional diversity of microbial communities and the identification of genes involved in the degradation of hydrocarbons and biosurfactants production.

In this research, we analyzed 46 public metagenomes, from 20 different biomes, water and terrestrial, to increase our understanding of the biogeographical distribution of biodegradation and biosurfactant-production genes. Additionally, a metagenomics pipeline that relies on BioSurfDB, to effectively and efficiently process a large amount of data, was developed and optimized.

Methods

All the computational processing was performed in a AMD server running Slackware version 14 in 64bits, with 64 CPUs and 258GB of RAM.

Metagenome sequences were downloaded from the SRA at NCBI website, the Metagenomic samples detailed information on SRA project and Run are available in Suplementary Material (Additional file 1: Table S1). The Metagenomes Summary table (Additional file 2: Table S2), summarizes the information regarding both soil and water metagenomes. Whenever possible, several samples from each biome were selected. There were a total of 71 DNA-seq metagenomes with a heterogeneous set of possible environmental samples with worldwide representation. Sample Geography figure (Additional file 3: Figure S1) presents a geographic distribution of the metagenomes that have been analyzed. The pipeline presented in Fig. 1 was used to get a macro view of the taxonomic and functional differences between the metagenomes.

Filtering/Trimming

A filtering/trimming procedure was applied to all the metagenomes presenting low quality parameters in the FASTQC (http://www.bioinformatics.babraham.ac.uk/projects/fastqc) report. Based on the generated quality report, the trimming of k-mer contaminated and heterogeneous GC-content areas was performed using *Fastx toolkit trimmer*. Also *fastq_quality_filter* from the same toolkit was used to assure a minimum Phred-Score of 20 for at least 90% of the reads. This procedure revealed to be an iterative and supervised-dependent process, as it had to be repeated for some samples until the FASTQC reports showed acceptable quality. The final number of sequences was also analyzed and the metagenomes with less than 100.000 sequences were discarded. It was decided to use a more conservative approach, by using less samples but with higher quality per sample.

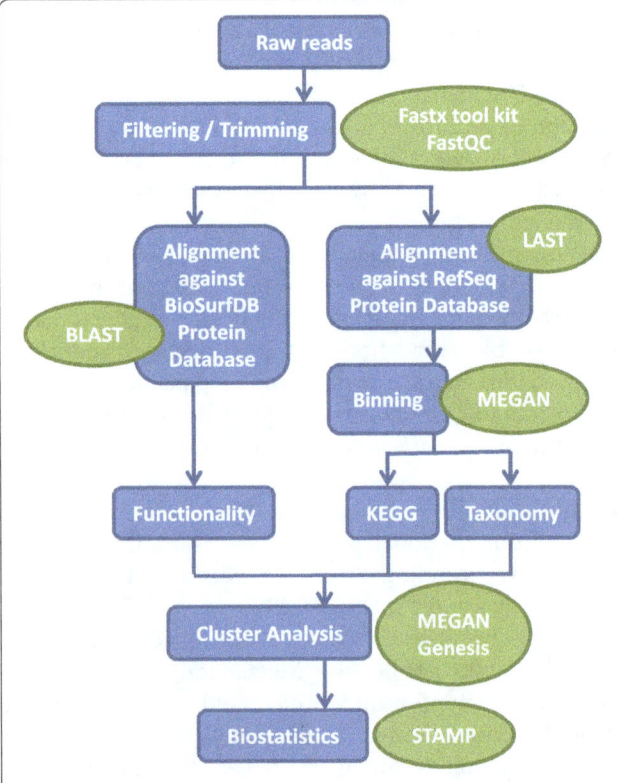

Fig. 1 Computational pipeline for taxonomic and functional analysis. The main processing steps are in *blue* and the software used is highlighted in *green*

Alignment

After the quality assessment, two parallel alignment steps were performed: (i) an alignment against BioSurfDB, a domain-based database, and (ii) an alignment against the RefSeq, a generic sequence database [16].

At this stage we should stress that the alignment was carried out using all the reads in the datasets and no assembling was performed to obtain contigs. This decision was significant and was based on the following observed during a preliminary study that had evaluated the impact of using contigs when abundance analysis is performed:

1. if the goal is to compare gene abundances between metagenomes, the use of contigs instead of reads will significantly reduce the abundance of information leading to inaccurate results;
2. in metagenomics, the organism diversity is so high that it is very difficult for assemblers to distinguish a repeated read from a homolog one, thus masking the real number of organisms present in the datasets;
3. when dealing with a large amount of heterogeneous sequencing data, average read length, coverage or quality a consistently high quality assembly step might not be possible because of the sequencing technology used.

RefSeq

RefSeq is a non-redundant database integrating sequences from many sources. The full set of non-redundant protein sequences (9.5 GBs) was downloaded. The selected sequence alignment program was LAST [17], an aligner optimized for repeat-rich datasets that performs much faster than the traditional BLAST [18]. This algorithm is very useful in situations where the size of the data hampers the alignment. Each metagenome was aligned to the RefSeq database using the default parameters for the LAST aligner. Taxonomic and Functional binning was performed by MEGAN (version 5) using its respective RefSeq and KEGG maps databases.

BioSurfDB

BioSurfDB is a curated information system with a focus on biodegradation and biosurfactant production organisms. It was developed to support research in the bioremediation field. This information system includes tools for the alignment of metagenomes against a number of genomic or protein sequences. One sample of each group of metagenomes, in a total of 46 samples, was uploaded to the BioSurfDB system and the BLASTx tool. Nucleotide query versus protein database with an E-value of $1e^{-4}$ was used for sequence alignment. Currently, the BioSurfDB database includes 3956 protein sequences from different pathways. The list of pathways available in BioSurfDB at the time of this study is shown in the BioSurfDB Pathways table (Additional file 4: Table S3). Following the alignment, the BioSurfDB system automatically performs taxonomic and functional binning. However, as BioSurfDB is a domain-specific database, its taxonomic prediction might be biased and therefore, we decided not to use it for taxonomic classification.

Cluster analysis

Alignment results from all the analyzed metagenomes, from both BioSurfDB and RefSeq analysis, were uploaded to MEGAN to compute UPGMA trees and PCoA (Principal Coordinates Analysis).

The metagenomics computational pipeline used includes scripts that cross the BLASTx results and the database tree, creating hit-count tables for taxonomy, proteins and metabolic pathways. These pathway tables were uploaded to Genesis [19], where normalization was applied, followed by the calculation of hierarchical clustering for both metagenomes and pathways, using a complete link approach.

Statistics

Results from the BLAST alignment using the BioSurfDB as database were grouped in a metadata file, according to the functional clusters obtained in the previous step. These tables were uploaded to STAMP [20] to perform the statistical tests between metagenomes and to Graphpad Prism to test the correlation between the surfactant production and hydrocarbon degradation.

To calculate the correlation coefficient between the diversity, i.e., the number of different blast alignments mapped, of biosurfactant and degradation genes in the environment, a Pearson parametric test was used, with a confidence interval of 0.95 and a P-value <0.0001.

A preliminary data analysis, automatically performed by STAMP, decides which is the best statistical test to be performed. A two-sided Welch's t-test with a confidence interval of 0.95 and Benjamini-Hochberg multiple test correction was performed to identify significant differences between groups. Two filters were used: a minimal q-value of 0.05 and a minimum difference of proportions of 1 (program defaults).

Results

Quality assessment

From the initial dataset of 71 metagenomics samples, 24 samples were discarded by failing the quality assessment, and 46 samples from the several biomes, shown in Table 1 were used for further analysis. At this stage of the data analysis, it was not possible to guarantee a uniform number of samples per biome, because for many of the projects the samples were not of acceptable quality.

Taxonomy annotation using RefSeq

For the 46 samples, the metagenomes annotations were obtained by using the alignment program LAST to compare the metagenomic sequences with the RefSeq protein database. The obtained results were grouped using a hierarchical clustering algorithm available in MEGAN. Unfortunately, due to the large size of the metagenomes, our server could not process 8 of these samples in MEGAN. Therefore, and solely in the hierarchical cluster step, only 39 samples, corresponding to 17 biomes were analyzed. The results in Fig. 2 show the formation of distinct taxonomic clusters. From the dendrogram analysis we have considered three different clusters. In cluster 1, it is possible to see water metagenomes, mainly samples from the Atlantic and Pacific oceans and grouped into distinct cluster extensions. The second cluster includes only terrestrial metagenomes. However, it is possible to verify the grouping of terrestrial metagenomes by similar climatic regions. The third cluster is also formed by water metagenomes, but from tropical regions.

These results validate the samples for consistency, as the samples from the same metagenomes are in the same clusters. Based on this clustering result, we decided to use just one sample dataset as a representative of each metagenome for further analysis. Consequently, it was possible to optimize the use of computational resources.

Furthermore, we computed a rarefaction curve in the MEGAN tool, to assure that the metagenomic datasets included a significant number of reads to cover most taxons. As seen in the Rarefaction Curve figure (Additional file 5: Figure S2), the number of leaves in taxonomy reaches a plateau in all samples and this confirms the acceptable sample size of the data under analysis.

BioSurfDB cluster analysis

Using the 46 samples, a functional clustering was carried out examining the data obtained from a BLAST compared with the databases included in the BioSurfDB information system using the Genesis software tool. Figure 3 shows the resulting hierarchical clusters when only the degradation genes are considered, see Fig. 3a, and when considering only the biosurfactants production genes, see Fig. 3b. K-means clustering was also used and revealed clusters like those obtained by the hierarchical clustering algorithm.

These results highlight two important clusters: (1) a cluster of tropical or near-equatorial terrestrial metagenomes (represented by the red square in Fig. 3 (a) and (b)) that show the highest values of reads mapping both for degradation and biosurfactant genes, showing the similarity of the microorganism communities; and (2) metagenomes from

Table 1 Analyzed Biomes, classified by soil or water type, with information about the region, number of reads, average read length, sequencing technology used and sequencing project SRA code and link

	Regions	Number of reads	Read length (bp)	Seq. Tech.	SRA Link
Soil					
Tundra	Siberia & Canada	1.31E + 07	183.5	Illumina	SRP047512
Temp. Woodland	Australia	1.23E + 07	290	Illumina	ERP008551
Arid Grassland	Australia	1.92E + 07	299	Illumina	ERP008551
Saline Desert	India	2.07E + 06	124	Ion	SRP041239
Atlantic Forest	Brazil	9.62E + 04	380	Illumina	SRP004544
Tropical Forest	French Guiana	4.04E + 05	384	454	ERP002426
Temp. Coniferous Forest	Canada	2.18E + 07	136	Illumina	ERP009498
Mangrove	Brazil	5.26E + 05	418	454	SRP004544
Caatinga	Brazil	2.31E + 05	426	454	SRP004544
Paddy Soil	China	2.16E + 06	190	Illumina	SRP039858
Temp. Plantation Soil	Australia	3.32E + 07	299	Illumina	ERP008551
Grassland Soil	Oklahoma	9.43E + 06	169	Illumina	SRP029969
Terrestrial Subsurface	South Africa	1.11E + 07	186	Illumina	SRP049336
Water					
Sea Water	North Pacific	2.67E + 07	187	Illumina	ERP003628
Sea Water	South Pacific	2.66E + 07	188	Illumina	ERP003628
Sea Water	Indian Ocean	1.63E + 07	185	Illumina	ERP001736
South Atlantic	Brazil	2.46E + 07	184	Illumina	ERP003708
North Atlantic	Iceland	8.39E + 05	460	Illumina	ERP009703
North Atlantic	Portugal	3.32E + 06	293	Illumina	ERP009703
River Plume	Amazon	5.23E + 06	286	Illumina	SRP039390
Adriatic / Ionian Sea	Mediterranean	9.62E + 07	193	Illumina	ERP003628
River Estuary	Brazil	1.00E + 05	438	454	SRP004544

All data and metadata can be retrieved from the link provided

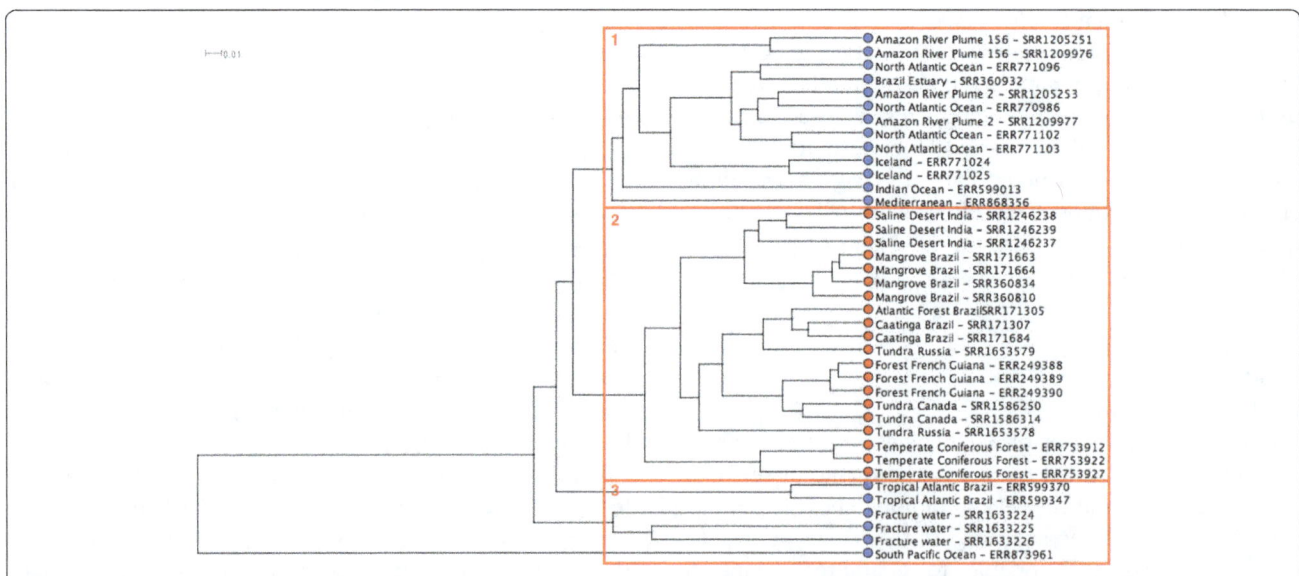

Fig. 2 UPGMA Tree computed by MEGAN with RefSeq data. The distance between the clusters is based on pairwise distance among taxa. The soil samples are represented with a *red dot* and the water samples with *blue*. The *red squares* show the proposed cluster division

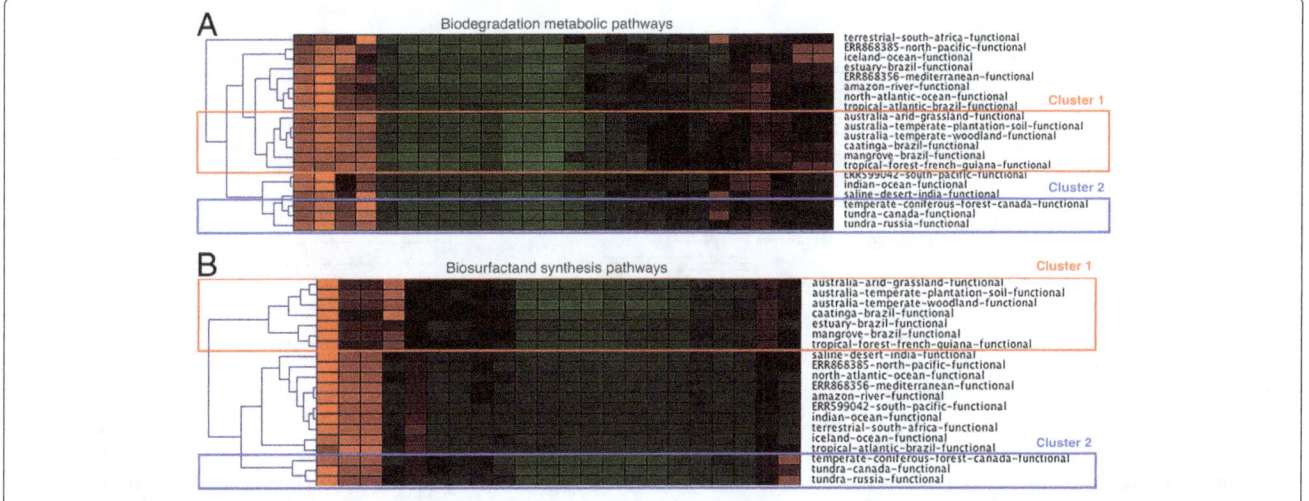

Fig. 3 Hierarchical clusters obtained from the BioSurfDB functional data through Genesis software, for degradation (**a**) and biosurfactants (**b**). Inside the *red borders* the "equatorial region clusters" can be seen whilst inside the *blue borders* are the "cold region clusters". Each column represents a specific pathway and the colour schema for their relative abundances is *green* for low and *red* for a high number of blast hits

cold regions in Russia and Canada (blue square) that have a low abundance of microorganisms involved in the degradation of biosurfactant production processes.

A global analysis of the 46 samples resulted in an important correlation between the diversity of biosurfactant genes when compared with the existence of degradation genes in the environment (Fig. 4). The parametric Pearson-correlation test showed a positive linear correlation, with an R^2 of 0.9, suggesting that both biosurfactant and biodegradation genetic diversity are related. This and the observations presented by the *BioSurfDB Cluster Analysis* underlined the importance of analyzing both biosurfactant and biodegradation genes at the same time.

Statistics

According to the results presented in Fig. 5, from the comparison of the two most distant clusters: *Cluster 2*,

non-tropical metagenomes and *Cluster 1*, tropical metagenomes, it is possible to identify significant differences of more than 3% in the abundance of the microorganisms' genus. The abundance of *Mycobacterium* is significantly higher in *Cluster 2*, while Streptomyces is more abundant in *Cluster 1*.

Regarding the comparison of functional data, see Fig. 5b, the results demonstrate a significant prevalence of degradation of aromatic hydrocarbon genes in Cluster 1, composed of tropical metagenomes. These genes are associated with xylene and aromatic degradation, the metabolism of xenobiotics by cytochrome P450 and alnumycin biosynthesis.

From a different perspective, the soil and water metagenomes were also compared (Fig. 6). The *Alcanivorax* and *Escherichia* genera are more abundant in water metagenomes, while *Streptomyces* is more abundant in terrestrial metagenomes.

Functional comparison of terrestrial and water samples revealed that some cyclic hydrocarbon degradation pathways, namely toluene, chlorocyclohexane, chlorobenzene and nitrotoluene degradation are significantly more abundant in terrestrial metagenomes, while linear hydrocarbon degradation pathways, as alkane degradation and cytochrome P450 metabolism are significantly more abundant in water ecosystems. In addition, streptomycin and polyketides biosynthesis pathways are more representative of the water biomes, while Alnumycin biosynthesis is more abundant in terrestrial biomes. Methane metabolism is also significantly higher in terrestrial biomes.

Discussion

In this article we have focused on the biogeographical distribution of hydrocarbon degrading and biosurfactant producing genetic diversity, in the environment.

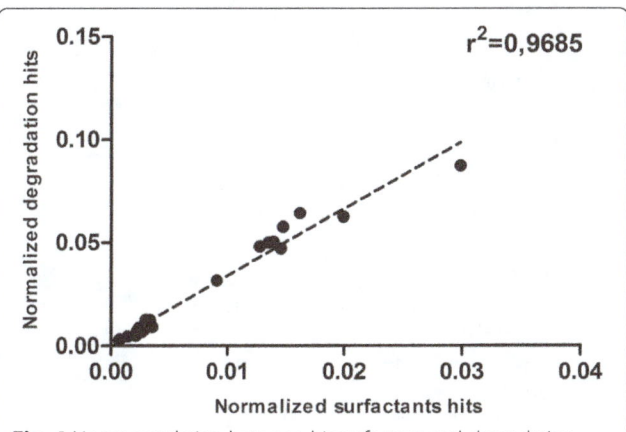

Fig. 4 Linear correlation between biosurfactant and degradation gene diversity

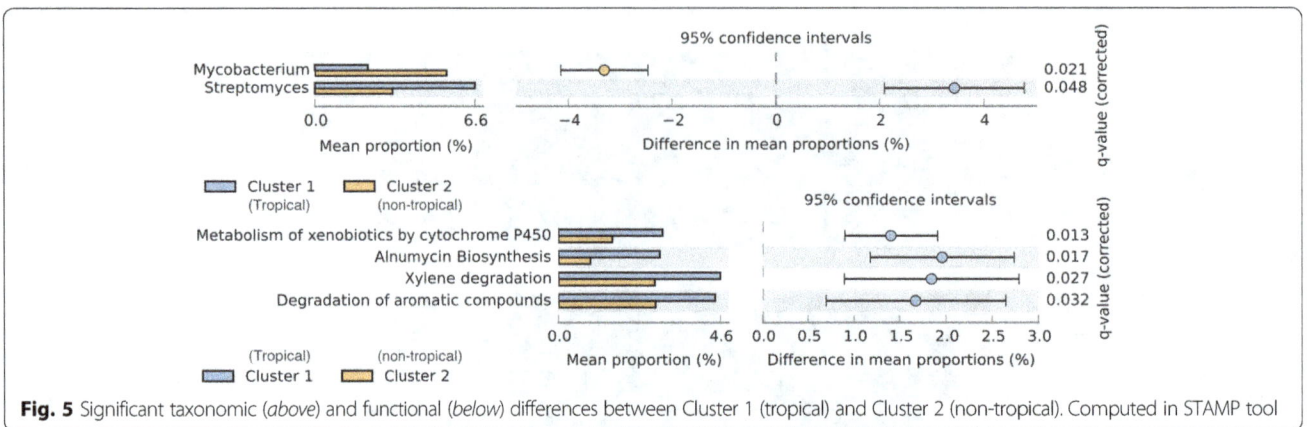

Fig. 5 Significant taxonomic (*above*) and functional (*below*) differences between Cluster 1 (tropical) and Cluster 2 (non-tropical). Computed in STAMP tool

Taxonomic analysis using RefSeq

Using Refseq databases, the formation of clusters from water and terrestrial metagenomes are in accordance with previous studies suggesting that the principle factor influencing the microbiota is if the substrate is terrestrial or water [21]. On a second level, metagenomes subjected to similar abiotic and biotic conditions such as sunlight, temperature, oxygen supply, osmotic and redox potential, pH and nutrient supply should have a similar bacterial community in their environments [22]. Therefore, these factors possibly determine the formation of the clusters observed in this paper.

Functional analysis using BioSurfDB

In the functional analysis performed using BioSurfDB, we analyzed all the available genes involved with the hydrocarbon degradation pathways along with the genes of the biosurfactant synthesis (Cluster 1 in Fig. 3). One of the main reasons for this analysis was the fact that

biodegradation is favored by the biosurfactant miscibility effect on hydrophobic material in order to assure its biodisponibility for bacteria.

Temperature is another factor that directly affects hydrocarbon biodegradation [23, 24]. Low temperatures are an important limitation to hydrocarbon biodegradation because they generate suboptimal environmental conditions for biodegradation such as increased viscosity, retarded volatilization of short-chain alkanes that are <C10, insolubility of long-chain alkanes, limited availability of water and nutrients; specifically, nitrogen, and extremes in pH and salinity [25]. In contrast, higher temperatures increase the rates of hydrocarbon metabolism to a maximum, typically in the range of 30 to 40 °C [23]. Moreover, in tropical areas there are high incidences of light and high average temperatures that favor photoautotrophic organisms, such as plants, algae and cyanobacteria, which can naturally synthesize linear or aromatic hydrocarbons [26–28]. Therefore, the higher

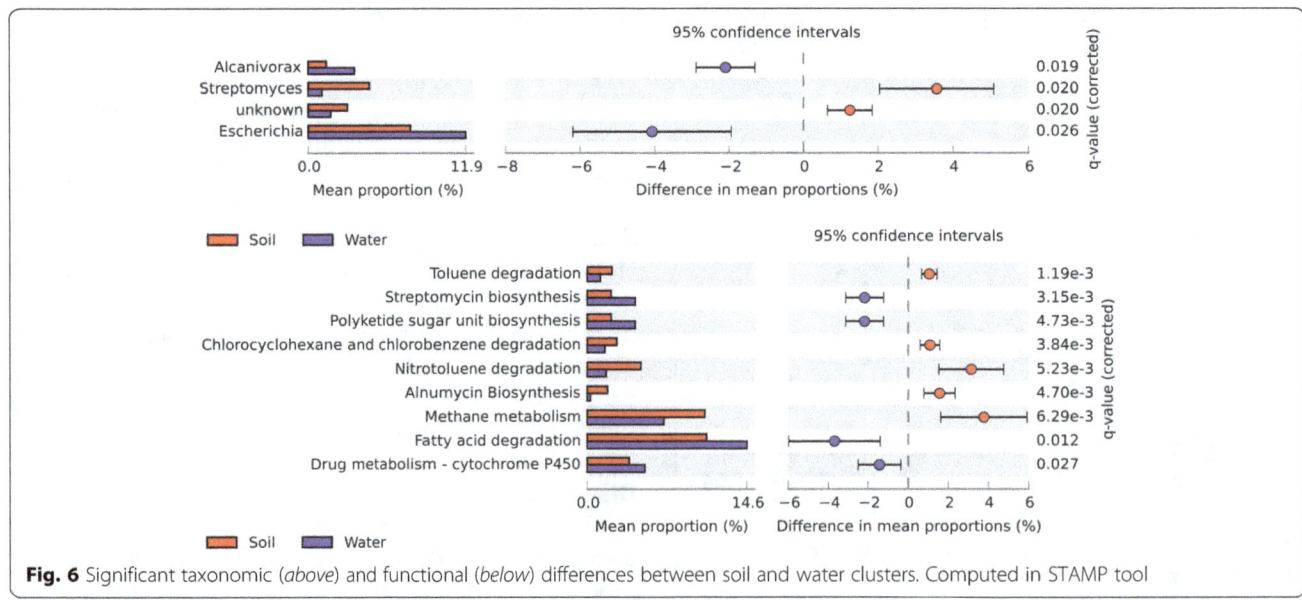

Fig. 6 Significant taxonomic (*above*) and functional (*below*) differences between soil and water clusters. Computed in STAMP tool

occurrence of degrading organisms and producers of biosurfactants in tropical areas see cluster 1 in Fig. 3) is probably favored by the documented higher bioavailability of hydrocarbons in these regions when compared with cold regions, see cluster 2 in Fig. 3 [4, 29, 30].

Correlation between biosurfactant production and biodegradation

The strong correlation (0.9) between degradation genes and genes involved in the biosynthesis of biosurfactants, observed in this study, reinforces the need for more research on biogeography distribution of both degradation and biosurfactants synthesis genes, to increase our understanding of their integrated action in the environment. This evidence is an important contribution to this knowledge, as most of the existing biogeographical studies on degradation and surfactant gene abundance analyze those pathways separately [1–3].

Statistical comparisons
Tropical vs. Non-tropical regions

Streptomyces and *Mycobacterium* are the most represented genus in tropical areas (Cluster 1 in Fig. 5) and non-tropical (Cluster 2 in Fig. 5), respectively. Both genera are described as capable of degrading hydrocarbons and produce biosurfactants [4]. In fact, hydrocarbon-degrading microorganisms are ubiquitous in several ecosystems, although they constitute less than 0.1% of the microbial community. However, in oil-polluted environments, they can represent up to 100% of the viable microorganisms [31]. Therefore, when we analyze the abundance of these genes in contaminated environments we are not only observing the natural dynamic or abundance of the bacterial community.

In this study, *Mycobacterium*, included in the *Actinobacteria* phylum, was the most representative genera in non--tropical (Cluster 2 in Fig. 5). Similarly, the first metagenomic analysis of permafrost samples showed Actinobacteria as a dominant phylum in accordance with the community composition reported from other polar soils [32]. Biofilm formation has been suggested to optimize the bioavailability of the substrate necessary for the growth of *Mycobacterium* under low concentrations of anthracene (PAH) [33]. However, biosurfactant production was not observed for *Mycobacterium* [33], which can explain the low abundance of surfactants in our results.

Soil vs. Water metagenomes

In this study, *Escherichia* and *Alcanivorax* genus were predominant in water metagenomes while *Streptomyces* was shown to be abundant in terrestrial metagenomes. *Escherichia* belongs to the *Enterobacteriaceae* family, which is not expected to show extracorporeal existence. However, the success of *E. coli* in the gut ecosystem, an example of a harsh environment, is thought to reflect its abilities to occupy different ecological niches. Corroborating this hypothesis, recent studies reporting the isolation of indigenous *E. coli* able to degrade hydrocarbon from contaminated soils [34, 35] showed the property of another bacterium from the *Escherichia* genus, the *E. fergusonii* KLU01, isolated from oil contaminated soil, as a hydrocarbon degrading, heavy metal tolerant and a potent producer of biosurfactant using diesel oil as the sole carbon and energy source [36]. Similarly, Sarma et al. 2004, isolated an enteric strain *Leclercia adecarboxylata* PS4040 from soil samples, collected from an oily sludge contaminated site that had had a contamination history of over 100 years, which is genotypically different from a clinical strain of *L. adecarboxylata* and showed that it can degrade other two- and three-benzene-ring PAH [37].

In water metagenomes, the *Alcanivorax* hydrocarbonoclastic genus is predominant when compared to those in soil. Despite being predominantly marine and described as almost exclusively linear alkane degrading and being up to 90% present in seawater contaminated with petroleum [38], it has also been found in some saline terrestrial environments contaminated with hydrocarbons [39]. Alkanes are open-chain hydrocarbons, which may represent up to 50% of the crude oil [40], and may also be synthesized by cyanobacteria [41], being rapidly degraded in marine environments [42]. Furthermore, the functional analysis of this study shows the predominance of the linear alkanes degradation pathway (fatty acid degradation) in water metagenomes and the predominant degradation genes of P450. This is probably due to the high incidence of the *Alcanivorax* genus that has a highly restricted genome of catabolic enzyme, since this organism uses predominantly aliphatic hydrocarbons as a source of carbon and energy and has several well-annotated genes encoding for AlkB1 and AlkB2 and Cytochrome P450 [43]. Furthermore, *Alcanivorax* and *Streptomyces*, are significantly abundant in clusters with a prevalence of genes involved in biosurfactant synthesis and hydrocarbon degradation which have also already been reported as biosurfactant producers [43–45].

Moreover, other studies noted the predominance of aromatic compound degradation genes in soil [46] when compared to alkane degradation genes AlkB. We observed the predominance of aromatic degradation genes in soil when compared with water metagenomes. This is possibly justified by the fact that polycyclic aromatic compounds are released into the atmosphere due to the use of fossil fuels and are subjected to chemical and physical degradation. Consequently, soils are the primary repository of aromatic compounds due to their capacity for retaining hydrophobic compounds [47]. *Streptomyces* are also typical soil bacteria already described as capable of utilizing

PAH and petroleum as carbon and energy sources [36, 37]. Our results are in accordance with this, as they showed significant predominance of *Streptomyces* in soil metagenomes.

Computational challenges

One of the main challenges in this research was investigating the possibility of obtaining new knowledge from the analysis of heterogeneous and publicly available metagenomics datasets. Advanced analytics, associated with high-performance computing, has made possible a more comprehensive analysis of many metagenomes. However, data integration often revealed deficiencies in data quality, e.g. inconsistency, redundancy, poor annotations and incompleteness. It was also clear that although the proposed bioinformatics pipeline could produce very interesting results, additional types of data should be considered to improve the knowledge regarding gene diversity. A more comprehensive analysis of these datasets should include DNA-Seq and RNA-Seq data to understand the ultimate activity of the identified genes.

One important result of this study is that the metagenomics data that is publicly available still needs to be improved in terms of its quality. Most of the available datasets are of poor quality, limiting the statistical significance of further analysis. In this research we have faced a 34% reduction in the size of the datasets when compared with the raw data.

Conclusion

From our research It was possible to see that: (1) surfaction and degradation are correlated events; (2) terrestrial biomes have more degradation genes, especially cyclic compounds, and less surfaction genes when compared to water biomes; and (3) latitude has a significant influence on the diversity of genes involved in biodegradation and biosurfactant production, suggesting that microbiomes near the equator have richer genes that have a role in these processes.

This information can be used in the application of bioremediation techniques, by taking into considering the biostimulation, optimization or manipulation of microbial consortia from these areas.

Additional files

Additional file 1: Table S1. Metagenomes Summary. Country, number of samples and sequencing technology for each biome.

Additional file 2: Table S2. Samples Information. Feature, location, run and project SRA information for each sample.

Additional file 3: Figure S1. Sample Geography. Geographical distribution of the metagenome samples.

Additional file 4: Table S3. BioSurfDB Pathways. Name and KEGG Map ID for Alkane biodegradation and surfactant biosynthesis pathways analyzed.

Additional file 5: Figure S2. Rarefaction Curve. Rarefaction Curves performed in MEGAN.

Abbreviations

CPU: Central processing unit; DNA: Deoxyribonucleic acid; DNA-Seq: DNA sequencing; GC: Guanine-cytosine; NCBI: National Center for Biotechnology Information; PAH: Poly-aromatic-hydrocarbons; PCoA: Principal coordinates analysis; RAM: Random-access-memory; RNA: Ribonucleic acid; RNA-Seq: RNA sequencing; UPGMA: Unweighted pair group method with arithmetic mean

Acknowledgements

JO and AV were funded by FAPERJ. RF was funded by FCT under project EXCL/EEI-ESS/0257/2012. JO, AB and SA were funded by CAPES. CM was funded by FUNPEC. ATF was funded by FCT under project UID/CEC/50021/2013. LL, WA, RP and AC were funded by CNPq and CAPES.

Funding

This work was supported by the Conselho Nacional de Desenvolvimento Científico e Tecnológico (CNPq-Brazil), Coordenação de Aperfeiçoamento de Pessoal de Nível Superior (CAPES-Brazil), Fundação Norte-Rio-Grandense de Pesquisa e Cultura (FUNPEC), Fundação de Amparo à Pesquisa do Rio de Janeiro (FAPERJ) and Fundação para a Ciência e a Tecnologia (FCT) with references UID/CEC/50021/2013 and EXCL/EEI-ESS/0257/2012.

Authors' contributions

AV, AF and LL have made substantial contributions to conception and design; JO, WA and JF were responsible for acquisition of data and analysis; and JO, WA, RP, AG, SA, CM and AC were responsible for the interpretation of data; All the authors have been involved in drafting the manuscript or revising it critically for important intellectual content; All the authors have given final approval of the version to be published.

Competing interests

The authors declare that they have no competing interests.

Author details

[1]Laboratório de Biologia Molecular e Genômica, Departamento de Biologia Celular e Genética, Centro de Biociências, Universidade Federal do Rio Grande do Norte, Natal, RN, Brazil. [2]INESC-ID/IST Instituto de Engenharia de Sistemas e Computadores/Instituto Superior Técnico, Universidade de Lisboa, Rua Alves Redol, 9, 1000-029 Lisbon, Portugal. [3]Laboratório de Bioinformática, Laboratório Nacional de Computação Científica, Petrópolis, RJ, Brazil.

References

1. Kurata N, Vella K, Hamilton B, Shivji M, Soloviev A, Matt S, Tartar A, Perrie W. Surfactant-associated bacteria in the near-surface layer of the ocean. Sci Rep. 2014;6:19123.

2. Jan BVB, Li Z, Wouter D, Last BW. Diversity of alkane hydroxylase systems in the environment. Oil Gas Sci Technol. 2003;58(4):427–40. –rev. IFP,

3. Bodour AA, Drees KP, Maier RM. Distribution of biosurfactant producing bacteria in undisturbed and contaminated arid southwestern soils. Appl Environ Microbiol. 2003;69:3280–7.

4. Jitendra DD, Ibrahim MB. Microbial production of surfactants and their commercial potential. Microbiol Mol Biol Rev. 1997:61(1):47–64.

5. Hassanshahian M, Zeynalipour MS, Musa FH. Isolation and characterization of crude oil degrading bacteria from the Persian Gulf (Khorramshahr provenance). Mar Pollut Bull. 2014;82(1–2):39–44.

6. Wallisch S, Gril T, Dong X, Welzl G, Bruns C, Heath E, Engel M, Suhadolc M, Schloter M. Effects of different compost amendments on the abundance and composition of alkB harboring bacterial communities in a soil under industrial use contaminated with hydrocarbons. Front Microbiol. 2014;5:96.

7. Gilbert JA, Jansson JK, Knight R. The earth microbiome project: successes and aspirations. BMC Biol. 2014;12:69.

8. Delmont TO, Robe P, Cecillon S, Clark IM, Constancias F, Simonet P, Hirsch PR, Vogel TM. Terra genome: a consortium for the sequencing of a soil metagenome. Nat Rev Microbiol. 2009;7:252.

9. Pylro VS, Roesch L, Ortega JM, do Amaral AM. Brazilian microbiome project: revealing the unexplored microbial diversity— challenges and prospects. Microb Ecol. 2014;67:237–41.

10. Nesme J, Achouak W, Agathos SN, Bailey M, Baldrian P, Brunel D, Frostegård A, Heulin T, Jansson JK, Jurkevitch E, Kruus KL, Kowalchuk GA, Lagares A, Lappin-Scott HM, Lemanceau P, Paslier DL, Mandic-Mulec I, Murrell JC, Myrold DD, Nalin R, Nannipieri P, Neufeld JD, Gara FO, Parnell JJ, Pühler A, Pylro V, Ramos JL, Roesch LFW, Schloter M, Schleper C, Sczyrba A, Sessitsch A, Sjöling S, Sørensen J, Sørensen SJ, Tebbe CC, Topp E, Tsiamis G, JDV E, Keulen GV, Widmer F, Wagner M, Zhang T, Zhang X, Zhao L, Zhu YG, Vogel TM, Simonet P. Back to the future of soil metagenomics. Front Microbiol. 2016;10(7):73.

11. Meyer F, Paarmann D, D'Souza M, Olson R, Glass EM, Kubal M, Paczian T, Rodriguez A, Stevens R, Wilke A, Wilkening J, Edwards RA. The metagenomics RAST server – a public resource for the automatic phylogenetic and functional analysis of metagenomes. BMC Bioinf. 2008;9(386):1–8.

12. Huson DH, Mitra S, Ruscheweyh H, Weber N, Schuster SC. Integrative analysis of environmental sequences using MEGAN4. Genome Res. 2011;21:1552–60.

13. Wood DE, Salzberg SL. Kraken: ultrafast metagenomic sequence classification using exact alignments. Genome Biol. 2014;15:R46.

14. Treangen TJ, Koren S, Sommer DD, Liu B, Astrovskaya I, Ondov B, Aaron BO. MetAMOS: a modular and open source metagenomic assembly and analysis pipeline. Genome Biol. 2013;14(1):R2.

15. Oliveira JS, Araújo W, Sales AIL, Guerra AB, Araújo SCS, Vasconcelos ATR, Agnez-Lima LF, Freitas AT. BioSurfDB: knowledge and algorithms to support biosurfactants and biodegradation studies. Database. 2015;2015:1–8. https://doi.org/10.1093/database/bav033.

16. Tatusova T, Ciufo S, Fedorov B, O'Neill K, Tolstoy I. RefSeq microbial genomes database: new representation and annotation strategy. Nucleic Acids Res. 2014;42(1):D553–9.

17. Kiełbasa SM, Wan R, Sato K, Horton P, Frith MC. Adaptive seeds tame genomic sequence comparison. Genome Res. 2011;21(3):487–93.

18. Altschul SF, Gish W, Miller W, Myers EW, Lipman DJ. Basic local alignment search tool. J Mol Biol. 1990;215:403–10.

19. Sturn A, Quackenbush J, Trajanoski Z. Genesis: Cluster analysis of microarray data. Bioinformatics. 2002;18(1):207–8.

20. Parks DH, Tyson GW, Hugenholtz P, Beiko RG. STAMP: Statistical analysis of taxonomic and functional profiles. Bioinformatics. 2014;30:3123–4.

21. Jeffries TC, Seymour JR, Gilbert JA, Dinsdale EA, Newton K, Leterme SSC, Roudnew B, Smith RJ, Seuront L, Mitchell JG. Substrate type determines metagenomic profiles from diverse chemical habitats. PLoS One. 2011; 6(9):e25173.

22. Standing D, Killham K. 2006. Modern soil microbiology, 2nd Edition. Chapter 1, International standard book number-13: 978-1-4200-1520-1.

23. Bossert I, Bartha R. The fate of petroleum in soil ecosystems, p. 434-476. In: Atlas RM, editor. Petroleum microbiology. New York: Macmillan Publishing Co; 1984.

24. Davis SJ, Gibbs CF. The effect of weathering on crude oil residue exposed at sea. Water Res. 1975;9:275–85.

25. Margesin R. Potential of cold-adapted microorganisms for bioremediation of oil-polluted Alpine soils. Int Biodeterior Biodegrad. 2000;46:3–10.

26. Pattanaik B, Lindberg P. Terpenoids and their biosynthesis in cyanobacteria. Life. 2015;5(1):269–93.

27. Winters K, Parker PL, van Baalen C. Hydrocarbons of blue-green algae: geochemical significance. Science. 1969;163(3866):467–8.

28. Timmis KN, McGenity TJ, van der Meer, JR, de Lorenzo V. 2010. Handbook of hydrocarbon and lipid microbiology.

29. Leahy JG, Colwell RR. Microbial Degradation of Hydrocarbons in the Environment. Microbiol Rev. 1990;54(3):305–15.

30. Eliora ZR, Eugene R. Natural roles of biosurfactants. Environ Microbiol. 2001; 3(4):229–36.

31. Atlas RM. Microbial degradation of petroleum hydrocarbons: an environmental perspective. Microbiol Rev. 1981;45(1):180–209.

32. Barabas G, Vargha G, Szab IM, Penyige A, Damjanovich S, Szöllösi J, Matk J, Hirano T, Matyus A, Szab I. n-Alkane uptake and utilisation by Streptomyces strains. Antonie Van Leeuwenhoek. 2001;79:269–76.

33. Wick LY, de Munain AR, Springael D, Harms H. Responses of *Mycobacterium sp.* LB501T to the low bioavailability of solid anthracene. Appl Microbiol Biotechnol. 2002;58:378–85.

34. Yergeau E, Hogues H, Whyte LG, Greer CW. The functional potential of high Arctic permafrost revealed by metagenomic sequencing, qPCR and microarray analyses. ISME J. 2010;4:1206–14.

35. Ferradji FZ, Fodil D, Mnif S, Eddouaouda K, Badis A, Rebbani S, Sayadi S. Naphthalene and crude oil degradation by biosurfactant producing *Streptomyces spp.* isolated from Mitidja plain soil (North of Algeria). Int Biodeterior Biodegrad. 2014;86:300–8.

36. Shekhar SK, Godheja J, Modi DR. Hydrocarbon bioremediation efficiency by five indigenous isolated from contaminated soils. Int J Curr Microbiol Appl Sci. 2014;4(3):892–905.

37. Sriram MI, Kalishwaralal K, Deepak V, Gracerosepat R, Srisakthi K, Gurunathan S. Biofilm inhibition and antimicrobial action of lipopeptide biosurfactant produced by heavy metal tolerant strain *Bacillus cereus* NK1. Colloids Surf B: Biointerfaces. 2011;85:174–81.

38. Sarma PM, Bhattacharya D, Krishnan S, Lal B. Degradation of polycyclic aromatic hydrocarbons by a newly discovered enteric bacterium, *Leclercia adecarboxylata*. Appl Environ Microbiol. 2004;70(5):3163–6.

39. Harayama S, Kishira H, Kasai Y, Shutsubo K. Petroleum biodegradation in marine environments. J Molec Microbiol Biotechnol. 1999;1(1):63–70.

40. Yakimov MM, Timmis KN, Golyshin PN. Obligate oil-degrading marine bacteria. Curr Opin Biotechnol. 2007;18:257–66.

41. Rojo F. Degradation of alkanes by bacteria. Environ Microbiol. 2009; 11(10):2477–90.

42. McGenity TJ, Folwell BD, McKew BA, Sanni GO. Marine crude-oil biodegradation: a central role for interspecies interactions. Aquat Biosyst. 2012;8:10.

43. Schneiker S, Santos VAP, Bartels D, Bekel T, Brecht M, Buhrmester J, Chernikova TN, Denaro R, Ferrer M, Gertler C, Goesmann A, Golyshina OV, Kaminski F, Khachane AN, Lang S, Linke B, AC MH, Meyer F, Nechitaylo T, Pühler A, Regenhardt D, Rupp O, Sabirova JO, Selbitschka W, Yakimov MM, Timmis KN, Vorhölter F, Weidner S, Kaiser O, Golyshin PN. Genome sequence of the ubiquitous hydrocarbon-degrading marine bacterium *Alcanivorax borkumensis*. Nat Biotechnol. 2006;24:997–1004.

44. Batista SB, Mounteer A, Amorim FR, Totola MR. Isolation and characterization of biosurfactant/bioemulsifier-producing bacteria from petroleum contaminated sites. Bioresour Technol. 2006;97:868–75.

45. Wang L, Wang W, Lai Q, Shao Z. Gene diversity of CYP153A and AlkB alkane hydroxylases in oil-degrading bacteria isolated from the Atlantic Oceanemi. Environ Microbiol. 2010;12(5):1230–42.

46. Liu Q, Tang J, Bai Z, Hecker M, Giesy JP. Distribution of petroleum degrading genes and factor analysis of petroleum contaminated soil from the Dagang Oilfield, China. Sci Rep. 2015;5:11068.

47. Wild SR, Jones KC. Polynuclear aromatic hydrocarbons in the united kingdom environment: a preliminary source inventory and budget. Environ Pollut. 1995;88:91–108.

12

YPTB3816 of *Yersinia pseudotuberculosis* strain IP32953 is a virulence-related metallo-oligopeptidase

Ali Atas[1], Alan M. Seddon[1], Donna C. Ford[3], Ian A. Cooper[3], Brendan W. Wren[2], Petra C. F. Oyston[3] and Andrey V. Karlyshev[1*]

Abstract

Background: Although bacterial peptidases are known to be produced by various microorganisms, including pathogenic bacteria, their role in bacterial physiology is not fully understood. In particular, oligopeptidases are thought to be mainly involved in degradation of short peptides e.g. leader peptides released during classical protein secretion pathways. The aim of this study was to investigate effects of inactivation of an oligopeptidase encoding gene *opdA* gene of *Yersinia pseudotuberculosis* on bacterial properties in vivo and in vitro, and to test dependence of the enzymatic activity of the respective purified enzyme on the presence of different divalent cations.

Results: In this study we found that oligopeptidase OpdA of *Yersinia pseudotuberculosis* is required for bacterial virulence, whilst knocking out the respective gene did not have any effect on bacterial viability or growth rate in vitro. In addition, we studied enzymatic properties of this enzyme after expression and purification from *E. coli*. Using an enzyme depleted of contaminant divalent cations and different types of fluorescently labelled substrates, we found strong dependence of its activity on the presence of particular cations. Unexpectedly, Zn^{2+} showed stimulatory activity only at low concentrations, but inhibited the enzyme at higher concentrations. In contrast, Co^{2+}, Ca^{2+} and Mn^{2+} stimulated activity at all concentrations tested, whilst Mg^{2+} revealed no effect on the enzyme activity at all concentrations used.

Conclusions: The results of this study provide valuable contribution to the investigation of bacterial peptidases in general, and that of metallo-oligopeptidases in particular. This is the first study demonstrating that *opdA* in *Yersinia pseudotuberculsosis* is required for pathogenicity. The data reported are important for better understanding of the role of OpdA-like enzymes in pathogenesis in bacterial infections. Characterisation of this protein may serve as a basis for the development of novel antibacterials based on specific inhibition of this peptidase activity.

Keywords: *Yersinia pseudotuberculosis*, Proteases, Proteolysis, Oligopeptidases, Metallopeptidases, Virulence

Background

Proteases are ubiquitous enzymes found in both eukaryotes and prokaryotes and play a pivotal role in many biological functions [1, 2]. Oligopeptidases, such as OpdA, OpdB and Dcp, specifically cleave short peptides and are inactive against full-length large proteins such as casein [1]. OpdA (or PrlC [3]) is a cytoplasmic zinc-dependent oligopeptidase belonging to the M3A subfamily of proteases containing a highly conserved alpha-helix forming zinc binding motif HEXXH located in a channel restricting access of proteins [4]. The enzymatic activity requires binding of the histidine residues with a zinc cation, and the glutamic acid residue carries out the catalytic role [5]. OpdA was first identified in *Salmonella enterica* serovar Typhimurium as an enzyme capable of hydrolysing *N*-acetyl-L-alanyl-L-alanyl-L-alanyl-L-alanine (AcAla$_4$) [6], and subsequently was also identified in *Escherichia coli*

* Correspondence: a.karlyshev@kingston.ac.uk
[1]School of Life Sciences, Pharmacy and Chemistry; Faculty of Science, Engineering and Computing, Kingston University, Penrhyn Road, Kingston upon Thames KT1 2EE, UK
Full list of author information is available at the end of the article

[1, 7, 8]. A tyrosine residue Y607 was identified as a key residue in substrate recognition [7].

OpdA and some eukaryotic oligopeptidases, such as thimet oligopeptidase (TOP) and neurolysin, share the same zinc binding motif HEFGH and some amino acid sequence similarity and can be differentiated by substrate specificity [8]. In contrast to mammalian oligopeptidases, the biological role of OpdA is not well understood and has only been investigated for *E. coli* and *S.* Typhimurium. Diverse functions have been suggested for the enzyme, such as signal peptide break down [1, 9], downstream hydrolysis of peptides for amino acid recycling [1], and the roles in the development of phage P22 in *Salmonella* Typhimurium [10] and in a heat shock response [11].

Apart from *S.* Typhimurium and *E. coli* there have been no reports on the characterization of OpdA in other bacterial species. The aim of this study was to characterize an OpdA homologue in a further member of the Enterobacteriaceae, *Yersinia pseudotuberculosis*. Hosts infected with this pathogen show symptoms of mesenteric lymphadenitis, gastroenteritis and septicemia in a disease termed yersiniosis [12]. We show here that inactivation of a putative OpdA encoding gene in *Y. pseudotuberculosis* IP32953 (YPTB3816, also annotated as *prlC*, GenBank accession number CAH23054.1) resulted in attenuation in a murine model of infection, suggesting a role for OpdA in the pathogenesis of *Y. pseudotuberculosis*. The derived amino acid sequence of this protein revealed 81% identity (100% coverage) with that of *E. coli* OpdA (GenPept accession number NP_417955) suggesting some similarity in their functions. The recombinant OpdA of *Y. pseudotuberculosis* was purified, and its catalytic activity was characterized.

Methods
Bacterial strains and growth conditions
E. coli XL1 Blue and *Y. pseudotuberculosis* IP32953 strains were grown on Luria Bertani plates or in LB broth at 37 or 28 °C respectively. When required, the media were supplemented with kanamycin (50 µg/ml) or chloramphenicol (10 µg/ml).

Construction of the *Y. pseudotuberculosis* ΔopdA mutant
Mutagenesis of *Y. pseudotuberculosis* was performed as described previously [13, 14]. Briefly, a PCR product, containing the kanamycin resistance gene from plasmid pUC4K and flanking regions corresponding to the 5′ and 3′ proximal parts of *opdA*, was generated using the following primers: Yptb3816_kan_for (CCGTTCTCCCTG CCACCGTTTTCTGCTATTCGGCCTGAAGATATCGT GCCCACAGGAAACAGCTATGACC) and Yptb3816_- kan_rev (GCAACATGGCATCTAACTGCGGTTCACG GCCACGGAAGCGTTTGAACAGTCAAGTCAGCGT AATGCTCTGC). The PCR product was transformed

into *Y. pseudotuberculosis* IP32953/pAJD434 by electroporation. Transformants were verified by PCR using screening primers Yptb3816_for (ATGACAAACCCGC TGTTGACT) and Yptb3816_rev (TTAGCCCTTAATA CCGTAATGAC) (Additional file 1: Figure S1). The mutant was cured of the helper plasmid, and the presence of the virulence plasmid pYV was confirmed using primers yscU-for (TCTGTACTGTTGGCTTTGTGC) and yscU-rev (TTGCGCACAGTCTGAACTTGG). The procedure resulted in a deletion of 98% of the *opdA* gene.

Effect of *opdA* mutation on bacterial fitness in vivo
Six to eight week old female BALB/c mice were obtained from a commercial supplier (Charles River, United Kingdom). On arrival, mice were housed in groups of 5 in polypropylene solid bottom cages with a wire mesh lid, integral diet hopper and water bottle holder (M3, NKP cages, Coalville, UK) within a UK Advisory Committee on Dangerous Pathogens (ACDP) level 3 isolator and allowed to acclimatize before experimental use. Mice were provided with ad libitum irradiated water and ad libitum irradiated diet (5002 Certified Rodent Diet, LabDiet, St Louis, Missouri, USA). Mice were provided with corn cob bedding (1014 Corn Cob, IPS Product Supplies Ltd, London, UK) with enrichment provided as a dome home (LBS Biotech, Crawley UK), aspen wood wool (LBS Biotech, Crawley, UK) and hemp fiber mat (Happi-Mat, Marshall Bio-Resources, Hull, UK). Lighting cycle was 12 h light, 12 h dark with environmental temperatures and humidity maintained within the specified range for rodents under ASPA. Mice were checked a minimum of twice daily, with clinical signs observed, scored and recorded and used to apply the humane endpoint specified in the project license for mice challenged with *Yersinia* spp. Mice were observed at least twice daily for end-point criteria, including loss of appetite, hunched posture, gait and righting difficulty, prostration, ruffled fur and gummy eyes. The animals that reached end-point criteria and animals that survived through the end of the experiment were humanly euthanized by cervical dislocation.

In vivo competitive index (CI) studies were performed as described in [13]. Briefly, mutant and wild type strains were grown separately to exponential phase, cells deposited by centrifugation and the pellet washed once with sterile PBS. The bacteria were re-suspended in PBS and the OD_{600} adjusted to 0.55–0.60. Wild type and mutant bacterial suspensions were then mixed in a 1:1 ratio and serially diluted with sterile PBS to produce bacterial suspension with approximately 1×10^3 cfu/ml. Groups of six mice were then dosed with 0.1 ml of bacterial suspension by the intravenous (i.v.) route. Retrospective viable counts were determined by plating out dilutions (in triplicate) on LB agar and LB agar supplemented with kanamycin to determine the input ratio. Mice were killed by cervical dislocation on day 5, with spleens collected after confirmation of death. Spleens were passed through 70 µm sieves (Becton Dickinson) to produce a cell suspension in 3 ml of PBS. Cell suspensions were serially diluted in sterile PBS and plated onto LB agar and LB agar supplemented with kanamycin to determine the output ratio. The CI is defined as the output ratio (mutant/wild type) divided by the input ratio (mutant/wild type) [15].

Testing effect of *opdA* mutation on bacterial growth rate in vitro

For growth curves, bacteria were suspended to an OD_{590} of 0.05 in 50 ml LB broth and incubated with shaking (200 rpm) overnight at 28 °C. Both wild-type and mutant strains grew to a similar density during overnight incubation. The bacteria were pelleted by centrifugation and washed once with LB broth before being re-suspended to an OD_{590} of 0.05 in L-broth. Growth curves were performed in a 96 well microtitre plate format. Outer wells were filled with 200 µl distilled water to reduce evaporation and test wells with 200 µl of each test culture. Each strain was tested three times each with six technical replicates. A sterile gas permeable membrane (Breathe-Easy, Diversified Biotech) was used to seal the 96-well plates. Growth curves were generated using a microplate reader (Multiskan FC, Thermo Scientific) housed in a class II biological safety cabinet. The plate was incubated at 28 °C with shaking at 5 Hz, amplitude 15 mm and the OD_{595} recorded every 15 min for 24 h.

Expression and purification of OpdA

Genomic DNA of *Y. pseudotuberculosis* IP32953 was used as a template to amplify the *opdA* gene using PCR. The primers used to amplify the gene were: GAT<u>TCTAGA</u>A GAAGGAGATATACCATGCATCATCATCATCATCACA CAAACCCGCTGTTGACTCCGTTCTCCCTG (forward) and TTAGCCCTTAATACCGTAATGACGCAAC (reverse). Nucleotides underlined indicate the introduced restriction site for *Xba*I required for sub-cloning. The amplified 2.1 kb fragment was cloned into the pGEM-T Easy vector (Promega) using the manufacturer's instruction. After sequence verification, the *Xba*I/*Sph*I fragment was subcloned into the expression vector pBAD33 [16] to produce plasmid pBAD33opdA-His. The derived recombinant plasmid encoded a full copy of OpdA protein with an N-terminal 6xHis tag.

E. coli XL1 Blue (Stratagene) strain carrying pBAD33opdA-His was grown at 37 °C in LB broth (Sigma) supplemented with 10 µg/ml chloramphenicol (Sigma). Expression of OpdA fusion protein (referred to as OpdA in this article) was induced at an optical density at 600 nm of 0.6–0.7 by addition of 0.1% *w/v* L-arabinose (Acros Organics) followed by incubation for 1 h at 37 °C. The protein was purified according using Ni^{2+}-NTA Fast

Start Kit (Qiagen) according to manufacturer's protocol. Briefly, the cells were harvested by centrifugation and incubated for 30 min on ice with lysis buffer (Qiagen) containing 50 mM monosodium phosphate, pH 8.0, 300 mM NaCl, 10 mM imidazole and 3 µg lysozyme. After removal of the cellular debris by centrifugation at 4000 rpm for 30 min at 4 °C, the cleared lysate was loaded onto a Ni^{2+}-NTA Fast Start Kit column (Qiagen), and the enzyme was purified under native conditions following the manufacturer's protocol. Proteins were analyzed using sodium dodecyl sulfate polyacrylamide gel electrophoresis (SDS-PAGE). The eluate (1 ml) was dialyzed at room temperature against phosphate buffered saline (PBS) (1000 ml) using 3.5 kDa dialysis tubing (Sigma) to remove imidazole that may have interfered with the enzyme assays. Protein concentration was determined using bicinchoninic acid assay (BCA) kit (Fisher Scientific). The dialyzed protein was stored at 4 °C.

Enzymatic assays

In this study we used internally quenched fluorogenic substrates, containing Abz (fluorophore) and EDDnp (quencher) at the N-terminal and C-terminal ends respectively. Fluorescence is induced upon digestion of the peptide, followed by separation of the fluorophore from the quencher. Fluorescence of the products of hydrolysis of Abz-NKPRRPQ-EDDnp and Abz-AAL-EDDnp substrates (LifeTein) by OpdA was measured at 37 °C in 50 mM Tris–HCl buffer, pH 7.0, using a FLUOstar Optima plate reader (BMG LABTECH) with filters $\lambda ex = 320$ nm and $\lambda em = 420$ nm. The reaction was initiated by the addition of 20 µM substrate. Unless stated otherwise, the incubation time was 5 min. The relative activity was estimated using a formula: $(T-R)/(C-R)*100\%$, where T, C and R are test, control and reference samples respectively. The C sample refers to activity of the enzyme with no additive, and R refers to fluorescence background of the sample containing peptide only (no enzyme).

In order to analyze the effects of cations, the purified OpdA was subjected to treatment with 10 mM ethylene glycol tetraacetic acid (EGTA) for 10 min at room temperature followed by dialysis against PBS buffer for 2 h at room temperature. This treatment was repeated twice if required. The samples were incubated in 50 mM Tris–HCl buffer, (pH 7.0) supplemented with metal cations at required concentrations, at room temperature for 5 min.

To determine the effect of inhibitors on OpdA, the enzyme was first incubated with the inhibitor in 50 mM Tris–HCl buffer, pH 7.0 at room temperature for 5 min and the reaction initiated with 20 µM fluorogenic substrate. The inhibitors evaluated were 74 µM antipain, 1 mM EGTA, 0.1 mM chymostatin, 0.13 mM bestatin, 0.01 mM N-[N-(N-acetyl-L-leucyl)-L-leucyl]-L-norleucine

(ALLN), 0.01 mM leupeptin, 1 mM PMSF, 1 mM 4-(2-aminoethyl) benzenesulfonyl fluoride (AEBSF), 28 μM E-64 and 0.01 mg/ml phosphoramidon according to manufacturer's protocol (G-Biosciences). In order to study effects of pH on the enzymatic activity, the pH of the test buffer was adjusted with hydrochloric acid.

Statistical analysis of the results was carried out using a one way ANOVA. P value less than 0.05 was used to demonstrate statistically significant difference. Data were expressed as means ± SD of three readings from each of two independent experiments.

Results

The gene encoding OpdA of Y. pseudotuberculosis is required for virulence

In order to determine whether OpdA plays a role in pathogenesis in *Yersinia*, the gene encoding OpdA was disrupted in *Y. pseudotuberculosis* IP32953 (Additional file 1: Figure S1). Mice were then infected with mutant and wild-type *Y. pseudotuberculosis* and the CI was calculated. A CI value of 0.2 or less indicates that the locus is attenuating. The CI of the OpdA-defective mutant was 0.05, identifying OpdA as a potential virulence-associated

protein. In contrast, no growth defect in vitro could be detected (Additional file 2: Figure S4).

Purified OpdA protein reveals oligo-peptidase activity

OpdA was expressed as an N-terminal 6xHis fusion protein and purified from *E. coli*. Position of the recombinant protein on the gel (Fig. 1) fully corresponds to its molecular weight (77 kDa) estimated from the amino acid sequence. The protein was highly pure, stable and produced at a yield of 2 mg/l (Fig. 1).

No activity of the purified enzyme could be detected employing commercial protease assay (Protease Screening™ kit, Geno-Technology Inc) using casein as a substrate (data not shown). However, a very strong activity with Abz-NKPRRPQ-EDDnp substrate confirmed that the enzyme is an oligopeptidase. It was also found that, although EGTA inhibited the activity of the native enzyme (indicating that this was a metallopeptidase), a full inhibition of activity could not be achieved (Additional file 3: Figure S2). We reasoned that this could be due to extremely high affinity of the enzyme for metal cations, leaving residual amount of bound cations after a single treatment with EGTA. This hypothesis was confirmed by repeated treatment and dialysis, which resulted in a significant reduction

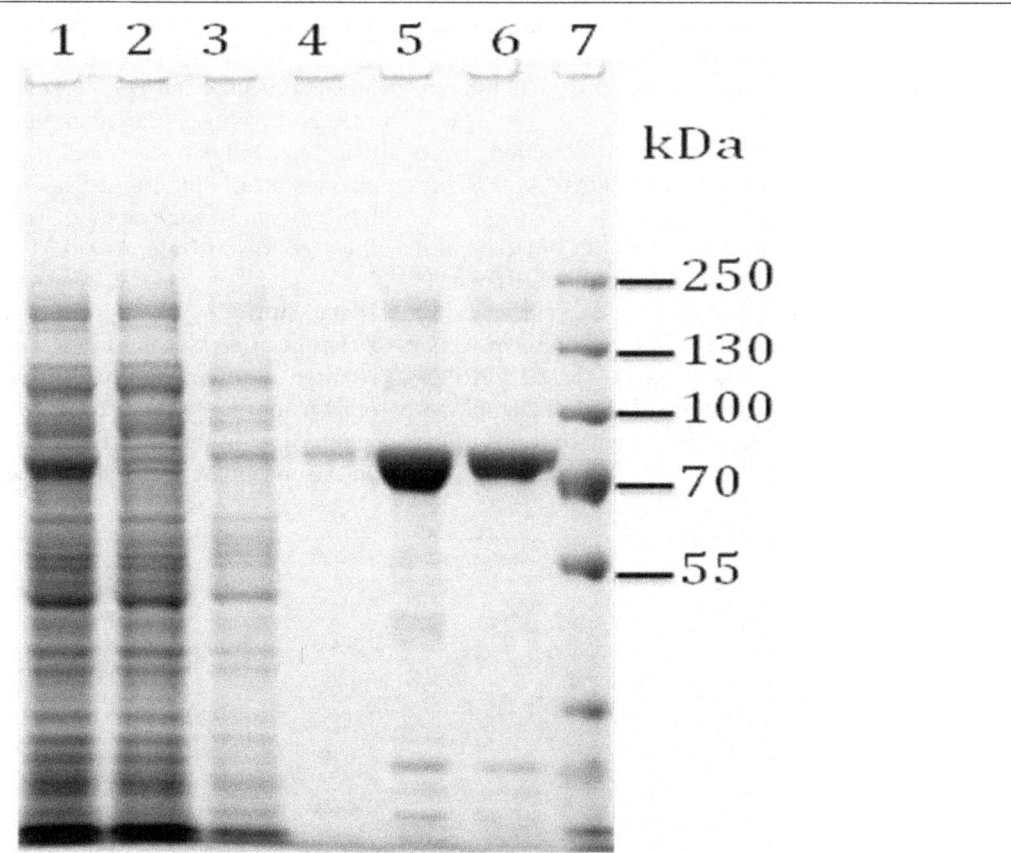

Fig. 1 Purification of OpdA protein from *E. coli* XL1/pBAD33opdA-His strain after induction with L-arabinose. *Lane 1*, cell lysate; *lane 2*, flow-through; *lane 3*, wash 1; *lane 4*. wash 2, *lanes 5* and *6*, eluates; *lane 7* protein size markers (Fisher Scientific)

of activity. Unless stated otherwise, the experiments described in this study were conducted with this enzyme 'fully depleted' of metal cations following dialysis treatment.

Effect of pH on the activity of OpdA

Optimal pH range for OpdA activity with Abz-NKPRRPQ-EDDnp substrate was found to be between 6.0 and 8.0 (Fig. 2). The activities at pH 6.0 and pH 8.5 were 50 and 60% respectively when compared with the highest activity observed at pH 6.5.

Effect of divalent cations on the hydrolytic activity of OpdA

The effect of different cations on the activity of OpdA and its ability to hydrolyze the substrate Abz-NKPRRPQ-EDDnp was evaluated (Fig. 3). Addition of Ca^{2+} and Mn^{2+} did not produce any statistically significant change in activity at low concentration, but higher concentrations stimulated hydrolysis (Fig. 3a and b respectively). Addition of Co^{2+} stimulated hydrolysis at all concentrations (Fig. 3c) while Mg^{2+} and Cu^{2+} did not show any stimulatory effect on activity of OpdA at any of the concentrations tested (data not shown). Zn^{2+}, the proposed metal cofactor of OpdA, increased activity of the enzyme in the assay at low concentrations, but there appeared to be a limit to the concentration of Zn^{2+} the protein could tolerate, as the stimulation was abrogated at higher concentrations (Fig. 3e).

To determine whether the effects observed were substrate-specific, a second substrate, Abz-AAL-EDDnp was evaluated in the same way (Fig. 4).

Similar to a reaction with Abz-NKPRRPQ-EDDnp, the hydrolysis of this substrate required much higher concentrations of Ca^{2+} and Mn^{2+} compared to Co^{2+} (Fig. 4a-c).

The enzyme was unaffected by the addition of Mg^{2+} and Cu^{2+} (data not shown). As with Abz-NKPRRPQ-EDDnp, the hydrolysis of Abz-NKPRRPQ-EDDnp was stimulated at low, but inhibited at high concentrations of Zn^{2+} (Fig. 4e).

To ascertain whether the inhibitory effect of zinc was due to steric changes in the active site following zinc binding, EGTA-treated OpdA was first incubated with 0.01 mM Zn^{2+} for 5 min, and then with the substrate Abz-NKPRRPQ-EDDnp for another 5 min at room temperature, to allow the reaction to proceed. This was followed by the addition of 0.1 mM Zn^{2+} in an attempt to inhibit the enzyme. The reaction was transferred to 37 °C and the hydrolysis of the peptide was analyzed for an hour. The results demonstrated the inhibition of hydrolysis after the addition of 0.1 mM Zn^{2+} (Fig. 5). To determine whether the inhibition was reversible, non-EGTA treated OpdA was first incubated with 0.1 mM Zn^{2+} for 5 min, then with 1 mM EGTA. The reaction was initiated with 20 μM Abz-NKPRRPQ-EDDnp substrate and incubated at 37 °C. EGTA was able to reverse the inhibitory effect of excess Zn^{2+} (Fig. 6).

Effect of inhibitors on the hydrolytic activity of OpdA

The effect of inhibitors on OpdA was evaluated. Protein preparations were used that had been purified from *E. coli* as above, but not treated with EGTA. Enzyme activity was evaluated after 5 min pre-incubation with the inhibitor, using Abz-NKPRRPQ-EDDnp as the substrate (Table 1). The activity of OpdA was not affected by phosphoramidon, E-64, bestatin, PMSF, AEBSF and leupeptin, but was reduced in the presence of antipain, chymostatin, ALLN and EGTA. The activity of OpdA was not affected by phosphoramidon, bestatin, PMSF, AEBSF and leupeptin, but was decreased in the presence of antipain, chymostatin, ALLN and EGTA. A similar effect of inhibitors on enzyme activity was observed when using the second substrate, Abz-AAL-EDDnp (Table 2).

Discussion

Deletion of *opdA* gene in *Y. pseudotuberculosis* IP32953 resulted in attenuation of this strain in a mouse model of infection, suggesting a possible role of the OpdA protein in infection. In this study we confirmed that this protein as an oligopeptidase and tested effects of pH, various cations

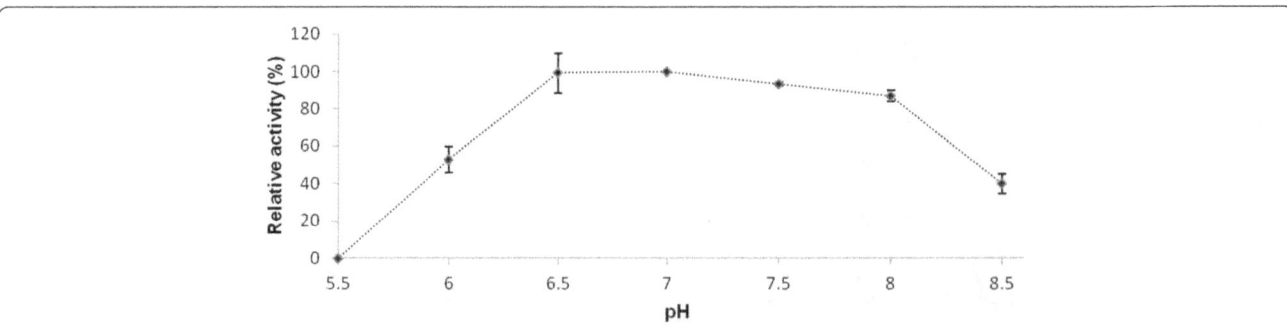

Fig. 2 Effect of pH on OpdA activity. Abz-NKPRRPQ-EDDnp was used as the substrate. The activity at different pH values relative to that at pH 6.5 was determined. Mean values ± SD of three readings from each of two independent experiments are shown

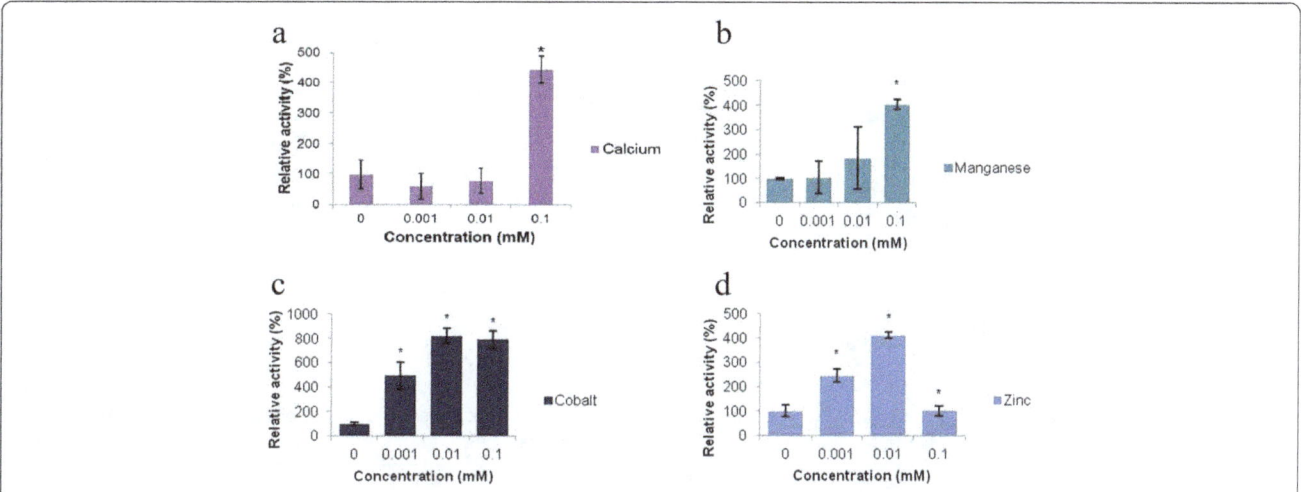

Fig. 3 Effects of divalent cations on the hydrolysis of Abz-NKPRRPQ-EDDnp (20 µM) by OpdA. A *star* denotes a statistically significant difference compared to the test conducted in the absence of a cation. Mean values ± SD of three readings from each of two independent experiments are shown

and peptidase inhibitors on its activity using two substrates. OpdA from *S.* Typhimurium and *E. coli* were shown to be able to use a broad range of oligopeptide substrates, but no preference was reported [1, 7, 8, 10, 17]. Although OpdA belongs to the M3A subfamily of Zn-dependent metallo-proteases [7, 8, 10], in the current study the enzyme was able to fully hydrolyze the substrates in the absence of exogenously added Zn^{2+} possibly due to trace cations present in the OpdA active site.

Previous reports demonstrated inhibition of *S.* Typhimurium OpdA by Zn^{2+} at 0.1 mM [3, 17]. Our data support this result. We investigated this phenomenon in more detail by using OpdA depleted of naturally bound divalent cations by repeated treatment with EGTA followed by dialysis (in

order to remove EGTA and soluble EGTA-complexes). In these samples we found stimulation of OpdA activity by low Zn^{2+} concentrations (10 µM), whilst there was complete inhibition at 100 mM of this cation. As observed previously with OpdA from *S. typhimurium* [6, 17], cobalt, calcium and manganese cations stimulated the activity of OpdA. The ability of OpdA to utilize different metal ions for hydrolysis is possibly due to the co-ordination geometries of these metals and the flexibility of the active center [18].

A previous study suggested a possibility of OpdA from *E. coli* having two active sites [3]. In particular, that study showed stimulation of OpdA activity by Co^{2+} when hydrolysing Z-AALpNA, but slight inhibition when hydrolysing

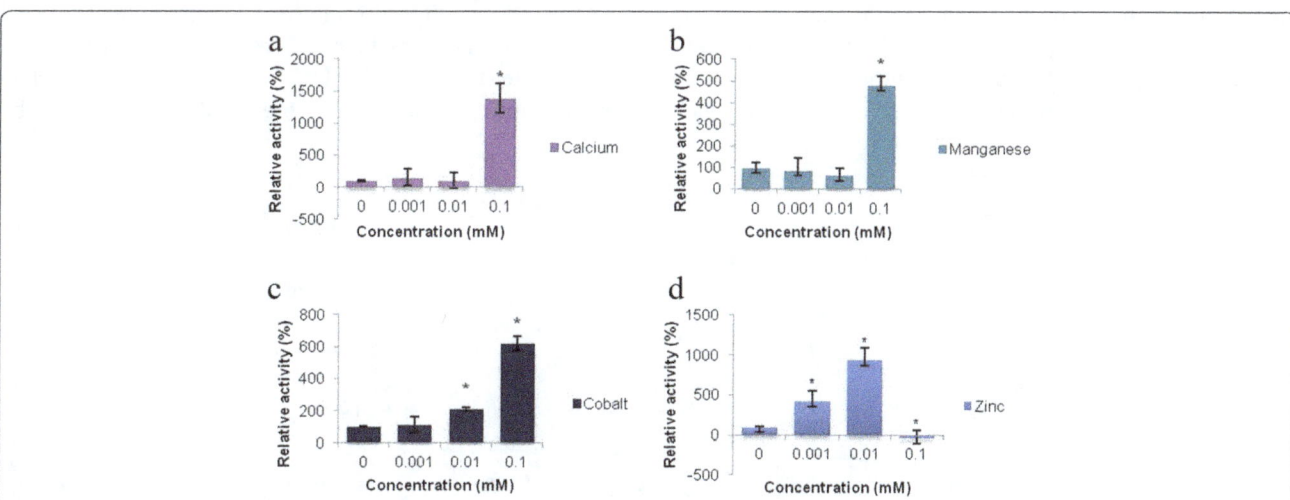

Fig. 4 Effect of divalent cations on the hydrolysis of 20 µM Abz-AAL-EDDnp by double EGTA treated OpdA after 5 min. A *star* denotes a statistically significant difference compared to the test conducted in the absence of a cation. Mean values ± SD of three readings from each of two independent experiments are shown. Fluorescence (Y axis) is represented by arbitrary units

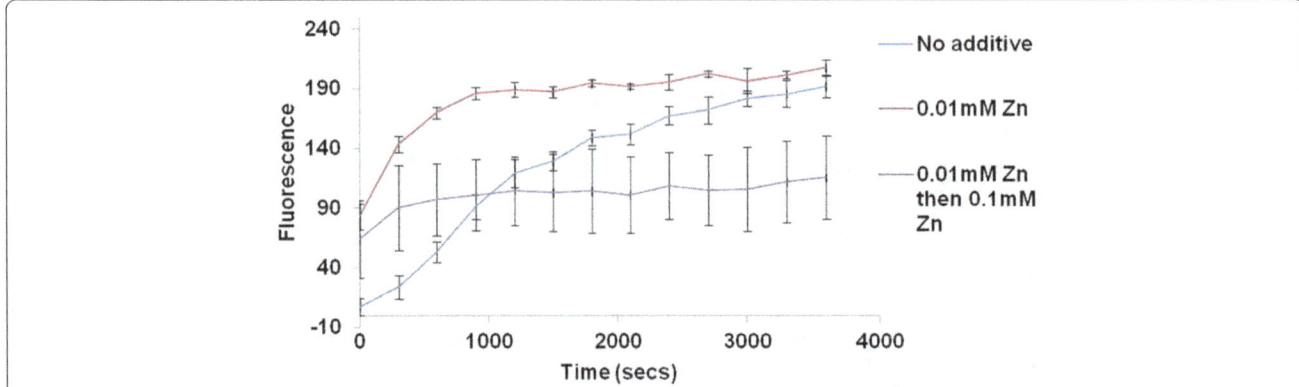

Fig. 5 Effect of 0.1 mM Zn^{2+} on the activity of OpdA (0.4 µg) pre-treated with 0.01 mM of Zn^{2+} Mean values ± SD of three readings from each of two independent experiments are shown. Fluorescence (Y axis) is represented by arbitrary units

Boc-Val-Pro-Arg-NH-Mec. In addition, inhibition assays from the study showed that the hydrolysis of Z-AALpNA was not affected by antipain or 4-guanidino benzoic acid 4- tert-butylphenyl ester, whereas the hydrolysis of Boc-Val-Pro-Arg-NH-Mec was completely inhibited by these two inhibitors.

Following these data reported by Jiang et al. [3], in the current study we decided to examine the 'trypsin-like' activity of OpdA by using Abz-NKPRRPQ-EDDnp as it has two arginine residues, which trypsin could cleave [19]. However, we found that the hydrolysis of Abz-NKPRRPQ-EDDnp by OpdA was not completely inhibited by trypsin specific inhibitor antipain and the enzyme was able to fully hydrolyze the substrate within an hour. In contrast to data by Jiang et al. [3] describing *E. coli* OpdA-like protein (named PrlC, or rPrlC for its 'recombinant' form), in our experiments an addition of Co^{2+} to partially depleted OpdA of *Y. pseudotuberculosis* stimulated the hydrolysis of both Abz-NKPRRPQ-EDDnp and Abz-AAL-EDDnp. The discrepancies in the results may be due to different properties of OpdA proteins extracted from different sources. Neither our data with two different substrates, nor more recent studies provide

any evidence for the presence of two active centers in OpdA and similar enzymes [1, 7, 8]. Putative amino acid residues involved in catalytic activity of *Y. pseudotuberculosis* OpdA by similarity to *E. coli* OpdA are showing in Additional file 4: Figure S3.

Previous studies showed the inhibitory effect of EDTA on the hydrolysis of $AcAla_4$ by OpdA [6, 17]. In contrast, our initial experiments with the hydrolysis of Abz-NKPRRPQ-EDDnp using native (untreated) OpdA showed only partial inhibition of activity in the presence of 1 mM EDTA (data not shown). Increasing concentration of the latter to 5 mM did not increase the inhibitory effect. No statistically valid difference in the latter was observed when EDTA was replaced with EGTA. This could be due to much higher affinity of divalent cations to OpdA than to EDTA or EGTA. We managed to alleviate the problem with partial inhibition by using sequential double treatment of the enzyme with EGTA followed by dialysis.

Despite confirming that OpdA is a metallopeptidase, unexpectedly we found inhibitory effects of antipain and chymostatin. This could be a result of disturbances to the flexible loop i.e. [600]SHIFAGGYAAGYYSY[614]. The flexible loop contains two serine residues that could be

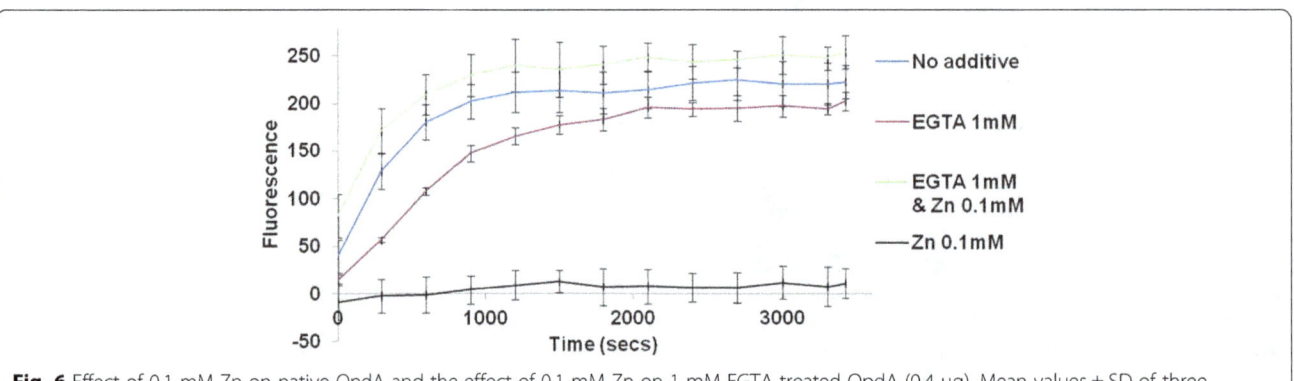

Fig. 6 Effect of 0.1 mM Zn on native OpdA and the effect of 0.1 mM Zn on 1 mM EGTA treated OpdA (0.4 µg). Mean values ± SD of three readings from each of two independent experiments are shown. Fluorescence (Y axis) is represented by arbitrary units

Table 1 Effect of inhibitors on the hydrolysis of Abz-NKPRRPQ-EDDnp by OpdA

Inhibitor	Relative activity (%)	P Value
None	100	
Phosphoramidon	92 ± 9	0.5759
E-64	101 ± 3	1.0000
Bestatin	95 ± 22	0.7791
PMSF	92 ± 14	0.4685
AEBSF	99 ± 10	0.9252
Leupeptin	87 ± 8	0.4590
Antipain	31 ± 2	0.0002*
ALLN	45 ± 9	0.0127*
Chymostatin	39 ± 11	0.0028*
EGTA	73 ± 9	0.0228*

A star (*) denotes statistically valid difference (P < 0.05) compared to control (no inhibitors)

important for enzyme activity and may be the site of action by antipain and chymostatin. It is also noteworthy that the loop also contains the important substrate recognition residue Tyr607 [7].

OpdA and Dcp are thought to be the major peptidases involved in the hydrolysis of peptides in the cytoplasm since bacterial lysates lacking both of these enzymes have no catalytical activity against AcAla$_3$ and AcAla$_4$ [6]. Incubation of both Abz-GFSPFR-EDDnp and Abz-GFSPFRQ-EDDnp by OpdA with *E. coli* bacterial lysate produced the Abz-GF product suggesting that OpdA is the major oligopeptidase in *E. coli* [8]. Although another study also demonstrated intracellular location of OpdA [9], one can't exclude a possibility of extracellular release of this enzyme during infection.

Table 2 Effect of inhibitors on the hydrolysis of Abz-AAL-EDDnp by OpdA

Inhibitor	Relative activity (%)	P Value
None	100	-
Phosphoramidon	150 ± 66	0.3515
E-64	212 ± 129	0.2361
Bestatin	103 ± 31	1.0000
PMSF	148 ± 61	0.3053
AEBSF	82 ± 12	0.3262
Leupeptin	73 ± 7	0.1690
Antipain	28 ± 5	0.0008*
ALLN	1 ± 10	0.0004*
Chymostatin	4 ± 34	0.0077*
EGTA	3 ± 16	0.0175*

A star (*) denotes statistically valid difference (P < 0.05) compared to control (no inhibitors)

A role of OpdA protein in pathogenesis remains to be elucidated. One attractive hypothesis is a possibility of this enzyme destroying short positively charged antimicrobial peptides involved in innate host immunity. Remarkably, the substrate used in this study is derived from bradykinin peptide known for its antimicrobial activity [20]. One of the mechanisms of bacterial resistance to antimicrobial peptides is degradation of these compounds by oligopeptidases [21].

Overall, the current study has demonstrated the effect of various metal ions and inhibitors on the hydrolysis of two fluorogenic substrates by OpdA. The study showed concentration dependent inhibitory effect of zinc on OpdA activity. Further research could be aimed at determining the residues responsible for the zinc inhibition via site directed mutagenesis. Furthermore, it will be interesting to determine if replacement of the wild type copy of the *opdA* gene with that encoding a modified protein (e.g. not inhibited by elevated Zn^{2+} concentrations) would have any effect on the virulence of *Y. pseudotuberculosis*.

Conclusions

In this study we report the identification and enzymatic properties of *Yersinia pseudotuberculosis* oligopeptidase (OpdA) required for bacterial pathogenicity. The enzyme was found to be specific to short oligopeptides, with its peptidase activity dependent on the presence of particular divalent cations. The activity of this enzyme towards positively charged peptides mimicking those produced by host innate immune system may explain the reason for attenuation of the *opdA* mutant of this bacterium.

The data generated by this study are novel and important as they:

- present results of extensive studies on the effects of different cations of the activity of this enzyme.
- explain unexpected limited effects of chelators on the native enzyme predicted to be a metallopeptidase.
- describe a novel approach used for 'depletion' of the native enzyme of divalent cation(s) by repeated treatment with a chelator and dialysis.
- show unusual and unexpected inhibitory effect of higher concentrations of Zn^{2+} on the activity.
- demonstrate that the latter could be observed even after initial activation of the enzyme at lower Zn^{2+} concentrations.
- show that the inhibitory effect is reversible due to unexpected restoration of activity after addition of EGTA.

In addition, this is the first study suggesting a role of OpdA enzyme in pathogenicity. The results reported provide valuable contribution to the investigation of bacterial metallopeptidases in general, and that of oligopeptidases in particular, and are important for better understanding of the role of OpdA-like enzymes in pathogenesis of bacterial infections. Furthermore, characterisation of this protein may serve as a basis for the development of novel antibacterials based on specific inhibition of its peptidase activity.

Additional files

Additional file 1: Figure S1. PCR analysis of *Y. psedutuberculosis* YPIII wild type (lane 2), and three clonal isolates of the ΔopdA mutant (lanes 3–5) using gene specific primers Yptb3816_for (ATGACAAACCCGCTGTTGACT) and Yptb3816_rev (TTAGCCCTTAATACCGTAATGAC). The expected product size after deletion is 1.2 kb, corresponding to a deletion of almost the entire gene sequence (1.9 out of 2.1 kb) plus 1 kb corresponding to the inserted *kan'* resistance gene. Lane 1, DNA size ladder 1 kb plus (Life Technologies).

Additional file 2: Figure S4. Comparison of growth rates of the wild strain of *Y. psedotuberculosis* IP32953 and its opdA mutant (opdA). Y axis, optical density (OD$_{595}$); X axis, time (hours); $n = 18$ (3 biological and 6 technical replicates.

Additional file 3: Figure S2. A representative experiment demonstrating the lack of complete inhibition of the activity of native OpdA enzyme (0.4 μg, green triangles) in the presence of EGTA (1 mM, blue crosses).

Additional file 4: Figure S3. Multiple alignment of amino acid sequences of OpdA proteins found in different bacteria: *Y. pseudotuberculosis* (accession number CAH23054.1), *E. coli* (synonym name PrlC, accession number P27298.3), *Salmonella enterica* serovar Typhimurium (accession number P27237.1) and *Haemophilus influenzae* (accession number P44573.1). Amino acid residues corresponding to Zn^{2+} binding motif (by similarity to *E. coli* OpdA) are highlighted in red. A Tyr residue corresponding to Tyr607 of *E. coli* OpdA shown to be essential for enzyme activity and specificity is also highlighted (in green).

Abbreviations

Abz: O-aminobenzoyl, fluorophore; EDDnp: Ethylenediamine 2,4-dinitrophenyl, quencher; OpdA: Oligopeptidase A

Funding

This study was not supported by any external funding. It was supported by internal DSTL and Kingston University research funds.

Authors' contributions

AVK, conceived and supervised the study; AMS, BWW and PCFO, supervised the study and provided tools and reagents; AVK and AA designed and performed protein expression, purification and analysis experiments; DCF and IAC performed in vivo experiments; AA, AMS, DCF, IAC, BWW, PCFO and AVK analysed the data and wrote the manuscript. All authors read and approved the final manuscript.

Competing interests

The authors declare that they have no competing interests.

Author details

[1]School of Life Sciences, Pharmacy and Chemistry; Faculty of Science, Engineering and Computing, Kingston University, Penrhyn Road, Kingston upon Thames KT1 2EE, UK. [2]Department of Pathogen Molecular Biology, London School of Hygiene and Tropical Medicine, Keppel Street, London WC1E 7HT, UK. [3]Biomedical Sciences, DSTL Porton Down, Salisbury, Wiltshire SP4 0JQ, UK.

References

1. Jain R, Chan MK. Support for a potential role of *E. coli* oligopeptidase A in protein degradation. Biochem Biophys Res Commun. 2007;359(3):486–90.
2. Gottesman S. Proteolysis in bacterial regulatory circuits. Annu Rev Cell Dev Biol. 2003;19:565–87.
3. Jiang XY, Zhang MY, Ding Y, Yao J, Chen H, Zhu DX, Muramatu M. *Escherichia coli* prIC gene encodes a trypsin-like proteinase regulating the cell cycle. J Biochem. 1998;124(5):980–5.
4. Brown CK, Madauss K, Lian W, Beck MR, Tolbert WD, Rodgers DW. Structure of neurolysin reveals a deep channel that limits substrate access. Proc Natl Acad Sci U S A. 2001;98(6):3127–32.
5. Hooper NM. Families of zinc metalloproteases. FEBS Lett. 1994;354(1):1–6.
6. Vimr ER, Green L, Miller CG. Oligopeptidase-deficient mutants of *Salmonella* Typhimurium. J Bacteriol. 1983;153(3):1259–65.
7. Lorenzon RZ, Cunha CEL, Marcondes MF, Machado MFM, Juliano MA, Oliveira V, Travassos LR, Paschoalin T, Carmona AK. Kinetic characterization of the *Escherichia coli* oligopeptidase A (OpdA) and the role of the Tyr(607) residue. Arch Biochem Biophys. 2010;500(2):131–6.
8. Paschoalin T, Carmona AK, Oliveira V, Juliano L, Travassos LR. Characterization of thimet- and neurolysin-like activities in *Escherichia coli* M3A peptidases and description of a specific substrate. Arch Biochem Biophys. 2005;441(1):25–34.
9. Novak P, Dev IK. Degradation of a signal peptide by protease Iv and Oligopeptidase A. J Bacteriol. 1988;170(11):5067–75.
10. Conlin CA, Vimr ER, Miller CG. Oligopeptidase A is required for normal phage-P22 development. J Bacteriol. 1992;174(18):5869–80.
11. Conlin C, Miller C. opdA, a *Salmonella enterica* serovar Typhimurium gene encoding a protease, is part of an operon regulated by heat shock. J Bacteriol. 2000;182(2):518–21.
12. Long C, Jones TF, Vugia DJ, Scheftel J, Strockbine N, Ryan P, Shiferaw B, Tauxe RV, Gould LH. *Yersinia pseudotuberculosis* and *Y. enterocolitica* infections, FoodNet, 1996–2007. Emerg Infect Dis. 2010;16(3):566–7.
13. Stubben CJ, Duffield ML, Cooper IA, Ford DC, Gans JD, Karlyshev AV, Lingard B, Oyston PCF, de Rochefort A, Song J, Wren BW, Titball RW, Wolinsky M. Steps toward broad-spectrum therapeutics: discovering virulence-associated genes present in diverse human pathogens. BMC Genomics. 2009;10:501.
14. Derbise A, Lesic B, Dacheux D, Ghigo JM, Carniel E. A rapid and simple method for inactivating chromosomal genes in Yersinia. FEMS Immunol Med Microbiol. 2003;38(2):113–6.
15. Taylor RK, Miller VL, Furlong DB, Mekalanos JJ. Use of phoA gene fusions to identify a pilus colonization factor coordinately regulated with cholera-toxin. Proc Natl Acad Sci U S A. 1987;84(9):2833–7.
16. Guzman LM, Belin D, Carson MJ, Beckwith J. Tight regulation, modulation, and high-level expression by vectors containing the arabinose P-BAD promoter. J Bacteriol. 1995;177(14):4121–30.
17. Conlin CA, Miller CG. Cloning and nucleotide sequence of opdA, the gene encoding oligopeptidase A in *Salmonella* Typhimurium. J Bacteriol. 1992;174(5):1631–40.
18. Fukasawa KM, Hata T, Ono Y, Hirose J. Metal preferences of zinc-binding motif on metalloproteases. J Amino Acids. 2011;2011:574816.
19. Olsen JV, Ong SE, Mann M. Trypsin cleaves exclusively C-terminal to arginine and lysine residues. Mol Cell Proteomics. 2004;3(6):608–14.
20. Kowalska K, Carr DB, Lipkowski AW. Direct antimicrobial properties of substance P. Life Sci. 2002;71(7):747–50.
21. Gruenheid S, Le Moual H. Resistance to antimicrobial peptides in Gram-negative bacteria. FEMS Microbiol Lett. 2012;330(2):81–9.

13

Dominant bacterial phyla in caves and their predicted functional roles in C and N cycle

Surajit De Mandal[1], Raghunath Chatterjee[2] and Nachimuthu Senthil Kumar[1*]

Abstract

Background: Bacteria present in cave often survive by modifying their metabolic pathway or other mechanism. Understanding these adopted bacteria and their survival strategy inside the cave is an important aspect of microbial ecology. Present study focuses on the bacterial community and geochemistry in five caves of Mizoram, Northeast India. The objective of this study was to explore the taxonomic composition and presumed functional diversity of cave sediment metagenomes using paired end Illumina sequencing using V3 region of 16S rRNA gene and bioinformatics pipeline.

Results: Actinobacteria, Proteobacteria, Verrucomicrobia and Acidobacteria were the major phyla in all the five cave sediment samples. Among the five caves the highest diversity is found in Lamsialpuk with a Shannon index 12.5 and the lowest in Bukpuk (Shannon index 8.22). In addition, imputed metagenomic approach was used to predict the functional role of microbial community in biogeochemical cycling in the cave environments. Functional module showed high representation of genes involved in Amino Acid Metabolism in (20.9%) and Carbohydrate Metabolism (20.4%) in the KEGG pathways. Genes responsible for carbon degradation, carbon fixation, methane metabolism, nitrification, nitrate reduction and ammonia assimilation were also predicted in the present study.

Conclusion: The cave sediments of the biodiversity hotspot region possessing a oligotrophic environment harbours high phylogenetic diversity dominated by Actinobacteria and Proteobacteria. Among the geochemical factors, ferric oxide was correlated with increased microbial diversity. In-silico analysis detected genes involved in carbon, nitrogen, methane metabolism and complex metabolic pathways responsible for the survival of the bacterial community in nutrient limited cave environments. Present study with Paired end Illumina sequencing along with bioinformatics analysis revealed the essential ecological role of the cave bacterial communities. These results will be useful in documenting the biospeleology of this region and systematic understanding of bacterial communities in natural sediment environments as well.

Keywords: Cave, Illumina sequencing, Functional diversity, KEGG pathways, Biospeleology

Background

Bacteria constitute the major portion of the cave biodiversity and plays a key role in maintaining cave ecosystem [1]. Limited nutrient and energy sources create an oligotrophic environment inside the caves, wherein the primary production is carried out by auto-trophic bacteria which inturn supports the growth of several chemo-organotrophic microbes [2]. Bacteria present under this oligotrophic environment often survive by modifying their metabolic pathway or other

mechanism [3]. Understanding these adopted bacteria and their survival strategy inside the oligotrophic environment is an important aspect of microbial ecology.

Geomicrobial Investigations in nutrient limited caves are sparse and most of them have been carried out using culture based techniques. Such approach can only detect a minute portion of the total community. Such limitation is solved by the introduction of next generation sequencing (NGS) and expands our knowledge on uncultured microbes [4]. Although the cost of amplicon sequencing (16S rDNA) used for the bacterial community composition studies has rapidly decreased, the functional study using the Shotgun approach or Geochip still remains expensive and thus, is restricted for selected

* Correspondence: nskmzu@gmail.com
[1]Department of Biotechnology, Mizoram University, Aizawl, Mizoram 796004, India
Full list of author information is available at the end of the article

studies [5]. An indirect approach is to compare the uncultured bacterial sequences with closely related and well studied microbes to predict the functional role in the ecosystem. This is also useful to understand the unknown energy source required for metabolism [6, 7]. A computational approach, PICRUSt (phylogenetic investigation of communities by reconstruction of unobserved states) based on the relationship between phylogeny and function was developed to predict functional diversity using 16S rDNA data and a reference database and has been used to study in diverse environments [8].

Cave microorganisms contain a wide range of bacterial groups influenced by the geology, soil or sediment and other factors [9]. Geochemistry parameter often drives the diversity and bacterial community composition inside the caves [10]. Present study focuses on the bacterial community and geochemistry in five unexplored and unknown caves of Mizoram, Northeast India falling under the less- known biodiversity hotspot zone of the eastern Himalayan belt. The objective of this study was to explore the taxonomic composition and to understand how the bacterial communities respond to the cave oligotropic environments. This study was based on the hypothesis that the undisturbed and nutrient- limited cave habitats will host specific bacterial species and the cave geochemical parameters might favour species diversity and richness.

Methods
Sample collection and community DNA extraction
Cave sediment samples were collected from different sites of the caves – Bukpuk (CBP V3), Lamsialpuk (CLP V3) and Reiekpuk (CRP V3) followed by sieved and preserved at 4 °C (Fig. 1). The geochemical and molecular data of the sediment sample Lamsialpuk (CLP V3) and Khuangcherapuk (CKP V3) were collected from our previous study [11, 12]. All sites were not subjected to any human disturbances, except CLPV3 [4]. The elevation, pH and other geochemical parameters of the caves are given in Table 1. The pH of the sediment samples was analysed using pH meter (Eutech, pH 510, USA). Major oxides and trace elements were measured using X-ray Fluorescence (XRF) (Bruker AXS, S4 Pioneer, Germany) at IIT Rookie, India.

DNA was extracted from the cave sediment samples using the Fast DNA spin kit (MP Biomedical, Solon, OH, USA) and the V3 hypervariable region of the 16S rRNA gene was amplified using 10 pmol/µl of each forward 341F (5′-CCTACGGGAGGCAGCAG-3′) and reverse 518R (5′-ATTACCGCGGCTGCTGG-3′) primer. PCR Master Mix will contain 2 µL each primers, 0.5 µL of 40 mM dNTP (NEB, USA), 5 µL of 5X Phusion HF reaction buffer (NEB, USA), 0.2 µL of 2 U/ µL F-540Special Phusion HS DNA Polymerase (NEB,

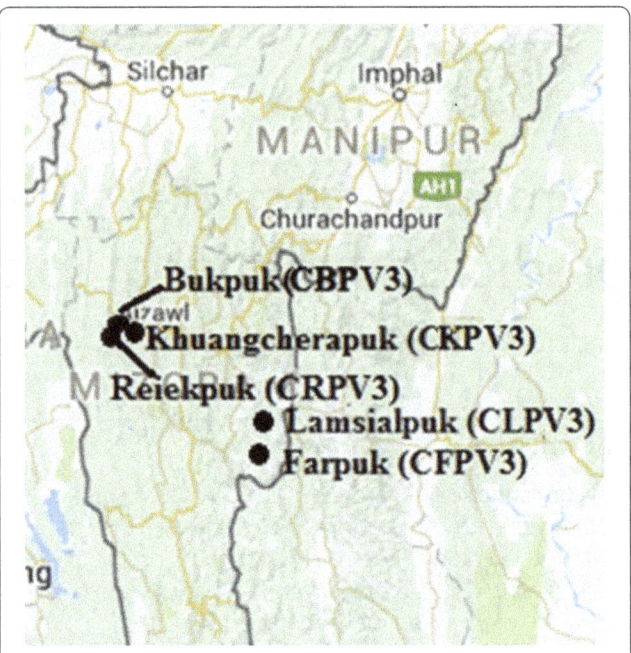

Fig. 1 Geographical location of the sampling sites in Mizoram, Northeast India. The figure has been adopted from Google Map and modified

USA), 5 ng input DNA and water to make up the total volume to 25 µL. The PCR conditions were 98 °C for 30 s followed by 30 cycles of 98 °C for 10 s; 72 °C for 30 s and a final extension at 72 °C for 5 s followed by 4 °C hold.

Pre-processing and sequence analysis
Paired end Illumina sequencing (2 × 150 bp) was carried out at Scigenome Lab, Cochin, India. Raw sequence data for the two cave sediment samples, Farpuk (CFPV3) and Khuangcherapuk (CKPV3), were derived from our previous study [11, 12]. Raw fastq sequences were processed using the QIIME software package v.1.8.0 [13, 14]. Poor quality (quality score < 25) and smaller reads (read length < 100 bp) were filtered out using the split_libraries command. Pre-processed sequence reads were clustered to operational taxonomic units (OTU's) using UCLUST method with similarity threshold of 97% [15] and were taxonomically classified using Greengenes database. Relative abundance of the bacterial phyla was calculated using QIIME. Statistical analysis was performed after rarefying the OTU table to 50,000 sequences per sample. Alpha and beta diversity plots were also generated using QIIME. Beta diversity between five bacterial cave communities was measured using unweighted UniFrac approach [16]. Pearson correlations between soil characteristics and bacterial major phylum were estimated using PASW Statistics 18 (SPSS Inc., Chicago, IL, USA). Additionally, we performed imputed metagenomic analysis by the genome prediction

Table 1 Geochemical parameters of the cave samples

Sample ID	GPS Coordinates	Elevation (MSL)	pH	Na$_2$O %	MgO%	Al2O3%	SiO$_2$%	P$_2$O5%	SO$_3$%	K$_2$O%	CaO%	Fe2O3%	Cr2O3%	MnO%	NiO%	CuO%	ZnO%	Rb2O%	SrO%	ZrO2%	BaO%	Cl%	V2O5%
CBPV3 (Bukpuk)	N23.69, E93.29	4003	7.2	0.08	1.04	4.59	12.3	7.53	11.9	2.9	8.85	3.21	0.03	0.20	0.03	0.05	0.071	0.03	0.08	0	0.00	0.06	0
CFPV3 (Farpuk)	N23.11, E93.53	4645	7.3	31	0.92	11.8	44.2	0.16	0.16	2.5	0.64	5.15	0.04	0.12	0.02	0.01	0.01	0.01	0.03	0.12	0.07	0	0
CLPV3 (Lamsialpuk)	N23.13, E93.29	4446	7.5	42.4	0.92	13.2	37.9	0.71	0.23	2.7	1.23	6.35	0.03	0.09	0.01	0.01	0.013	0.01	0.03	0.08	0.07	0.06	0.001
CRPV3 (Reiekpuk)	N23.69, E92.60	4312	6.8	0.6	1.67	11.7	39.1	0.16	1.92	2.8	3.23	5.72	0.04	0.09	0.03	0.01	0.013	0.01	0.02	0.07	0.07	0	0
CKPV3 (Khuangcherapuk)	N23.69, E92.61	4900	6.7	0.37	1.04	10.5	33.4	0.84	0.82	2.5	0.92	5.64	0.08	0.05	0.01	0.01	0.012	0.03	0.04	0.07	0.07	0	0

All the samples were collected during March 2014

MSL meters above sea level

software PICRUSt [8]. The input used here was normalized OTU prepared by closed reference based approach. OTU's were assigned at 97% similarity and were mapped to the Greengenes ver.13.5 database for functional prediction.

Statistical analysis

Multivariate principal component analysis (PCA) of 20 physicochemical parameters i.e., pH, Na_2O, MgO, Al_2O_3, SiO_2, P_2O_5, SO_3, K_2O, CaO, Fe_2O_3, Cr_2O_3, MnO, NiO, CuO, ZnO, Rb_2O, SrO, ZrO_2, BaO and Cl was carried out to determine which environmental variables best explained the observed community patterns using the PAST v3.02 software [17].

Results

Geochemical characteristics of the cave sediment samples

The pH of the five cave sediment were recorded in the range of 6.7–7.5. The highest pH was recorded at CLPV3 (7.5) followed by CFPV3 (7.3) and CBPV3 (7.2), whereas the lowest pH was recorded at CKPV3 (6.7). The concentration of the oxides such as Na_2O, MgO, Al_2O_3, SiO_2, P_2O_5, SO_3, K_2O, CaO, CuO, ZnO, Fe_2O_3, Cr_2O_3, MnO, NiO, CuO, ZnO, Rb_2O, SrO, ZrO_2, BaO and Cl varied among the samples (Table 1). Soil samples from both CKPV3 and CRPV3 had similar, but relatively lower pH compared to the other three cave samples. Similarly, CLPV3 and CFPV3 were also geochemically similar with high concentration of Na_2O (Additional file 1: Figure S1). CBPV3 showed the highest concentration of P_2O_5, SO_3, CaO, MnO, CuO, ZnO and SrO, whereas the lowest concentration of Al_2O_3, SiO_2, Na_2O and Fe_2O_3 compared to other caves. Interestingly, the elevation of CBPV3 was lower than the other four caves under study. A principal component analysis (PCA) of the physicochemical parameters showed that the five caves were separated into four geochemically distinct habitats. The first two principal components explained 88.06% of the total variance. The sample CKPV3 and CRPV3 were found geochemically similar and were grouped together in the 2-dimensional PCA plot. The key influencing parameters for the geochemical diversity were Na_2O and P_2O_5, while Cl and SO_3 were the other influencing parameters in component 1 and component 2, respectively.

Analysis of bacterial community composition

The high throughput sequencing effort yielded a total of 54,90,239 paired end reads with an average of 9,15,040 paired end reads per sample. After assembly and quality assessment of the reads, a total of 54,88,530 high quality reads were obtained. A total of 48 phyla (AC1, Acidobacteria, Actinobacteria, AD3, Armatimonadetes, Bacteroidetes, BHI80–139, BRC1, Caldithrix, Chlorobi,

Chloroflexi, Cyanobacteria, Deferribacteres, Elusimicrobia, FCPU426, Fibrobacteres, Firmicutes, Fusobacteria, GAL15, Gemmatimonadetes, GN02, GN04, MVP-21, NC10, Nitrospirae, NKB19, OD1, OP1, OP11, OP3, OP8, OP9, Planctomycetes, Proteobacteria, SBR1093, SC4, Spirochaetes, SR1, Synergistetes, Tenericutes, Thermi, TM6, TM7, Verrucomicrobia, WPS-2, WS2, WS3 and ZB3) were detected from different cave sediments (Fig. 2). The total bacterial community analysis showed that the phylum Actinobacteria was the most dominant contributing up to 65.1%, followed by Proteobacteria (24.8%), Acidobacteria (4.2%) and Firmicutes (3.6%) and the top ten phyla present in individual cave is shown in Fig. 3.

Actinobacteria

In the present study, the identified class under this phylum were Actinobacteria, Acidimicrobiia, Thermoleophilia, Rubrobacteria, MB-A2–108, Coriobacteriia, Nitriliruptoria,

OPB41 and KIST-JJY01. High abundance of dominant family (>0.01%) under Actinobacteria were *Nocardiaceae*, *Streptomycetaceae*, *Micrococcaceae*, *Frankiaceae*, *Gaiellaceae*, *Pseudonocardiaceae*, *Streptomycetaceae*, *EB1017*, *Mycobacteriaceae*, *Actinosynnemataceae*, *Corynebacteriaceae*, *Rubrobacteraceae*, *Nocardioidaceae*, *Micromonosporacea*, *Geoderma tophilaceae*, *Sporichthyaceae*, *Actinosynnemataceae*, *Nakamurellaceae*, *Pseudonocardiaceae*, *Cryptosporangiaceae*, *Kineosporiaceae and Ruaniaceae*. Other dominant genus under Actinomycetes was *Mycobacterium*, *Corynebacterium*, *Rubrobacter*, *Actinoplanes*, *Saccharothrix* and *Pseudonocardia*.

Proteobacteria

Within the Proteobacteria, most phylotypes were classified under the class Alphaproteobacteria and Gammaproteobacteria. Other identified class were Betaproteobacteria, Deltaproteobacteria, Gammaproteobacteria, TA18, Epsilonproteobacteria and Zetaproteobacteria. Abundant genera (≥0.01%) under this phylum were *Rhodoplanes*, *Kaistobacter*, *Sphingomonas*, *Bradyrhizobium*, *Alteromonas*, *Acidiphilium* and *Halomona*. Under the class Alphaproteobacteria, two families (Hyphomicrobiaceae and Sphingomonadaceae) and three abundant genus (*Rhodoplanes*, *Kaistobacter* and *Sphingomonas*) were identified. Other detected genera, present in low abundance, under this class were *Candidatus entotheonella*, *Plesiocystis*, *Desulfococcus*, *Nannocystis*, *Anaeromyxobacter*, *Sorangium*, *Haliangium*, *Geobacter*, *Cystobacter* and *Syntrophus*. The dominant genera under the class Gammaproteobacteria were Alteromonas and Halomonas. Other genera (<0.01%) were *Acinetobacter*, *Alcanivorax*, *Aquicella*, *Cronobacter*, *Dickeya*, *Dokdonella*, *Enhydrobacter*, *Enterobacter*, *Enterovibrio*, *Erwinia*, *Fulvimonas*, *Glaciecola*, *Hafnia*, *Halorhodospira*, *Idiomarina*, *Klebsiella*, *Legionella*, *Luteibacter*, *Luteimonas*, *Marinobacter*, *Marinobacterium*,

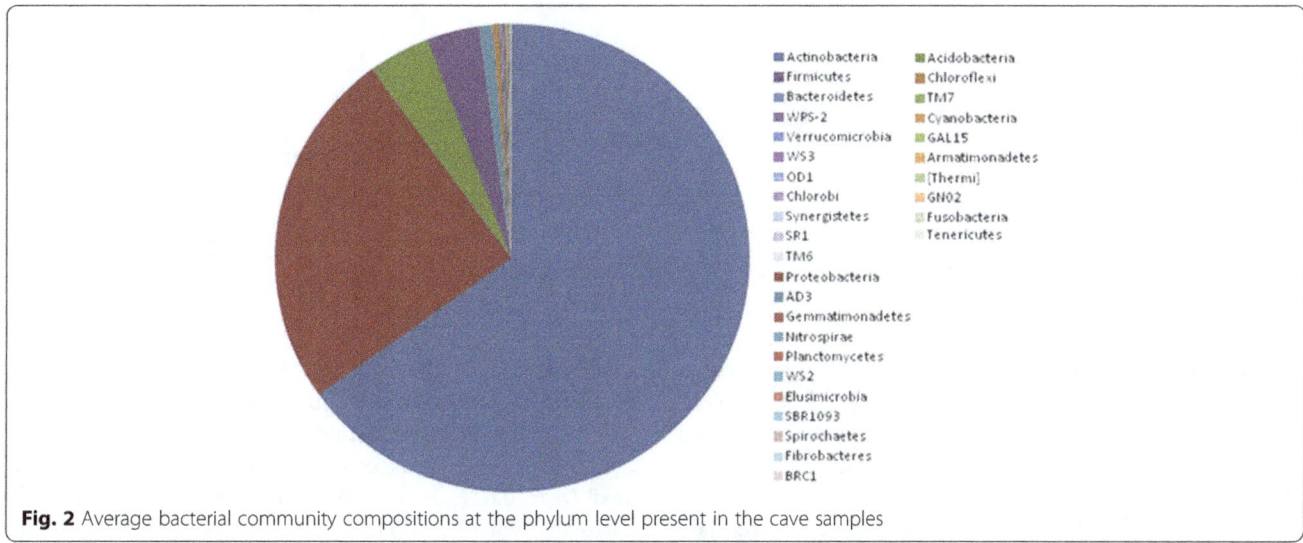

Fig. 2 Average bacterial community compositions at the phylum level present in the cave samples

Oceanospirillum, Providencia, Psychrobacter, Pseudoxanthomonas, Pseudomonas, Pseudoalteromonas, Rheinheimera, Rhodanobacter, Salinisphaera, Salinivibrio, Serpens, Serratia, Shewanella, Stenotrophomonas, Steroidobacter and *Thermomonas.* The class Epsilonproteobacteria was present in low abundance which consisted three families (*Helicobacteraceae, Campylobacteraceae* and *Helicobacteraceae*) and two genus (*Arcobacter* and *Sulfurimonas*). However, no genus was identified under the class Zetaproteobacteria and TA18.

Acidobacteria

Acidobacteria was the third dominant phyla with eight families and 10 identified genera. Dominant families under this phylum were *Solibacteraceae, Koribacteraceae* and *Acidobacteriaceae.* Assigned genera under the family *Acidobacteriaceae* were *Acidobacterium, Edaphobacter, Terriglobus, Acidicapsa* and *Acidopila.*

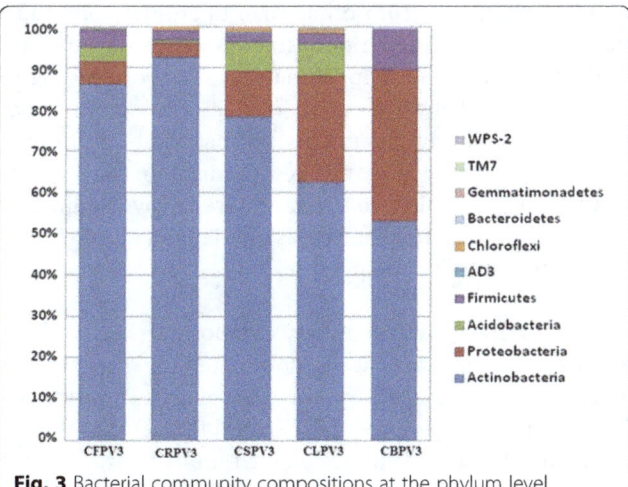

Fig. 3 Bacterial community compositions at the phylum level present in the individual cave samples

Diversity estimates of the cave bacterial community

Based on the Shannon index, a high bacterial diversity was observed in CLPV3 (12.50) and low in CBPV3 (8.22) (Table 2). The principal coordinate analysis plot of the UniFrac distance matrix distinguish CBPV3 from rest of the samples suggesting the presence of different composition of the bacterial communities, whereas other four cave samples had similar community composition (Fig. 4).

Function prediction using PICRUSt

Analysis revealed five functional modules (i.e. metabolism, genetic information processing, environmental information processing, cellular process and organismal systems) where metabolism was the most represented, accounting for about 60% of the entire data set. A deeper analysis of the terms encompassed by the metabolism functional module showed high representation of Amino Acid Metabolism (24%), Carbohydrate Metabolism (23%), Energy Metabolism (10%), Lipid Metabolism (10%), Metabolism of Cofactors and Vitamins (8%), Metabolism of Terpenoids and Polyketides (6%), Nucleotide Metabolism (6%), Metabolism of Other Amino Acids (4%), Enzyme Families (3.0%), Glycan Biosynthesis and Metabolism (3%) and Biosynthesis of Other Secondary Metabolites (3%) (Additional file 1: Figure S2).

With the carbon metabolism, three reactions were involved (carbon degradation, carbon fixation, and methane metabolism). The identified carbon degradation enzymes included genes encoding alpha-amylase, glucoamylase, neopullulanase, and pullulanase (involved in starch degradation); alpha-glucosidase, endoglucanase; beta-glucanase, beta-glucosidase (involved in cellulose degradation); arabinofuranosidase, xylanase, and mannanase (involved in hemicellulose degradation); Chitinase; beta-hexosaminidase; alpha-mannosidase and beta-mannosidase acetyl-glucosaminidase, polygalacturonase (involved in chitin,

Table 2 Alpha diversity index of the cave samples

	Observed species	Simpson reciprocal	Shannon	Simpson	PD whole tree
CFPV3	87,179	62.04	9.97	0.001	2914.7
CRPV3	72,638	86.33	10.25	0.001	2357.9
CKPV3	89,805	89.17	11.35	0.001	3020.0
CLPV3	83,136	316.81	12.50	0.004	2873.1
CBPV3	22,004	57.32	8.22	0.003	827.8

All the diversity index is calculated using QIIME
PD Phylogenetic Diversity

pectin degradation) and other carbohydrate degradation enzymes. A rare fraction of the predicted metagenomes sequence was classified as 4-hydroxybutyryl-CoA dehydratase. The predicted carbohydrate degrading enzymes were shown in Additional file 1: Table S1.

Gene's codes for the enzymes methenyltetrahydrofolate cyclohydrolase is also detected in our study. Predicted genes and enzymes show the prevalence of methane cycle in the caves (Additional file 1: Table S2). Analysis also revealed nitrogen cycling genes involved in nitrification, nitrate reduction and ammonia assimilation. Genes codes for the enzyme involved nitronate monooxygenase; nitrile hydratase; nitrate reductase; nitrilase; nitric oxide dioxygenase; nitric oxide reductase; nitric-oxide synthase; nitrite reductase; nitric-oxide reductase; nitrogenase; nitric nitrogen fixation protein; nitroreductase/dihydropteridine reductase; nitrous-oxide reductase; nitroreductase; nitrate reductase; nitrogenase; nitric oxide reductase; nitrogenise (Additional file 1: Table S3).

Association between bacterial communities with geochemical parameters

A correlation analysis was performed to study the association between the most abundant phyla identified (AD3,

Acidobacteria, Actinobacteria, Bacteroidetes, Chloroflexi, Firmicutes, Gemmatimonadetes, Proteobacteria, TM7 and WPS-2) and the geochemical parameters. Analysis revealed that Al_2O_3 was positively correlated with Chloroflexi ($r = 0.627$, $p = 0.060$); and MnO was negatively correlated with Acidobacteria ($r = -0.790$, $p = -0.060$). No other relationship between geochemical parameters and the relative abundance of the major phyla was significant different among sampling sites. Within the candidate phyla, MgO was correlated with the relative abundance of the AD3 ($r = 0.978$, $p = 0.001$), TM7 ($r = 0.974$, $p = -0.001$); and WPS-2 ($r = 0.938$, $p = -0.006$) (Additional file 1: Table S4). Furthermore, the content of Fe_2O_3 showed highest positive correlation with the Shannon diversity index ($r = 0.926$, $p = 0.001$), followed by Al_2O_3, NiO and negative correlation with SO_3 and MnO (Additional file 1: Table S5).

Discussion

Speleological studies with NGS approaches are now becoming an important approach for analyzing the concealed microbial diversity in belowground ecosystems [18]. Adaptation of the microorganism in cave ecosystem mostly involves interaction with the minerals, mobilizing inorganic phosphate, oxidizing methane and hydrogen, and deriving energy by hydrolyzing macromolecules derived from other cave microbial communities [19]. High competition for resources in nutrient limited environment helps in natural selection leading to innovation and diversification of bacterial communities [20]. Present study documents the bacterial community composition along with the geochemical analysis of the bacterial community from five different cave sediments in Mizoram, a state of northeast India, situated in Indo-Burma biodiversity hotspot zone.

Analysis of bacterial community composition

All the cave samples were dominated by the phylum Actinobacteria as seen by our previous study using V4 hypervariable region of 16SrRNA [4].The three most abundant bacterial phyla detected in this study were Actinobacteria a common cave inhabitant has been isolated in rock walls and biofilm of various caves [21].

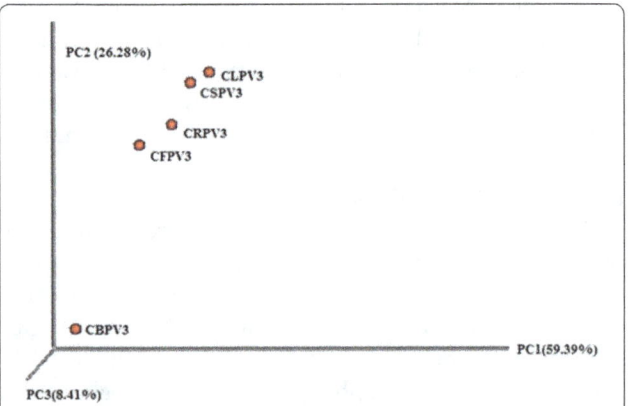

Fig. 4 Principal coordinate analysis (PCoA) plot of samples using the unweighted UniFrac distance metric. The variance explained by each principal coordinate axis is shown in parentheses. Datasets were subsample to equal depth prior to the UniFrac distance computation

Isolation of rare and novel Actinobacteria from unexplored environment is an important area of research [22]. Members of the dominant family *Nocardiaceae* have previously been reported in cave ecosystem, are oligotrophic, and can metabolize various substrates such as toluene, herbicides, naphthalene and PCBs [23–27]. The genus *Streptomyces* was the second highest genus falling under the family Streptomycetaceae. Members of this group can metabolize different compounds including alcohols, sugars, amino acids and aromatic compounds and capable of synthesizing clinically useful antibiotics [28].

Proteobacteria was dominated by Alphaproteobacterial species and Gammaproteobacteria. Some species under this subphylum can survive under extreme environment by using ABC (ATP-Binding Cassettes) and TRAP: (Tripartite ATP-independent periplasmic transporters) mechanism [29]. The genus Rhodoplanes under the subphylum Alphaproteobacteria accounts for 0.15% of the total bacterial community and possesses Photo- and chemo-organ heterotrophic growth [30]. They can also produce hopanoids and carotenoids [31, 32]. Another identified genus under Alphaproteobacteria- *Sphingomona* (0.133%), a group commonly found in nutrient-limited subsurface environments can metabolize a large number of different aromatic compounds [33].

The most abundant genus under Gammaproteobacteria was Alteromonas, a gram negative heterotrophic bacteria capable of degrading aromatic carbon rings introduced through oil spill [34]. Another dominant genus under this subphylum was Halomonas known to resist extreme conditions and also involve in sandstone formations [35]. Among the Betaproteobacteria, the most abundant genera were *Thiobacillus*, *Burkholderia* and *Delftia*, but they were present in less number (<0.002%). *Thiobacillus* can obtain energy by oxidizingo sulfur and ferrous iron compounds [36]. Most of the members under the genus *Burkholderia* were diazotrophs and degrades a variety of xenobiotic compounds [37].

The unique characteristics of the genus *Bdellovibrio* are that they can enter into the periplasmic space of other bacteria and feed on the biopolymers and thereby used as biocontrol purposes [38]. The abundant genus under Acidobacteria was *Candidatus Solibacter*, an aerobic, chemoorganotrophic bacteria having a large number of anion: cation symporters which helps them to survive in nutrient limited condition [39]. Other abundant genus *Candidatus koribacter* was primarily considered as heterotroph [40].

Metabolic prediction using PICRUSt
The cave environment is a diverse habitat harbouring organisms from all hierarchies starting from prokaryotes to higher eukaryotes [4]. Phylogenetic analysis using 16S

SSU-rDNAs data were also applied to assume the metabolic role of the identified bacterial in cave ecosystem by aligning the sequence information to its next nearest culturable representatives [41, 42]. More recently, PICRUSt software package was developed used to infer the potential functional role of bacterial communities in the cave sediment samples using 16S SSU-rDNAs data [8].

Microbial communities are well known key players of biogeochemical cycles and mainly contribute to the global biogeochemical cycling of carbon and nitrogen [29]. Present study detected enzyme 4-hydroxybutyryl-CoA dehydratase involved in the CO_2-fixation of Archea and fermentation in bacteria which supports the hypothesis that autotrophic archaea contribute to carbon assimilation in cave and other environments [43–45]. Analysis also detected methenyltetrahydrofolate cyclohydrolase which is involved in reverse methanogenesis prevalent in anaerobic methanotrophic archaea [46, 47]. The presence of genes encoding proteins for the phosphate recycling mechanism, such as phosphonate transpoters (PhnB, PhnG, PhnH, PhnI, PhnJ, and PhnM) in the cave samples suggest that they form carbonphosphorus lyase complex which is involved in methane production from methyl phosphonate [48].

Role of bacteria in nitrogen cycle have been well studied in soil and aquatic habitats, but information on cave sediment is limited. Some reports are available where microbes can accrue energy as well as nutrients in oligotrophic environments through nitrogen cycling processes. Most of the genes involved in nitrogen cycle were detected in the present study. Presence of the genes codes for hydroxylamine oxidase indicates the presence of a key ammonia oxidizing bacteria (AOB) [49]. Presence of AOB and sulfur-oxidizing bacteria were also reported in chemolithotrophic Cave [48] and thus lithochemotrophy might be a survival strategy of the bacterial communities present in the cave sediments. Identified genus, *Nitrospira* and *Nitrosospira* were reported to perform autotrophic nitrification which is an indication of CO_2-fixation-coupled ammonia oxidation process in the studied cave ecosystems [50].

Association between bacterial communities with geochemical parameters
Bacterial community structure is greatly influenced by the mineral substrates present in an environment [51]. Present study observed the positive relationship between Fe_2O_3, Al_2O_3 and NiO with the Shannon diversity index. Fe (II) is produced on the subsurface under anoxic conditions by dissimilatory iron (III) reducing bacteria (DIRB) coupled with biotic/abiotic weathering of minerals. Reduced metals inside the cave serve as a source of electron donor for bacterial growth [52, 53]. Only certain organisms can survive in the presence of oligotrophic forces

and a high-metal environment, and the natural selection favours adaptations in microbial communities to sustain in these environments.

Conclusion

Present study used Illumina sequencing to examine the taxonomical diversity of bacterial communities present in cave sediment samples, which were collected from Mizoram, an Indo-Burma Biodiversity Hotspot. These oligotrophic cave harbours a high phylogenetic diversity, including organisms from all hierarchies as well as a higher proportion of unclassified sequences indicating the possibility of novel species. The cave sediments were dominated by Actinobacteria and Proteobacteria. Fe_2O_3 content was correlated with increased microbial diversity in these cave environments. Bioinformatics analysis detected genes involved in various metabolic pathways which are essential for the survival of the community in nutrient limited cave environments. Further research by cultivating the uncultured communities or whole genome sequencing is needed to illustrate the actual survival strategies in the cave environments.

Additional file

Additional file 1: Table S1. List of the genes codes for enzymes involved in carbohydrate degradation identified using PICRUSt. **Table S2.** List of the homologs of methanogenesis-associated genes that were identified from the five cave sediments using PICRUSt. **Table S3.** List of the genes coding for enzymes involved in nitrogen cycle identified using PICRUSt. **Table S4.** Pearson correlation (PC) between physiochemical factors with the dominant bacterial phyla. **Table S5.** Pearson correlation (PC) between physiochemical factors with the bacterial diversity. **Figure S1.** Bioplot generated for the Principal Component Analysis (PCA) of 20 geochemical variables. Cave samples are shown as colored symbols and physicochemical variables are represented by green lines. **Figure S2.** Relative abundance of the functional genes present in the cave samples.

Abbreviations

ABC: ATP-Binding Cassettes; AOB: ammonia oxidizing bacteria; HMP: human microbiome project; NGS: next generation sequencing; PCA: principal component analysis; PICRUSt: phylogenetic investigation of communities by reconstruction of unobserved states; TRAP: tri-partite ATP-independent periplasmic transporters

Acknowledgements
Not applicable.

Funding

This research was supported by a grant from the Bioinformatics Infrastructure Facility sponsored by Department of Biotechnology, Govt. of India, New Delhi. The grant has enabled us to establish computational facility to work on Metagenomics pipeline, and the Advanced State Biotech Hub grant has helped in sampling and generate NGS data. The DeLCON facility has enabled for study design, analysis and interpretation of data.

Authors' contributions

SDM performed the sample collection and drafted the manuscript. SDM and RC carried out bioinformatics and statistical analyses. NSK and RC supervised the Illumina sequencing and edited the manuscript. All authors read and approved the final manuscript.

Competing interests

The authors declare that they have no competing interests.

Author details

[1]Department of Biotechnology, Mizoram University, Aizawl, Mizoram 796004, India. [2]Human Genetics Unit, Indian Statistical Institute, Kolkata 700108, India.

References

1. Ortiz M, Neilson JW, Nelson WM, Legatzki A, Byrne A, Yu Y, Wing RA, Soderlund CA, Pryor BM, Pierson LS, Maier RM. Profiling bacterial diversity and taxonomic composition on Speleothem surfaces in Kartchner caverns, AZ. Microb Ecol. 2013;65:371–83.

2. Canaveras JC, Sanchez-Moral S, Soler V, Saiz-Jimenez C. Microorganisms and microbially induced fabrics in cave walls. Geomicrobiol J. 2001;18:223–40.

3. Ivanova V, Tomova I, Kamburov A, Tomova A, Vasileva-Tonkova E, Kambourova M. High phylogenetic diversity of bacteria in the area of prehistoric paintings in Magura Cave,Bulgaria. J Caves Karst Stud. 2013; 75(3):218–28.

4. De Mandal S, Zothansanga PAK, Bisht SS, Kumar NS. MiSeq HV4 16S rRNA gene analysis of bacterial community composition among the cave sediments of Indo-Burma biodiversity hotspot. Environ Sci Pollut Res. 2016; 23:12216–26.

5. Gilbert JA, Dupont CL. Microbial Metagenomics: beyond the genome. Annu Rev Mar Sci. 2011;3:347–71.

6. Barns SM, Fundyga RE, Jeffries MW, Pace NR. Remarkable archaeal diversity detected in a Yellowstone National Park hot spring environment. Proc Natl Acad Sci U S A. 1994;91(5):1609–13.

7. Hugenholtz P, Goebel BM, Pace NR. Impact of culture-independent studies on the emerging phylogenetic view of bacterial diversity. J Bacteriol. 1998; 180:4765–74.

8. Langille MG, Zaneveld J, Caporaso JG, McDonald D, Knights D, Reyes JA, Clemente JC, Burkepile DE, Thurber RL, Knight R, Beiko RG. Predictive functional profiling of microbial communities using 16S rRNA marker gene sequences. Nat Biotechnol. 2013;31(9):814–21.

9. Adetutu EM, Thorpe K, Bourne S, Cao X, Shahsavari E, Kirby G, Ball AS. Phylogenetic diversity of fungal communities in areas accessible and not accessible to tourists in Naracoorte caves. Mycologia. 2011;103:959–68.

10. Barton HA, Taylor NM, Kreate MP, Springer AC, Oehrle SA, Bertog JL. The impact of host rock geochemistry on bacterial community structure in oligotrophic cave environments. Int J Speleol. 2007;36:93–104.

11. De Mandal S, Panda AK, Lalnunmawii E, Bisht SS, Kumar NS. Illumina-based analysis of bacterial community in Khuangcherapuk cave of Mizoram, Northeast India. Genomics data. 2015a;5:13–4.

12. De Mandal S, Sanga Z, Kumar NS. Metagenome sequencing reveals Rhodococcus dominance in Farpuk cave, Mizoram, India, an eastern Himalayan biodiversity hot spot region. Genome announcements. 2015b; 3(3):e00610–5.

13. Caporaso JG, Bittinger K, Bushman FD, DeSantis TZ, Andersen GL, Knight R. PyNAST: a flexible tool for aligning sequences to a template alignment. Bioinformatics. 2010a;26:266–7.

14. Caporaso JG, Kuczynski J, Stombaugh J, Bittinger K, Bushman FD, Costello EK, Fierer N, Peña AG, Goodrich JK, Gordon JI, Huttley GA, Kelley ST, Knights D, Koenig JE, Ley RE, Lozupone CA, McDonald D, Muegge BD, Pirrung M, Reeder J, Sevinsky JR, Turnbaugh PJ, Walters WA, Widmann J, Yatsunenko T, Zaneveld J, Knight R. QIIME allows analysis of high-throughput community sequencing data. Nat Methods. 2010b;7:335–6.

15. Edgar RC. Search and clustering orders of magnitude faster than BLAST. Bioinformatics. 2010;26(19):2460–1.

16. Lozupone C, Knight R. UniFrac: a new phylogenetic method for comparing microbial communities. Appl Environ Microbiol. 2005;71(12):8228–35.

17. Hammer O, Harper DAT, Ryan PD. PAST: Paleontological statistics software package for education and data analysis. Palaeontologia Electronica. 2001;4:1–9.

18. Margulies M, Egholm M, Altman WE, Attiya S, Bader JS, Bemben LA, Berka J, Braverman MS, Chen YJ, Chen Z, Dewell SB. Genome sequencing in microfabricated high-density picolitre reactors. Nature. 2005;437(7057):376–80.

19. Barton HA, Jurado V. What's up down there? Microbial diversity in caves. Microbe-American Society for Microbiology. 2007;2(3):132–8.

20. Schluter D. Ecological causes of adaptive radiation. Am Nat. 1996;148:S40.

21. Chelius MK, Moore JC. Molecular phylogenetic analysis of archaea and bacteria in wind cave, South Dakota. Geomicrobiol J. 2004;21(2):123–34.

22. Wu Y, Tan L, Liu W, Wang B, Wang J, Cai Y, Lin X. Profiling bacterial diversity in a limestone cave of the western loess plateau of China. Front Microbiol. 2015;6:244.

23. Vander Geize R, Dijkhuizen L. Harnessing the catabolic diversity of rhodococci for environmental and biotechnological applications. Curr Opin Microbiol. 2004;7(3):255–61.

24. Burkowski A. Corynebacteria – Genomics and Molecular Biology. Norfolk: Caister Academic Press; 2008.

25. McLeod MP, Warren RL, Hsiao WW, Araki N, Myhre M, Fernandes C, Miyazawa D, Wong W, Lillquist AL, Wang D, Dosanjh M. The complete genome of Rhodococcus sp. RHA1 provides insights into a catabolic powerhouse. PNAS. 2006;103(42):15582–7.

26. Groth I, Saiz-Jimenez C. Actinomycetes in hypogean environments. Geomicrobiol J. 1999;16:1–8.

27. Yoon JH, Cho YG, Lee ST, Suzuki KI, Nakase T, Park YH. Nocardioides nitrophenolicus sp. nov., a p-nitrophenol-degrading bacterium. Int J Systematic Bacteriol. 1999;49:675–80.

28. Madigan M, Martinko J. Brock biology of microorganisms. 11th ed; 2005. p. 149–52.

29. Kumbhare SV, Dhotre DP, Dhar SK, Jani K, Apte DA, Shouche YS, Sharma A. Insights into diversity and imputed metabolic potential of bacterial communities in the continental shelf of Agatti Island. PLoS One. 2015;10(6):e0129864.

30. Lakshmi KVNS, Sasikala C, Ramana CV. Rhodoplanes pokkaliisoli sp. nov., a phototrophic alphaproteobacterium isolated from a waterlogged brackish paddy soil. Int J Syst Evol Microbiol. 2009;59:2153–7.

31. Lodha TD, Srinivas A, Sasikala C, Ramana CV. Hopanoid inventory of Rhodoplanes spp. Arch Microbiol. 2015;197(6):861–7.

32. Takaichi S, Sasikala C, Ramana Ch V, Okamura K, Hiraishi A. Carotenoids in Rhodoplanes species: variation of compositions and substrate specificity of predicted carotenogenesis enzymes. Curr Microbiol. 2012;65:150–5.

33. Balkwill DL, Drake GR, Reeves RH, Fredrickson JK, White DC, Ringelberg DB, Chandler DP, Romine MF, Kennedy DW, Spadoni CM. Taxonomic study of aromatic-degrading bacteria from deep-terrestrial-subsurface sediments and description of Sphingomonas aromaticivorans sp. nov., Sphingomonas subterranea sp. nov., and Sphingomonas stygia sp. nov. Int J Syst Bacteriol. 1997;47(1):191–201.

34. Jin HM, Kim JM, Lee HJ, Madsen EL, Jeon CO. Alteromonas as a key agent of polycyclic aromatic hydrocarbon biodegradation in crude oil-contaminated coastal sediment. Environ Sci Technol. 2012 Jun 27; 46(14):7731–40.

35. Dong Y, Kumar CG, Chia N, Kim PJ, Miller PA, Price ND, Cann IK, Flynn TM, Sanford RA, Krapac IG, Locke RA. Halomonas sulfidaeris dominated microbial community inhabits a 1.8 km deep subsurface Cambrian sandstone reservoir. Environ Microbiol. 2014;16(6):1695–708.

36. Frankel RB, Bazylinski DA. Biologically induced mineralization by bacteria. Rev Mineral Geochem. 2003;54(1):95–114.

37. Rusch A, Islam S, Savalia P, Amend JP. Burkholderia insulsa sp. nov., a facultatively chemolithotrophic bacterium isolated from an arsenic-rich shallow marine hydrothermal system. Int J Syst Evol Microbiol. 2015; 65(1):189–94.

38. Yair S, Yaacov D, Susan K, Jurkevitch E. Small eats big: ecology and diversity of Bdellovibrio and like organisms, and their dynamics in predator-prey interactions. Agronomie. 2003;23:433–9.

39. Rowe OF, Sánchez España J, Hallberg KB, Johnson DB. Microbial communities and geochemical dynamics in an extremely acidic, metal rich stream at an abandoned sulfide mine (Huelva, Spain) underpinned by two functional primary production systems. Environ Microbiol. 9(7):1761–71.

40. Sait M, Davis KE, Janssen PH. Effect of pH on isolation and distribution of members of subdivision 1 of the phylum Acidobacteria occurring in soil. Appl Environ Microbiol. 2006;72(3):1852–7.

41. Pace NR. A molecular view of microbial diversity and the biosphere. Science. 1997;276:734–40.

42. Barton HA, Northup DE. Geomicrobiology in cave environments: past, current and future perspectives. J Cave Karst Stud. 2007;69(1):163–78.

43. Berg IA, Kockelkorn D, Buckel W, Fuchs G. A 3-hydroxypropionate/4-hydroxybutyrate autotrophic carbon dioxide assimilation pathway in archaea. Science. 2007;318:1782–6.

44. Zhang LM, Offre PR, He JZ, Verhamme DT, Nicol GW, Prosser JI. Autotrophic ammonia oxidation by soil thaumarchaea. Proc Natl Acad Sci U S A. 2010; 107:17240–5.

45. Ortiz M, Legatzki A, Neilson JW, Fryslie B, Nelson WM, Wing RA, Soderlund CA, Pryor BM, Maier RM. Making a living while starving in the dark: metagenomic insights into the energy dynamics of a carbonate cave. ISME J. 2014;8:478–91.

46. Williams TJ, Cavicchioli R. Marine metaproteomics: deciphering the microbial metabolic food web. Trends Microbiol. 2014;22(5):248–60.

47. Wang DZ, Xie ZX, Zhang SF. Marine metaproteomics: current status and future directions. J Proteome. 2014;97:27–35.

48. Chen Y, Wu L, Boden R, Hillebrand A, Kumaresan D, Moussard H, et al. Life without light: microbial diversity and evidence of sulfur- and ammonium-based chemolithotrophy in Movile cave. ISME J. 2009;3:1093–104.

49. Kim BK, Jung MY, Yu DS, Park SJ, Oh TK, Rhee SK, Kim JF. Genome sequence of an ammonia-oxidizing soil archaeon, 'Candidatus Nitrosoarchaeum koreensis' MY1. J Bacteriol. 2011;193:5539–40.

50. Sarbu SM, Kane TC, Kinkle BK. A chemoautotrophically based cave ecosystem. Science. 1996;272:1953–5.

51. Carson JK, Campbell L, Rooney D, Clipson N, Gleeson DB. Minerals in soil select distinct bacterial communities in their microhabitats. FEMS Microbiol Ecol. 2009;67(3):381–8.

52. Cunningham KI, Northup DE, Pollastro RM, Wright WG, LaRock EJ. Bacteria, fungi and biokarst in Lechuguilla cave, Carlsbad caverns National Park, New Mexico. Environ Geol. 1995;25:2–8.

53. Northup DE, Dahm CN, Melim LA, Spilde MN, Crossey LJ, Lavoie KH, Mallory LM, Boston PJ, Cunningham KI, Barns SM. Evidence for geomicrobiological interactions in Guadalupe caves. J Cave Karst Stud. 2000;62:80–90.

14

Intra- and inter-isolate variation of ribosomal and protein-coding genes in *Pleurotus*: implications for molecular identification and phylogeny on fungal groups

Xiao-Lan He[1†], Qian Li[1,2,3†], Wei-Hong Peng[1], Jie Zhou[1], Xue-Lian Cao[1], Di Wang[1], Zhong-Qian Huang[1], Wei Tan[1], Yu Li[2] and Bing-Cheng Gan[1*]

Abstract

Background: The internal transcribed spacer (ITS), RNA polymerase II second largest subunit (RPB2), and elongation factor 1-alpha (EF1α) are often used in fungal taxonomy and phylogenetic analysis. As we know, an ideal molecular marker used in molecular identification and phylogenetic studies is homogeneous within species, and interspecific variation exceeds intraspecific variation. However, during our process of performing ITS, RPB2, and EF1α sequencing on the *Pleurotus* spp., we found that intra-isolate sequence polymorphism might be present in these genes because direct sequencing of PCR products failed in some isolates. Therefore, we detected intra- and inter-isolate variation of the three genes in *Pleurotus* by polymerase chain reaction amplification and cloning in this study.

Results: Results showed that intra-isolate variation of ITS was not uncommon but the polymorphic level in each isolate was relatively low in *Pleurotus*; intra-isolate variations of EF1α and RPB2 sequences were present in an unexpectedly high amount. The polymorphism level differed significantly between ITS, RPB2, and EF1α in the same individual, and the intra-isolate heterogeneity level of each gene varied between isolates within the same species. Intra-isolate and intraspecific variation of ITS in the tested isolates was less than interspecific variation, and intra-isolate and intraspecific variation of RPB2 was probably equal with interspecific divergence. Meanwhile, intra-isolate and intraspecific variation of EF1α could exceed interspecific divergence. These findings suggested that RPB2 and EF1α are not desirable barcoding candidates for *Pleurotus*. We also discussed the reason why rDNA and protein-coding genes showed variants within a single isolate in *Pleurotus*, but must be addressed in further research.

Conclusions: Our study demonstrated that intra-isolate variation of ribosomal and protein-coding genes are likely widespread in fungi. This has implications for studies on fungal evolution, taxonomy, phylogenetics, and population genetics. More extensive sampling of these genes and other candidates will be required to ensure reliability as phylogenetic markers and DNA barcodes.

Keywords: Edible mushroom, Intra-isolate polymorphism, Specific variation

* Correspondence: bcgan918@163.com
†Equal contributors
[1]Soil and Fertilizer Institute, Sichuan Academy of Agricultural Sciences, Chengdu 610066, China
Full list of author information is available at the end of the article

Background

The internal transcribed spacer (ITS) region is the most extensively used nuclear ribosomal gene for specific identification and phylogenetic analysis in fungal groups, and has been declared as the DNA barcode for fungi [1]. The RNA polymerase II second largest subunit (RPB2) and elongation factor 1 alpha (EF1α) have also rapidly become the most prevalent single copy protein-coding genes in fungal identification and phylogenetic reconstructions at the species level [2, 3]. Whether a molecular marker could be used to achieve authentic fungal identification depends very much on the sequence homogeneity within individual and species. However, ITS sequences could show variation within individuals, and sequence heterogeneity has been reported in many fungal groups, such as *Ganoderma*, *Sclerotium*, *Laetiporus*, *Wolfiporia*, and *Trichaptum abietinum* [4–17]. A few reports have also found that sequence polymorphism within isolates is also present in EF1α; however, no study has focused on this in depth [18]. Furthermore, intra-isolate variation of RPB2 remains unknown in fungal groups.

Pleurotus represents one of the most diverse groups of cultivated mushrooms, including some edible species with high commercial values and that are cultivated worldwide. Phylogenetics, taxonomy, and population genetics on this group have been widely studied using molecular markers, including ITS, RPB2, and EF1α sequences [3, 19]. During our process of performing ITS, RPB2, and EF1α sequencing on the *Pleurotus* spp., direct sequencing of PCR products failed in some isolates. Especially in EF1α sequences, most direct sequencing results showed ambiguous sequences, and double or multiple peaks were frequently observed in sequencing chromatograms. We assumed that intra-isolate sequence heterogeneities are present in these genes of *Pleurotus* spp. The presence of intra-isolate variability renders the use of ITS, RPB2, and EF1α for barcoding *Pleurotus* and other fungal groups questionable. However, in many phylogenetic studies on fungal groups, ribosomal and protein-coding genes without assessment of intra-isolate or intraspecific divergence were directly used for phylogeny reconstruction or specific identification [3, 20–22]. Under such circumstances, specific divergence might have been underestimated in some groups, while species diversity overestimated.

In this study, ITS, partial RPB2 and partial EF1α regions of different *Pleurotus* isolates were cloned and analyzed to (1) examine differences of intra-isolate polymorphism levels between genes, isolates, and species, and (2) to test the intraspecific and interspecific divergence of the 3 genes in *Pleurotus*. The present study has implications for molecular identification and phylogenetic analyses in fungal groups.

Methods

Collections and DNA extraction

We collected fruiting bodies of *Pleurotus*, and then performed isolations on them. The dried fruiting bodies were identified by morphological and molecular evidence. The isolated *Pleurotus* strains used in this study were deposited in the Soil and Fertilizer Institute, Sichuan Academy of Agricultural Sciences (SAAS). One hundred and 22 *Pleurotus* isolates were collected, and then ITS, RPB2, and EF1α of these isolates were amplified and sequenced. Twenty-two isolates were selected for cloning and sequence polymorphism analyses (Table 1). Fungal genomic DNAs were extracted from mycelia, according to the procedures described by Peng et al. [23]. Identifications of the fruiting bodies and corresponding isolates were compared to confirm the purification of the tested materials.

PCR, cloning, and sequencing

Universal primers ITS4 and ITS5 [24] were used for ITS amplification, primers fRPB2 5F (5′ GAYGAYMG WG ATCAYTTYGG 3′) and bRPB2 7.1R (5′ CCCATR GCYTGYTTMCCCATDGC 3′) were used for RPB2 amplification, and EF116OR (5′ CCGAT CTTGTA

Table 1 List of strains used for polymorphism analysis in this study

Taxa	Isolates	Origin
Pleurotus citrinopileatus	p145	China: Sichuan, Chengdu
P. citrinopileatus	p146	China: Jilin, Changbai Mountains
P. citrinopileatus	p147	China: Jilin, Changbai Mountains
P. ostreatus	p019	China: Sichuan, Chengdu
P. ostreatus	p021	China: Sichuan, Chengdu
P. ostreatus	p024	China: Sichuan, Zhongjiang,
P. ostreatus	p026	China: Sichuan, Qingbaijiang,
P. ostreatus	p027	China: Sichuan, Qingbaijiang,
P. ostreatus	p028	China: Sichuan, Qingbaijiang,
P. ostreatus	p053	China: Sichuan, Pengxi
P. ostreatus	p055	China: Sichuan, Pengxi
P. ostreatus	p057	China: Sichuan, Zhongjiang
P. ostreatus	p058	China: Sichuan, Jintang
P. ostreatus	p069	China: Sichuan, Leshan
P. ostreatus	p079	China: Sichuan, Dazhou
P. ostreatus	p082	China: Sichuan, Zhongjiang,
P. pulmonarius	p073	China: Jilin, Changbai Mountains
P. pulmonarius	p077	China: Sichuan, Dazhou
P. pulmonarius	p078	China: Sichuan, Dazhou
P. pulmonarius	p003	China: Jilin, Changbai Mountains
P. pulmonarius	p038	China: Sichuan, Chengdu
P. pulmonarius	p041	China: Sichuan, Chengdu

GACGT CCTG 3′) and EF595F (5′ CGTGACTTCAT CAAGAAC ATG 3′) were used for the amplification of a part of the EF1α gene [3]. PCR conditions of ITS were as follows: 94 °C/5 min, 35 cycles of 94 °C/1 min, 55 °C/ 1 min, 72 °C/90 s, and a final extension step of 72 °C/ 10 min. PCRs of RPB2 and EF1α were performed as described by Rodriguez Estrada et al. [3]. For each isolate, PCR amplification was performed at least twice to exclude error of the polymerase. Both Taq polymerase and *pfu* polymerase were used for amplification.

Cloning was carried out with the Qiagen PCR Cloning Kit (Qiagen, Hilden, Germany) according to the manufacturer's instructions. All PCR products were sequenced with primers M13F and M13R. At least 30 clones for each isolate and gene were sequenced to statistically substantiate results.

Analysis of sequences

Sequences obtained in this study were deposited in GenBank. Sequences were manually checked and edited with Bioedit 7.0.9.0 [25]. Sequences from 1 gene of the same isolates and species were aligned in Mega 4.0 [26] to display polymorphisms. For the ITS region, only ITS1, 5.8S, and ITS2 were analyzed. RPB2 and EF1α sequences were trimmed for analyses. Both nucleotides and corresponding amino acid sequences were examined. Single occurrence sequence variants (singletons) in the multiple clones within a single isolate were assumed PCR or cloning errors and corrected. Nucleotide diversity (π-values) was calculated using the software DnaSP based on Nei and Li [27, 28].

Results
Polymorphism of ITS sequences

For *P. ostreatus*, 492 ITS sequences of 10 isolates were generated and the total length varied within species (588–591 bp). A total 32 "singletons (single occurrence sequence variants)" were supposed to be PCR errors, and 8 nucleotide substitutes were recognized as polymorphisms between these sequences (Additional file 1: Figure S1). All variants were nested in the non-coding region (5 in ITS1 and 3 in ITS2); two variants are T/C transitions, four are indels (insertions-deletions), and two are A/T transversions. In different isolates, the polymorphisms could occur in the same sites, or different sites. Only 1 variant was detected in isolate P027, while

8 were detected in the other 9 isolates (Additional file 1: Figure S1). Twenty-two ITS types were found within *P. ostreatus* (Additional file 2: Figure S2), and up to 13 types were detected within a single isolate P021 (Additional file 3: Figure S3). ITS sequence divergence within a single isolate was up to 1.35%, and intraspecific variation was also up to 1.35% in *P. ostreatus* (Table 2).

For *P. pulmonarius*, 93 ITS sequences of P038, P041, and P073 were obtained. The total length of the ITS was identical in the 3 different isolates (581 bp); 5 variable sites were recognized as singletons, and totally 10 polymorphic sites were found between these isolates (Additional file 4: Figure S4). One variant (306) was located in 5.8S, 3 in ITS1, and 6 in ITS2. Six ITS types and 5 polymorphic sites were observed in P038, 4 types and 2 variants in P041, and 4 types and 3 substitutes in P073. ITS sequence similarity was 98.70–100% within *P. pulmonarius*, and 99.31% within a single isolate (Table 2).

For *P. citrinopileatus*, 27 sequences of 3 isolates (P145, P146, and P147) were obtained. Expect for 2 "singletons", no intra-isolate polymorphism site was observed in these isolates, while 1 inter-isolate variable site was found between them (27 sequences; Additional file 5: Figure S5). Sequence similarity within individuals was 100%, and intraspecific divergence was 0.20% (Table 2).

Of the 18 polymorphisms in the 3 *Pleurotus* species, 3 were caused by transversions, 9 by T/C transitions, 2 by A/G transitions, and 5 by indels (Fig. 1). The highest intraspecific divergence was 1.35%, and the lowest interspecific variation between the 3 tested species was 2.90%.

Polymorphism of RPB2 sequences

For *P. ostreatus*, a total 115 sequences of P027, P028, and P053 were obtained; 11 variable sites were supposed to be singletons, and 101 variants between these sequences were considered polymorphisms (Additional file 6: Figure S6). In isolate P019, 10 RPB2 types and 20 variable sites were recovered. The 20 variable sites were composed of 12 T/C transitions, 3 A/G transitions, 5 transversions. One variant was in the intron region, and 19 in the exon region of 17 were synonymous mutations and 2 (sites 467, 687) were non-synonymous. In isolate P027, 6 sequence types and 10 variable sites (4 T/C transitions, 3 A/G transitions and 3 transversions) were observed. Among these variable bases, 4 were located in the intron region, and 6 in the exon

Table 2 Polymorphic sites, variation and nucleotide diversity (π) of ITS sequences in the tested *Pleurotus* species

Species	Nr. of clones	Nr. of polymorphic sites		Variation		Nucleotide diversity (π)	
		Intra-isolate	Intraspecies	Intra-isolate	Intraspecies	Intra-isolate	Intraspecies
P. ostreatus	289	8	8	1.35%	1.35%	0.0035	0.0034
P. pulmonarius	93	5	10	0.69%	1.30%	0.0025	0.0029
P. ciltrinopileatus	27	0	1	0.00%	0.20%	0.0000	0.0000

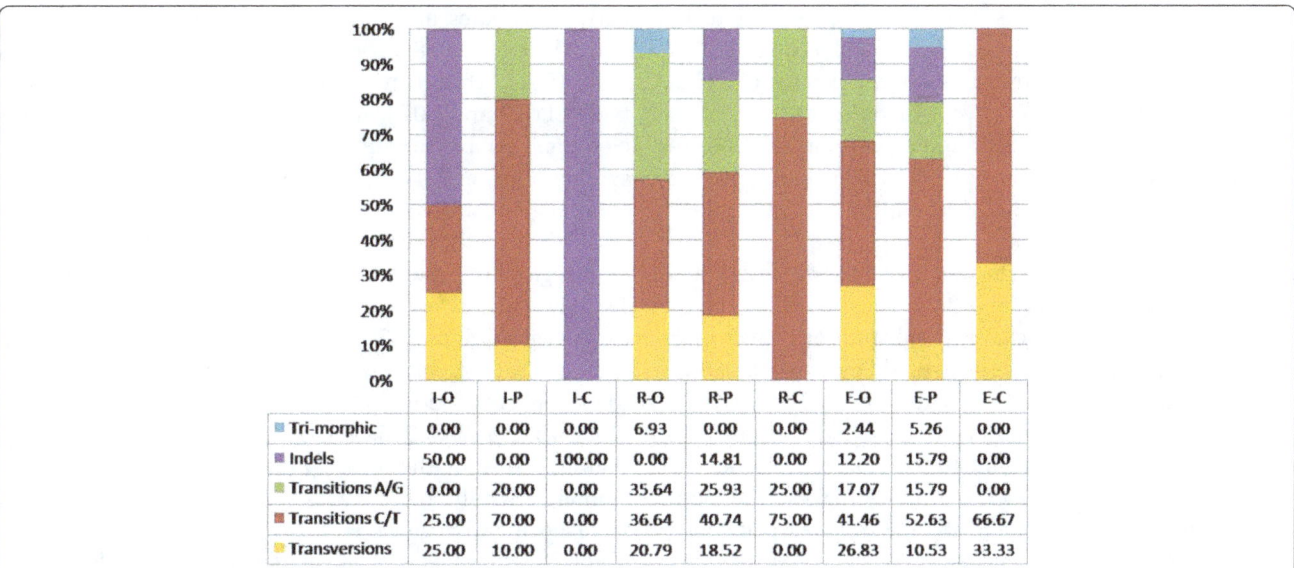

Fig. 1 Proportion (percentage) of indels, transversions, and transitions for each gene and species. Transitions are further split into substitutions between A and G, and between C and T. I: ITS; R: RPB2; E: EF1α; O: *P. ostreatus*; P: *P. pulmonarius*; C: *P. citrinopileatus*

region; 2 (486, 571) of them were non-synonymous. In isolate P028, 5 RPB2 types and 61 variable sites were found, among which 24 bases were T/C transitions, 20 were A/G transitions, and 17 were transversions. Four of these variable sites were located in the intron region, and 57 in the exon region. Only 1 variant in the exon was non-synonymous. In isolate P053, 9 sequence types and 61 variable sites were detected. Among these variable sites, 24 bases were T/C transitions, 22 were A/G transitions, 14 were transversions, and 1 was a tri-morphic site of T, C, and G. Four of these variable sites were located in the intron region, and 2 (sites 480, 828) of the 57 mutations in the exon region were non-synonymous. Sequence divergence within 1 individual as high as 6.03%, and sequence similarity within *P. ostreatus* was 93.40–100% (Table 3).

For *P. pulmonarius*, 83 sequences of P003, P077, and P078 were obtained; 27 polymorphisms (Additional file 7: Figure S7) and 16 "singletons" were recognized between these sequences. Twenty-one polymorphisms were found between P003 and P077, 6 variants between P077 and P078, and 24 between P003 and P078. In isolate P003, 6 sequence types and 8 polymorphism sites were observed. Among the variable sites, we found 3 T/

C transitions, 4 were A/G transitions, and 1 transversion. All polymorphism bases were in the exon region, and 2 (492, 712) were non-synonymous. Isolate P077 possessed 4 sequence types and 17 variable sites, of which 10 sites were T/C transitions, 2 were A/G transitions, 1 was a transversion, and 4 were indels. Eight of these variable sites were in the exon region and synonymous; meanwhile, 8 were in the intron region. Isolate P078 had 3 sequence types and 20 variable sites. Among these variable sites, 10 were T/C transitions, 3 were A/G transitions, 3 were transversions, and 4 were indels; furthermore, 8 sites were located in the intron region, 12 were in the exon region, and only 1 was non-synonymous. Sequence similarity within a single isolate was as low as 98.03%, and 98.00–100% within *P. pulmonarius* (Table 3).

For *P. citrinopileatus*, 86 sequences of P145, P146, and P147 were obtained; 8 polymorphisms and 5 "singletons" between these sequences were observed (Additional file 8: Figure S8). There was 1 variable site between P145 and P146, 4 between P145 and P147, and 3 between P146 and P147 (Additional file 8: Figure S8). In isolate P145, 7 sequence types and 5 variables were observed. Four of the polymorphisms were T/C transitions between, and 1 was A/G transition; 2 were in the intron region, and 3

Table 3 Polymorphic sites, variation, and nucleotide diversity (π) of RPB2 sequences in the tested *Pleurotus* species

Species	Nr. of clones	Nr. of polymorphic sites		Variation		Nucleotide diversity (π)	
		Intra-isolate	Intraspecies	Intra-isolate	Intraspecies	Intra-isolate	Intraspecies
P. ostreatus	115	61	102	6.03%	6.60%	0.0294	0.0312
P. pulmonarius	83	20	27	1.97%	2.00%	0.0029	0.0046
P. ciltrinopileatus	86	8	9	0.49%	0.50%	0.0025	0.0021

synonymous mutation sites were in the exon region. In isolate P146, 2 sequence types and 4 variable sites (3 T/C transitions and 1 G/A transition) were detected, and 3 synonymous mutation sites were in the exon region. In isolate P147, 5 sequence types and 8 polymorphisms were found. Six polymorphisms were T/C transitions, and 2 were A/G transitions; 3 (467, 686, 948) of the 6 variations in exon region were non-synonymous. Sequence divergence within a single isolate was up to 0.49%, and 0.50% within *P. citrinopileatus* (Table 3).

Of the total 136 nucleotide substitutions in the 3 *Pleurotus* species, 26 were caused by transversions, 54 T/C by transitions, 48 by A/G transitions, 5 by indels, and 7 by tri-morphic sites. In each species, transitions from T to C were seen most frequently, followed by transitions from A to G (Fig. 1). The highest intraspecific variation was 6.60%, and the lowest interspecific divergence between the three species was also 6.60%.

Polymorphism of EF1α Sequences

For *P. ostreatus*, 71 sequences of P019, P027, and P053 were obtained. With the exception of 23 "singletons", 41 polymorphisms were found between these sequences (Additional file 9: Figure S9). Polymorphic sites were varied in different isolates. Thirty-one different sites were observed between P019 and P027; 39 between P019 and P053; and 19 between P027 and P053. In isolate P019, 9 sequence types and 27 variations of 14 T/C transitions, 4 G/A transitions, 4 indels, and 5 transversions were demonstrated. Seven of these variations were located in the exon region, and all of them were synonymous. In isolate P027, 5 sequence types and 5 variations of 3 T/C transitions, 1 G/A transitions and 1 transversion were found. Two polymorphisms were in the exon region, and 1 (187, 359) were non-synonymous. In isolate P053, 9 sequence types and 15 variations of 5 T/C transitions, 3 G/A transitions, 1 indel, and 6 transversions were found. Five variations were in the exon region, and only 1 (439) was non-synonymous. Sequence divergence within 1 individual was as low as 95.57%, and sequence similarity within *P. ostreatus* was 95.20–100% (Table 4).

For *P. pulmonarius*, a total 85 sequences (559–563 bp) of P003, P077, and P078 were obtained; 19 polymorphisms and 11 "singletons" were found (Additional file 10: Figure S10). As in *P. ostreatus*, polymorphisms

between isolates in *P. pulmonarius* could also be different. There were 18 variable sites between isolates P003 and P077, 19 between P003 and P078, and only one between P077 and P078. In isolate P078, we found 12 nucleotide substitutions; furthermore, all were synonymous, and 4 (27, 45, 105, 141) were nested in the exon region, while 8 (204, 215, 217, 399, 400, 401, 402, 413) were in the intron region. Additionally, in the intron region, there was a 4 base indel (399–402). In P077, 11 polymorphisms were detected: 3 (27, 105, 141) in the exon region, and 8 (204, 215, 217, 399, 400, 401, 402, 413) in the intron region. In P003, 8 variants were found: 3 substitutions (48, 150, 548) were in the exon region, while 5 (203, 206, 211, 352, 401) were in the intron region. None of them was non-synonymous. Sequence similarity within 1 individual was as low as 97.87%, and intraspecific divergence was up to 2.70% (Table 4).

For *P. citrinopileatus*, 90 sequences of P145, P146, and P147 were obtained; 3 nucleotide substitutions and 5 "singletons" were found between these sequences (Additional file 11: Figure S11). Two of the 3 nucleotide substitutions (214: A/C; 223: T/C) were in the intron region, and 1 synonymous variation was in the exon region (189: T/C). Two variants were observed within P145, while no polymorphisms were found within P146 or P147. The lowest similarity within the same isolate was 99.65%, and sequence similarity within *P. citrinopileatus* was 99.40%–100% (Table 4).

Of the total 63 nucleotide substitutions in the 3 *Pleurotus* species, 14 were caused by transversions, 29 by transitions from T to C, 10 by transitions from A to G, 8 by indels, and 2 by tri-morphic variants. Proportions of transversions, transitions, indels for each gene and species are shown in Fig. 1. The highest intraspecific variation was up to 4.80%, but the lowest interspecific divergence between the three species was only 3.10%.

Discussion

Intra- and inter-isolate variation of ribosomal and protein-coding genes has an appreciable effect on molecular identification and phylogenetic analysis of fungal groups. In the present study, we demonstrated that intra-isolate sequence variation is commonly present within both ribosomal and protein-coding genes of *Pleurotus*. Among the 3 tested markers, ITS showed the lowest intra-isolate and intraspecific variation level, and

Table 4 Polymorphic sites, variation, and nucleotide diversity (π) of EF1α sequences in the tested *Pleurotus* species

Species	Nr. of clones	Nr. of polymorphic sites		Variation		Nucleotide diversity (π)	
		Intra-isolate	Intraspecies	Intra-isolate	Intraspecies	Intra-isolate	Intraspecies
P. ostreatus	71	27	42	4.43%	4.80%	0.0163	0.0181
P. pulmonarius	85	12	19	2.13%	2.70%	0.0073	0.0092
P. ciltrinopileatus	90	2	3	0.35%	0.60%	0.0018	0.0017

intra-isolate or intraspecific divergence was often less than that of interspecific divergence. However, RPB2 and EF1α sequence divergence within isolate or species was high, and could be identical or surpass the variation between species in the tested *Pleurotus* species.

ITS has been proposed as a universal DNA barcode marker for fungi [1]. As we know, a requirement of molecular markers used in taxonomic and phylogenetic studies is that all copies within the genome are constant and identical, and that intraspecific variation is lower than interspecific variation. However, intra-isolate variation in ITS regions has been reported in many fungal groups, and this variation could be observed in an unexpectedly high level in some groups. Woo et al. [16] reported 144 (19.4%) nucleotide differences between different ITS types in a single isolate of *Rhizopus microsporus*. As a general rule, ITS sequence similarities less than 90% should be considered to represent different species undoubtedly, and such a high level of variation within an individual is unbelievable. These results suggested that ITS -isolate variation level in fungal groups might be much higher than previously expected. ITS polymorphisms in *Pleurotus* have been reported in the literature. Huang et al. [29] observed a base pair indel in ITS sequences of *P. eryngii* var. *tuoliensis* (as *P. nebrodensis*), and intra-isolate variation has also been detected in *P. abolonus* [30]. In the present study, we found that the intra-isolate heterogeneity level of ITS was variable between isolates within the same species, and was different between species. On the other hand, intraspecific variation of ITS in *Pleurotus* was usually lower than variation between species in this genus. It seems that ITS could be used for specific identification in *Pleurotus*. However, it is premature to conclude that ITS is an ideal DNA barcode for *Pleurotus* due to the scarcity of samples in the present study, and care should be taken when using ITS sequences in taxonomy and phylogenetic studies in *Pleurotus* or other fungal groups for the presence of intra-isolate polymorphisms.

RPB2 has a higher specific recognition in many fungal groups, and has been used as a de facto barcode in identification and phylogenetic studies of some fungal groups, including *Pleurotus* [3, 19]. Contrary to the nuclear ribosomal genes, no intra-isolate variations were reported prior to this study in any fungal RPB2 gene sequences. In our study, RPB2 showed an unexpectedly high sequence heterogeneity level within single isolates. In the tested isolates, polymorphisms varied from 2 to 61 if "singletons" are excluded within individuals, and also differed between isolates or species. The EF1α gene was also regarded as an ideal candidate for studying the relationships of fungi. However, intra-isolate heterogeneity has also been observed in the EF1α gene of some fungal species [18] while no evidence of EF1α intra-

isolate polymorphism has been found in *Pleurotus*. In this study, EF1α sequences also demonstrated high intra-isolate variation in *Pleurotus*. Even though at least 30 clones of each isolate were sequenced, genetic variation was not sampled exhaustively. Additionally, polymorphisms counts were probably underestimated since the quality checks were extremely stringent in the present study, and the actual heterogeneity level was likely to be much higher. For another protein-coding gene, *BiP*, Kuhn et al. [31] reported that sequence divergence in a single isolate of *Glomus irregulare* could be as high as 8%. Such high variation within individuals will influence the results of identification and phylogenetic analyses. We also detected inter- and intraspecific divergence of RPB2 and EF1α in the tested species. Results indicated that polymorphic sites of different isolates within 1 species could be different. Furthermore, divergence of RPB2 within *P. ostreatus* could be identical to that between *P. ostreatus* and *P. pulmonarius*, and intraspecific heterogeneity of EF1α in *P. ostreatus* could surpass the interspecific variability between *P. ostreatus* and *P. pulmonarius*.

These results suggested that high intra-isolate polymorphisms of RPB2 and EF1α are also likely widespread in other fungal groups, and interspecific variation might be less than or equal to intraspecific variation. This posed challenges to the molecular identification and phylogenetic studies on fungi. As we know, an ideal molecular marker used in molecular identification and phylogenetic studies is homogeneous within species, and interspecific variation exceeds intraspecific variation [1]. It could be speculated that the RPB2 and EF1α are too polymorphic to use for identification and phylogenetic studies in *Pleurotus* at the species level, and this hold true for other fungal groups. Ideally, these genes could be used for specific identification or phylogeny analyses only when high intra-isolate variation level is excluded by sequencing at least 10 clones of each sample. As already pointed out by Álvarez and Wendel [32], relying only on direct sequencing of a single PCR product may yield incorrect results due to the numerical inequality among several repeat types within one genome and the possible preferential amplification of one type. In the study of Lindner and Banik [8], they also indicated that phylogenetic analyses with the cloned sequences produced different trees relative to analyses with consensus sequences, and cloned sequences of a single isolate felled into more than one species clades or entirely new clades. If analyzed on their own, these cloned sequences of a single isolate most likely would be recognized as different "undescribed" or "novel" taxa [8]. However, screening for intra-isolate polymorphisms is absent in the routine use of RPB2 and EF1α in phylogenetic studies of fungal groups. Under such circumstances, specific divergence might have been underestimated in some groups, while species diversity overestimated.

Among the variants observed in RPB2 and EF1α, most were synonymous and would not cause amino acid change. On the other hand, very few indel polymorphisms were observed, and they were located in the intron region. This was largely due tomissense variations and indels that affected translation products were tended to be flushed out by natural selection. Of all detected variants, transitions between T and C showed the highest frequency. Additionally, most detected variants were single nucleotide polymorphisms, and there were some differences between isolates. These differences could probably be used to discriminate different isolates.

The reasons for intra-isolate polymorphisms could be manifold. Taq polymerase misreading is a potential contribution to sequence polymorphisms. In the present study, a repeated PCR with *pfu* polymerase and cloning was performed to evaluate the frequency of Taq misreadings. Results showed that Taq polymerase errors were in a very low rate or absent. Sequence deviation of the clone sequences of ITS, RPB2, and EF1α from different *P. ostreatus* isolates, as well as from *P. pulmonarius* and *P. citrinopileatus* isolates, were far out of the range of a misincorporation error of Taq polymerase. Additionally, we presumed that the polymorphisms occurring only in 1 of the 30 clones were Taq misreadings, and these sites were corrected. Therefore, Taq polymerase errors should not be considered as the source of the variations reported in our study.

The protein-coding genes RPB2 and EF1α usually present as single copy genes in fungi; however, high intraisolate polymorphism levels indicated that this is likely not always the case in *Pleurotus*. The RPB2 gene has been found to be multicopy in certain taxa in plants and trypanosomes [33], and multiple copies of EF1α have been reported in humans and in the spider genus *Habronattus* [34–36]. Multicopy genes are assumed to evolve under the mechanism of concerted evolution to maintain homogeneity of all copies [37]. However, intra-isolate sequence variability will accumulate if the conversion rate is lower than the rate of mutation, or if concerted evolution is slower than speciation. It is also possible that differences between mutation rates and DNA repair rates lead to an accumulation of sequence variants in genes with multiple copies [38]. As we know, heterokaryosis is present in fungi, and genetically differentiated nuclei might cause intra-isolate variation. However, the internal mechanism behind why rDNA and protein-coding genes show variants within single isolates in *Pleurotus* is still unclear and must be addressed in further research.

Conclusions

In conclusion, we reported the polymorphisms of rDNA sequences and protein-coding genes of the tested species, which may also be present in other fungal groups. When inter- or intra-isolate variations of these markers are high, molecular identification or phylogeny analyses might cause opposing results. Especially when intra-isolate or intraspecific variation exceeds that of interspecific variation, individuals within 1 species might be considered as different species based on sequence comparisons. Consequently, identification and quantification of intra-isolate and intraspecific heterogeneity is of real importance, and more extensive sampling of these genes may be required to ensure reliability as desirable phylogenetic markers and DNA barcodes for specific identification.

Additional files

Additional file 1: Figure S1. ITS polymorphic sites within *P. ostreatus*. Variable sites of 9 *P. ostreatus* are shown; polymorphisms could occur in the same sites, or different sites.

Additional file 2: Figure S2. Twenty-two ITS types observed within *P. ostreatus*. Two hundred and eighty-nine sequences of 10 *P. ostreatus* isolates were detected.

Additional file 3:Figure S3. Thirteen ITS types detected within a single isolate of *P. ostreatus* (isolate P021).

Additional file 4: Figure S4. Polymorphic sites of ITS sequences in the 3 *P. pulmonarius* isolates. Polymorphisms differed between isolates.

Additional file 5: Figure S5. ITS polymorphic sites in *P. citrinopileatus* isolates. No intra-isolate variant was detected, and 1 nucleotide substitution was observed between the 3 isolates.

Additional file 6: Figure S6. Polymorphisms of RPB2 sequences in *P. ostreatus* isolates. Variation of RPB2 within individuals in *P. ostreatus* was unexpectedly high.

Additional file 7: Figure S7. Variants of RPB2 sequences in *P. pulmonarius* isolates. Indels were located in the intron region.

Additional file 8: Figure S8. Polymorphisms of RPB2 sequences in *P. citrinopileatus* isolates. Intra-isolate variations were found in the 3 isolates.

Additional file 9: Figure S9. Polymorphic sites of EF1α sequences in *P. ostreatus* isolates. Variation levels were differed markedly between isolates.

Additional file 10: Figure S10. Variants of EF1α sequences in *P. pulmonarius* isolates. T/C transitions had the highest frequency.

Additional file 11: Figure S11. Sequence polymorphisms of EF1α in *P. citrinopileatus* isolates. Intra-isolate variation was observed only in isolate P145.

Abbreviations

EF1α: elongation factor 1-alpha; ITS: Internal transcribed spacer; PCR: Polymerase chain reaction; rDNA: Ribosomal deoxyribonucleic acid; RPB2: RNA polymerase II second largest subunit

Acknowledgments
We are grateful to Mr. Miao Ren-Yun for assisting with *Pleurotus* isolates collection. We thank the anonymous reviewers for their helpful suggestions and critical advice.

Funding
This study was partly funded by the National Public Welfare (Agriculture) Science and Technology Project (201503137), Sichuan Provincial Innovation Ability Promotion Engineering (2015LWJJ-004, 2016ZYPZ-028), and Sichuan Provincial Infrastructure of Microbial Resources, Science and Technology Department of Sichuan Province (2016TJPT0002).

Authors' contributions
XLH designed the experiment; QL, JZ, XLC and DW performed the molecular experiments; WHP, ZQH, WT, YL and BCG collected isolates and improved the experiment; XLH and QL analyzed the data; XLH drafted the manuscript. All authors polished the manuscript.

Authors' information
Xiao-Lan He and Qian Li contributed equally to this work, and share the first author.

Competing interests
The authors declare that they have no competing interests.

Author details
[1]Soil and Fertilizer Institute, Sichuan Academy of Agricultural Sciences, Chengdu 610066, China. [2]Jilin Agricultural University, Changchun 130118, China. [3]Mianyang Institute of Agricultural Sciences, Mianyang 621023, China.

References
1. Schoch CL, Seifert K, Huhndorf S, Robert V, Spouge JL, Levesque CA, et al. Nuclear ribosomal internal transcribed spacer (ITS) region as a universal DNA barcode marker for fungi. Proc Natl Acad Sci U S A. 2012;109:6241–6.
2. Froslev TG, Matheny PB, Hibbett DS. Lower level relationships in the mushroom genus *Cortinarius* (Basidiomycota, Agaricales): a comparison of RPB1, RPB2, and ITS phylogenies. Mol Phylogenet Evol. 2005;37:602–18.
3. Rodriguez Estrada AE, Jimenez-Gasco M d M, Royse DJ. *Pleurotus eryngii* species complex: sequence analysis and phylogeny based on partial EF1α and RPB2 genes. Fungal Biol. 2010;114:421–8.
4. Belbahri L, McLeod A, Paul B, Calmin G, Moralejo E, Spies CFJ, et al. Intraspecific and within-isolate sequence variation in the ITS rRNA gene region of *Pythium mercuriale* sp. nov. (Pythiaceae). FEMS Microbiol Lett. 2008;284:17–27.
5. Fatehi J, Bridge P. Detection of multiple rRNA-ITS regions in isolates of *Ascochyta*. Mycol Res. 1998;102:762–6.
6. Kageyama K, Senda M, Asano T, Suga H, Ishiguro K. Intra-isolate heterogeneity of the ITS region of rDNA in *Pythium helicoides*. Mycol Res. 2007;111:416–23.
7. Ko KS, Jung HS. Three nonorthologous ITS1 types are present in a polypore fungus *Trichaptum abietinum*. Mol Phylogenet Evol. 2002;23:112–22.
8. Lindner D, Banik MT. Intragenomic variation in the ITS rDNA region obscures phylogenetic relationships and inflates estimates of operational taxonomic units in genus *Laetiporus*. Mycologia. 2011;103:731–40.
9. O'Donnell K, Cigelnik E. Two divergent intragenomic rDNA ITS2 types within a monophyletic lineage of the fungus *Fusarium* are nonorthologous. Mol Phylogenet Evol. 1997;7:103–16.
10. Okabe I, Arawaka M, Matsumoto N. ITS polymorphism within a single strain of *Sclerotium rolfsii*. Mycoscience. 2001;42:107–13.
11. Pannecoucque J, Höfte M. Detection of rDNA ITS polymorphism in *Rhizoctonia solani* AG 2–1 isolates. Mycologia. 2009;101:26–33.
12. Simon UK, Weiss M. Intragenomic variation of fungal ribosomal genes is higher than previously thought. Mol Biol Evol. 2008;25:2251–4.
13. Vydryakova GA, Van DT, Shoukouhi P, Psurtseva NV, Bissett J. Intergenomic and intragenomic ITS sequence heterogeneity in *Neonothopanus nambi* (*Agaricales*) from Vietnam. Mycology. 2012;3:89–99.
14. Yao CL, Fredriksen RA, Magill CW. Length heterogeneity in ITS2 and the methylation status of CCGG and GCGC sites in the rRNA genes of the genus *Peronosclerospora*. Curr Genet. 1992;22:219–29.
15. Wang DM, Yao YJ. Intrastrain internal transcribed spacer heterogeneity in *Ganoderma* species. Can J Microbiol. 2005;51:113–21.
16. Woo PCY, Leung S-Y, To KKW, Chan JFW, Ngan AHY, Cheng VCC, et al. Internal transcribed spacer region sequence heterogeneity in *Rhizopus microsporus*: implications for molecular diagnosis in clinical microbiology laboratories. J Clin Microbiol. 2010;48:208–14.
17. Zhao Y, Tsang C, Xiao M, Cheng J, Xu Y, Lau SKP, et al. Intra-genomic internal transcribed spacer region sequence heterogeneity and molecular diagnosis in clinical microbiology. Int J Mol Sci. 2015;16:25067–79.
18. Hasegawa E, Ota Y, Hattori T, Kikuchi T. Sequence-based identification of Japanese *Armillaria* species using the elongation factor-1 alpha gene. Mycologia. 2010;102:8–910.
19. He XL, Wu B, Li Q, Peng WH, Huang ZQ, Gan BC. Phylogenetic relationship of two popular edible *Pleurotus* in China, Bailinggu (*P. eryngii* Var. *tuoliensis*) and Xingbaogu (*P. eryngii*), determined by ITS, RPB2 and EF1α sequences. Mol Biol Rep. 2016;43:573–82.
20. Han ML, Chen YY, Shen LL, Song J, Vlasák J, Dai YC, et al. Taxonomy and phylogeny of the brown-rot fungi: *Fomitopsis* and its related genera. Fungal Divers. 2016;80:343–73.
21. Song J, Chen JJ, Wang M, Chen YY, Cui BK. Phylogeny and biogeography of the remarkable genus *Bondarzewia* (Basidiomycota, Russulales). Sci Rep. 2016;6:34568.
22. Matheny PB, Wang Z, Binder M, Curtis JM, Lim YW, Nilsson RH, et al. Contributions of rpb2 and tef1 to the phylogeny of mushrooms and allies (Basidiomycota, fungi). Mol Phylogenet Evol. 2007;43:430–51.
23. Peng W, He X, Wang Y, Zhang Y, Ye X, Jia D, et al. A new species of *Scytalidium* causing slippery scar on cultivated *Auricularia polytricha* in China. FEMS Microbiol Lett. 2014;359:72–80.
24. White TJ, Bruns T, Lee S, Taylor J. Amplification and direct sequencing of fungal ribosomal RNA genes for phylogenies. In: Innis MA, Gelfand DH, Sninsky JJ, White TJ, editors. PCR protocols, a guide to methods and applications. San Diego: Academic; 1990.
25. Hall TA. Bioedit: a user-friendly biological sequence alignment editor and analysis program for windows 95/98/NT. Nucleic Acids Symp Ser. 1999;41: 95–8.
26. Tamura K, Dudley J, Nei M, Kumar S. MEGA4: molecular evolutionary genetics analysis (MEGA) software version 4.0. Mol Biol Evol. 2007;24:1596–9.
27. Rozas J, Sanchez-DelBarrio JC, Messeguer X, Rozas R. DnaSP, DNA polymorphism analyses by the coalescent and other methods. Bioinformatics. 2003;19:2496–7.
28. Nei M, Li WH. Mathematical model for studying genetic variation in terms of restriction endonucleases. Proc Natl Acad Sci U S A. 1979;94: 7799–806.
29. Huang C, Xu J, Gao W, Chen Q, Wen H, Zhang J. A reason for overlap peaks in direct sequencing of rRNA gene ITS in *Pleurotus nebrodensis*. FEMS Microbiol Lett. 2010;305:14–7.
30. Zervakis GI, Moncalvo JM, Vilgalys R. Molecular phylogeny, biogeography and speciation of the mushroom species *Pleurotus cystidiosus* and allied taxa. Microbiology. 2004;150:715–26.
31. Kuhn G, Hijri M, Sanders IR. Evidence for the evolution of multiple genomes in arbuscular mycorrhizal fungi. Nature. 2001;414:745–8.
32. Álvarez I, Wendel JF. Ribosomal ITS sequences and plant phylogenetic inference. Mol Phylogenet Evol. 2003;29:417–34.

33. Goetsch LA, Eckert AJ, Hall BD. The molecular systematics of rhododendron (Ericaceae): a phylogeny based upon RPB2 gene sequences. Sys Bot. 2005; 30:616–26.
34. Hedin MC, Maddison WP. Phylogenetic utility and evidence for multiple copies of elongation factor-1α in the spider genus *Habronattus* (Araneae: Salticidae). Mol Biol Evol. 2001;18:1512–21.
35. Madsen HO, Poulsen K, Dahl O, Clark BFC, Hjorth JP. Retropseudogenes sconstitute the major part of the human elongation factor 1α gene family. Nucleic Acids Res. 1990;18:1513–6.
36. Opdenakker G, Cabeza-Arvelaiz Y, Fiten P, Dijkmans R, Damme JV, Volckaert G, et al. Human elongation factor 1α: a polymorphic and conserved multigene family with multiple chromosomal localizations. Hum Genet. 1987;75:339–44.
37. Liao DQ. Concerted evolution: molecular mechanism and biological implications. Am J Hum Genet. 1999;64:24–30.
38. Elder JF, Turner BJ. Concerted evolution of repetitive DNA-sequences in eukaryotes. Q Rev Biol. 1995;70:297–320.

15

Clonality, outer-membrane proteins profile and efflux pump in KPC- producing *Enterobacter sp.* in Brazil

Juliana Ferraz Rosa[1], Camila Rizek[1], Ana Paula Marchi[1], Thais Guimaraes[1], Lourdes Miranda[2], Claudia Carrilho[3], Anna S Levin[1] and Silvia F Costa[4*]

Abstract

Background: Carbapenems resistance in *Enterobacter* spp. has increased in the last decade, few studies, however, described the mechanisms of resistance in this bacterium. This study evaluated clonality and mechanisms of carbapenems resistance in clinical isolates of *Enterobacter* spp. identified in three hospitals in Brazil (Hospital A, B and C) over 7-year.

Methods: Antibiotics sensitivity, pulsed-field gel electrophoresis (PFGE), PCR for carbapenemase and efflux pump genes were performed for all carbapenems-resistant isolates. Outer-membrane protein (OMP) was evaluated based on PFGE profile.

Results: A total of 130 isolates of *Enterobacter* spp were analyzed, 44/105 (41, 9%) *E. aerogenes* and 8/25 (32,0%) *E. cloacae* were resistant to carbapenems. All isolates were susceptible to fosfomycin, polymyxin B and tigecycline. KPC was present in 88.6% of *E. aerogenes* and in all *E. cloacae* resistant to carbapenems. The carbapenems-resistant *E. aerogenes* identified in hospital A belonged to six clones, however, a predominant clone was identified in this hospital over the study period. There is a predominant clone in Hospital B and Hospital C as well. The mechanisms of resistance to carbapenems differ among subtypes. Most of the isolates co-harbored *bla*KPC, *bla*TEM and /or *bla*CTX associated with decreased or lost of 35–36KDa and or 39 KDa OMP. The efflux pump AcrAB-TolC gene was only identified in carbapenems-resistant *E. cloacae*.

Conclusions: There was a predominant clone in each hospital suggesting that cross-transmission of carbapenems-resistant *Enterobacter* spp. was frequent. The isolates presented multiple mechanisms of resistance to carbapenems including OMP alteration.

Keyword: *E. aerogenes*, *E. cloacae*, Resistance, Carbapenems, Efflux Pump, Outer Membrane Proteins, β-lactamases and Activity efflux pump with inhibitor Carbonyl-cyanide-m-chlorophenylhydrazone (CCCP)

Background

Healthcare associated infections caused by *Enterobacter* spp. have increased in the last decade all over the world [1, 2]. Carbapenems are frequently used to treat serious infections caused by multi-resistant Gram-negative bacilli, especially those caused by over production of AmpC cephalosporinases or extended-spectrum β-lactamases (ESBL), such as infections caused by *Enterobacter* spp. Thus, the emergence of carbapenems resistance, defined as resistance to ertapenem, imipenem and/or meropenem, is becoming a therapeutic challenge [1, 2].

To date, carbapenemase is the most frequent mechanism of carbapenems resistance reported in *Enterobacter* spp [2, 3]. Studies have shown the presence of carbapenemases (*bla*KPC, *bla*IMP, *bla*VIM and *bla*NDM) in association with ESBL (*bla*TEM, *bla*SHV and *bla*CTX-M) in isolates of *E. aerogenes* and *E. cloacae* resistant to carbapenems [2, 3]. Although, *E. aerogenes* and *E. cloacae* carbapenems-resistant isolates can decrease and or loss OmpK 35–36 and 39KDa outer membrane proteins (OMPs), which lead to alteration of permeability

* Correspondence: costasilviaf@ig.com.br

[4]LIM-54, Faculdade de Medicina da Universidade de São Paulo, São Paulo, Brazil
Full list of author information is available at the end of the article

and the induction of active drug efflux AcrAB-tolc, that contribute to resistance to carbapenems [2–6]. However, few studies demonstrated the importance of OMPs and efflux pump on carbapenems resistance in *Enterobacter* spp [1, 2, 7].

Therefore, the role of mechanisms of carbapenems resistance, such as OMP and efflux pump in *Enterobacter* spp, needs to be better addressed. The present study was conducted to investigate the clonality and mechanisms of carbapenems resistance in *E. aerogenes* and *E. cloacae* identified in three hospitals in Brazil.

Methods

Bacteria collection

One hundred and thirty *Enterobacter* spp. clinical isolates (105 *E. aerogenes* and 25 *E. cloacae*) identified in two hospitals (Hospital A and Hospital B whit 30 km of distance between them) in São Paulo in the state of São Paulo, and one hospital (Hospital C) 530 km distant, in Londrina in the state of Paraná, Brazil, were evaluated over a 7-year period, from 2005 to 2011.

Although, located in another state, the strains of *E. cloacae* identified in Hospital C in Londrina were evaluated in order to investigate the mechanism of carbapenems resistance and clonality in this species as well.

The identification of species was performed by API20 E (bioMérieux, France) and additional tests (Modify Rugai, Motility and Lysine).

Clinical data

The following clinical and demographic data from the medical records of patients hospitalized in Hospital A and Hospital B, were registered: age, gender, underlying diseases, site of infection, length of stay in the Intensive Care Unit and death. Definitions (CDC) for the infections were those used by the Centers for Disease Control and Prevention. An Epi Info™ database was built, and results were expressed as means (standard deviation) or median (interquartile range), depending on normality. All data were analyzed anonymously and confidentially, with approval by the Research Ethics Committee of the three hospitals.

Ethics statement

The study was performed in two hospitals located in São Paulo, Brazil, the Central Institute of Hospital das Clínicas of University of São Paulo (ICHC-FMUSP) and Hospital Itapecerica da Serra and one hospital in Paraná, the Universitary Hospital in Londrina. It was approved by the ethics committee of the hospitals. The approval number is 007/11.

Susceptible testing

The minimal inhibitory concentrations (MICs) of imipenem (Merck & Co. Inc., Elkton, EUA), meropenem (Astra Zeneca), ertapenem (Sigma Chemical, St. Louis, Mo.), cefepime (Bristol-Myers Squibb, Guayaquil, Equador), polymyxin B (Sigma Chemical, St. Louis, Mo.) and tigecycline (Sigma Chemical, St. Louis, Mo.), was performed by broth microdilution with Mueller-Hinton broth. In addition, the MIC of fosfomycin (Sigma Chemical, St. Louis, Mo.) was performed using agar dilution as described in Clinical and Laboratory Standards Institute (CLSI). *P. aeruginosa* ATCC 27853, *E. coli* ATTC 25922, *S. aureus* ATCC 29213 and *E. faecalis* ATCC 29212, were used as control for all isolates (105 *E. aerogenes* and 25 *E. cloacae*). Carbapenems resistance was defined as: resistance to one or more carbepenems (ertapenem, imipenem and or meropenem) according with CLSI breakpoint.

Carbapenemase genes and efflux pump

The presence of genes encoding ESBL (*bla*TEM, *bla*SHV, *bla*CTX-M), carbapenemases Class A (*bla*KPC, *bla*IMI and *bla*GES), Class B (*bla*IMP-1, *bla*VIM-2, *bla*GIM –1, *bla*SPM, *bla*NDM-1) and Class D (*bla*OXA-48), was investigated in all isolates (105 *E. aerogenes* and 25 *E. cloacae*) by PCR as described elsewhere (Table 1) [8–11], genbanks accession numbers: KF285575-KF285585, **KY524253 and MTZP00000000**.

PCR for all isolates (105 *E. aerogenes* and 25 *E. cloacae*) was performed to detect genes of efflux pump acrART using primers according to the study of Perez et al. [12].

DNA sequencing of genes of resistance was performed using the MegaBACE 1000, DNA Analysis System (Amersham Biosciences, UK. England), using DYEnamic ET Dye Terminator Kit (with Thermo Sequence™ DNA Polymerase II) US81090 code. The sequences were analyzed using the Sequence Analyzer software using the Cimarron Base Caller 3.12. The genetic sequence was compared with the database available on the Internet (BLAST - http://www.ncbi.nlm.nhi.gov/blast/).

Pulsed-field gel electrophoresis

DNA analysis of all carbapenems-resistant isolates, 44 *E. aerogenes* and 8 *E. cloacae* were performed by PFGE, after digestion with XbaI Fast (Invitrogen) and the electrophoretic run was made with the following parameters: 23 h with pulse times ranging from 5 to 60s at 6 V/cm, using the CHEF-DR III System (Bio-Rad Laboratories, Richmond, CA, USA). DNA relatedness was computationally analyzed using BioNumerics v.7.1 software (Applied Maths, Sint-Martens-Latem, Belgium). The banding patterns were compared by using the unweighted pair-group method with arithmetic averages (UPGMA), with the Dice similarity coefficient required to be >80% for the pattern to be considered as belonging to the same PFGE type (dendrogram).

Table 1 Primers of all carbapenems resistance genes studied and PCR annealing temperature

Primers	Sequences (5'-3')	Annealing temperature	Size (pb)
blaTEM - F	TCGCCGCATACACTATTCTCAGAATGA	55	420
blaTEM - R	ACG CTC ACC GGC TCC AGA TTT AT		
blaCTXM - F	GCT CTAGAATTATTGCATCAGAAA CCGTG	55	893
blaCTXM - R	CGGAATTCATGATGACTCAGAGCATTGG		
bla$_{OSHV}$ - F	TGCTTTGTTAATTCGGGCCAA	55	730
bla$_{SHV}$ - R	ATGCGTTATATTCGCCTGTG		
bla$_{OXA48}$ - F	GTAACAATGCTTGGTTCG	55	177
bla$_{OXA48}$ - R	TGTTTTTGGTGGCATCGA		
blakpc - F	GTTACGCCAAAGGACGAAC		893
blakpc - R	TTTTCAGAGCCTTACTGCCC		
bla$_{SPM}$ -F	CCTTTTCCGCGACCTTGATC	59	798
bla$_{SPM}$ - R	ATGCGCTTCATTCACGCAC		
bla$_{SIM}$ - F	GTACAAGGGATTCGGCATCG	58	569
bla$_{SIM}$ - R	GTACAAGGGATTCGGCATCG		
bla$_{IMP}$ -F	TTGGAAAATTATATAATCCC	47	188
bla$_{IMP}$ - R	CCAAACCACTAGGTTATC		
bla$_{VIM}$ - F	TTTGGTCGCATATCGCAAAG	60	382
bla$_{VIM}$- R	CCATTCAGCCCAGATCGGCAT		
bla$_{NDM}$ - F	GGCGGAATGGCTCATCACGA	60	375
bla$_{NDM}$ - R	CGCAACACAGCCTGACTTTC		
AcrART - F	GAT TAT GAT TCT GCC TTG GCCG	60	130
AcrART - R	CAA TGC GAC CGC TGA TAG GGG		

Outer membrane protein profile

Based on clonality, the OMP of 22 *E. aerogenes* isolates and 5 *E. cloacae* isolates were analyzed. Bacterial outer membrane proteins (OMPs) were purified by treatment of the cell envelops with 2% sodium-N-lauryl sarcosinate (Sigma Chemical.St Louis., MO). The proteins were quantified by the Bradford method with the aid of Bradford reagent (Bio-Rad Laboratories, Brazil) by spectrophotometry at 595 nm (GeneQuant Pro, GE- Healthcare Life Sciences) and subsequently treated with beta-mercapto-ethanol at a ratio of 2 µl acid beta-mercaptoethanol acid to 10 µl protein. The proteins studied were applied to polyacrylamide gels prior manufactured 12.5% (GE Healthcare) at a concentration of 20 ng. Electrophoreses run was performed on the Multiphor II device (SG) at 600 V, 50 mA, and 30 W for about 60 min. The following molecular weights were used as markers: 97.4 KDa (phosphorylase B from rat muscle), 66,2KDa (bovine serum albumin), 45 kDa (egg albumin), 31,0KDa (bovine carbonic anhydrase), 21,51KDa (trypsin inhibitor) and 14,4KDa (lysozyme). *Enterobacter aerogenes* ATCC13048 was used as control. After electrophoresis, the gel was stained with Silver Staining Kit plus one Kit, Protein (GE Healthcare) according to the manufacturer's instructions. We classified the protein profile based on intensity of band in without lack or loss of protein (++++), very little reduction of protein (+++), reduction (++), major reduction (+) and absent of protein based on previously described by Mostachio et al. [13].

Efflux pump activity

The efflux pump activity was analyzed based on carbapenems's MIC in 5 *E. cloacae* that harbored efflux pump gene and 5 in *E. aerogenes* belonged to different clones. MICs of imipenem, meropenem and ertapenem with 50 and 100 mg/mL and without the efflux pump inhibitors Carbonyl-cyanide-m-chlorophenylhydrazone (CCCP-Sigma Chemical.St Louis., MO) were determined by agar dilution to investigate the role of efflux pump on carbapenems-resistant *E. aerogenes* and *E. cloacae* isolates. The influence of an efflux pump on the carbapenems's MIC for a given bacterial strain was determined by a reduction of at least four-fold of the respective MIC in the presence of CCCP [2].

Results

A total of 44/105 (41, 9%) *E. aerogenes* and 8/25 (32, 0%) *E. cloacae* were resistant to carbapenems. Thirty-nine isolates of *E. aerogenes* resistant to carbapenems were identified in Hospital A, 5 isolates of *E. aerogenes* in Hospital B and 8 isolates of *E. cloacae* in

Table 2 In vitro activity of 7 antibiotics against 44 *E. aerogenes* and 8 *E. cloacae* carbapenems-resistant strains isolated in three Brazilians hospitals using microdilution and agar dilution

Antibiotics	Range	CIM 50 (mg/mL)	CIM 90 (mg/mL)	% resistance	CIM 50 (mg/mL)	CIM 90 (mg/mL)	% resistance
Imipenem	0,25–128	8	32	93,2	16	64	100
Meropenem	0,25–128	8	32	93,2	16	16	87,5
Ertapenem	0,25–128	64	128	86,4	64	64	100
Tigecycline	0,03–16	0,25	2	0	1	2	0
Fosfomycin[a]	0,25–256	16	64	0	16	32	0
Polymyxin B	0,003–16	1	2	0	1	2	0
Cefepime	0,25–128	≥128	≥128	100	≥128	≥128	100

[a]MIC by agar dilution

Hospital C. *E. aerogenes* isolates presented MIC ranging from 2 to 128 mg /ml for imipenem, 4 to 64 mg /ml for meropenem and 1 to ≥128 mg /mL for ertapenem. *E. cloacae* isolates showed MIC ranging from 8 to 64 mg/ml for imipenem, 2 to 16 mg / ml for meropenem and 8 to 64 mg/mL for ertapenem. All isolates were susceptible to fosfomycin, polymyxin B and tigecycline (Table 2).

Demographic and clinical data of 39 patients with colonization and infection caused by *E. aerogenes* resistant to carbapenems in Hospital A and 5 in hospital B are shown in Table 2. Most of the isolates were from blood 12 (30.8%), followed by 9 (23%) from respiratory tract secretion and 6 (15.4%) from urine. Most of patients (*N* = 17) undergone surgery (4 liver and 2 kidney transplant), 2 patients received chemotherapy (1 acute leukemia and 1 bone marrow

transplant) and 1 patient was HIV positive (Table 3). There was no clinical information regarding isolates identified in Hospital C, although, all isolates were from surveillance swabs.

The PCR showed that 39/44 (88.6%) *E. aerogenes* isolates harbored *bla*KPC, other genes encoding carbapenemases were not identified in any isolate. Thirty-nine (88.6%) had *bla*TEM-1 gene and 41 (93.2%) *bla*CTX-M gene. Among the eight isolates of *E. cloacae*, all harbored *bla*KPC and *bla*TEM-1 genes; other carbapenemases studied were not identified.

The outer membrane proteins of 22 of the 44 isolates of *E. aerogenes* and 5 of 8 isolates of *E. cloacae* were analyzed. The intensity of the proteins of interest Omp F (39 KDa) OMP C (42 KDa), 35–36KDa were compared with the molecular weight and ATCC *E. aerogenes* 13048 was used as control (Table 4).

Table 3 Clinical and demographic characteristics of 39 patients with infection and colonization caused by carbapenems-resistant *E. aerogenes* from Hospital A and 5 patients with colonization and infection caused by carbapenems-resistant *E. aerogenes* from Hospital B

Patients variables	Hospital A N = 39 (%)	Hospital B N = 5 (%)
Age (range), mean	18–87 years old , 55.6 years old	24–88 years old, ,58.2 years old
Gender		
Female	13 (33,3%)	0
Male	22 (56,4%)	5 (100%)
Length of stay before identification of *E. aerogenes* (range), mean	6–190 days, 43.7 days	9–17 days, 16.2 days
Site of isolation		
Blood	12 (30,8%)	1 (20%)
Respiratory Tract secretion[a]	9 (23,1%)	0
Urine	6 (15,4%)	0
Peritoneal fluid	3 (8%)	0
Skin	2 (6%)	0
Others	7 (18%)	1 (20%)
Rectal swab	0	3 (60%)
Intensive care unit	23 (59%)	3 (60%)
Death	24 (61,5%)	1 (20%)

[a]Tracheal secretion and Bronco alveolar lavage (BAL)

Table 4 Minimum inhibitory concentration of carbapenems of 5 carbapenems-resistant *E. aerogenes* and 5 carbapenems-resistant *E. cloacae* isolates by agar dilution with or without efflux inhibitor CCCP

Antimicrobial agent	*E. aerogenes* (non-harboring acrART gene) N = 5						*E. cloacae* (harboring acrART gene) N = 5					
	50 mg/mL CCCP						100 mg/mL CCCP					
	MIC (mg/mL) SD						MIC (mg/mL)					SD
Imipenem	8	8	32	64	2	25,75	4	8	4	8	32	11,79
Imipenem + CCCP	≤0,125	≤0,125	≤0,125	≤0,125	0,25	0	≤0,125	≤0,125	≤0,125	≤0,125	≤0,125	0
Meropenem	4	4	8	32	8	11,79	4	4	2	8	16	5,58
Meropenem + CCCP	≤0,125	≤0,125	1	≤0,125	1	0	≤0,125	≤0,125	≤0,125	≤0,125	≤0,125	0
Ertapenem	32	64	32	256	32	93,38	16	16	8	32	32	10,73
Ertapenem + CCCP	≤0,125	2	≤0,125	2	4	1,61	8	8	≤0,125	≤0,125	≤0,125	4,35

All isolates harbored KPC
SD standart deviation

The protein profile was classified based on intensity of band in without lack or loss of protein (++++), very little reduction of protein (+++), reduction (++), great reduction (+) and absent (Table 4). All isolates resistant to carbapenems in this study showed a decrease in 35–36 KDa protein. We observed that the 42KDa protein in our analysis did not appear to be involved in carbapenems resistance.

E. aerogenes isolates with decrease of 39 KDa protein, presented CIM of 2–128 mg/mL to imipenem, 4–64 mg/mL to meropenem and 8 a ≥128 mg/mL to ertapenem. Isolates of *E. aerogenes*, which presented absence of 39 KDa protein showed MIC of 8 to 32 mg/ml for imipenem, 4 to 32 mg/ml for meropenem and 32–64 mg/mL to ertapenem.

E. aerogenes isolates with decrease of 35–36 KDa protein, presented MIC of 2–32 mg/ml for imipenem and 4 to 32 mg/ml for meropenem and 8–128 mg / mL to ertapenem. *E. aerogenes* isolates with absence of the protein of 35–36 KDa showed MIC of 128 mg/mL to imipenem, 64 mg/mL to meropenem and ≥128 mg/mL to ertapenem.

Moreover, all isolates of carbapenems-resistant *E. cloacae* showed decreased of 39KDa protein and absence of 35–36 KDa protein and their MICs to imipenem were 8 to 64 mg/ml, 2 to 16 mg/mL to meropenem and 8–64 mg/mL to ertapenem (Table 5).

The PFGE showed 6 clones (A, B, C, E, F and G) among 39 isolates of carbapenems-resistant *E. aerogenes* identified in Hospital A. Clone A was present only in the first two year of study (2005 and 2006), and it was replaced by clone C that circulated during the entirely study. Twenty-seven (69.2%) isolates belonged to this predominant clone C that harbored KPC, TEM-1 and CTXM, whose subtypes ranging from C1 to C19 over the 7-year study period, from 2005 to 2011 (Fig. 1). Two subtypes (C17 and C19) of clone C and clone F were KPC negative.

The 4 carbapenems-resistant *E. aerogenes* identified in Hospital B belonged to a predominant clone nominated

as clone D with subtypes from D1 to D3. This clone harbored KPC and TEM-1, and showed decreased of 39 and 35–36 KDa proteins (Fig. 1 and Table 5).

In addition, the 8 carbapenems-resistant *E. cloacae* identified in Hospital C in Londrina belonged to two clones, clone A (predominant clone) and clone B (Fig. 2). All carbapenems-resistant *E. cloacae* isolates harbored KPC, TEM-1 and efflux pump gene acrART (Fig. 2). These clones showed as well lost of 35–36KDa proteins (Table 5).

The efflux pump AcrAB-TolC gene was only identified in carbapenems-resistant *E. cloacae*. All 5 isolates positive for this efflux pump were resistant to imipenem, meropenem and ertapenem with MICs above 8 mg/mL for all carbapenems (Tables 4 and 5).

The 5 carbapenems-resistant *E. aerogenes* isolates belonged to different clones and showed efflux pump activity in presence of CCCP inhibitor. Among the carbapenems-resistant *E. cloacae*, 2 of the 5 isolates showed efflux pump activity for meropenem, imipenem and ertapenem, and 3 isolates showed no activity on presence of ertapenem (Tables 4 and 5).

Discussion

In the present study, we demonstrated that all *E. aerogenes* and *E. cloacae* isolates resistant to carbapenems were susceptible to polymyxin B, tigecycline and fosfomycin. KPC was the only carbapenemase identified. In contrast with previous studies [5, 10, 14], most of carbapenems-resistant *E. aerogenes* isolates co-harbored KPC and wide spectrum ESBL, such as *bla*TEM and blaCTX-M, in association with alteration of OMPs, mainly reduction or loss of 35–36 KDa and 39KDa proteins. In addition to alteration of OMPs, *E. cloacae* isolates harbored blaKPC; blaTEM and efflux pump gene acrART. *E. aerogenes* isolates were polyclonal, although a different predominant clone was finding in each hospital. On the other hand, *E. cloacae* belonged to two clones. The mechanisms of resistance to carbapenems

Table 5 Outer membrane proteins, *bla*KPC gene, *bla*TEM and *bla*CTX and pump efflux (AcrART) profile of 22 carbapenems-resistant isolates of *E. aerogenes* (hospital A and B) and 5 *E. cloacae* (hospital C)

PFGE	Hospital	Date	Imipenem MIC	Meropenem MIC	Ertapenem MIC	PCR *bla* KPC	*bla* CTX	*bla* TEM 1	acrART	SDS-PAGE 42 KDa	39 Kda	35 e 36 KDa
A2	A	13/05/2005	32	8	32	+	+	+	-	++++	+++	++
A2	A	10/12/2005	16	8	64	+	+	+	-	++++	+++	++
C11	A	18/05/2006	32	16	128	+	+	+	-	++++	+++	++
C11	A	13/07/2006	32	16	128	+	+	+	-	++++	++++	++
C8	A	14/07/2006	16	16	64	+	+	+	-	++++	++++	++
C7	A	02/01/2007	16	4	32	+	+	+	-	++++	Absent	++
C6	A	23/03/2007	16	8	64	+	+	+	-	++++	+++	++
C13	A	25/04/2007	8	4	32	+	+	+	-	+++	Absent	++
C14	A	21/09/2007	8	4	64	+	+	+	-	++++	+++	++
G1	A	10/11/2008	32	16	64	+	-	-	-	+++	Absent	++
C1	A	06/01/2009	16	32	64	+	-	+	-	+++	Absent	++
C2	A	31/08/2009	16	4	16	+	+	+	-	+++	++	++
C4	A	03/11/2009	32	16	64	+	+	+	-	+++	Absent	++
C5	A	29/11/2009	16	4	32	+	+	+	-	+++	+	++
C3	A	05/12/2009	4	4	16	+	+	+	-	+++	++	++
B1	A	16/03/2010	16	8	64	+	+	+	-	+++	Absent	++
F1	A	08/06/2011	128	64	≥128	+	-	+	-	+++	++	Absent
D2	B	26/12/2010	2	8	32	+	+	+	-	++++	+	++
D1	B	05/01/2011	4	8	32	+	+	+	-	++++	+	++
D1	B	05/01/2011	4	8	32	+	+	+	-	++++	+	++
D3	B	06/01/2011	4	8	8	+	+	+	-	+++	+	++
A2	C	19/11/2012	8	2	8	+	-	+	+	+++	+	Absent
B1	C	-	8	8	64	+	-	+	+	+++	+	Absent
B1	C	-	8	8	64	+	-	+	+	+++	+	Absent
A3	C	-	64	16	64	+	-	+	+	+++	+	Absent
A5	C	-	16	16	64	+	-	+	+	+++	+	Absent

differed among *E. aerogenes* subtypes. We identified isolates of *E. aerogenes* resistant to carbapenems that harbored only ESBL (TEM-1 and CTX-M) associated with decreased or loss of 35–36 kDa and 39 kDa proteins.

E. aerogenes and *E. cloacae* are important agents of healthcare-associated infections in several countries and resistance to carbapenems has been increasing in the last decade, thus there is a need for in vitro studies showing alternatives for the treatment of infections due to these microorganisms [3, 7, 14]. Since our susceptibility results showed that all carbapenems-resistant isolates were susceptible to tigecycline, fosfomycin, and polymyxin B, we consider these findings quite promising. However, it is noteworthy that despite the excellent results in vitro, pharmacokinetics and pharmacodynamics of tigecycline and fosfomycin, and the limited clinical experience in serious infections, especially bloodstream infections, are obstacles to the routine use of these antimicrobials

in the treatment of systemic infections. Despite of these drawbacks, these drugs have been used with success as combination therapy in the treatment of severe infections caused by Enterobacteriaceae resistant to carbapenems [15, 16].

The demographic and clinical data from the studied hospitals showed that the most frequent site of isolation of carbapenems-resistant Enterobacter was blood, most of our patients had undergone surgical procedures, and had a high overall mortality, similar to previous studies [14, 17]. The only carbapenemase identified in our study was KPC. Data regarding carbapenems resistance in *Enterobacter* spp. in Brazil are scarce, although KPC is the carbapenemase most often described in *Enterobacter* spp. isolates in Brazilians hospitals [5, 7, 14].

Regarding the clonality, we identified in Hospital A, in São Paulo, 6 clones of *E. aerogenes*, named A, B, C, E, F and G over the study period, from 2005 to 2011, with

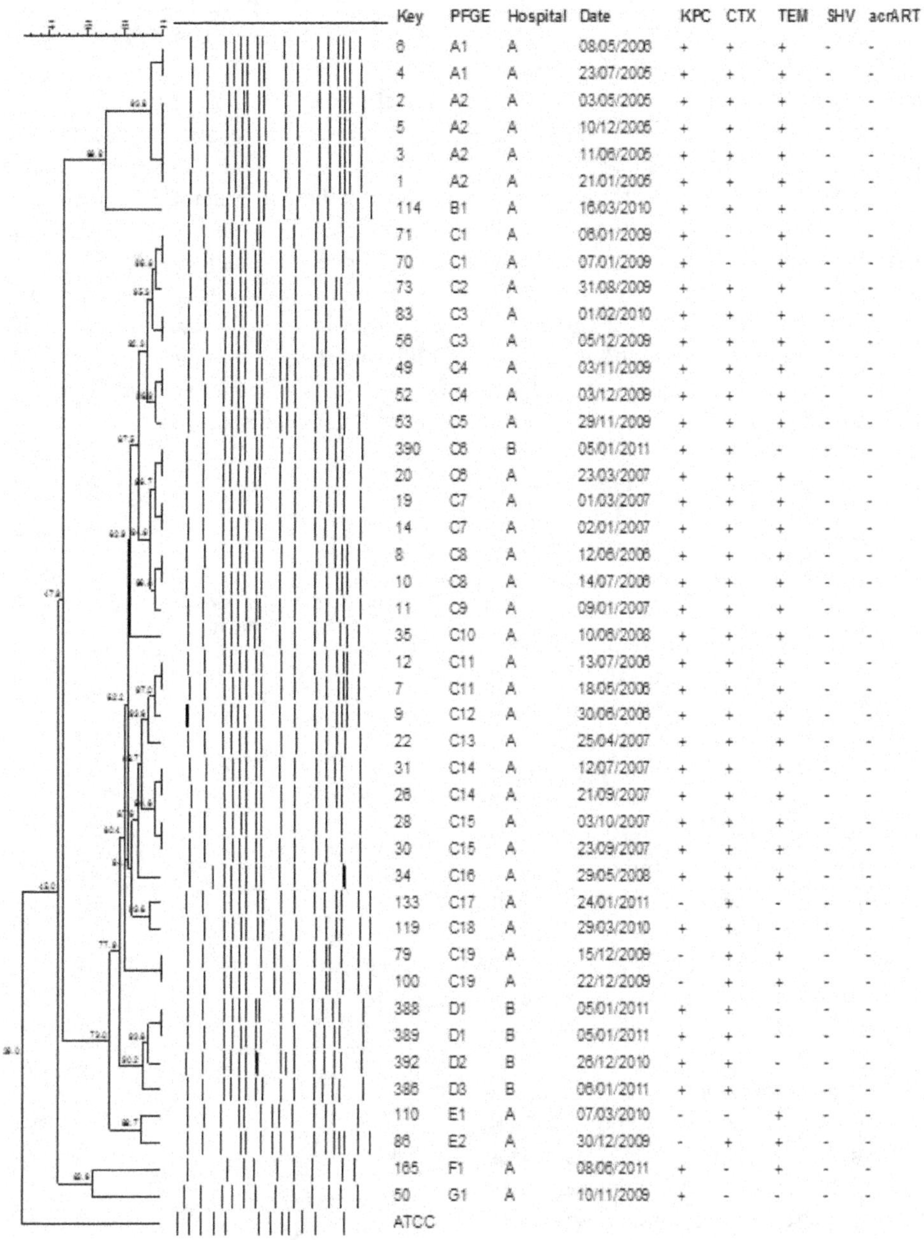

Key	PFGE	Hospital	Date	KPC	CTX	TEM	SHV	acrART
6	A1	A	08/05/2006	+	+	+	-	-
4	A1	A	23/07/2005	+	+	+	-	-
2	A2	A	03/05/2005	+	+	+	-	-
5	A2	A	10/12/2005	+	+	+	-	-
3	A2	A	11/06/2005	+	+	+	-	-
1	A2	A	21/01/2005	+	+	+	-	-
114	B1	A	16/03/2010	+	+	+	-	-
71	C1	A	06/01/2009	+	-	+	-	-
70	C1	A	07/01/2009	+	-	+	-	-
73	C2	A	31/08/2009	+	+	+	-	-
83	C3	A	01/02/2010	+	+	+	-	-
56	C3	A	05/12/2009	+	+	+	-	-
49	C4	A	03/11/2009	+	+	+	-	-
52	C4	A	03/12/2009	+	+	+	-	-
53	C5	A	29/11/2009	+	+	+	-	-
390	C6	B	05/01/2011	+	+	-	-	-
20	C6	A	23/03/2007	+	+	+	-	-
19	C7	A	01/03/2007	+	+	+	-	-
14	C7	A	02/01/2007	+	+	+	-	-
8	C8	A	12/06/2006	+	+	+	-	-
10	C8	A	14/07/2006	+	+	+	-	-
11	C9	A	09/01/2007	+	+	+	-	-
35	C10	A	10/06/2008	+	+	+	-	-
12	C11	A	13/07/2006	+	+	+	-	-
7	C11	A	18/05/2006	+	+	+	-	-
9	C12	A	30/06/2008	+	+	+	-	-
22	C13	A	25/04/2007	+	+	+	-	-
31	C14	A	12/07/2007	+	+	+	-	-
26	C14	A	21/09/2007	+	+	+	-	-
28	C15	A	03/10/2007	+	+	+	-	-
30	C15	A	23/09/2007	+	+	+	-	-
34	C16	A	29/05/2008	+	+	+	-	-
133	C17	A	24/01/2011	-	+	-	-	-
119	C18	A	29/03/2010	+	+	-	-	-
79	C19	A	15/12/2009	-	+	+	-	-
100	C19	A	22/12/2009	-	+	+	-	-
388	D1	B	05/01/2011	+	+	-	-	-
389	D1	B	05/01/2011	+	+	-	-	-
392	D2	B	26/12/2010	+	+	-	-	-
386	D3	B	06/01/2011	+	+	-	-	-
110	E1	A	07/03/2010	-	-	+	-	-
86	E2	A	30/12/2009	-	+	+	-	-
165	F1	A	08/06/2011	+	-	+	-	-
50	G1	A	10/11/2009	+	-	-	-	-
ATCC								

Fig. 1 Dendogram of 39 carbapenems-resistant *E. aerogenes* isolates from Hospital A and 5 carbapenems-resistant *E aerogenes* from Hospital B

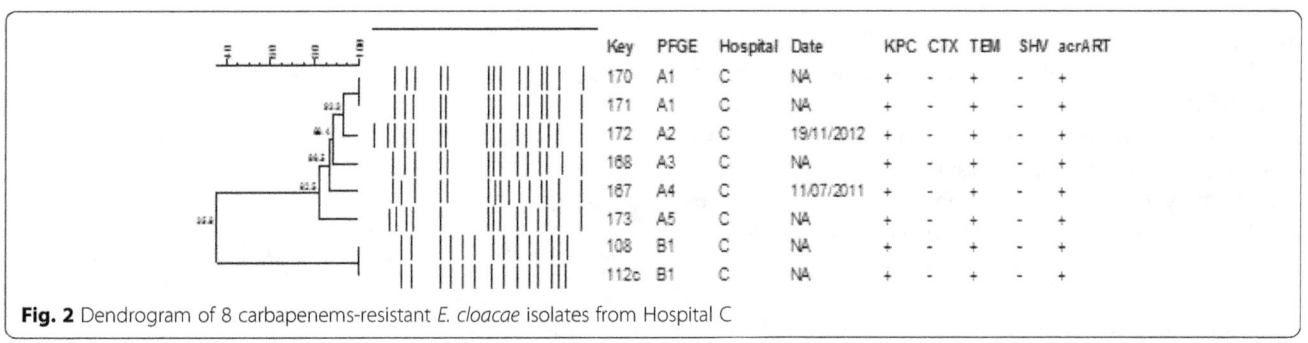

Key	PFGE	Hospital	Date	KPC	CTX	TEM	SHV	acrART
170	A1	C	NA	+	-	+	-	+
171	A1	C	NA	+	-	+	-	+
172	A2	C	19/11/2012	+	-	+	-	+
168	A3	C	NA	+	-	+	-	+
167	A4	C	11/07/2011	+	-	+	-	+
173	A5	C	NA	+	-	+	-	+
108	B1	C	NA	+	-	+	-	+
112c	B1	C	NA	+	-	+	-	+

Fig. 2 Dendrogram of 8 carbapenems-resistant *E. cloacae* isolates from Hospital C

the predominance of clone C. One predominant clone was also identified in hospital B in São Paulo. An intriguing finding is that one isolate of hospital B seems to be related to a clone of hospital A, pointed to a possible inter-hospital spread. Two clones A (subtypes A1 to A5) and B were identified among E. cloacae isolates resistant to carbapenems in Hospital C, in Londrina, Paraná. These findings suggest that cross-transmissions have occurred over the study period in all hospitals and highlight the importance of the stringent enforcement of hand hygiene and contact precautions to control the spread of this agent. Another interesting finding is that the mechanism of resistance to carbapenems in E. aerogenes from hospital A differs among subtypes and CTX-M was only identified in this hospital.

The profile of outer-membrane proteins of our isolates demonstrated that all isolates studied showed reduction of 35–36 KDa OMPs. These OMPs have been associated with β-lactam resistance in Enterobacter spp [18–21]. Studies in vivo and in vitro showed that diminished or loss of these OPMS were associated with increase on imipenem and meropenem MIC [18–22]. Other OMPs previously associated with carbapenems-resistance among E. aerogenes are OMPs of 42KDa and OMP of 39KDa [19]. However, in contrast with previous studies, we could not associate the alteration of the 42KDa OmpC with carbapenems resistance in our E. aerogenes and E. cloacae isolates. This finding could be explained perhaps by the clones that circulated in Brazil. Data regarding OMP in carbapenems–resistant Enterobacter in Brazil is limited. However, a previous Brazilian study demonstrated that OmpC and OmpF were present in only 6.6% of E. cloacae isolates resistant to ertapenem [5].

Few authors had evaluated the role of efflux pump on carbapenems resistance in Enterobacter spp [23, 24]. Efflux pump AcrAB system has been described as associated with resistance to meropenem and imipenem in E. cloacae [2, 22, 23]. In carbapenems-resistant E. aerogenes there were descriptions of efflux pump associated with KPC and/or porin loss as well [2, 23, 24]. In our study, the efflux pump AcrART was found only in E. cloacae. However, eventhough E. aerogenes isolates did not present the efflux pump AcrART gene, carbapenem efflux pump activity was observed in the presence of CCCP inhibitor. Possibly, these E. aerogenes isolates have other efflux pump, not described yet. These findings reinforce the importance of association of mechanism of resistance in carbapenems-resistant Enterobacter spp.

This study has several limitations: besides being retrospective, we evaluated only three hospitals and we could access the clinical data from only two hospitals. However, we were able to demonstrate that there was an association of mechanism of resistance such as β-lactamase, alteration of outer-membrane protein and efflux pump in carbapenems-resistant Enterobacter spp. in Brazil.

Conclusion

In conclusion, we observed that there was a predominant clone in each hospital, suggesting that cross-transmission of carbapenems-resistant Enterobacter spp. was frequent. The isolates presented multiple mechanisms of resistance to carbapenems, such as KPC, ESBL and alteration of the outer membrane protein. E. cloacae presented as well the efflux pump AcrART, and in vitro activity of efflux pump inhibitor. These findings can be useful to control the spread of carbapenems resistance among Enterobacter spp. in the hospital setting.

Abbreviations

ATCC: American Type Culture Collection; CCCP: Cyanide 3-chlorophenylhydrazone; DNA: Deoxyribonucleic acid; E. aerogenes: Enterobacter aerogenes; E. cloacae: Enterobacter cloacae; ESBL: Extended-spectrum β-lactamases; KPC: Klebsiella pneumonia carbapenemase; M-PCR: Multiplex PCR; NDM: New Deli metallo- β-lactamase; OMP: Outer-membrane protein; PCR: Polymerase Chain Reaction; PFGE: Pulsed-Field Gel Electrophoresis; VIM: Verona integrin metallo-β-lactamase

Acknowledgments

Not applicable.

Funding

The study was financial support by CNPQ (Conselho Nacional de Desenvolvimento Científico e Tecnológico) and FAPESP (Fundação de Amparo à pesquisa do Estado de São Paulo, Brazil.

Authors' contributions

JFR and SFC designed the study. JFR, LM, CC, TG, ASL and SFC collected the samples. JFR, CR and APM carried out the laboratory work. JFR, TG, ASL and SFC analyzed the data. JFR and SFC wrote the manuscript. All authors read and approved the final version of manuscript.

Competing interests

The authors declare that they have no competing interests.

Author details

[1]Department of Infectious Diseases, University of São Paulo, Laboratory of Medical Investigation 54 (LIM-54), Hospital Das Clínicas FMUSP, São Paulo, Brazil. [2]Hospital de Itapecerica da Serra, Itapecerica da Serra, SP, Brazil. [3]Hospital University of Londrina, Londrina, Paraná, Brazil. [4]LIM-54, Faculdade de Medicina da Universidade de São Paulo, São Paulo, Brazil.

References

1. Bratu S, Landman D, Alam M, Tolentino E, Quale J. Detection of KPC carbapenem-hydrolyzing enzymes in *Enterobacter spp.* from Brooklyn, New York. Antimicrob Agents Chemother. 2005;49:776–8.
2. Qin X, Yang Y, Hu F, Zhu D. Hospital clonal dissemination of *Enterobacter aerogenes* producing carbapenemase KPC-2 in a Chinese teaching hospital. J Med Microbiol. 2014;63:222–8.
3. Davin-Regli A, Pages JM. *Enterobacter aerogenes* and *Enterobacter cloacae*; versatile bacterial pathogens confronting antibiotic treatment. Front Microbiol. 2015;6:392.
4. Doumith M, Ellington MJ, Livermore DM, Woodford N. Molecular mechanisms disrupting porin expression in ertapenem-resistant *Klebsiella* and *Enterobacter spp.* clinical isolates from the UK. J Antimicrob Chemother. 2009;4:659–67.
5. Jaskulski MR, Medeiros BC, Borges JV, Zalewsky R, Fonseca ME, Marinowic DR, et al. Assessment of extended-spectrum beta-lactamase, KPC carbapenemase and porin resistance mechanisms in clinical samples of *Klebsiella pneumoniae* and *Enterobacter spp.* Int J Antimicrob Agents. 2013;1:76–9.
6. Pérez A, Canle D, Latasa C, Poza M, Beceiro A, Tomás M d, et al. Cloning, nucleotide sequencing, and analysis of the AcrAB-TolC efflux pump of *Enterobacter cloacae* and determination of its involvement in antibiotic resistance in a clinical isolate. Antimicrob Agents Chemother. 2007;51:3247–53.
7. Tavares CP, Pereira PS, Marques EA, Faria Jr C, de Almeida R, et al. Molecular epidemiology of KPC-2-producing Enterobacteriaceae (non-*Klebsiella pneumoniae*) isolated from Brazil. Diagn Microbiol Infect Dis. 2015;82:326–30.
8. Kim SY, Park YJ, Yu JK, Kim HS, Park YS, Yoon JB, et al. Prevalence and mechanisms of decreased susceptibility to carbapenems in *Klebsiella pneumoniae* isolates. Diagn Microbiol Infect Dis. 2007;1:85–91.
9. Chen Y, Zhou Z, Jiang Y, Yu Y. Emergence of NDM-1-producing *Acinetobacter baumannii* in China. J Antimicrob Chemother. 2011;6:1255–9.
10. Chen Z, Li H, Feng J, Li Y, Chen X, Guo X, et al. NDM-1 encoded by a pNDM-BJ01-like plasmid p3SP-NDM in clinical *Enterobacter aerogenes*. Front Microbiol. 2015;6:294.
11. Monteiro J, Widen RH, Pignatari AC, Kubasek C, Silbert S. Rapid detection of carbapenemase genes by multiplex real-time PCR. J Antimicrob Chemother. 2012;67:906–9.
12. Pérez A, Poza M, Fernández A, Fernández MC, Mallo S, Merino M, et al. Involvement of the AcrAB-TolC efflux pump in the resistance, fitness, and virulence of *Enterobacter cloacae*. Antimicrob Agents Chemother. 2012;56:2084–90.
13. Mostachio AK, Levin AS, Rizek C, Rossi F, Zerbini J, Costa SF. High prevalence of OXA-143 and alteration of outer membrane proteins in carbapenem-resistant Acinetobacter spp. isolates in Brazil. Int J Antimicrob Agents. 2012;5:396–401.
14. Tuon FF, Scharf C, Rocha JL, Cieslinsk J, Becker GN, Arend LN. KPC-producing *Enterobacter aerogenes* infection. Braz J Infect Dis. 2015;19:324–7.
15. Entenza JM, Moreillon P. Tigecycline in combination with other antimicrobials: a review of in vitro, animal and case report studies. Int J Antimicrob Agents. 2009;1:8.e1–9.
16. Datta S, Roy S, Chatterjee S, Saha A, Sen B, Pal T, Som T, Basu S. A five-year experience of carbapenem resistance in Enterobacteriaceae causing neonatal septicaemia: predominance of NDM-1. PLoS One. 2014;11:e112101.
17. De Gheldre Y, Struelens MJ, Glupczynski Y, De Mol P, Maes N, Nonhoff C, et al. National epidemiologic surveys of *Enterobacter aerogenes* in Belgian hospitals from 1996 to 1998. J Clin Microbiol. 2001;3:889–96.
18. Thiolas A, Bornet C, Davin-Regli A, Pages JM, Bollet C. Resistance to imipenem, cefepime, and cefpirome associated with mutation in Omp36 osmoporin of *Enterobacter aerogenes*. Biochem Biophys Res Commun. 2004;317:851–6.
19. Yigit H, Anderson GJ, Biddle JW, Steward CD, Rasheed JK, Valera LL, et al. Carbapenem resistance in a clinical isolate of *Enterobacter aerogenes* is associated with decreased expression of OmpF and OmpC porin analogs. Antimicrob Agents Chemother. 2002;46:3817–22.
20. Lavigne JP, Sotto A, Nicolas-Chanoine MH, Bouziges N, Pages JM, Davin-Regli A. An adaptive response of *Enterobacter aerogenes* to imipenem: regulation of porin balance in clinical isolates. Int J Antimicrob Agents. 2013;41:130–6.
21. Lavigne JP, Sotto A, Nicolas-Chanoine MH, Bouziges N, Bourg G, Davin-Regli A. Membrane permeability, a pivotal function involved in antibiotic resistance and virulence in *Enterobacter aerogenes* clinical isolates. Clin Microbiol Infect. 2012;18:539–45.
22. Wozniak A, Villagra NA, Undabarrena A, Gallardo N, Keller N, Moraga M, et al. Porin alterations present in non-carbapenemase-producing Enterobacteriaceae with high and intermediate levels of carbapenem resistance in Chile. J Med Microbiol. 2012;61:1270–9.
23. Philippe N, Maigre L, Santini S, Pinet E, Claverie JM, Davin-Régli AV, et al. In vivo evolution of bacterial resistance in two cases of *Enterobacter aerogenes* Infections during treatment with imipenem. PLoS One. 2015;10: e0138828.
24. Bornet C, Chollet R, Malléa M, Chevalier J, Davin-Regli A, Pagès JM, Bollet C. Imipenem and expression of multidrug efflux pump in *Enterobacter aerogenes*. Biochem Biophys Res Commun. 2003;301(4):985–90.

16

Gene expression of *Vibrio parahaemolyticus* growing in laboratory isolation conditions compared to those common in its natural ocean environment

Katherine García[1], Cristian Yáñez[2], Nicolás Plaza[1,2], Francisca Peña[3], Pedro Sepúlveda[3], Diliana Pérez-Reytor[1] and Romilio T. Espejo[2*]

Abstract

Background: *Vibrio parahaemolyticus* is an autochthonous marine bacterial species comprising strains able to grow in broth containing bile salts at 37 °C, a condition seldom found in the ocean. However, this condition is used for isolation in the laboratory because it is considered a necessary property for pathogenesis. In this context, revealing how gene expression enables *V. parahaemolyticus* to adapt to this particular condition –common to almost all *V. parahaemolyticus* isolates- will improve our understanding of the biology of this important pathogen. To determine the genes of *V. parahaemolyticus* differentially expressed when growing in isolation condition (37 °C, 0.9% NaCl, and 0.04% bile salts) referred to those at the temperature and salt concentration prevailing in ocean south of Chile (marine-like condition; 12 °C, 3% NaCl, and absence of bile salts) we used high-throughput sequencing of RNA.

Results: Our results showed that in the isolation condition, among the 5034 genes annotated in the *V. parahaemolyticus* RIMD2210633 genome, 344 were upregulated and 433 downregulated referred to the marine-like condition, managing an adjusted *P*-value (Padj) $< E^{-5}$. Between the 50 more highly expressed genes, among the small RNAs (sRNA), the three carbon storage regulators B (CsrB) were up four to six times, while RyhB, related to iron metabolism besides motility control, was down about eight times. Among proteins, BfdA, a hemolysin-co-regulated protein (Hcp1) secreted by T6SS1, one of the most highly expressed genes, was about 140 times downregulated in isolation condition. The highest changes in relative expression were found among neighboring genes coding for proteins related to respiration, which were about 40 times upregulated.

Conclusions: When *V. parahaemolyticus* is grown in conditions used for laboratory isolation 777 genes are up- or downregulated referred to conditions prevailing in the marine-like condition; the most significantly overrepresented categories among upregulated processes were those related to transport and localization, while secretion and pathogenesis were overrepresented among downregulated genes. Genes with the highest differential expression included the sRNAs CsrB and RhyB and the mRNAs related with secretion, nutritional upshift, respiration and rapid growing.

Keywords: RNA seq, *Vibrio parahaemolyticus*, Genome, Bioinformatics, sRNA, mRNA

* Correspondence: romilio.espejo@gmail.com
[2]Institute of Nutrition and Food Technology, Universidad de Chile, Av. El Líbano 5524, Macul, Santiago, Chile
Full list of author information is available at the end of the article

Background

Vibrio parahaemolyticus is an autochthonous ocean-dwelling bacterial species comprising strains that are widely disseminated in marine environments throughout the world (see by example review [1]). Some of these strains can cause severe diarrhea when present in seafood [1]. Since traditional isolation of *V. parahaemolyticus* is performed in broth containing bile salts at 37 °C (Bacteriological Analytical Manual of the US Food and Drug Administration [2]) almost all the known strains are able to grow in this condition though it is seldom found in the ocean. Insights into how regulation of gene expression enables *V. parahaemolyticus* to adapt from marine environment to laboratory isolation conditions will improve our understanding of the biology of this important pathogen. We studied this growth adaptation by determining the differential gene expression when growing at isolation (37 °C, 0.9% NaCl and 0.04% bile salt) versus marine-like condition (12 °C and 3% NaCl without bile salt). We used 12 °C and 3% NaCl for the second condition because these are the average superficial seawater temperature and salt concentration in the southern Chile littoral where the arrival of the pandemic strain of *V. parahaemolyticus* caused one of the world's largest outbreaks of seafood-related diarrhea [3, 4], while we used 37 °C, 0.9% NaCl and 0.04% bile salt because these are the isolation conditions. Differential gene expression upon change in growth temperature, salt concentration and addition of bile salts originally focused on selected genes and lately, with the advent of high-throughput RNA sequencing, on the whole genome. Several studies of gene expression has been compared in different conditions but not between those employed in this study. One of them explore the expression of three selected proteins, superoxide dismutase (SOD), catalase (CAT) and TDH by *V. parahaemolyticus* as influenced by heat shock. They showed that SOD and CAT activities are reduced but TDH protein synthesis is enhanced after heat shock at 42 °C [5]. Other studies have focused on secretion systems. They have received particular attention because of their involvement in pathogenesis. It was shown that Hcp2 of type VI secretion system chromosome II (T6SS2) but not Hcp1 of type VI secretion system chromosome I (T6SS1) is expressed at 37 °C [6]. On the other hand, Livny et al. [6] compared the transcriptomes of *V. parahaemolyticus* cultured under standard laboratory conditions in the presence of bile, and directly isolated from the ceca of infected infant rabbits, which mimic the human infection. They showed that expression of genes encoding type III secretion system chromosome II (T3SS2) and its effectors is markedly upregulated under bile and infection conditions. However, the expression of 277 genes was only altered under infection, suggesting that *V. parahaemolyticus* gene expression during infection is subject to significant regulatory influences in addition to bile induction. More recently, Urmersbach et al. [7] showed that 638 genes are differentially expressed between 15 and 37 °C but virulence-associated genes (*tdh1*, *tdh2*, *toxR*, *toxS*, *vopC*, T6SS1, T6SS2) remained mostly unaffected by temperature.

In this work, we try to understand which genes are involved in the ability of *V. parahaemolyticus* strains to grow in a condition commonly used for isolation in the laboratory, but seldom found in the marine environment. We show that about 15% (777 of 5034) of the genes are differentially expressed under the condition of isolation in the laboratory with respect those expected to prevail in the marine-like condition. We interpret and discuss the overall differences between growth in laboratory isolation and marine-like conditions and discuss in further detail some results that seemed most interesting, namely: Increased expression of the three carbon storage related sRNAs, downregulation of RyhB sRNA which translate into upregulation of genes involved in motility, downregulation of T6SS in chromosome I and II but differential regulation of the T3SS in chromosome I and II, and rise of cold shock proteins, CspA and VPA0552 at 37 °C.

Methods

Strains and culture conditions

Vibrio parahaemolyticus RIMD2210633, a strain maintained in our laboratory since 2002, was used in all experiments. The strain was grown in two conditions; Marine-like condition (E): Luria-Bertani (LB) with NaCl 3%, 12 °C and Isolation condition (I): LB with NaCl 0.9%, 37 °C, supplemented with bile salt 0.04% (Bile salt for microbiology, Sigma-Aldrich, MO, USA). LB was prepared using yeast extract 0.5% and Tryptone 1%, both from Becton Dickinson (NJ, USA). Preliminary growth studies showed that the generation times of *V. parahaemolyticus* when growing in conditions E and I were 12 and 0.5 h, respectively. Stationary phase was reached at OD_{600} of 4 and 5, respectively.

Total RNA isolation

RNA was isolated from three parallel independent cultures for both conditions. One and a half ml cultures were harvested in exponential phase at an OD_{600} of 1.0 and rinsed once with PBS. Total RNA was isolated from the pellet with Trizol Max Bacterial (ThermoFisher Scientific) according to the manufacturer's protocol. The quantity of total RNA was determined by Nanodrop 2000 (Thermo Scientific, Wilmington, DE, USA) and QuantiFluor-ST Fluorometer-Ribogreen (Promega, USA), and quality and integrity of the RNA was determined by electrophoresis in the Bioanalyzer 2100 (Agilent Technologies, USA). Yield of RNA per optical density was 0.6-times lower in condition E than condition I.

RNA integrity (RIN) was higher than 7.8 in all RNA samples. No differences in the electrophoretic patterns of the RNA in the two conditions were evident.

RNA library preparation and nucleotide sequencing

Four micrograms of total RNA was initially depleted of ribosomal RNA using the Ribo-Zero rRNA Removal Kit (Gram-Positive Bacteria, Illumina; catalog no. MRZGP126) according to the manufacturer's protocol, using Riboguard and purifying the RNA by precipitation with ethanol in the last step. Libraries were then prepared following the TruSeq Stranded mRNA Sample Preparation Guide (Part no. 15031047, Rev. E October 2013), following the low sample protocol (LS) but bypassing the mRNA purification. The following barcodes were used: sample I1, AR013 - AGTCAA(C); sample I2, AR014 - AGTTCC(G); sample I3, AR015 - ATGTCA(G); sample E1, AR016 - CCGTCC(C); sample E2, AR018 - GTCCGC(A); sample E3, AR019 - GTGAAA(C). RNA dissolved in re-suspension buffer was immediately used for first strand cDNA synthesis (beginning at step 12, page 20 of the protocol), with the following options: no control was used and re-suspension buffer was used instead in every occasion indicating addition of control. The clean-up PCR step (page 40) was performed using MagJET NGS Cleanup and Size Selection Kit (catalog no. K2821, Thermo Scientific) instead of the AMPure XP Beads. Libraries were validated using Bioanalyzer High Sensitivity Chip.

Created libraries were sequenced at Beijing Genomics Institute (BGI) on the Hiseq 2000 platform (Illumina, San Diego, USA) using single-end 50 bp, and 19.7 to 25.4 million reads per sample were received from BGI after processing and filtration. The number of reads was reduced by random selection to 19 million reads in each sample using ShortRead Bioconductor package [8]. Subsequent filtration by removing adaptors and quality control of reads reduced the number of reads to 17.4–17.5 million per sample using BBDuk (sourceforge.net/projects/bbmap). Reads of each sample were aligned against chromosome I and II of *V. parahaemolyticus* RIMD2210633 with Burrows-Wheeler Aligner (BWA) [9] using the algorithm BWA-BackTrack. Alignments were ordered by position in the reference genome using SAMtools [10]. Annotation of the *V. parahaemolyticus* RIMD2210633 genome was done using information in RefSeq [11] for open reading frames (ORF), tRNA, and rRNA, and BSRD [12] for sRNA with a custom script. Reads counting for each feature were determined using HTSeq-count with the following parameters: -f bam, −r pos, −s reverse, −a 10, −m union [13]. The overall fragments coverage of genomic regions corresponding to features, such as ORFs, tRNAs, rRNAs, sRNAs, and inter-genes were calculated according to the counts obtained and mapped in total alignment reads. Coefficient of Determination (R^2) was calculated using RPKM (*Reads Per Kilobase of transcripted gene per Million mapped reads with features*), calculated for the counts of the features of all samples using a custom script. The differential expression analysis was performed using the DESeq2 package of Bioconductor [14]. The parameters used for analysis were 3-fold change ($\log_2$FC: 1.58), Padj $1E^{-5}$, and the 12 °C growth condition as reference. Gene Ontology (GO) [15] terms were obtained from UniprotKB databases [16].

Clustering of GO terms for the differentially expressed genes was carried out with clusterProfiler Bioconductor [17] using R: A library was created using the locus tag and Entrez ID in the annotated genome described above and the GO terms id associated with each protein, using the *Annotation-Forge* Bioconductor package (*AnnotationDbi: Annotation Database Interface*, R package version 1.36.0. [https://bioconductor.org/packages/release/bioc/html/AnnotationForge.html]. Functional partners of the highly expressed genes were predicted using the STRING program [18] to identify networks involved in biological interactions. Overrepresentation of biological functions among up- or downregulated genes was calculated using EnrichGO of clusterProfiler using the same library created above in R, and Padj = 0.01 -calculated using Benjamini & Hochberg method- and q-value = 0.05.

Results

The natural habitat of pathogenic *V. parahaemolyticus* is seawater, which in southern Chile has average surface water temperature around 12 °C. However, almost every strain of the species is able to grow at 37 °C in bile salt. To compare gene expression between these two conditions RNA was single-end sequenced and raw reads were randomly selected and filtered to approximately 17 million reads for informatics analysis.

Total transcript abundance

Figure 1a shows the results of the counting of the reads aligned against the annotated genome, calculated with HTSeq-count. "With Feature" corresponds to reads aligned to annotated regions, which in this case includes the 43 sRNA described for *V. parahaemolyticus* [12]. 'Without Feature' corresponds to reads aligned to non-annotated regions and may include sRNA not yet described in *V. parahaemolyticus*, antisense in ORF, UTRs, etc. The percentage of reads corresponding to annotated genes ("With Feature") and to non-annotated intergenic regions ("Without Feature") roughly parallels the corresponding percentages of annotated genome; 88.2% annotated and 11.8% non-annotated. Figure 1b shows the percentage of reads aligned to regions with different features: the vast majority of the reads corresponded to mRNA, the sense strands of annotated ORFs, and second in abundance are reads aligning with regions annotated as sRNA. The percentage of sRNA reads observed

Fig. 1 Reads alignment against the annotated genome calculated with HTSeq-count using the following parameters: -f bam, –r pos, –s reverse, –a 10, –m union. E, Marine-like conditions. I, Isolation conditions. **a** Percentage of reads aligned with features and without features and reads not aligned because of lack of similarity. **b** Percentage of reads aligned with different gene classes

is more than 200 times the percentage of the genome annotated as such, which covers only 0.12% of the whole genome. This is because some sRNA are present in very large amounts.

Relative gene expression
The Coefficient of Determination (R^2) of normalized expression per gene, RPKM was between 0.985 and 0.992 for I samples, 0.952 and 0.974 for E samples, and 0.689 and 0.783 between I and E samples. RPKM values oscillated between 0 and 55,000 for mRNA, and 0 and 400,000 for sRNA. To define the transcriptome

(total genes with reads with features) expressed in each condition, we selected genes with RPKM >5 among the 5034 annotated loci. This selection included 3841 genes in I samples and 4143 genes in E samples. This implies that 76% and 82% of the total annotated genes in *V. parahaemolyticus* were significantly expressed in each condition or that 302 gene expressed in E samples were repressed in I samples (Additional file 1).

Differential expression
To identify genes that were differentially expressed in cultures growing at 37 °C, 0.9% NaCl and 0.04% bile

salts versus those growing at 12 °C with 3% NaCl, we compared the RNA-Seq data for these conditions using DESeq2, a differential expression analysis package for RNAseq data that presumes read abundance can be modeled by a negative binomial distribution [20]. Of the 5034 annotated genes in our *V. parahaemolyticus* genome file (Additional file 1), 777 genes in total were significantly affected, 344 (6.8%) and 433 (8.6%) were significantly up- or downregulated (>3-fold, Padj $<E^{-5}$), respectively. Nine of 43 sRNA (21%) were up- or downregulated. Of the 777 genes differentially expressed, 456 genes were from chromosome I and 321 from chromosome II, approximately the ratio of the chromosomes sizes.

Regulation of the highly expressed genes

Table 1 shows the RPKM, differential expression parameters, and some properties for the 21 genes differentially expressed among the 50 more highly expressed genes (RPKM) in each condition. Five sRNA were among these 21 genes. The three carbon-storage regulator (CsrB) sRNA described in *V. parahaemolyticus* were upregulated between four to six times. CsrB antagonize the activity of CsrA, an RNA-binding protein which coordinates a wide range of cellular physiology including regulation of carbon

metabolism, motility, biofilm formation, production of secondary metabolites and quorum sensing, thus affecting virulence [19]. The small 6S RNA was also upregulated about five times; this RNA binds to the σ70-holoenzyme form of RNA polymerase reducing its activity in stationary phase and hence repressing the expression of σ 70-dependent genes [20]. Small RNA svpa113.1, homologous to RyhB was downregulated about eight times; RyhB is expressed when cellular iron concentration is low [21]but also modulates the expression of several genes that control motility, chemotaxis and biofilm formation [22]. Among the highly expressed genes stands up the one encoding BfdA, a hemolysin-co-regulated protein (Hcp1) secreted by T6SS1 and involved in the structural tube formation [23, 24], which is downregulated more than 140 times Also, all the BfdA functional partners genes predicted by the STRING program [18]; VgrG protein, VP1401, VP1402, VP1403, VP1405, VP1406, IcmF-like protein, VP1409, VPA1027 (Hcp2) and VPA1034 were downregulated.

Oddly, the cold shock protein CspA is upregulated about 30 times at 37 °C, as well as other cold shock proteins (VPA0552). However, as cited in Discussion section, CspA can also function as a nutritional up-shift stress protein [25].

Table 1 Differentially expressed genes among the 50 more highly expressed in both growth conditions

Gene	RPKM I	RPKM E	log$_2$FC	p-value	Observations
Upregulated genes in isolation condition					
svpa172.1	157,710	37,082	2.7	7.2E^{-31}	Carbon storage regulator (CsrB)
svpa2439.1	121,815	43,795	2.1	2.2E^{-51}	CsrB
svpa3216.1	128,785	50,025	1.9	7.3E^{-20}	CsrB
svpa2734.1	429,216	166,980	2.3	1.6E^{-25}	6S RNA
VPt061	4867	1639	2.1	2.0E^{-16}	Arginine tRNA
VPt108	7708	2916	2.0	6.3E^{-28}	Ileucine tRNA
VPA1289	20,530	1102	4.8	4.8E^{-151}	CspA
VPA1476	4431	277	4.6	5.4E^{-106}	Hypothetical protein
VPA1475	3828	531	3.4	2.3E^{-46}	Purine nucleoside phosphorylase DeoD-type 2
VP0129	4086	613	3.3	2.0E^{-68}	Phosphoenolpyruvate carboxykinase
VPA0552	55,155	9280	3.1	1.7E^{-85}	Cold shock DNA-binding domain protein
VP2585	37,016	9182	2.6	2.2E^{-53}	Hypothetical protein
VP0256	4022	1260	2.2	1.4E^{-40}	30S ribosomal protein S10
VPA0466	7349	2418	2.2	8.1E^{-71}	Universal stress protein A (Usp)
VPA1428	3936	1310	2.2	6.6E^{-42}	Azurin: blue copper protein
VP0076	5271	2400	1.7	7.6E^{-19}	Usp
VPA1469	22,981	10,708	1.7	3.4E^{-26}	Outer membrane protein (OMP)
Downregulated in isolation condition					
svpa113.1	2014	24,669	-3.0	2.7E^{-66}	RyhB
VP1393	19	4134	-7.2	0	BfdA, Hcp1
VP0795	1150	11,616	-2.7	1.2E^{-116}	Histidine-containing phosphocarrier protein
VP2157	1169	8519	-2.3	2.6E^{-27}	Glyceraldehyde-3-phosphate dehydrogenase

Genes with the highest differential expression

Table 2 shows the RPKM and differential expression parameters for the 10 genes with highest upregulation and the 10 genes with greatest downregulation. Five neighboring genes VP1512-VP1516, probably all related to respiration, are among those showing the highest upregulation, close to 70 times. Two other neighboring genes (VP1634, VP1635) related to polyamine production are also highly upregulated. Among the most downregulated genes (about 140 times) is the gene encoding BfdA, the hemolysin-co-regulated protein (Hcp1) secreted by T6SS1, found also among the most highly expressed genes. Further, four genes close together in the genome (VP1777, VP1783, VP1794 and VP1796) and three (VP1067, VP1068 and VP1072) are more than 100 times downregulated; they are described as hypothetical proteins but might have related functions.

Overall differential expression

To compare the expression of all the differentially expressed genes, they were grouped by gene ontology term using enrichGO from clusterProfiler [17]. An overrepresentation analysis of gene ontology terms was performed for up- and downregulated genes and the results are shown in Fig. 2a and b, respectively.

Among the 777 differentially expressed genes, 753 were associated with a GO term. Small RNAs do not have GO terms and hence they are not included in this particular comparison. The most significantly overrepresented upregulated processes were those related to transport and localization. Among overrepresented processes with downregulated genes were secretion- and pathogenesis-related genes.

Discussion

Expression of 777 genes differed when the RNA of *V. parahaemolyticus* grown at 37 °C, 0.9% NaCl in the presence of bile salts (conditions for laboratory isolation) was compared with the same grown at marine like-conditions with temperature 12 °C and salt 3% NaCl, (more common in the marine environment). Expression of genes measured as RPKM varied from zero up to hundreds of thousands while the extent of change varied from 3 up to 200 times. Some of the more interesting changes when comparing the bacteria growing in isolation versus marine like conditions were:

- Increase in expression of the three CsrB sRNAs genes probably related to the faster carbon

Table 2 RPKM and differential expression parameters for the 25 genes with the highest differential expression between both conditions

Gene	RPKM I	RPKM E	log$_2$FC	p-value	Observations
Genes upregulated in isolation condition					
VP1512	1104	29	5.7	$5.0E^{-83}$	Hypothetical protein
VP1513	392	11	5.6	$1.1E^{-89}$	Formate dehydrogenase large subunit
VP1514	357	10	5.5	$1.9E^{-79}$	Formate dehydrogenase, iron-sulfur subunit. Interaction with 4Fe-4S
VP1515	216	5	5.9	$1.6E^{-141}$	Formate dehydrogenase, cytochrome b556 subunit. Membrane protein
VP1516	671	16	5.9	$8.2E^{-219}$	Hypothetical protein
VPA1634	2140	8	8.5	0	Putrescine transporter
VPA1635	3652	16	8.4	0	Ornithine decarboxylase, initial enzyme in polyamine synthesis
VPA0040	679	9	6.7	$1.5E^{-289}$	Hypothetical protein
VP0061	330	7	6.0	$4.1E^{-130}$	Multidrug transmembrane resistance signal peptide protein
VPA0475	1764	5	8.8	$6.1E^{-228}$	Hypothetical protein
Genes downregulated in isolation condition					
VP1393	19	4134	−7.2	0	BfdA, Hcp1
VP1777	0	14	-7.1	$4.1E^{-24}$	Aldehyde dehydrogenase Oxidation of acetaldehyde NAD or NADP reduction
VP1783	0	51	-7.7	$3.1E^{-29}$	Hypothetical protein
VP1794	0	29	-7.5	$6.6E^{-23}$	Hypothetical protein
VP1796	0	21	-7.2	$4.7E^{-25}$	Hypothetical protein
VP1067	0	10	-7.1	$6.6E^{-20}$	Hypothetical protein
VP1068	1	178	-7.0	$9.0E^{-114}$	Hypothetical protein
VP1400	1	207	-7.5	$3.5E^{-248}$	Hypothetical protein
VP1072	1	142	-7.2	$1.1E^{-186}$	Helicase
VPA1424	2	492	−7.2	0	Fructose transport

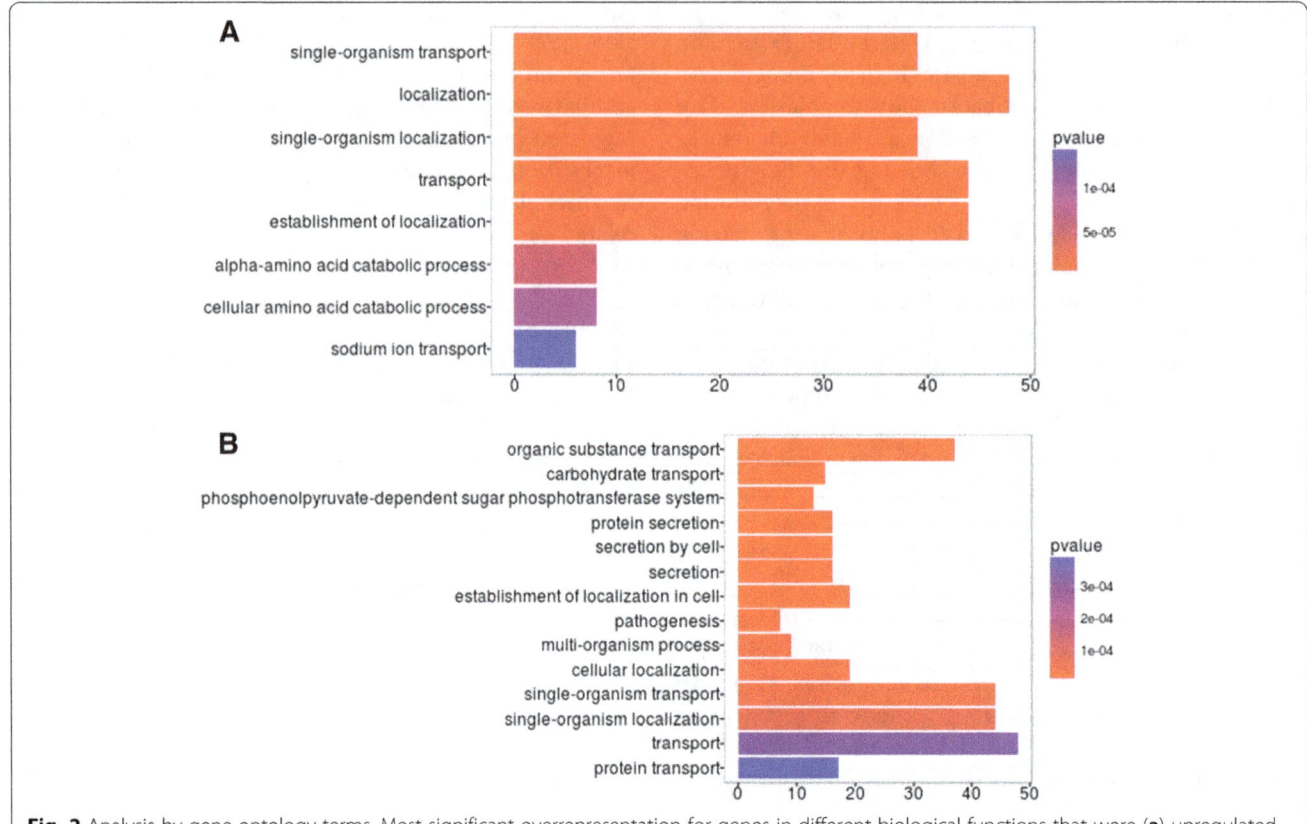

Fig. 2 Analysis by gene ontology terms. Most significant overrepresentation for genes in different biological functions that were (**a**) upregulated or (**b**) downregulated in the isolation condition

metabolism required for growing 24 times more rapid at 37C.

- Downregulation of RyhB sRNA which translate into upregulation of genes involved in motility.
- Downregulation of T6SS in chromosome I and II but differential regulation of the T3SS in chromosomes I and II.
- Non-induction of both *tdh*s and VopL, VopT, and VopC when using simple non-conjugated bile salt, and induction of 16 of the T3SS2 genes in absence of VtrB upregulation.
- Increase of cold shock protein CspA and other cold shock protein (VPA0552) by the nutritional upshift at 37C.

Increase in expression of the three CsrB genes was probably due to the faster carbon metabolism required for growing 24 times more rapid at 37 °C. In *E. coli* CsrB RNA functions as an antagonist of CsrA, which is a central component of the carbon storage regulator system and inhibits glycogen biosynthesis, catabolism and gluconeogenesis, among others [19]. Upregulation of CsrB probably increases these processes supporting rapid *Vibrio* growth.

RyhB modulates the expression of several genes that control motility, chemotaxis and biofilm formation, besides iron metabolism. In *Vibrio cholerae*, several genes involved in flagellar biosynthesis or chemotaxis are negatively regulated in a *ryhB* mutant, and therefore RyhB is necessary for the formation of biofilms and chemotactic motility [26]. By contrast, in *Salmonella Typhimurium*, RyhB2 is involved in the downregulation of the flagellar and chemotaxis genes *flgJ*, *cheY* and *fliF*, thus leading to an increased motility phenotype in a *ryhB2* mutant strain [27]. As observed in our results, there could be an association between the significant downregulation of RhyB and the upregulation of various genes involved in motility under the isolation condition.

All genes differentially expressed for both secretion system type VI(T6SS) were downregulated; 25 of the 29 genes of T6SS1 (VP1386-VP1414) including BfdA, the hemolysin-co-regulated protein (Hcp1), and 17 of the 22 genes of T6SS2, including Hcp2 protein (VPA1025-VPA1046). These results coincide with Salomon et al. in the idea that some of these T6SS play an important role in the adaptability to the environment and fitness of *V. parahaemolyticus* [23], because it is overexpressed in the E condition (12 °C and 3% NaCl). Conversely, there was differential regulation of the T3SS1 and T3SS2, both related to pathogenicity; 31 of the 42 genes that compose the T3SS in chromosome I (VP1656-VP1697) were significantly downregulated (Padj $< 1.3E^{-6}$) while 16 of the 50 genes of T3SS2 (VPA1321-VPA1370) were significantly upregulated (Padj $< 1.7E^{-7}$). This observation is partially in accordance with previous publications where it has been shown that bile salts also induce Vops, the needle-like secretion apparatus, including components such as translocon VopD2, and the expression of *tdh* genes, which belong to pathogenicity island 7 [28–30]. However, unlike these authors, we did not observe some important pathogenicity genes within this pathogenicity island increasing in the presence of bile salts, for example both *tdh* (*tdhA*, VPA1314 and *tdhS*, VPA1378), VPA1370 (VopL), VPA1327 (VopT) and VPA1321 (VopC). This apparent discrepancy could be due to the usage of different bile salts because they have different transcription-inducing activity. We used bile salt from Sigma-Aldrich, which consists of deoxycholate/cholate which according to Gotoh et al. [31] lack or show low transcription induction of *tdh* and Vp-PAI. Taurodeoxycholate and glycodeoxycholate are the bile salt with highest inducing activity. These last two bile salts consist of deoxycholate conjugated with glycine and taurine respectively. In *V. parahaemolyticus*, activation of T3SS2 by bile salts is regulated by VtrA (VPA1332), VtrB (VPA1348) and VtrC (VPA1333). VtrA and VtrC form a functional complex that binds bile salts to activate the cytoplasmic DNA binding domain of VtrA, which in turn induces T3SS2 via the downstream transcription factor VtrB [32]. Since we do not observe a significant upregulation of VtrB (log_2FC = 0.6), we speculate that upregulation of some genes of T3SS2 could also occur by an independent mechanism. Interestingly, the downregulation of pathogenesis-related genes in condition I (Fig. 2) suggests that besides pathogenesis in humans, these genes are also required for functions that increase *V. parahaemolyticus* fitness in the environment and hence are highly expressed in the condition E. This idea is supported by similar observations of Yang et al. [33] of increased expression of genes responsible for the general secretion pathway, type IV prepilin biogenesis, and pathogenesis when growing at 10 °C instead of 37 °C and by Urmersbach [7] who also observed increased expression of genes related to secretion and pathogenicity comparing growth at 15 °C with 37 °C.

Other mRNAs that changed significantly their expression were the cold shock protein CspA as well as other cold shock proteins (VPA0552). The upregulation of these proteins at 37 °C was surprising and unexpected since other authors has shown that CspA is upregulated in cold shock as expected [33, 34]. However, the observed upregulation of CspA in condition I could be explained by the high nutrients availability in this condition which has been shown increases these proteins [35].

Additionally, the large increase of the five neighbor genes VP1512-VP1516 could be related to the increased growth rate since three of them are related to respiration. The other two are hypothetical proteins that could also participate in this function. Finally, the increase of VP1634 and VP1635, and VP0061 involved in putrescine and spermidine production, commonly associated with putrefaction, may be related to other less known but important functions of these compounds in bacteria when the growth rate is increased, such as rapid growth itself and incorporation into the cell wall [36].

Within 777 genes up- or downregulated in *V. parahaemolyticus*, the most significantly overrepresented categories among upregulated processes were those related to transport and localization, while secretion and pathogenesis were overrepresented among downregulated genes.

Most of differences observed are independent of the availability of nutrients since we used the base medium LB reach in organic nutrients in both conditions. The availability of nutrients in seawater is probably orders of magnitude lower but it is likely that *V. parahaemolyticus* preferentially grow in the ocean when associated to a host supplying nutrients in abundance, specifically when in mollusks or in the intestine of higher animals. Attempts to grow *V. parahaemolyticus* at the low nutrient concentration prevailing in seawater were unsuccessful.

Conclusions

When *V. parahaemolyticus* is grown in conditions used for laboratory isolation 777 genes are up- or downregulated referred to conditions prevailing in the ocean when organic nutrients are in high supply; the most significantly overrepresented categories among upregulated processes were transport and localization, while secretion and pathogenesis were overrepresented among downregulated genes. Genes with the highest differential expression included the CsrB and RhyB sRNAs and the mRNAs related with secretion, nutritional upshift, respiration and rapid growing.

Abbreviations

BGI: Beijing Genomics Institute; BWA: Burrows-Wheeler Aligner; E: Marine-like environmental condition; FC: Fold Change; GO: Gene Ontology; I: Isolation condition; LB: Luria-Bertani; LS: Low protocol sample; ORF: Open reading frame; RPKM: Reads Per Kilobase of transcribed gene per Million mapped reads with features; T3SS: Type III secretion system; T6SS: Type VI secretion system

Acknowledgments
It is a pleasure to acknowledge Diego Riquelme for growing the bacteria, isolating the RNA and helping in the libraries preparation. KG and RE acknowledge to FONDECYT 11140257 and FONDECYT 1140732 respectively.

Funding
This work was funded by grant FONDECYT 11140257 and FONDECYT 1140732.

Authors' contributions
RE y KG: conception and design of the work; acquisition, analysis and interpretation of data and writing the manuscript. CY, NP, FP, PS, and DP-R: acquisition, analysis and interpretation of data. All authors read and approved the final manuscript.

Competing interests
The authors declare that they have no competing interests.

Author details
[1]Centro de Investigación Biomédica, Facultad de Ciencias de la Salud, Instituto de Ciencias Biomédicas, Universidad Autónoma de Chile, Av. El Llano Subercaseaux, 2801 Santiago, Chile. [2]Institute of Nutrition and Food Technology, Universidad de Chile, Av. El Líbano 5524, Macul, Santiago, Chile. [3]School of Bioinformatics Engineering, University of Talca, Talca, Chile.

References

1. Letchumanan V, Chan KG, Lee LH. *Vibrio parahaemolyticus*: a review on the pathogenesis, prevalence, and advance molecular identification techniques. Microbiol: Front; 2014.

2. Kayser, A; DePaola A. BAM:Vibrio. 2012. Available from: http://www.fda.gov/Food/ScienceResearch/LaboratoryMethods/BacteriologicalAnalyticManualBAM/ucm070830.htm.

3. González-Escalona N, Cachicas V, Acevedo C, Rioseco ML, Vergara JA, Cabello F, et al. *Vibrio parahaemolyticus* Diarrhea, Chile, 1998 and 2004. Emerg. Infect. Dis. Centers for Disease Control and Prevention. 2005;11: 2004–2006. Available from: https://www.fda.gov/food/foodscienceresearch/laboratorymethods/ucm070830.htm.

4. García K, Bastías R, Higuera G, Torres R, Mellado A, Uribe P, et al. Rise and fall of pandemic *Vibrio parahaemolyticus* serotype O3:K6 in southern Chile. Environ. Microbiol. 2013;15:527–34. [cited 2014 Mar 7]. Available from: http://www.ncbi.nlm.nih.gov/pubmed/23051148

5. Chiang M-L, Chou C-C. Expression of superoxide dismutase, catalase and thermostable direct hemolysin by, and growth in the presence of various nitrogen and carbon sources of heat-shocked and ethanol-shocked *Vibrio parahaemolyticus*. Int J Food Microbiol. 2008;121:268–74. [cited 2016 May 18]. Available from: http://www.sciencedirect.com/science/article/pii/S0168160507005715

6. Mandlik A, Livny J, Robins WP, Ritchie JM, Mekalanos JJ, Waldor MK. RNA-Seq-based monitoring of infection-linked changes in Vibrio cholerae Gene expression. Cell Host Microbe. 2011;10:165–74. Available from: http://dx.doi.org/10.1016/j.chom.2011.07.007.

7. Urmersbach S, Aho T, Alter T, Hassan SS, Autio R, Huehn S. Changes in global gene expression of *Vibrio parahaemolyticus* induced by cold- and heat-stress. BMC Microbiol. 2015;15:229. Available from: http://www.biomedcentral.com/1471-2180/15/229

8. Morgan M, Anders S, Lawrence M, Aboyoun P, Pagès H, Gentleman R. ShortRead: a bioconductor package for input, quality assessment and exploration of high-throughput sequence data. Bioinformatics. 2009;25:2607–8.

9. Li H, Durbin R. Fast and accurate short read alignment with Burrows-Wheeler transform. Bioinformatics. 2009;25:1754–60.

10. Li H, Handsaker B, Wysoker A, Fennell T, Ruan J, Homer N, et al. The sequence alignment/map format and SAMtools. Bioinformatics. 2009;25:2078–9.

11. Pruitt KD, Tatusova T, Maglott DR. NCBI reference sequences (RefSeq): a curated non-redundant sequence database of genomes, transcripts and proteins. Nucleic Acids Res. 2007;35:501–4.

12. Li L, Huang D, Cheung MK, Nong W, Huang Q, Kwan HS. BSRD: a repository for bacterial small regulatory RNA. Nucleic Acids Res. 2013;41:233–8.

13. Anders S, Pyl PT, Huber W. HTSeq-A python framework to work with high-throughput sequencing data. Bioinformatics. 2015;31:166–9.

14. Love MI, Huber W, Anders S. Moderated estimation of fold change and dispersion for RNA-seq data with DESeq2. Genome Biol. 2014;15:1–34.

15. Ashburner M, Ball CA, Blake JA, Botstein D, Butler H, Cherry JM, et al. The Gene ontology consortium. Gene ontology: tool for the unification of biology. Nat Genet. 2011;25:25–9.

16. Bateman A, Martin MJ, O'Donovan C, Magrane M, Apweiler R, Alpi E, et al. UniProt: a hub for protein information. Nucleic Acids Res. 2015;43:D204–12.

17. Yu G, Wang L-G, Han Y, He Q-Y. clusterProfiler: an R package for comparing biological themes among gene clusters. OMICS. 2012;16:284–7. Available from: http://www.pubmedcentral.nih.gov/articlerender.fcgi?artid=3339379&tool=pmcentrez&rendertype=abstract.

18. Szklarczyk D, Franceschini A, Wyder S, Forslund K, Heller D, Huerta-Cepas J, Simonovic M, Roth A, Santos A, Tsafou KP, Kuhn M, Bork P, Jensen LJ, von Mering C. STRING v10: protein-protein interaction networks, integrated over the tree of life. Nucleic Acids Res. 2015;43:447–52.

19. Suzuki K, Wang X, Weilbacher T, Pernestig A, Georgellis D, Babitzke P, et al. Regulatory circuitry of the CsrA/CsrB and BarA/UvrY systems of *Escherichia coli*. J Bacteriol. 2002;184:5130–40.

20. Wassarman KM, Storz G. 6S RNA regulates *E. coli* RNA polymerase activity. Cell. 2000;101:613–23.

21. Kanehisa M, Goto S, Sato Y, Kawashima M, Furumichi M, Tanabe M. Data, information, knowledge and principle: back to metabolism in KEGG. Nucleic Acids Res. 2014;42:D199–205. Available from: https://www.ncbi.nlm.nih.gov/pmc/articles/PMC333379/?tool=pmcentrez.

22. Pérez-Reytor D, Plaza N, Espejo RT, Navarrete P, Bastías R, Garcia K. Role of Non-coding Regulatory RNA in the Virulence of Human Pathogenic Vibrios. Front. Microbiol. 2017 [cited 2017 Mar 11];7. Available from: http://journal.frontiersin.org/article/10.3389/fmicb.2016.02160/full.

23. Salomon D, Gonzalez H, Updegraff BL, Orth K. *Vibrio parahaemolyticus* Type VI secretion system 1 is activated in marine conditions to target bacteria, and is differentially regulated from system 2. PLoS One. 2013;8:e61086. Available from: http://dx.plos.org/10.1371/journal.pone.0061086

24. Ho BT, Dong TG, Mekalanos JJ. A view to a kill: the bacterial type VI secretion system. Cell Host Microbe. 2014;15:9–21. Available from: http://dx.doi.org/10.1016/j.chom.2013.11.008%5Cnhttp://linkinghub.elsevier.com/retrieve/pii/S1931312813004095%5Cnhttp://www.ncbi.nlm.nih.gov/pubmed/24332978%5Cnhttp://www.pubmedcentral.nih.gov/articlerender.fcgi?artid=PMC3936019

25. Yamanaka K, Inouye M. Selective mRNA degradation by polynucleotide Phosphorylase in cold shock adaptation in *Escherichia coli* selective mRNA degradation by polynucleotide Phosphorylase in cold shock adaptation in *Escherichia coli*. J Bacteriol. 2001;183:2808–16.

26. Porcheron G, Dozois CM. Interplay between iron homeostasis and virulence: fur and RyhB as major regulators of bacterial pathogenicity. Vet Microbiol. 2015:2–14.

27. Kim JN, Kwon YM. Identification of target transcripts regulated by small RNA RyhB homologs in salmonella: RyhB-2 regulates motility phenotype. Microbiol Res. 2013;168:621–9.

28. Livny J, Zhou X, Mandlik A, Hubbard T, Davis BM, Waldor MK. Comparative RNA-Seq based dissection of the regulatory networks and environmental stimuli underlying *Vibrio parahaemolyticus* gene expression during infection. Nucleic Acids Res. 2014;42:12212–23. Available from: http://dx.doi.org/10.1016/j.chom.2013.11.008

29. Gotoh K, Kodama T, Hiyoshi H, Izutsu K, Park KS, Dryselius R, et al. Bile acid-induced virulence gene expression of *Vibrio parahaemolyticus* reveals a novel therapeutic potential for bile acid sequestrants. PLoS One. 2010; 5(10):e13365.

30. Broberg CA, Calder TJ, Orth K. *Vibrio parahaemolyticus* Cell biology and pathogenicity determinants. Microbes Infect. 2011;13(12–13):992–1001.

31. Kodama T, Hiyoshi H, Gotoh K, Akeda Y, Matsuda S, Park K-S, et al. Identification of two translocon proteins of *Vibrio parahaemolyticus* type III secretion system 2. Infect Immun. 2008;76:4282–9.

32. Li P, Rivera-Cancel G, Kinch LN, Salomon D, Tomchick DR, Grishin NV, et al. Bile salt receptor complex activates a pathogenic type III secretion system. elife. 2016;5 doi:10.7554/eLife.15718.

33. Yang L, Zhou D, Liu X, Han H, Zhan L, Guo Z, et al. Cold-induced gene expression profiles of *Vibrio parahaemolyticus*: a time-course analysis. FEMS Microbiol Lett. 2009;291:50–8.

34. Datta PP, Bhadra RK. Cold shock response and major cold shock proteins of Vibrio cholerae. Appl Environ Microbiol. 2003;69:6361–9. [cited 2017 Mar 11]. Available from: http://www.ncbi.nlm.nih.gov/pubmed/14602587

35. Yamanaka K, Inouye M. Induction of CspA, an *E. coli* Major cold-shock protein, upon nutritional upshift at 37 oC. Genes Cells. 2001;6:279–90. [cited 2016 Nov 16]. Available from: http://doi.wiley.com/10.1046/j.1365-2443.2001.00424.x

36. Wortham BW, Patel CN, Oliveira MA. Polyamines in bacteria: pleiotropic effects yet specific mechanisms. Adv Exp Med Biol. 2007;603:106–15. [cited 2016 Nov 18]. Available from: http://www.ncbi.nlm.nih.gov/pubmed/17966408.

17

Functional characterization of thiolase-encoding genes from *Xanthophyllomyces dendrorhous* and their effects on carotenoid synthesis

Nicole Werner, Melissa Gómez, Marcelo Baeza, Víctor Cifuentes and Jennifer Alcaíno*

Abstract

Background: The basidiomycetous yeast *Xanthophyllomyces dendrorhous* has been described as a potential biofactory for terpenoid-derived compounds due to its ability to synthesize astaxanthin. Functional knowledge of the genes involved in terpenoid synthesis would create opportunities to enhance carotenoid production. A thiolase enzyme catalyzes the first step in terpenoid synthesis.

Results: Two potential thiolase-encoding genes were found in the yeast genome; bioinformatically, one was identified as an acetyl-CoA C-acetyltransferase (*ERG10*), and the other was identified as a 3-ketoacyl Co-A thiolase (*POT1*). Heterologous complementation assays in *Saccharomyces cerevisiae* showed that the *ERG10* gene from *X. dendrorhous* could complement the lack of the endogenous *ERG10* gene in *S. cerevisiae*, thereby allowing cellular growth and sterol synthesis. *X. dendrorhous* heterozygous mutants for each gene were created, and a homozygous *POT1* mutant was also obtained. This mutant exhibited changes in pigment composition and higher *ERG10* transcript levels than the wild type strain.

Conclusions: The results support the notion that the *ERG10* gene in *X. dendrorhous* is a functional acetyl-CoA C-acetyltransferase essential for the synthesis of mevalonate in yeast. The *POT1* gene would encode a functional 3-ketoacyl Co-A thiolase that is non-essential for cell growth, but its mutation indirectly affects pigment production.

Keywords: Thiolase, Mevalonate, Astaxanthin, Sterols, Carotenoids, Functional complementation

Background

Isoprenoids are widely distributed in nature with multiple functions due to their structural diversity and represent an important resource for the biotechnology industry with uses ranging from aroma and flavor enhancers (terpenes) to anticarcinogenic molecules (taxol) [1]. All isoprenoids originate from isopentenyl pyrophosphate (IPP), containing five carbon units, which in most organisms is obtained through the mevalonate (MVA) pathway or the methyl-D-erythritol-4-phosphate (MEP) pathway, the latter of which occurs in plant plastids, algae and some bacteria [2].

Condensation between two acetyl-CoA molecules to form acetoacetyl-CoA has been recognized as the first step in the mevalonate pathway in eukaryotes [3]. This reaction is catalyzed through a non-decarboxylative Claisen-type condensation by an enzyme that belongs to the thiolase protein family known as acetyl-CoA C-acetyltransferase (ACAT) [4, 5]. The thiolase protein family comprises enzymes that have different subcellular localizations and expression patterns, depending on the pathway with which they are associated; in yeast, enzymes with cytoplasmic locations are mevalonate pathway related enzymes (ACAT), and 3-ketoacyl-CoA thiolase (ACAA), which is involved in the β-oxidation of fatty acids, is found in peroxisomal locations [6, 7].

The second step in mevalonate biosynthesis involves the addition of a third acetyl-CoA molecule to form

* Correspondence: jalcainog@u.uchile.cl

Departamento de Ciencias Ecológicas y Centro de Biotecnología, Facultad de Ciencias, Universidad de Chile, Las Palmeras 3425, Casilla 653, Ñuñoa, Santiago, Chile

3-hydroxy-3-methylglutaryl-CoA (HMG-CoA) by the HMG-CoA synthase (HMGS) enzyme, followed by the conversion of HMG-CoA to mevalonic acid via HMG-CoA reductase (HMGR), which is the most studied step of the mevalonate pathway and has been defined as the rate-limiting step in sterol biosynthesis in eukaryotes [8]. In *Saccharomyces cerevisiae*, two HMGR encoding genes have been identified, *HMG1* and *HMG2*, with *HMG1* contributing the majority of the enzymatic activity in the pathway [9]; however, only one ACAT encoding gene (*ERG10*) has been reported in this yeast [10]. The concluding steps in the mevalonate pathway include two sequential phosphorylation reactions performed by the enzymes mevalonate kinase and phosphomevalonate kinase, respectively, and a final decarboxylation step catalyzed by phosphomevalonate decarboxylase to produce IPP [11], which is the precursor of the wide variety of isoprenoid compounds and derivatives, including secondary metabolites of commercial value.

The basidiomycetous yeast *Xanthophyllomyces dendrorhous* has been extensively studied for its ability to produce large amounts of isoprenoids, primarily astaxanthin (3,3′-dihydroxy-b, b-carotene-4,4′-dione), a red carotenoid used as a feeding additive in aquaculture, making it appealing to the biotechnological industry. Most genetic studies with this yeast have focused on the characterization of genes involved in carotenogenesis [12, 13] and genes involved in the biosynthesis of carotenoids precursors [14]. It has been shown that astaxanthin production in *X. dendrorhous* is favored when cultures are supplemented with mevalonate [15] or in strains that have higher *HMGR* transcript levels [16]. These results reflect the critical role of the MVA pathway in carotenogenesis in *X. dendrorhous*, highlighting the importance of studying the MVA pathway and identifying the genes that are directly involved in the synthesis of IPP.

Regarding the *ERG10* gene of *X. dendrorhous*, a possible ACAT encoding gene was recently identified in a genomic study of this yeast [17], and its overexpression in *X. dendrorhous* led to an increase in total carotenoid production [18]. In the present study, another possible *X. dendrorhous* thiolase-encoding gene is identified, and additional functional evidence for the previously reported ACAT encoding gene and for the newly characterized ACAA encoding gene in this yeast is provided.

Results and discussion
Bioinformatic characterization and expression analysis
Gene identification was accomplished by running a local BLASTp search in the program CLC Genomics Workbench through the *X. dendrorhous* genomic and transcriptomic databases [19], using related *ERG10* sequences obtained from the GenBank database as queries. Two potential *ERG10* sequences were obtained and designated

ERG10A and *ERG10B*. The *ERG10A* gene is composed of 7 introns and 8 exons, and its cDNA has a length of 1,296 bp encoding a deduced protein of 431 amino acids with a total size of 44.38 kDa, while the *ERG10B* gene contains 5 introns and 6 exons, and its cDNA has a length of 1,212 bp encoding a 42.01-kDa deduced protein of 403 amino acids.

Within the thiolase family, degradative and biosynthetic enzymes share the same catalytic mechanism in which the catalytic site is conformed by four loops, each carrying conserved residues at the active site [4]. These residues correspond to Cys-125, Asn-357, His-389 and Cys-417 in the deduced thiolase encoded by the *ERG10A* gene and Cys-101, Asn-326, His-358 and Cys-388 in the thiolase deduced from *ERG10B*. Two reactions occur in the catalytic site of this enzyme family; first a conserved highly reactive Cys residue (Cys-89 in the *Zoogloea ramigera* biosynthetic thiolase [PDB: 1 dm3]) is acetylated, while a second Cys (Cys-378 in the *Z. ramigera* biosynthetic thiolase) protonates the CoA leaving group. The second reaction involves condensation, in which the negatively charged Cys-378 deprotonates a second acetyl-CoA, to which the acetyl moiety of Cys-89 is transferred [4, 5]. According to bioinformatic analyses, both putative *X. dendrorhous* thiolases contain all of the catalytic site amino acids, but no further information regarding their catalytic activity could be obtained.

To determine the possible function of the thiolase encoded by each gene, a phylogenetic tree was created using thiolase amino acid sequences described as having different functions: acetyl-CoA C-acetyltransferase or 3-ketoacyl-CoA thiolase. For multiple alignments, the program ClustalW 2.1 was used and a phylogenetic tree was built using the Neighbor Joining method with a 1,000-replica bootstrap with MEGA6 software [20]. As shown in Fig. 1, the thiolase encoded by the *ERG10A* gene groups with acetyl-CoA C-acetyltransferases, and the thiolase encoded by *ERG10B* groups with 3-ketoacyl-CoA thiolases. The multiple alignment used to generate this tree is shown in Additional file 1: Figure S1, where Cys-141, Asn-384, His-416 and Cys-447 correspond to the conserved active site residues. This result suggests that *ERG10A* most likely encodes the thiolase involved in the mevalonate pathway and that *ERG10B* is the thiolase involved in the β-oxidation of fatty acids in *X. dendrorhous*. Moreover, the *ERG10A* nucleotide sequence has a 98% identity with 91% coverage compared to the *X. dendrorhous acaT* reported nucleotide sequence available at GenBank [AB919149] [18].

To identify possible binding sites for transcriptional regulators, the promoter regions, which were considered to comprise 1,000 bp upstream of the first ATG codon of each gene, were analyzed using JASPAR (http://jaspar.genereg.net/) and TBfind (http://tfbind.hgc.jp/). Several

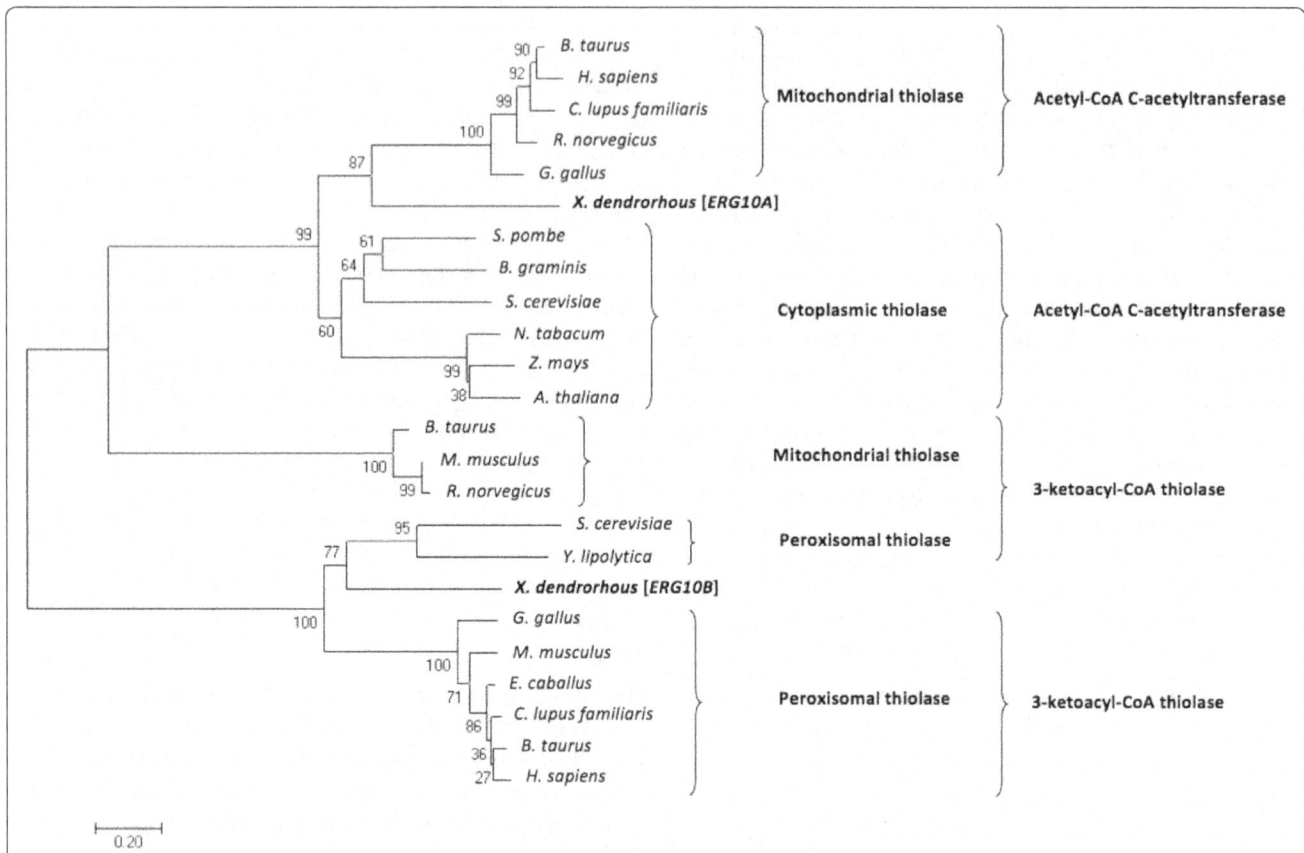

Fig. 1 Phylogenetic tree of *ERG10A* and *ERG10B* thiolases from *X. dendrorhous* compared to other organisms. The unrooted tree was created in MEGA 6.0 using the neighbor-joining method [20] with the following amino acid sequences. Mitochondrial thiolase/acetyl-CoA C-acetyltransferase: *B. taurus* [NP_001039540.1], *H. sapiens* [BAA01387.1], *C. lupus familiaris* [XP_546539.2], *R. norvegicus* [NP_058771.2], *G. gallus* [NP_001264708.1]. Cytoplasmic thiolase/acetyl-CoA C-acetyltransferase: *S. pombe* [Q9UQW6.1], *B. graminis f. sp. tritici 96224* [EPQ61678.1], *S. cerevisiae* [P41338.3], *N. tabacum* [AAU95618.1], *Z. mays* [NP_001266315.1], *A. thaliana* [Q9FIK7.1]. Mitochondrial thiolase/3-ketoacyl-CoA thiolase: *B. taurus* [NP_001030419.1], *M. musculus* [NP_803421.1], *R. norvegicus* [NP_569117.1]. Peroxisomal thiolase/3-ketoacyl-CoA thiolase: *S. cerevisiae* [CAA37472.1], *Y. lipolytica* [Q05493.1], *G. gallus* [NP_001184217.1], *M. musculus* [NP_570934.1], *E. caballus* [XP_001488609.1], *C. lupus familiaris* [XP_534222.2]. *B. taurus* [NP_001029491.1], *H. sapiens* [NP_001598.1]. *X. dendrorhous* [*ERG10A*]: thiolase encoded by the *ERG10A* gene. *X. dendrorhous* [*ERG10B*]: thiolase encoded by the *ERG10B* gene. Accession numbers are given in parentheses. Numbers at each node indicate the percentage support for a specific node after 1,000-replica bootstrap analysis

genes involved in the MVA pathway of *Schizosaccharomyces pombe* have been shown to be regulated by ergosterol levels through the Sterol Regulatory Element Binding Protein (SREBP), which has been described as a major regulator of genes related to the synthesis of cellular sterols [21]. In the promoter analysis, one possible SRE element was identified upstream of each *ERG10* gene-encoding region, suggesting possible regulation through SREBP. To gain further insight, the relative transcript levels of both genes were analyzed in a *X. dendrorhous* mutant strain (385-*cyp61*$^{(-/-)}$) that does not produce ergosterol but accumulates other sterols and demonstrates up-regulation of the *HMGR* gene involved in the MVA pathway in relation to the wild type strain UCD 67–385 [16]. Transcript levels for the *ERG10A* and *ERG10B* genes were evaluated in both strains after 24 h (exponential stage) and 120 h (stationary stage) of culture (Fig. 2a). At both time points, the transcript levels of *ERG10A* were higher in the mutant

strain in relation to the wild type strain. Maximum differences were observed for the *ERG10A* transcript levels after 120 h of culture (approximately 3-fold higher in the mutant strain in relation to wild type). For *ERG10B*, a significant change in transcript levels was observed only after 24 h. In the study made by Loto et al. using the same strain and growth conditions [16], transcripts for the *HMGR* gene were measured. The change in the expression of this gene was 23 times higher in the strain that doesn't produce ergosterol compared to wild type strain. As the SREBP pathway has been described as a regulatory mechanism that acts upon genes related to sterol biosynthesis [21], the change could be attributed to a SREBP-like mechanism. Compared to these results, the change in expression of *ERG10A* is minor and cannot be attributed to SREBP *a priori*, although the gene contains potential SRE sites in its promoter region. Further experiments should be performed in the future to identify if there is a functional SREBP-like

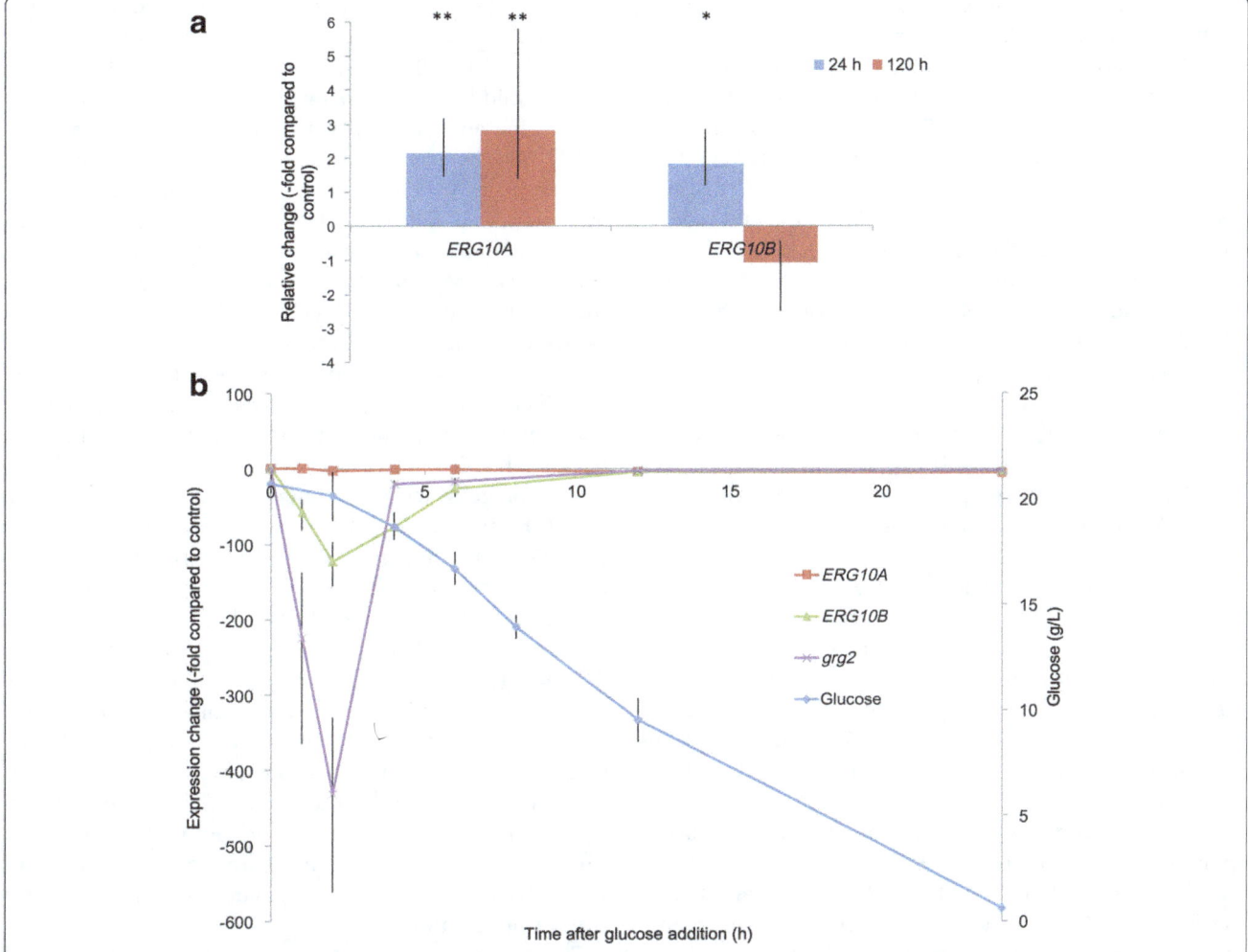

Fig. 2 *ERG10A* and *ERG10B* relative transcript levels. **a** In strain 385-*cyp61*$^{(-/-)}$, which does not produce ergosterol. Level of change was determined by comparison to wild type UCD 67–385 (control). Cultures were grown in YM liquid media for 24 h (blue) or 120 h (red). Each bar represents an average of three independent cultures. Black lines indicate standard deviation. *$p < 0.05$, **$p < 0.01$. **b** Effect of glucose addition. Gene expression kinetics and glucose concentration were quantified in strain UCD 67–385 after adding glucose (20 g/l final concentration). Error bars correspond to the standard deviation ($n = 3$). The negative values on the y-axis denote decrease relative to control

mechanism in *X. dendrorhous* that may be regulating the aforementioned genes.

According to Marcoleta et al. [22], carotenogenesis in *X. dendrorhous* is repressed by glucose. Moreover, the *MIG1* gene encoding the catabolic repressor Mig1, which mediates transcriptional glucose-dependent repression in other yeasts, was recently described and its mutation alleviated the glucose-mediated repression of carotenogenesis in *X. dendrorhous* [23]. Considering these findings, the possible glucose-dependent regulation of *ERG10A* and *ERG10B* gene expression was studied to evaluate whether catabolite repression could act over pathways upstream of carotenogenesis. For this, the wild type *X. dendrorhous* strain was cultured in YM media without glucose supplementation until the stationary phase of growth was reached; then, the culture was divided into two flasks: one was supplemented with glucose to a final concentration of

20 mg/ml and the other was left as a control without glucose supplementation. Cultures were incubated at 22 °C with constant agitation, and samples were taken after 1, 2, 4, 6, 12 and 24 h of treatment to extract total RNA and evaluate *ERG10A* and *ERG10B* transcript levels by RT-qPCR (Fig. 2b). The *ERG10A* gene transcript levels did not show any significant changes after the addition of glucose, but they were reduced approximately 5-fold compared to the control when glucose consumption began, which could be attributable to the by-products of glucose metabolism affecting *ERG10A* expression. In contrast, *ERG10B* transcript levels decreased approximately 120-fold when glucose was added to the culture compared to the untreated control, in a similar manner to the glucose repressed gene *grg2* [GenBank: JN043364], used as a glucose repression control (Fig. 2b) [22]. However, this decrease was only temporary as the transcript

levels normalized to control levels when the glucose in the media was consumed. Considering these results, only *ERG10B* is repressed by glucose in a manner similar to catabolite repression. Glucose repression, mediated by the transcription factors UME6, ABF1 and RP-A, has been reported previously for the *S. cerevisiae* gene *POT1/FOX3*, which encodes the yeast ACAA thiolase [24]. This evidence also supports the notion that *ERG10B* could indeed encode the ACAA thiolase in *X. dendrorhous*.

The results presented above support the idea that *ERG10A* encodes an acetyl-CoA C-acetyltransferase and *ERG10B* encodes a 3-ketoacyl-CoA thiolase, each participating in a different metabolic pathway (Additional file 2: Figure S2). Thus, hereafter each gene denomination was assigned according to the gene name given in *S. cerevisiae*, *ERG10* to *ERG10A* and *POT1* to *ERG10B*, and the sequences were uploaded to the GenBank database [KX267759 and KX26758, respectively].

Functional complementation in *S. cerevisiae*

To functionally assess the bioinformatic results, we performed heterologous expression of each gene cDNA in *S. cerevisiae* for complementation assays. For this analysis the *S. cerevisiae* diploid strain Meyen ex E.C Hansen YPL028W BY4743 (*Sc-ERG10/erg10*), a heterozygous mutant for the *ERG10* gene, was acquired from the ATCC collection.

The *S. cerevisiae* *ERG10* gene and the cDNAs corresponding to the two potential *X. dendrorhous* thiolase-encoding genes (*ERG10* and *POT1*) were inserted into the *S. cerevisiae* expression vector YEpNP (Table 1) and used to independently transform *Sc-ERG10/erg10*. From each transformation, two random colonies were selected and analyzed by PCR to confirm the presence of the tested gene in the plasmid.

Acetyl-CoA C-acetyltransferase activity is essential for cell viability [10]. In *S. cerevisiae*, this activity is only performed by the enzyme encoded by *ERG10*, so loss of this gene leads to unviable mutants. To assess if the gene expressed by the plasmid has the specified function, Random Spore Analysis was performed. After sporulation of the transformant strains and asci breaking followed by haploid selection, total DNA was extracted to confirm the lack of the endogenous *ERG10* gene and presence of the complementing gene in the plasmid by PCR analyses (Fig. 3). These could only be confirmed in strains carrying the YEpNP-10sc and YEpNP-c10xd plasmids (corresponding to strains Sc-ERG10sc and Sc-cERG10xd) that contain the *ERG10* genes from *S. cerevisiae* and *X. dendrorhous*, respectively. As shown in Fig. 3a, the chromosomal *ERG10* gene from *S. cerevisiae* was amplified only from strains S288c and Sc-dERG10, as haploid strains do not have this gene; instead, the band corresponding to the KanMX resistance gene was amplified from that locus

(Fig. 3b). The *S. cerevisiae* *ERG10* gene could be fully detected in the control strain S288c, the diploid mutant strain Sc-dERG10 and strain Sc-ERG10Sc, which carry the plasmid harboring this gene (Fig. 3c). A fragment of approximately 1,250 bp was amplified from strain Sc-cERG10xd, which corresponds to the cDNA of gene *ERG10* from *X. dendrorhous* (Fig. 3d); the same primer pair amplified a fragment of approximately 2,200 bp when genomic DNA from strain UCD 67−385 of *X. dendrorhous* was used as template, corresponding to the genomic version of *ERG10*. In the complementation assays with the *X. dendrorhous* *POT1* gene, no haploid colonies lacking the endogenous *ERG10* gene could be identified, even though more than 500 colonies were analyzed by replica plating and 30 potential candidates were analyzed by PCR.

Heterologous complementation in haploid strains was further analyzed by constructing growth curves (Additional file 3: Figure S3) and performing total sterol quantification/composition analyses. Samples for sterol extraction were recovered after 48 h of growth. The total amounts of sterols recovered from 10-ml samples were 3.3 ± 0.2 (mg/g dry yeast) for strain Sc-*ERG10/erg10*, 4.67 ± 0.01 (mg/g dry yeast) for strain Sc-ERG10sc and 6.4 ± 0.7 (mg/g dry yeast) for strain Sc-cERG10xd. Differences in the amounts of sterols produced between the parental diploid strain and the complemented strains may be attributable to the fact that in the latter, the gene is carried in an expression plasmid from which transcription is not regulated, leading to higher transcription rates.

These results support the notion that *ERG10* indeed encodes the *X. dendrorhous* acetyl-CoA C-acetyltransferase as it complements the *erg10-* mutation in *S. cerevisiae*.

ERG10 and *POT1* gene mutations in *X. dendrorhous*

To gain further knowledge regarding the functions and effects on carotenogenesis of the potential thiolase-encoding genes identified in *X. dendrorhous*, deletion mutants were generated. Plasmids pPHT-ERG10xd and pPHT-POT1xd were constructed and linearized to independently transform the diploid wild type strain UCD 67−385 to replace the corresponding genes with a hygromycin B resistance cassette by homologous recombination. PCR analyses were performed to confirm the gene replacement events on the heterozygous mutant strains obtained for genes *ERG10* (385-*erg10*[(+/−)]) and *POT1* (385-*pot1*[(+/−)]). To the naked eye, no pigmentation differences between both heterozygous mutant strains and the wild type could be appreciated. This observation was confirmed by pigment extraction and quantification from triplicate samples grown in liquid YM media at 22 °C for three days. The total amount of carotenoids extracted was 191 ± 5 μg/g dry weight for 385-*erg10*[(+/−)] and 175 ± 8 μg/g dry weight

Table 1 Strains and plasmids used in this work

Strain/Plasmid	Genotype or relevant features	Reference
Strain:		
X. dendrorhous		
UCD 67–385	Wild type, diploid strain.	ATCC 24230
385-*cyp61*(−/−)	Homozygote transformant derived from UCD 67–385 with both *CYP61* alleles interrupted, one with a hygromycin B resistance cassette and the other with a zeocin resistance cassette.	[16]
385-*erg10*(+/−)	Heterozygous mutant of UCD 67–385 with one allele of *ERG10* replaced by a hygromycin B resistance cassette.	This work
385-*pot1*(+/−)	Heterozygous mutant of UCD 67–385 with one allele of *POT1* replaced by a hygromycin B resistance cassette.	This work
385-*pot1*(−/−)	Homozygous mutant of UCD 67–385 with both *POT1* alleles replaced by a hygromycin B resistance cassette.	This work
385-*ERG10*	Heterozygote transformant derived from UCD 67–385 containing an additional *ERG10* allele and a hygromycin B resistance cassette integrated at *locus int*.	This work
385-*POT1*	Heterozygote transformant derived from UCD 67–385 containing an additional *POT1* allele and a hygromycin B resistance cassette integrated at *locus int*.	This work
385-Vexp2	Heterozygote transformant derived from UCD 67–385 containing an empty over-expressing cassette (without an inserted ORF) and a hygromycin B resistance cassette integrated at *locus int*.	This work
S. cerevisiae		
s288C	MATa, SUC2, gal2, mal, mel, flo1, flo8-1, hap1, ho, bio1, bio6	[38]
ATCC 4022800 (Sc-ERG10/erg10)	MATa/MATalpha his3delta1/his3delta1 leu2delta0/leu2delta0 lys2delta0/+ met15delta0/+ ura3delta0/ura3delta0 deltaERG10	[39]
Sc-ERG10sc	Strain Sc-ERG10/erg10 carrying plasmid YEpNP-10sc.	This work
Sc-cERG10xd	Strain Sc-ERG10/erg10 carrying plasmid YEpNP-c10Xd.	This work
Sc-cPOT1xd	Strain Sc-ERG10/erg10 carrying plasmid YEpNP-cPOT1Xd.	This work
Plasmid:		
pBluescript SK- (pBS)	ColE1 replication origin, Amp^R, LacZ for blue-white colony screening.	Stratagene
pBS-PT-ERG10xd	pBS with 620 bp of DNA upstream of *ERG10* gene and 412 bp downstream with a *Sma*I site between them.	This work
pBS-PT-POT1xd	pBS with 559 bp of DNA upstream of *POT1* gene and 560 bp downstream with a *Sma*I site between them.	This work
pMN-*hph*	Plasmid containing an hygromycin B resistance cassette for *X. dendrorhous*.	[12]
pPHT-Erg10xd	pBS -PT-ERG10Xd with a hygromycin B resistance cassette in the *Sma*I site.	This work
pPHT-POT1xd	pBS-PT-POT1Xd with a hygromycin B resistance cassette in the *Sma*I site.	This work
pPZT-Erg10xd	pBS-PT-ERG10Xd with a zeocin resistance cassette in the *Sma*I site.	This work
YEp-ACT4	pBR322 and 2 micron replication origins, Amp^R, *LEU2*, promoter *ACT4*.	[32]
YEp-NP	YEpACT4 with a *TDH3* terminator next to pACT4.	[28]
YEpNP-10sc	YEp-NP with *ERG10* DNA from *S. cerevisiae* between pACT4 and tTDH3.	This work
YEpNP-c10xd	YEp-NP with *ERG10* from *X. dendrorhous* cDNA between pACT4 and tTDH3.	This work
YEpNP-cPOT1xd	YEp-NP with *POT1* from *X. dendrorhous* cDNA between pACT4 and tTDH3	This work
pXdVexp2	*X. dendrorhous* expression vector: pBS bearing the *X. dendrorhous* ubiquitin promoter [GenBank: KJ140285] and GPD terminator [Genbank:Y08366] with a *Bam*HI site between them to insert the gene to express and the hygromycin B cassette for selection, flanked by non-encoding genomic [GenBank: KJ140286] regions to target the construction integration in the genome.	[14]
pXdVexp2-cERG10xd	pXdVexp2 containing the cDNA version of the *ERG10* gene from *X. dendrorhous*	This work
pXdVexp2-cPOT1xd	pXdVexp2 containing the cDNA version of the *POT1* gene from *X. dendrorhous*	This work

for 385-*pot1*$^{(+/−)}$ compared to 177 ± 11 µg/g dry weight from wild-type strain.

Among the properties that differentiate ACAT from ACAA encoding genes is the fact that ACAA null mutants can be obtained [25], whereas ACAT null mutants cannot unless there is more than one gene encoding enzymes with the same activity. We attempted to obtain *X.* *dendrorhous* double mutants for both strains using the double recombinant method (DRM, [26]). Briefly, the method consists of growing a heterozygous mutant in liquid media with increasing concentrations of the antibiotic corresponding to the selection marker considering that strains that become homozygous (by mitotic recombination) for the antibiotic marker would be able to grow

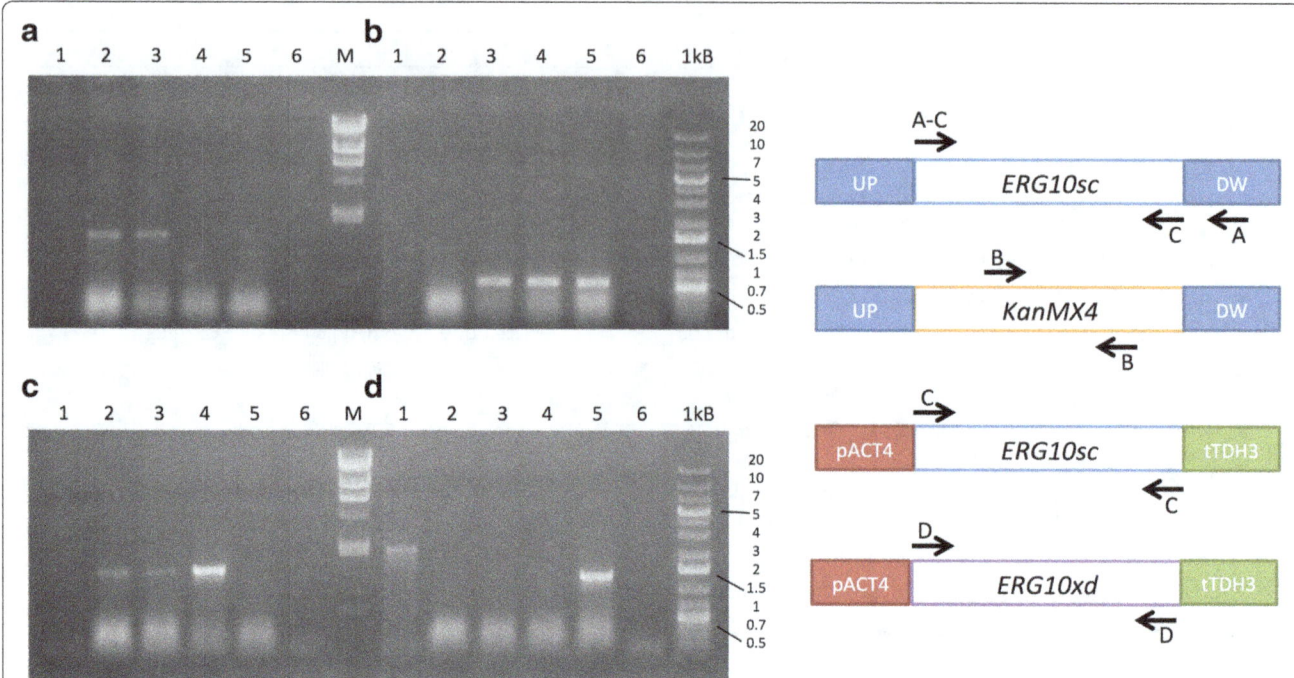

Fig. 3 PCR analyses of *S. cerevisiae* haploid strains. *S. cerevisiae* strains Sc-ERG10sc (carrying plasmid YEpNP-10sc) and Sc-cERG10xd (carrying plasmid YEpNP-c10xd) were analyzed by PCR to confirm the expected genotype. As controls, *X. dendrorhous* UCD 67–385 strain (Lane 1), *S. cerevisiae* strain S288c (Lane 2), *S. cerevisiae* diploid strain *Sc-ERG10/erg10* (Lane 3) and a no-template control (Lane 6), were included. *S. cerevisiae* Sc-ERG10sc haploid strain (Lane 4) and *S. cerevisiae* Sc-cERG10xd haploid strain (Lane 5) were analyzed to assess: **a** absence of chromosomal *ERG10* from *S. cerevisiae* (primers erg10scF and erg10scDWR); **b** presence of geneticin resistance cassette in *S. cerevisiae* (primers KanMXF2 and KanMXR2); **c** presence of *ERG10* from *S. cerevisiae* (primers erg10scF and erg10scR); and **d** presence of *ERG10* from *X. dendrorhous* (primers Thio2Fw and Thio2Rv). The molecular size markers Lambda DNA/*Hind*III (Lane M; 23.1, 9.4, 6.6, 4.4, 2.3, 2 and 0.6 kb) and GeneRuler 1 kb Plus (Lane 1kB, band size in kb is indicated) were used. On the right side of the picture, a schematic diagram of the amplification products is included; arrows represent primer sets with a letter indicating in which panel they were used. UP and DOWN (in blue) correspond to chromosomal regions located approximately 300 bp upstream and downstream of the *S. cerevisiae ERG10* gene, respectively, KanMX4 corresponds to the geneticin (G418) resistance module, and pACT4 (in red) and tTDH3 (in green) correspond to promoter and terminator regions in the vector YEpNP, respectively

at a higher concentration of the antibiotic than heterozygous strains. Antibiotic concentration was increased until a phenotypic color difference in colonies grown in plates was observed. For 385-*erg10*$^{(+/-)}$, antibiotic concentration could only be augmented from 15 µg/ml up to 100 µg/ml as higher concentrations inhibited cell growth; however, no differences in color phenotype were observed in the seeded colonies. Growth of strain 385-*pot1*$^{(+/-)}$ could still be accomplished at the maximum concentration of antibiotic used (400 µg/ml), and seeded colonies began to exhibit a paler phenotype when the antibiotic concentration in liquid media reached 200 µg/ml.

For both heterozygous mutant strain DRM assays, a few colonies selected at the maximum possible antibiotic concentration were randomly chosen and analyzed by PCR to determine if both alleles of the studied genes were lost after the treatment. No homozygous mutants deriving from strain 385-*erg10*$^{(+/-)}$ were found. Although this is an expected result if *ERG10* encodes the thiolase involved in the MVA pathway, a second approach was

attempted to try to obtain a homozygous mutant. A second plasmid for transformation, pPZT-ERG10xd, was constructed and used to transform 385-*erg10*$^{(+/-)}$; transformant selection was performed using plates supplemented with zeocin and hygromycin B. No colonies resistant to both antibiotics were obtained after several attempts of transformation, suggesting that a homozygous mutant for gene *ERG10* may not be viable as expected considering that *ERG10* is an essential gene.

On the other hand, it was confirmed by PCR analyses that paler colonies deriving from strain 385-*pot1*$^{(+/-)}$ after applying the DRM had lost both *POT1* alleles (homozygous mutants). For further analyses, only one of the analyzed colonies was used and designated 385-*pot1*$^{(-/-)}$.

To compare phenotypic changes between 385-*pot1*$^{(+/-)}$, 385-*pot1*$^{(-/-)}$ mutants and the wild type strain, each was grown in YM media at 22 °C with constant agitation for four days in triplicate. Samples were taken after 96 h of culture (stationary phase of growth) to analyze carotenoid and sterol content and composition. Within the growth curve, no significant differences were observed, either for total sterol or in carotenoid quantification. Carotenoid samples were analyzed by RP-HPLC and, as expected from visual inspection, strain 385-$pot1^{(-/-)}$ showed major differences regarding carotenoid composition in relation to the parental and heterozygous mutant strains, with a reduced proportion of astaxanthin and an increase in carotenogenesis intermediaries (Table 2). These results are similar to what has been previously observed in strains in which genes leading to sterol synthesis and carotenogenic pathways precursors were deleted to obtain heterozygous mutants [14]. In those mutants, decreases in the transcript levels of the genes crtS and crtR, which control the synthesis of astaxanthin from beta-carotene, were reduced, which could partially explain their differential carotenoid compositions compared to the wild type strain. Considering this background, total RNA was extracted from each sample, and RT-qPCR analyses were performed to compare the relative amounts of transcripts of the genes ERG10, POT1, crtS and crtR (Fig. 4). In the 385-$pot1^{(+/-)}$ strain, only the POT1 transcript showed significant changes, demonstrating half the amount of transcript when compared to the parental strain UCD 67–385. In the null 385-$pot1^{(-/-)}$ mutant, POT1 transcripts were not detected, confirming that the strain does not have the functional gene. In this strain, the ERG10 transcript level was increased 8-fold compared to the wild type, suggesting that some compensation for POT1 gene loss might be occurring. Although ACAA enzymes (encoded by POT1) are not directly related to the carotenogenic pathway, they might influence substrate availability as they catalyze the final steps in fatty acid β-oxidation, which could alter the acetyl-CoA pool in the peroxisome that can be transported to the mitochondria for energy production through the TCA cycle [27]. Then, if no acetyl-CoA is obtained from peroxisomal fatty acid β-oxidation, more cytoplasmic acetyl-CoA could be transported to the mitochondria; thus, a higher ERG10 transcript level could help to ensure the cytoplasmic acetyl-CoA flux towards the MVA pathway to maintain the sterol production necessary for cell replication. Although no significant changes were found in the crtR transcript level in the mutant 385-$pot1^{(-/-)}$, the

crtS transcript level was reduced 5-fold, which could explain, at least in part, the differences observed in carotenoid composition. The crtS gene encodes a cytochrome P450 enzyme, and it has been shown that in X. dendrorhous, the crtR gene that encodes the cytochrome P450 reductase is essential for the synthesis of astaxanthin in this yeast [13]. Then, considering that two cytochrome P450 enzymes are involved in ergosterol biosynthesis [16, 28], the reduced crtS transcript levels could be related to reduced carotenogenic activity to maintain the flux towards ergosterol biosynthesis.

ERG10 and POT1 gene overexpression in X. dendrorhous

To evaluate if overexpression of the studied genes affects the biosynthesis of secondary metabolites, strains that carry an extra copy of each gene were constructed. cDNAs of genes ERG10 and POT1 were inserted into the plasmid pXdVexp2 [14] to obtain pXdVexp2-cERG10xd and pXdVexp2-cPOT1xd, in which each cDNA is located between the ubiquitin promoter and the GPD terminator from X. dendrorhous. These plasmids also bear a hygromycin B resistance module for transformant selection.

Wild type strain UCD 67–385 was transformed using linearized plasmids to obtain the transformant strains 385-ERG10 and 385-POT1, in which the gene of interest was inserted by homologous recombination into a nonencoding single copy site named int [GenBank: KJ140286] under control of the ubiquitin promoter [14]. A control strain resistant to hygromycin B (385-Vexp2), transformed with the empty plasmid, was also obtained. The correct insertion of each DNA into the target locus was confirmed by PCR.

Relative expression for each gene in the overexpression mutants was compared to control strain 385-Vexp2 to determine if indeed there was an increase in transcript levels caused by the additional copy of each cDNA. RT-qPCR was performed on samples taken after 120 h of growth in liquid YM-hygromycin B media to analyze transcript levels of genes ERG10 and POT1. For strain 385-ERG10, a 3.2-fold ERG10 transcript level increment was observed without any significant changes in the levels of POT1 transcripts. On the other hand, in strain

Table 2 Total sterol and carotenoid content in X. dendrorhous POT1 mutant strains

Strain	UCD 67–385	385-pot1(+/−)	385-pot1(−/−)
Total Sterols (mg/g dry weight)	5.7 ± 0.3	4.6 ± 0.2	5.6 ± 0.6
Total Carotenoids (µg/g dry weight)	207.8 ± 21.5	201.5 ± 3.8	174 ± 19.5
% Astaxanthin	82.0 ± 3.4	78.6 ± 0.7	49.4* ± 2.4
% Beta-carotene	2.3 ± 0.7	2.8 ± 0.1	10.0* ± 0.8
% Phoenicoxanthin	8.0 ± 1.3	9.9 ± 0.4	16.5* ± 0.6
% Other carotenoids	7.7 ± 1.7	8.6 ± 0.3	23.0* ± 2.0

Each value corresponds to the average of three independent samples ± standard deviation. Other carotenoids include hydroxy-keto-γ-carotene, hydroxy-equinenone, hydroxy-keto-torulene and canthaxanthin. *$p < 0.01$, Student's T-test compared to wild type

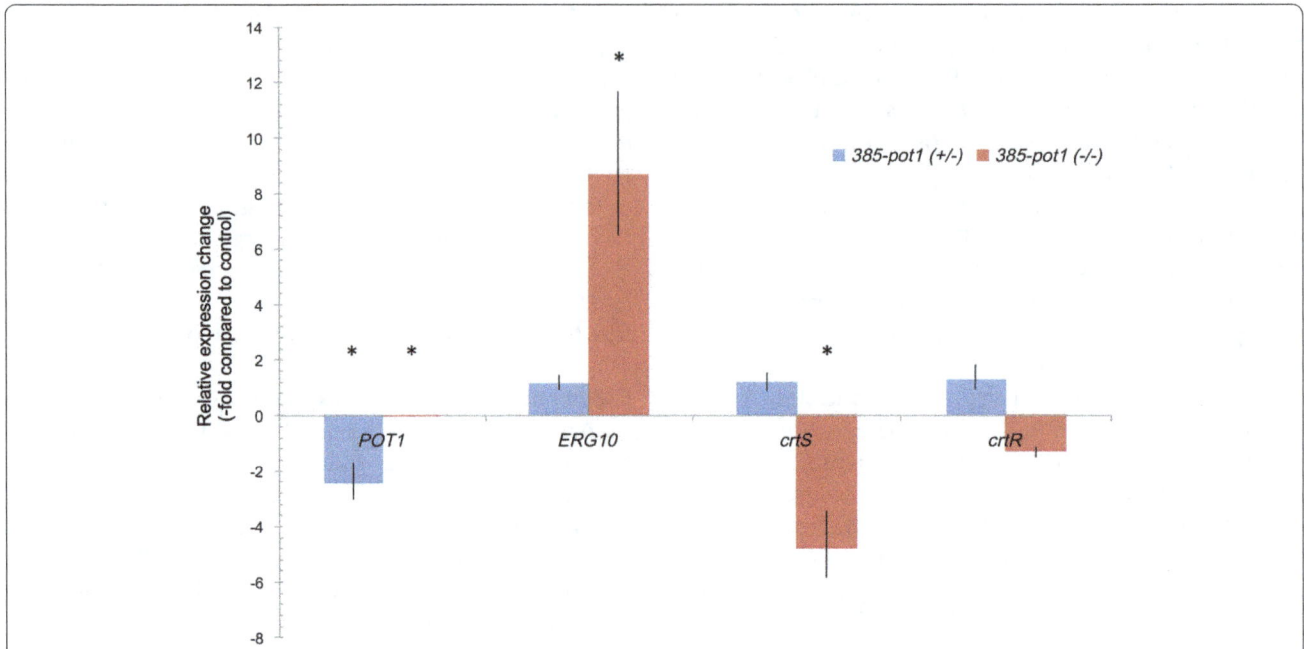

Fig. 4 Changes in transcript levels for *POT1* mutant strains. Relative transcript levels of genes *POT1*, *ERG10*, *crtS* and *crtR* of *X. dendrorhous* were analyzed in samples obtained after 96 h of growth in liquid YM media for strains UCD 67–385, 385-*pot1*$^{(+/-)}$ and 385-*pot1*$^{(-/-)}$. The results were analyzed using the $2^{-\Delta\Delta Ct}$ method using actin as the normalizer gene and UCD 67–385 as the control strain compared to 385-*pot1*$^{(+/-)}$ (blue) and 385-*pot1*$^{(-/-)}$ (red). For *POT1* transcript analysis in strain 385-*pot1*$^{(-/-)}$, no transcript was detected. Each bar represents an average of three independent samples. Black lines indicate standard deviation. *$p < 0.01$

385-*POT1*, a 4.9-fold increase in *POT1* transcript levels as well as no significant changes in *ERG10* transcripts in relation to the control strain was observed. These results confirm that the integration *locus* is appropriate for inserting additional gene copies for overexpression.

Total carotenoids and sterols were extracted from these strains to determine if gene overexpression had any phenotypic effects, as no differences could be observed to the naked eye. As shown in Table 3, the total amounts of carotenoids and sterols among these strains indeed did not demonstrate any significant differences. Previously, it was reported that *ERG10* (named as *acaT*) overexpression by integration in the rDNA using the *gdh* (glutamate

dehydrogenase) promoter led to a slight augmentation in the intracellular astaxanthin content produced by *X. dendrorhous* (approximately 1.3-fold higher than the control strain) [18]. The observed differences in this study may be attributable to several differences in the experimental strategy, including the use of a different integration *locus* [29] and promoter to regulate gene expression [30], gene copy number and the different genetic background of the strain used as the host for gene overexpression.

Our results indicate that the overexpression of either the *ERG10* or *POT1* genes does not lead to significant changes in the amount or composition of carotenoids and sterols in *X. dendrorhous*.

Table 3 Total sterol and carotenoid contents in *X. dendrorhous* thiolase overexpression strains

Strain	UCD 67–385	385-Vexp2	385-*POT1*	385-*ERG10*
Gene overexpression (–fold compared to control)	N/A	1.0 (0.9–1.1)	4.9*(3.7–6.4)	3.2*(2.2–4.7)
Total sterols (mg/g dry weight)	3.5 ± 0.4	3.8 ± 0.1	3.9 ± 0.2	3.9 ± 0.1
Total carotenoids (µg/g dry weight)	359.0 ± 16.3	401.7 ± 26.8	394.1 ± 15.9	349.7 ± 24.8
% Astaxanthin	75.4 ± 1.2	74.4 ± 0.2	68.7 ± 2.0	73.8 ± 1.4
% Beta-carotene	2.8 ± 0.7	2.6 ± 0.1	4.1 ± 0.9	2.8 ± 0.4
% Phoenicoxanthin	11.6 ± 0.3	10.8 ± 0.7	12.2 ± 0.6	10.9 ± 0.6

Each value corresponds to the average value of three independent samples ± standard deviation. Overexpression data are presented as an average with standard deviation in parentheses. The relative expression of the studied genes in 385-Vexp2 was compared to wild type strain UCD 67–385, and the results indicated no significant differences in the relative expression of these genes; therefore, relative expression for each gene in the overexpression mutants was compared to control strain 385-Vexp2. N/A = Not Analyzed. *$p < 0.01$, Student's *T*-test

Conclusions

The *X. dendrorhous* ERG10 gene encodes a functional acetyl-CoA C-acetyltransferase (ACAT) that acts in the first step towards mevalonate production, which was supported by a heterologous complementation assay in *S. cerevisiae*. The provided results suggest this gene may be regulated at a transcriptional level through an ergosterol-dependent mechanism. This observation corresponds with previous observations suggesting the possibility of a feedback regulatory mechanism mediated by ergosterol that regulates synthesis of sterols and carotenoids.

A second thiolase-encoding gene (*POT1*) was identified, and its potential involvement in carotenogenesis was tested by constructing *X. dendrorhous* mutants. According to the results, this second gene is not essential for the viability of the yeast and most likely encodes the *X. dendrorhous* 3-ketoacyl-CoA thiolase, which is important for supplying acetyl-CoA for yeast metabolism through β-oxidation of fatty acids. This study provides the first functional evidence that by altering the acetyl-CoA flux in different pathways and cell compartments, the overall synthesis of carotenoids in *X. dendrorhous* is affected.

Methods

Strains and culture conditions

All strains and plasmids used in this work are shown in Table 1.

S. cerevisiae strains were grown either in YPD media (1% yeast extract, 2% peptone, 2% glucose) or minimal synthetic media SD (0.67% Yeast Nitrogen Base w/o amino acids w/o ammonium sulfate, 0.5% ammonium sulfate, 2% glucose) supplemented with the corresponding amino acids. Geneticin (G418) was used when necessary at a concentration of 50 μg/ml.

X. dendrorhous was grown in YM media (0.3% yeast extract, 0.3% malt extract, 0.5% peptone) supplemented with 1% glucose. Hygromycin B at 15 μg/ml or zeocin at 20 μg/ml were supplemented in YM agar (1.5%) plates for transformant selection. All liquid cultures were grown in orbital shakers at 160 rpm at 22 °C for *X. dendrorhous* and at 30 °C for *S. cerevisiae*.

Plasmid propagation was performed in *Escherichia coli* strain DH5α, which was grown in LB media (1% tryptone, 0.5% yeast extract, 0.5% NaCl). For transformant selection, ampicillin was used at a concentration of 100 μg/ml and X-Gal (5-bromo-4-chloro-3-indolyl-β-D-galactopyranoside) at 80 μg/ml.

Nucleic Acid Extraction

Yeast DNA from *X. dendrorhous* and *S. cerevisiae* was obtained by mechanical cell rupture in 600 μl of TE buffer (25 mM Tris-HCl, 10 mM EDTA, pH 8.0) and 100 μl of 0.5-mm glass beads (BioSpec Products Inc., Bartlesville, OK, USA) using a Mini-beadbeater-16 (BioSpec Products

Inc., Bartlesville, OK, USA) for 1 min, followed by centrifugation for 5 min at 4,000 x g to recover the aqueous phase. DNA was extracted by adding 1 volume of a phenol:chloroform:isoamilic alcohol mixture (25: 24: 1, v/v/v), vortexing and centrifuging for 1 min at 20,000 x g. The aqueous phase was recovered and washed with 1 volume of chloroform:isoamilic alcohol (24: 1, v/v) to remove traces of phenol. DNA was precipitated with 1 ml of cold absolute ethanol and incubated at −20 °C for 1 h. Then, it was centrifuged for 10 min at 20,000 x g, and the supernatant was eliminated. The pellet was allowed to dry at 37 °C for 5 min and then suspended in 50 μl of water.

Total RNA was extracted from 2 to 5 ml of *X. dendrorhous* liquid cultures. Cells were harvested by centrifugation, and RNA was obtained using a RiboPure Yeast RNA purification kit (Life Technologies, Carlsbad, CA, USA).

cDNA synthesis and qPCR

cDNA was synthesized according to the enzyme manufacturer's protocols in a final volume of 20 μl with 5 μg of total RNA as template using M-MLV reverse transcriptase from Invitrogen (Carlsbad, CA, USA).

Quantitative real-time PCR was performed in an Mx3000P real-time PCR system (Agilent, Santa Clara, CA, USA) using the Sensimix SYBR Green I kit (Bioline, London, UK). Each sample was prepared in triplicate in a final volume of 20 μl using 1 μl of reverse transcription product, 0.25 μM of each primer and 10 μl of kit reaction mixture. To normalize Ct values, the corresponding value for the actin gene of *X. dendrorhous* [GenBank: X89898.1] was employed. In each experiment, the control condition was defined as the untreated culture or the wild type strain. Data were analyzed using the $2^{-\Delta\Delta Ct}$ method, giving an asymmetric distribution of standard deviation values due to the conversion of a exponential process to a linear comparison [31]. Decimal values obtained were converted to negative values according to [31].

S. cerevisiae heterologous complementation assays

All primers used in this work are listed in Additional file 4: Table S1 and were purchased from Integrated DNA Technologies (Coralville, IA, USA). PCR was performed with *Pfu* DNA polymerase (Promega, Madison, WI, USA) using 1 μl of total *X. dendrorhous* RT reaction obtained from the RNA of strain UCD 67–385 to obtain the cDNAs corresponding to each gene. The amplified products were inserted into the plasmid YEp-NP [28]; this plasmid corresponds to a modified YEp-ACT4 plasmid [32] in which a sequence corresponding to 300 bp downstream of the *TDH3* gene of *S. cerevisiae* was inserted into the *Hin*dIII recognition site of the plasmid (Additional file 5: Figure S4, A). Gene orientation in the plasmid was confirmed by colony PCR in a reaction mixture of 25 µl containing 2 U of *Taq* DNA polymerase, *Taq* buffer, 0.2 mM dNTPs, 2 mM $MgCl_2$ and 1 µM of each primer.

S. cerevisiae was transformed by electroporation. To prepare electrocompetent cells, a 30-ml liquid culture of *S. cerevisiae* that was grown in YPD media overnight at 22 °C was diluted by adding 30 ml of YPD media and incubated for 3 h. Cells at the exponential phase of growth were collected by centrifuging for 5 min at 4,000 x g and washing three times with 40 ml of distilled sterile water. The last wash was performed with 5 ml of 1 M sorbitol. Finally, cells were suspended in 0.2 ml of 1 M sorbitol and divided in 40-µl aliquots, which were stored at 4 °C until use. Before electroporation, 4 µl of plasmid DNA was added to the cells, and the mixture was placed in an electroporation cuvette. The electric pulse conditions were 1.5 kV, 25 µF, and 200 Ω using a GenePulser Xcell ™ (BioRad, Hercules, CA, USA). Cells were suspended in 1 ml of YEP media and incubated at 30 °C for 2 h. Finally, cells were plated on SD plates supplemented with 0.002% uracil and 1% histidine.

For sporulation induction, *S. cerevisiae* transformants were streaked on pre-sporulation agar plates (5% glucose, 0.8% yeast extract, 0.3% peptone, 2% agar) and incubated for 2 days at 30 °C. Cells were transferred to a sporulation agar plate (1% potassium acetate, 2% agar) and incubated at 22 °C until asci were observed by optical microscope (3 to 5 days). To recover the asci from the plate, 1 ml of sterile water was added onto the plate surface, cells were removed using a standard spreader and recovered into an Eppendorf tube. Ascospore isolation was performed using a combined diethyl ether and zymolyase treatment according to [33]. The cell suspension was spread on YEP agar plates supplemented with G418 to confirm the presence of the *S. cerevisiae* erg10⁻ mutant allele. Randomly selected colonies were replica plated on SD agar plates supplemented with uracil and histidine, or with uracil, histidine, lysine and methionine, to sustain all possible auxotrophies in the resulting haploid strains. Then, haploid strains were selected according to their methionine and/or lysine auxotrophy, both of which are heterozygous markers in the parental diploid strain, as auxotrophic strains because these amino acids should be haploid.

Plasmid construction for *X. dendrorhous* mutation

All plasmids used in this work are presented in Table 1. To eliminate the *ERG10* (Ex: *ERG10A*) and *POT1* (Ex: *ERG10B*) genes from the *X. dendrorhous* genome, the plasmids pPHT-ERG10xd and pPHT-POT1xd were constructed (Additional file 5: Figure S4, B). For this, the upstream and downstream regions of each gene of interest were PCR-amplified using genomic DNA from *X.*

dendrorhous strain UCD 67–385 as a template. Then, the upstream region was attached to the downstream region using overlap extension PCR [34], introducing a *Sma*I recognition site between them according to the primer design. The fragment was inserted in pBluescript SK at a *Sma*I site to obtain plasmids pBS-PT-ERG10xd and pBS-PT-POT1xd. Each plasmid was digested with *Sma*I to insert either a hygromycin B or zeocin resistance module, which were obtained via digestion from plasmids pMN-*hph* [12] or pIR-zeo [16], respectively. To overexpress the *ERG10* and *POT1* genes in *X. dendrorhous*, plasmids pXdVexp2-cERG10xd and pXdVexp2-cPOT1xd were constructed (Additional file 5: Figure S4, C). The *X. dendrorhous* ERG10 and POT1 cDNAs were amplified from plasmids YEpNPc10xd and YEpNPcPOT1xd, respectively, and were independently inserted at the *Hpa*I site of pXdVexp2 [14].

X. dendrorhous electrocompetent cells were prepared from an exponentially growing culture at an O.D.$_{600}$ of 2.0 according to [35]. Electroporation was performed using a GenePulser Xcell ™ (BioRad, Hercules, CA, USA), employing a square wave protocol of 5 pulses of 10 ms of 450 V each with 2 ms rest between pulses. For each transformation, 5 µg of linear DNA was used.

Sterol and carotenoid extraction and RP-HPLC analysis

Sterols and carotenoids were extracted, quantified spectrophotometrically and normalized to the dry weight of the yeast.

Sterols were extracted according to [36]. Briefly, cell pellets were saponified adding 4 g of KOH and 16 ml of 60% (v/v) ethanol/water at 80 °C for 2 h. Then, the mixture was cooled and sterols were extracted using 10 ml of petroleum ether. Total sterols were quantified spectrophometrically at 280 nm using an absorption coefficient of $A_{1\%}$ = 11,500. Ether was evaporated and sterols were suspended in 200 µl of methanol and separated by RP-HPLC using a C-18 column and methanol:water (97: 3, v/v) as the mobile phase with a 1 ml/min flux under isocratic conditions. Sterols were visualized with the 280 nm channel, and the elution spectra were recovered using a diode array detector. Standard ergosterol was acquired from Sigma (Saint Louis, MI, USA).

Carotenoids were extracted following [37]. Cell pellets were disrupted using glass beads and acetone for carotenoid extraction. Carotenoids were quantified spectrophotometrically at 465 nm using an absorption coefficient of $A_{1\%}$ = 2,100. For carotenoid identification, samples were run in an RP-HPLC using a C-18 column and acetonitrile: methanol:isopropanol (85: 10: 5, v/v/v) as the mobile phase with a 1 ml/min flux under isocratic conditions and compared to standards according to retention time and absorption spectra.

Additional files

Additional file 1: Figure S1. Protein sequence alignment used for phylogenetic tree construction. The deduced protein sequences for *ERG10A* and *ERG10B* from *X. dendrorhous* were aligned against sequences of diverse thiolases using ClustalW 2.1 and visualized in UGENE. Fully conserved residues are colored in blue. Mitochondrial thiolase/acetyl-CoA C-acetyltransferase: *B. taurus* [NP_001039540.1], *H. sapiens* [BAA01387.1], *C. lupus familiaris* [XP_546539.2], *R. norvegicus* [NP_058771.2], *G. gallus* [NP_001264708.1]. Cytoplasmic thiolase/acetyl-CoA C-acetyltransferase: *S. pombe* [Q9UQW6.1], *B. graminis f. sp. tritici 96224* [EPQ61678.1], *S. cerevisiae* [P41338.3], *N. tabacum* [AAU95618.1], *Z. mays* [NP_001266315.1], *A. thaliana* [Q9FIK7.1]. Mitochondrial thiolase/3-ketoacyl-CoA thiolase: *B. taurus* [NP_001030419.1], *M. musculus* [NP_803421.1], *R. norvegicus* [NP_569117.1]. Peroxisomal thiolase/3-ketoacyl-CoA thiolase: *S. cerevisiae* [CAA37472.1], *Y. lipolytica* [Q05493.1], *G. gallus* [NP_001184217.1], *M. musculus* [NP_570934.1], *E. caballus* [XP_001488609.1], *C. lupus familiaris* [XP_534222.2], *B. taurus* [NP_001029491.1], *H. sapiens* [NP_001598.1]. *X. dendrorhous* [*ERG10A*]: thiolase encoded by the *ERG10A* gene. *X. dendrorhous* [*ERG10B*]: thiolase encoded by the *ERG10B* gene.

Additional file 2: Figure S2. Pathways involving genes studied in this work. A schematic representation of (A) Sterol and carotenoid synthesis pathways and (B) β-oxidation of fatty acids is shown. The mevalonate pathway is highlighted inside the dashed rectangle. Steps catalyzed by the enzymes encoded by the genes studied in this work (*ERG10* and *POT1*) are highlighted in red.

Additional file 3: Figure S3. *S. cerevisiae* complementation strain growth curves. Strains Sc-*ERG10/erg10* (blue), Sc-ERG10sc (red) and Sc-cERG10xd (green) were grown in YEP media for 48 h with constant agitation at 30 °C. Each point represents the average of three independent cultures. Black bars indicate standard deviation.

Additional file 4: Table S1. Primers designed and used for this work.

Additional file 5: Figure S4. Scheme of plasmids constructed in this work. Schematic representation of plasmids used for yeast transformation. Selection markers are represented by blue arrows, promoters and terminators, by white arrows; yeast genomic sequences used as a platform for homologous recombination are shown in green and cDNA sequences that correspond to the gene of interest (GOI) are represented by red arrows. Plasmid size in bp is shown considering *POT1* as the GOI. (A) Representation of plasmid YEp-NP, used for heterologous complementation assays, where the cDNA of the gene GOI is represented by a red arrow. (B) Representation of plasmid pPHT-GOI, used to obtain DNA fragments to mutate *X. dendrorhous*. Green arrows represent the positions of the upstream and downstream regions of the GOI. (C) Representation of plasmid pXdVexp2 used for genomic insertion of genes for overexpression in *X. dendrorhous*. Abbreviations; Hygromycin resistance gene (HygR), ampicillin resistance gene (AmpR), *S. cerevisiae LEU2* gene (LEU2).

Abbreviations
ACAA: 3-ketoacyl-CoA thiolase; ACAT: Acetyl-CoA C-acetyltransferase; ATCC: American type culture collection; DRM: Double recombinant method; HMG-CoA: 3-hydroxy-3-methylglutaryl-CoA; HMGR: HMG-CoA reductase; HMGS: HMG-CoA synthase; IPP: Isopentenyl pyrophosphate; MEP: Methyl-D-erythritol-4-phosphate; MVA: Mevalonate; SREBP: Sterol regulatory element binding protein

Funding
FONDECYT 11121200 and FONDECYT 1160202 financially supported this work. NW was supported by a CONICYT fellowship 21110701.

Authors' contributions
NW and MG carried out the bioinformatic analyses. NW performed expression analyses, *S. cerevisiae* complementation assays and *X. dendrorhous* mutant transformation and analyses. MG performed transformation and analyses for gene overexpression studies. MB and VC contributed to the study design and analysis of the results. JA conceived the study and participated in the experimental design and coordination. NW and JA drafted the manuscript. All authors read and approved the final manuscript.

Competing interests
The authors declare that they have no competing interests.

References
1. Chemler JA, Yan Y, Koffas MA. Biosynthesis of isoprenoids, polyunsaturated fatty acids and flavonoids in *Saccharomyces cerevisiae*. Microb Cell Fact. 2006;5:20.
2. Hunter WN. The non-mevalonate pathway of isoprenoid precursor biosynthesis. J Biol Chem. 2007;282:21573–7.
3. Fox AR, Soto G, Mozzicafreddo M, Garcia AN, Cuccioloni M, Angeletti M, Salerno JC, Ayub ND. Understanding the function of bacterial and eukaryotic thiolases II by integrating evolutionary and functional approaches. Gene. 2014;533:5–10.
4. Haapalainen AM, Meriläinen G, Wierenga RK. The thiolase superfamily: condensing enzymes with diverse reaction specificities. Trends Biochem Sci. 2006;31:64–71.
5. Jiang C, Kim SY, Suh DY. Divergent evolution of the thiolase superfamily and chalcone synthase family. Mol Phylogenet Evol. 2008;49:691–701.
6. Peretó J, López-García P, Moreira D. Phylogenetic analysis of eukaryotic thiolases suggests multiple proteobacterial origins. J Mol Evol. 2005;61:65–74.
7. Wentzinger L, Gerber E, Bach TJ, Hartmann M-A. Occurrence of Two Acetoacetyl-Coenzyme A Thiolases with Distinct Expression Patterns and Subcellular Localization in Tobacco. In: Bach TJ, Rohmer M, editors. Isoprenoid Synthesis in Plants and Microorganisms. New York: Springer New York; 2013. p. 347–65.
8. Goldstein JL, Brown MS. Regulation of the mevalonate pathway. Nature. 1990;343:425–30.
9. Basson ME, Thorsness M, Rine J. *Saccharomyces cerevisiae* contains two functional genes encoding 3-hydroxy-3-methylglutaryl-coenzyme A reductase. Proc Natl Acad Sci. 1986;83:5563–7.
10. Hiser L, Basson ME, Rine J. *ERG10* from *Saccharomyces cerevisiae* encodes acetoacetyl-CoA thiolase. J Biol Chem. 1994;269:31383–9.
11. Miziorko HM. Enzymes of the mevalonate pathway of isoprenoid biosynthesis. Arch Biochem Biophys. 2011;505:131–43.
12. Niklitschek M, Alcaíno J, Barahona S, Sepulveda D, Lozano C, Carmona M, Marcoleta A, Martinez C, Lodato P, Baeza M, Cifuentes V. Genomic organization of the structural genes controlling the astaxanthin biosynthesis pathway of *Xanthophyllomyces dendrorhous*. Biol Res. 2008;41:93–108.
13. Alcaíno J, Barahona S, Carmona M, Lozano C, Marcoleta A, Niklitschek M, Sepulveda D, Baeza M, Cifuentes V. Cloning of the cytochrome p450 reductase (*crtR*) gene and its involvement in the astaxanthin biosynthesis of *Xanthophyllomyces dendrorhous*. BMC Microbiol. 2008;8:169.
14. Alcaíno J, Romero I, Niklitschek M, Sepúlveda D, Rojas MC, Baeza M, Cifuentes V. Functional characterization of the *Xanthophyllomyces dendrorhous* farnesyl pyrophosphate synthase and geranylgeranyl pyrophosphate synthase encoding genes that are involved in the synthesis of isoprenoid precursors. PLoS One. 2014;9:e96626.
15. Calo P, de Miguel T, Velázquez JB, Villa TG. Mevalonic acid increases trans-astaxanthin and carotenoid biosynthesis in *Phaffia rhodozyma*. Biotechnol Lett. 1995;17:575–8.
16. Loto I, Gutierrez MS, Barahona S, Sepulveda D, Martinez-Moya P, Baeza M, Cifuentes V, Alcaíno J. Enhancement of carotenoid production by disrupting

the C22-sterol desaturase gene (*CYP61*) in *Xanthophyllomyces dendrorhous*. BMC Microbiol. 2012;12:235.

17. Sharma R, Gassel S, Steiger S, Xia X, Bauer R, Sandmann G, Thines M. The genome of the basal agaricomycete *Xanthophyllomyces dendrorhous* provides insights into the organization of its acetyl-CoA derived pathways and the evolution of Agaricomycotina. BMC Genomics. 2015;16:233.

18. Hara KY, Morita T, Mochizuki M, Yamamoto K, Ogino C, Araki M, Kondo A. Development of a multi-gene expression system in *Xanthophyllomyces dendrorhous*. Microb Cell Fact. 2014;13:175.

19. Baeza M, Alcaíno J, Barahona S, Sepúlveda D, Cifuentes V. Codon usage and codon context bias in *Xanthophyllomyces dendrorhous*. BMC Genomics. 2015;16:293.

20. Tamura K, Stecher G, Peterson D, Filipski A, Kumar S. MEGA6: Molecular Evolutionary Genetics Analysis version 6.0. Mol Biol Evol. 2013;30:2725–9.

21. Bien C, Espenshade P. Sterol Regulatory Element Binding Proteins in Fungi: Hypoxic Transcription Factors Linked to Pathogenesis. Eukaryot Cell. 2010;9:352–9.

22. Marcoleta A, Niklitschek M, Wozniak A, Lozano C, Alcaíno J, Baeza M, Cifuentes V. Glucose and ethanol-dependent transcriptional regulation of the astaxanthin biosynthesis pathway in *Xanthophyllomyces dendrorhous*. BMC Microbiol. 2011;11:190.

23. Alcaíno J, Bravo N, Córdova P, Marcoleta AE, Contreras G, Barahona S, Sepúlveda D, Fernández-Lobato M, Baeza M, Cifuentes V. The involvement of Mig1 from *Xanthophyllomyces dendrorhous* in catabolic repression: an active mechanism contributing to the regulation of carotenoid production. PLoS One. 2016;11:e0162838.

24. Einerhand AW, Kos W, Smart WC, Kal AJ, Tabak HF, Cooper TG. The upstream region of the *FOX3* gene encoding peroxisomal 3-oxoacyl-coenzyme A thiolase in *Saccharomyces cerevisiae* contains ABF1-and replication protein A-binding sites that participate in its regulation by glucose repression. Mol Cell Biol. 1995;15:3405–14.

25. Igual JC, Matallana E, Gonzalez Bosch C, Franco L, Pérez Ortin JE. A new glucose repressible gene identified from the analysis of chromatin structure in deletion mutants of yeast *SUC2* locus. Yeast. 1991;7:379–89.

26. Niklitschek M, Baeza M, Fernández-Lobato M, Cifuentes V. Generation of astaxanthin mutants in Xanthophyllomyces dendrorhous using a double recombination method based on hygromycin resistance. In: Barredo JL, editor. Microbial Carotenoids From Fungi - Methods and Protocols. Heidelberg: Springer; 2012. p. 219–34.

27. Tang X, Lee J, Chen WN. Engineering the fatty acid metabolic pathway in *Saccharomyces cerevisiae* for advanced biofuel production. Metab Eng Commun. 2015;2:58–66.

28. Leiva K, Werner N, Sepulveda D, Barahona S, Baeza M, Cifuentes V, Alcaíno J. Identification and functional characterization of the *CYP51* gene from the yeast *Xanthophyllomyces dendrorhous* that is involved in ergosterol biosynthesis. BMC Microbiol. 2015;15:89.

29. Wery J, Gutker D, Renniers ACHM, Verdoes JC, van Ooyen AJJ. High copy number integration into the ribosomal DNA of the yeast *Phaffia rhodozyma*. Gene. 1997;184:89–97.

30. Hara KY, Morita T, Endo Y, Mochizuki M, Araki M, Kondo A. Evaluation and screening of efficient promoters to improve astaxanthin production in *Xanthophyllomyces dendrorhous*. Appl Microbiol Biotechnol. 2014;98:6787–93.

31. Schmittgen TD, Livak KJ. Analyzing real-time PCR data by the comparative CT method. Nat Protoc. 2008;3:1101–8.

32. Sánchez-Torres P, González-Candelas L, Ramón D. Heterologous expression of a *Candida molischiana* anthocyanin-β-glucosidase in a wine yeast strain. J Agric Food Chem. 1998;46:354–60.

33. Bahalul M, Kaneti G, Ysff K. Ether-zymolyase ascospore isolation procedure: an efficient protocol for ascospores isolation in *Saccharomyces cerevisiae* yeast. Yeast. 2010;27:999–1003.

34. Ho SN, Hunt HD, Horton RM, Pullen JK, Pease LR. Mevalonic acid increases trans-astaxanthin and carotenoid biosynthesis in *Phaffia rhodozyma*. Biotechnol Lett. 1989;77:51–9.

35. Adrio JL, Veiga M. Transformation of the astaxanthin-producing yeast *Phaffia rhodozyma*. Biotechnol Tech. 1995;9:509–12.

36. Shang F, Wen S, Wang X, Tan T. Effect of nitrogen limitation on the ergosterol production by fed-batch culture of *Saccharomyces cerevisiae*. J Biotechnol. 2006;122:285–92.

37. An G-H, Schuman DB, Johnson EA. Isolation of *Phaffia rhodozyma* mutants with increased astaxanthin content. Appl Environ Microbiol. 1989;55:116–24.

38. Mortimer RK, Johnston JR. Genealogy of principal strains of the yeast genetic stock center. Genetics. 1986;113:35–43.

39. Giaever G, Chu AM, Ni L, Connelly C, Riles L, Véronneau S, Dow S, Lucau-Danila A, Anderson K, André B, Arkin AP, Astromoff A, El-Bakkoury M, Bangham R, Benito R, Brachat S, Campanaro S, Curtiss M, Davis K, Deutschbauer A, Entian KD, Flaherty P, Foury F, Garfinkel DJ, Gerstein M, Gotte D, Güldener U, Hegemann JH, Hempel S, Herman Z, Jaramillo DF, Kelly DE, Kelly SL, Kötter P, LaBonte D, Lamb DC, Lan N, Liang H, Liao H, Liu L, Luo C, Lussier M, Mao R, Menard P, Ooi SL, Revuelta JL, Roberts CJ, Rose M, Ross-Macdonald P, Scherens B, Schimmack G, Shafer B, Shoemaker DD, Sookhai-Mahadeo S, Storms RK, Strathern JN, Valle G, Voet M, Volckaert G, Wang CY, Ward TR, Wilhelmy J, Winzeler EA, Yang Y, Yen G, Youngman E, Yu K, Bussey H, Boeke JD, Snyder M, Philippsen P, Davis RW, Johnston M. Functional profiling of the *Saccharomyces cerevisiae* genome. Nature. 2002;418:387–91.

18

Occurrence, diversity and community structure of culturable atrazine degraders in industrial and agricultural soils exposed to the herbicide in Shandong Province, P.R. China

Dmitry P. Bazhanov[1*†], Chengyun Li[2†], Hongmei Li[2], Jishun Li[2], Xinjian Zhang[1], Xiangfeng Chen[3] and Hetong Yang[2]

Abstract

Background: Soil populations of bacteria rapidly degrading atrazine are critical to the environmental fate of the herbicide. An enrichment bias from the routine isolation procedure prevents studying the diversity of atrazine degraders. In the present work, we analyzed the occurrence, diversity and community structure of soil atrazine-degrading bacteria based on their direct isolation.

Methods: Atrazine-degrading bacteria were isolated by direct plating on a specially developed SM agar. The atrazine degradation genes *trzN* and *atzABC* were detected by multiplex PCR. The diversity of atrazine degraders was characterized by enterobacterial repetitive intergenic consensus-PCR (ERIC-PCR) genotyping followed by 16S rRNA gene phylogenetic analysis. The occurrence of atrazine-degrading bacteria was also assessed by conventional PCR targeting *trzN* and *atzABC* in soil DNA.

Results: A total of 116 atrazine-degrading isolates were recovered from bulk and rhizosphere soils sampled near an atrazine factory and from geographically distant maize fields. Fifteen genotypes were distinguished among 56 industrial isolates, with 13 of them representing eight phylogenetic groups of the genus *Arthrobacter*. The remaining two were closely related to *Pseudomonas alcaliphila* and *Gulosibacter molinativorax* and constituted major components of the atrazine-degrading community in the most heavily contaminated industrial plantless soil. All isolates from the adjacent sites inhabited by cogon grass or common reed were various *Arthrobacter* spp. with a strong prevalence of *A. aurescens* group. Only three genotypes were distinguished among 60 agricultural strains. Genetically similar *Arthrobacter ureafaciens* bacteria which occurred as minor inhabitants of cogon grass roots in the industrial soil were ubiquitous and predominant atrazine degraders in the maize rhizosphere. The other two genotypes represented two distant *Nocardioides* spp. that were specific to their geographic origins.

(Continued on next page)

* Correspondence: bazhdp@outlook.com
†Equal contributors
[1]Key Laboratory for Applied Microbiology of Shandong Province, Ecology Institute (Biotechnology Center) of Shandong Academy of Sciences, Jinan, Shandong Province, People's Republic of China
Full list of author information is available at the end of the article

(Continued from previous page)

Conclusions: Direct plating on SM agar enabled rapid isolation of atrazine-degrading bacteria and analysis of their natural diversity in soil. The results obtained provided evidence that contaminated soils harbored communities of genetically distinct bacteria capable of individually degrading and utilizing atrazine. The community structures of culturable atrazine degraders were habitat-specific. Bacteria belonging to the genus *Arthrobacter* were the predominant degraders of atrazine in the plant rhizosphere.

Keywords: Atrazine-degrading bacteria, Diversity, Soil bacterial communities, *Arthrobacter*, *Gulosibacter*, *Nocardioides*, *Pseudomonas*, *trzN*, *atz* genes

Background

First registered in Switzerland and the United States in 1958, atrazine has soon become one of the world's best-selling and heavily applied herbicides [1]. Nowadays, atrazine is a commonly detected contaminant of soils, underground and surface streams and basins [2–9]. Research performed during the first three decades of atrazine application indicated that the herbicide, like other chlorinated *s*-triazines, was poorly biodegradable through N-dealkylation of side chains by microbial hydrolases with low specific activities and subsequent dechlorination of the intermediates [10]. Atrazine was considered moderately persistent in soils; with a half-life estimate of about 2 months [11, 12]. However, atrazine residues and metabolites are detectable in soil for years [11, 13–15] and even decades [16] after the herbicide application.

Since the mid-nineties of the 20th century, a rapid degradation of atrazine has been revealed in soils continuously exposed to the herbicide at various geographical locations [15]. The half-life of atrazine in such adapted soils can be as short as 1–3.5 days [17], causing reduced efficacy of weed control. From the ecological point of view, the enhanced degradation substantially decreases the harmful consequences of atrazine application by reducing both its conversion to stable dealkylated metabolites and the leaching of atrazine and its metabolites to deep soil horizons [15].

The enhanced atrazine degradation in soils was linked with the abundance of bacteria that had acquired novel metabolic abilities [17, 18] first discovered in *Pseudomonas* sp. strain ADP [19]. This bacterium mineralized atrazine through its dechlorination and further conversion of hydroxyatrazine to cyanuric acid by cleavage of the alkylamino side chains [20]. The genes *atzABC*, coding enzymes for the 3-step conversion were found to be highly conserved in phylogenetically distant bacteria [21]. While AtzB and AtzC are still the only known hydrolases for the specific transformation of hydroxyatrazine to cyanuric acid, an alternative, coded by the gene *trzN*, atrazine chlorohydrolase has been discovered in *Nocardioides* sp.190 [22] and later detected in some other bacteria [15]. Since cyanuric acid has been found to be readily degraded by a large number of microorganisms under a wide variety of natural conditions

[10, 23, 24], it is arguable that the presence of bacteria harboring functional genes *atzABC* or *trzN* and *atzBC* is critical for the enhanced mineralization of atrazine in soils.

According to the review of Krutz et al. [15], atrazine-utilizing bacteria bearing the genes *atzA* or *trzN* occur worldwide (except Antarctica) and belong to four genera within the phylum *Actinobacteria* (*Arthrobacter*, *Clavibacter*, *Nocardia*, *Nocardioides*) and ten genera within the phylum *Proteobacteria* (*Agrobacterium*, *Alcaligenes*, *Herbaspirillum*, *Pseudaminobacter*, *Pseudomonas*, *Polaromonas*, *Ralstonia*, *Rhizobium*, *Sinorhizobium*, *Stenotrophomonas*). Among the listed bacteria, several *Arthrobacter* sp. strains [25–28] and *Pseudomonas* sp. AD39 [28] were isolated from sewage systems or heavily polluted soils at atrazine manufacturing plants in China. Similar atrazine degraders were later isolated from Chinese agricultural soils at several geographically distant locations, and all the taxonomically characterized agricultural isolates, except *Pseudomonas stutzeri* SA1 [29], belonged to the genus *Arthrobacter* [26, 30–34]. Such a relatively narrow taxonomic range of atrazine degraders isolated in China might be explained by the fact that their isolation targeted the selection of highly effective strains for treatment of wastewaters or contaminated soils rather than the analysis of atrazine-degrading microbial communities. The geographical occurrence, abundance and diversity of atrazine degrading bacteria remained poorly characterized. Among the known atrazine degrading bacteria, including the Chinese strains, a vast majority was isolated after procedures of repeated subculturing in media containing atrazine as a sole nitrogen or carbon source, principally similar to those first described by Mandelbaum et al. [35]. Selection of the isolates during the enrichment and probable elimination of many other atrazine-degrading members of microbial communities prevented studying the natural diversity and community structure of atrazine degraders in soils. Also, the isolation results may be affected by horizontal transfer of atrazine degradation genes during repeated enrichment. Owing to the genetic rearrangements, atrazine-mineralizing strains can be recovered even from the mixed cultures originated from soils which harbor no bacteria utilizing atrazine individually [19].

Table 1 Characteristics of sampling sites

Sites	Geographic coordinates	Atrazine history[a]	Plants/Crop rotation	Sowing dates[a]	Atrazine treatment date[a]	Atrazine (active ingredient) rate[a, b], g ha^{-1}	Atrazine residues, μg kg^{-1}	Sampling date	Plant growth stage
S1	N: 36.66361 E: 117.26151	No treatments for ≥ 20 years	Miscellaneous wild herbs	n/a	n/a	n/a	0.21 ± 0.03	April 08, 2013	dormancy
D3	N: 37.12418 E: 119.07188	7 years, industrial contamination	Plantless soil	n/a	n/a	n/a	2091.3 ± 60.0	May 29, 2013	n/a
D5	N: 37.12544 E: 119.07162	7 years, industrial contamination	Cogon grass (*Imperata cylindrica* (Linn.) Beauv.)	n/a	n/a	n/a	NA[c]	May 29, 2013	3–5 leaves
D6	N: 37.12539 E: 119.07087	7 years, industrial contamination	Common reed (*Phragmites australis* (Cav.) Trin. ex Steud.)	n/a	n/a	n/a	1.05 ± 0.13	May 29, 2013	2–3 leaves
TD(a)	N: 35.96921 E: 117.06004	≥20 years	Maize/ annual rotation with winter wheat	June 05–10	June 25–30	780	0.43 ± 0.06	July 22, 2013	9 leaves
TD(b)	N: 35.96980 E: 117.05999	≥20 years	Maize/ annual rotation with winter wheat	June 01–05	June 20–25	780	TA[c]	July 22, 2013	11 leaves
DnW	N: 35.07492 E: 115.64059	≥5 years	Maize/ annual rotation with winter wheat	June 01–05	June 20–25	450	0.65 ± 0.08	July 22, 2013	10 leaves
DnL	N: 35.09441 E: 115.66002	≥5 years	Maize/ annual rotation with winter wheat	June 01–05	June 20–35	450	0.08 ± 0.02	July 23, 2013	11 leaves
GD	N: 36.85381 E: 116.37587	≥10 years	Maize/ annual rotation with winter wheat	June 18	July 04–05	950	TA	August 14, 2013	pollen shed beginning
WS	N: 37.23022 E: 116.11851	≥15 years	Maize/ annual rotation with winter wheat	June 18	June 20	540	0.10 ± 0.02	August 14, 2013	pollen shed beginning

[a]The information was received from specialists of the local agriculture bureaus, managers of the farms, atrazine factory and academy campus

[b]In 2013

[c]NA not analyzed, TA trace amount (0.01 < TA < 0.05 μg kg^{-1}), n/a not applicable

Shandong Province has been a region of intensive atrazine application for more than 20 years. Moreover, atrazine factories of several companies are located in Shandong Province, causing soil exposure to the herbicide of various rates, duration and periodicity, thus favoring the differential development of atrazine-degrading microbial communities in the surrounding area. The present work aimed to analyze the occurrence, phylogenetic diversity and community structure of atrazine-degrading bacteria in industrially contaminated and agricultural soils in Shandong Province based on their isolation by direct plating on a selective agar medium. Additionally, the occurrence of atrazine degraders was assessed by a direct detection of the genes for enhanced atrazine degradation in soil DNAs by conventional PCR.

Results and discussion
Soil properties
Soil samples were collected near an atrazine factory in Weifang Prefecture and from maize fields in four prefectures of Shandong Province during spring and summer of 2013 (Table 1, Additional file 1: Figure S1). Additionally, soil not previously treated with atrazine was collected from non-arable hillside area at the Eastern Campus of Shandong Academy of Sciences, Jinan. The detailed soil characteristics were analyzed in the Laboratory of Environmental Analysis of the Shandong Provincial Analysis and Test Center (Additional file 2: Table S1). Among three industrial sites, a high concentration of chloride (3200 ± 5 mg kg^{-1}), increased contents of available nitrogen (437 ± 22 mg kg^{-1}) and sodium (942 ± 94 mg kg^{-1}), and the highest atrazine concentration ($2091.3 \pm 60.$-0 μg kg^{-1}, Table 1) were detected in D3 soil, sampled near a control well of the factory wastewater pipeline. In soil taken at D6 site, a distance of about 150 m from D3, the atrazine content was nearly 2000 times lower (Table 1). All agricultural soils had a history of atrazine use for at least 5 years and received atrazine treatment from 3 to

7 weeks before sampling dates (Table 1). Low or trace amounts of atrazine were found in all agricultural soils tested and, surprisingly, in the "wild" S1 soil.

Detection, enumeration and isolation of atrazine-degrading bacteria by direct plating
The enrichment procedure [35] was believed to be essential for isolation of atrazine-degraders from soil [15]. The analyses of the literature allowed us to find four examples of a direct isolation of atrazine-utilizing bacteria [26, 36–38]. In three of the cases, soils used as isolation sources were previously exposed to extremely high (29 g kg^{-1} of soil [36]) or moderate (1.5 mg kg^{-1} [37, 38]) doses of atrazine, that could be deemed an alternative method of enrichment. Three-week incubation was required before colonies with clearing halos indicating degradation of atrazine were observed [26, 37, 38]. It was supposed that selective advantage for culturable atrazine-degrading bacteria in all isolation media used so far was reduced due to the hydrophobicity of atrazine. Therefore, Tween 80 was included in SM agar aiming to improve atrazine bioavailability. Addition of the surfactant provided perfect visual homogeneity to atrazine suspension in SM agar and enhanced colony growth of the reference atrazine-degrading strain *Arthrobacter* sp. SD41.

The occurrence of atrazine degraders was studied in industrial and agricultural soils that were differentially exposed to atrazine. The maize fields were normally treated with a low atrazine rate (Table 1) only once a year, therefore low densities of atrazine degraders were expected. Krutz et al. [18] demonstrated that the rhizosphere population of atrazine degraders was nearly twice that in a bulk soil. In order to increase the probability of detection the occurrence of atrazine-utilizing bacteria was analyzed in rhizosphere fractions of the agricultural soils sampled.

Direct plating on SM agar revealed atrazine-degrading bacteria in all samples of industrially contaminated and

Table 2 Detection and enumeration of atrazine-degrading bacteria in soils

Sampling site	Soil[a]	Min-max population densities, CFU g^{-1} of dry soil		Percentage of atrazine degraders in the culturable population
		Culturable bacteria on TY agar	Atrazine-degrading bacteria on SM agar	
D3	I, B	5.8×10^8–6.4×10^8	2.9×10^7–3.5×10^7	4.5–6.0 %
D5	I, B	8.2×10^6–8.6×10^6	1.6×10^6–2.4×10^6	18.6–29.3 %
	I, R	8.8×10^7–1.2×10^8	1.1×10^7–2.9×10^7	9.1–33.0 %
D6	I, B	1.6×10^7–2.0×10^7	5.7×10^4–7.4×10^4	0.29–0.46 %
	I, R	6.1×10^8–9.9×10^8	3.6×10^5–2.2×10^6	0.05–0.23 %
TD(a)	A, R	6.5×10^8–9.9×10^8	1.3×10^3–1.3×10^4	0.0002–0.0014 %
TD(b)	A, R	NE[b]	1.8×10^3–5.2×10^3	NE
DnW	A, R	NE	1.3×10^3–2.8×10^3	NE
DnL	A, R	NE	8.0×10^2–4.5×10^3	NE
GD	A, R	4.8×10^8–7.9×10^8	1.4×10^4–7.5×10^4	0.0035–0.0095 %
WS	A, R	NE	3.0×10^2–2.0×10^3	NE

[a] *I* industrial, *A* agricultural, *B* bulk, *R* rhizosphere. [b] *NE* not evaluated

agricultural soils (Table 2). Colonies with typical clearing zones were visible by the end of the 3rd incubation day (Additional file 3: Figure S2). Additional atrazine-degrading colonies appeared in most samples during further incubation, while some of the old colonies overlapped.

The highest population densities of atrazine degrading bacteria were detected in D3 and D5 soils (Table 2). In the latter, both in bulk soil and in the rhizosphere of cogon grass, atrazine degraders accounted for nearly 1/3 of the recoverable bacterial population. The densities of atrazine degrading bacteria in the rhizosphere of cogon grass (sample D5) and common reed (sample D6) were about ten times higher than those in the bulk soils.

Populations of atrazine-degrading bacteria in the maize rhizosphere varied from 10^2 to 10^4 CFU g^{-1} soil (Table 2), indicating that they constituted a minor component of the microbial communities. In general, the population densities of atrazine degraders calculated from direct plate counts were in agreement with the data obtained by a radiological most probable number method for rhizosphere of maize growing in atrazine-adapted soils [18].

In sum, 116 strains of atrazine degraders were isolated by direct plating of soil dilutions. Among these, 56 strains were derived from samples of industrially contaminated soil, and 60 strains originated from agricultural soils. Six to 26 strains were isolated from each of the sampling sites. The isolates varied in streak and colony morphologies and colony growth rate (Fig. 1), indicating that direct plating on SM agar allowed the isolation of diverse bacteria showing different culturability.

ERIC-PCR genotyping of atrazine-degrading strains
All the 116 atrazine degrading isolates were genotyped by Rep-PCR using primer ERIC2. As a result, 17 ERIC types

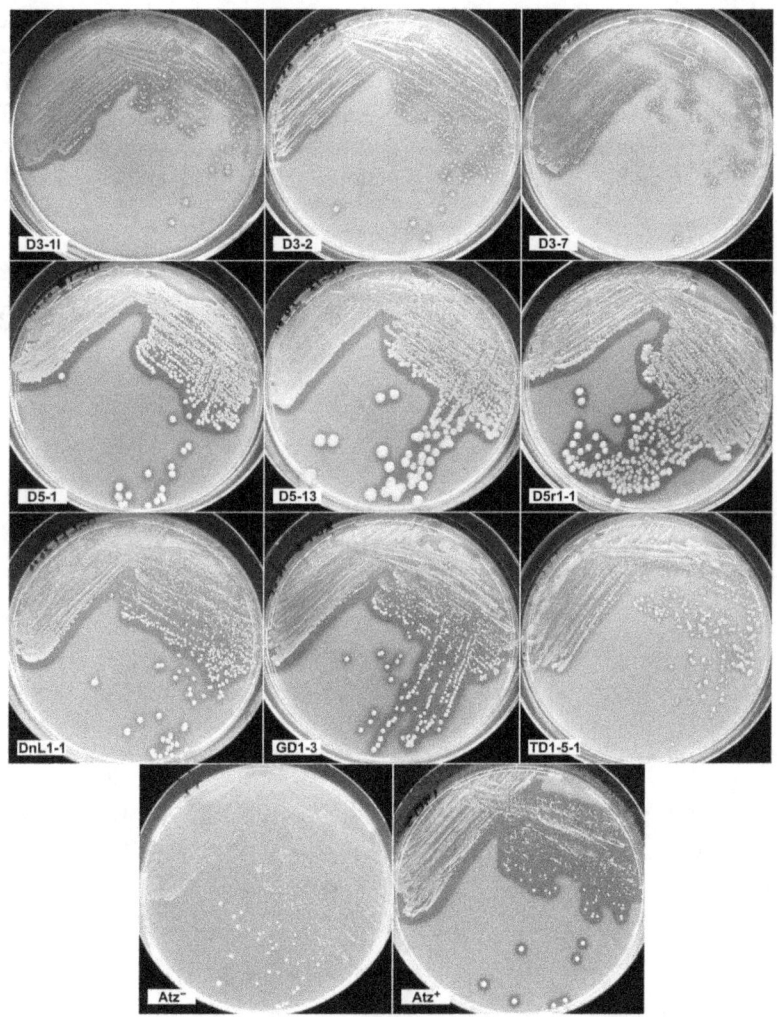

Fig. 1 Cultures of some atrazine degrading isolates on SMY agar. Age of the cultures was 5 days (D3-1l, D3-2, D5-1, D5-13, D5r1-1, DnL1-1) or 7 days (D3-7, GD1-3, TD1-5-1). (Atz$^-$) *A. ureafaciens* CGMCC 1.1897^T, a negative control strain; (Atz$^+$) the reference atrazine-degrading strain *Arthrobacter* sp. SD41

designated A-Q (Table 3, Additional file 4: Figure S3) were distinguished. Analysis of 56 cultures isolated from the industrially contaminated soils (D samples) allowed discrimination between 15 different ERIC types. Among these, ERIC type A was strongly predominant at sites D5 and D6, both in the rhizosphere and bulk soil (Table 3). Besides type A, two other ERIC types of atrazine-utilizing bacteria were found at both sites. The second largest, ERIC type L, comprised bacteria isolated from bulk soil. In contrast, bacteria of ERIC type E were isolated from the rhizosphere (1 isolate from each of the sites). In addition to the ERIC types common for these two sites, seven minor types (B, I, J, K, M, N and Q) of atrazine degraders were found only at site D5. Three ERIC types (F, G and H) were detected among nine isolates from D3 soil. No ERIC types found among isolates from D5 and/or D6 samples were detected among the bacteria isolated from D3 soil.

Only three ERIC types were discriminated among 60 strains isolated from the maize rhizosphere soils, collected from six geographically distant fields. Forty seven strains fitted into ERIC type B, which was originally detected among isolates associated with cogon grass roots in D5 soil. ERIC type B atrazine-degrading bacteria were found in all replicate samples of maize rhizosphere soils from all field sites. The reference bacterium *Arthrobacter* sp. SD41 also belonged to ERIC type B.

The other two ERIC types of atrazine-utilizing bacteria found in the maize rhizosphere were represented by seven type C and six type D isolates. Strains of ERIC type C were isolated form all replicate samples taken at TD(a) and TD(b) sites. Strains of ERIC type D were found in all GD replicate samples and in two replicate samples from WS site. Bacteria of ERIC types C and D were not found among atrazine degraders originating from other sampling sites, suggesting a localized distribution.

Thus, ERIC-PCR genotyping of atrazine-degrading bacteria directly isolated from industrially contaminated soils revealed communities of genetically distinct strains individually utilizing atrazine as a sole nitrogen source. ERIC type A bacteria were identified as dominant in atrazine-degrading communities at sites D5 and D6 both in bulk and rhizosphere soils, while most isolates from site D3 were ERIC type G bacteria. Along with the absence of any common ERIC type, this indicated a general dissimilarity between the structures of atrazine degrading communities in soil D3 and soils D5-D6, despite the sampling sites being separated by only 150 m. Population densities of atrazine degraders in all three soils were high (Table 2), indicating high rates of atrazine inflow. Among the populations in bulk soils the highest density was detected at site D3, where atrazine content was nearly 2000 times higher than in D6 soil (Table 1). Assuming similar

Table 3 ERIC types of atrazine-degrading isolates and their geographic occurrence

ERIC types	Number of atrazine degrading strains isolated from soils													
	D3	D5			D6			TD(a)	TD(b)	DnW	DnL	GD	WS	Total
	I, B[a]	I, B	I, R	Total	I, B	I, R	Total	A, R	A, R	A, R	A, R	A, R	A, R	
A	0	5	9	14	3	12	15	0	0	0	0	0	0	29
B	0	0	2	2	0	0	0	10	6	6	8	9	8	49
C	0	0	0	0	0	0	0	3	4	0	0	0	0	7
D	0	0	0	0	0	0	0	0	0	0	0	4	2	6
E	0	0	1	1	0	1	1	0	0	0	0	0	0	2
F	2	0	0	0	0	0	0	0	0	0	0	0	0	2
G	5	0	0	0	0	0	0	0	0	0	0	0	0	5
H	2	0	0	0	0	0	0	0	0	0	0	0	0	2
I	0	1	0	1	0	0	0	0	0	0	0	0	0	1
J	0	1	0	1	0	0	0	0	0	0	0	0	0	1
K	0	1	0	1	0	0	0	0	0	0	0	0	0	1
L	0	3	0	3	3	0	3	0	0	0	0	0	0	6
M	0	1	0	1	0	0	0	0	0	0	0	0	0	1
N	0	1	0	1	0	0	0	0	0	0	0	0	0	1
O	0	0	0	0	0	1	1	0	0	0	0	0	0	1
P	0	0	0	0	0	1	1	0	0	0	0	0	0	1
Q	0	1	0	1	0	0	0	0	0	0	0	0	0	1
Total strains	9	14	12	26	6	15	21	13	10	6	8	13	10	116
Total ERIC types	3	8	3	10	2	4	5	2	2	1	1	2	2	17

[a] *I* industrial, *A* agricultural, *B* bulk, *R* rhizosphere

leaching rates at sites D3 and D6, the difference between the measured atrazine concentrations would result mainly from the balance between atrazine inflow and degradation. Therefore, the high concentration of atrazine in D3 soil indicated that the contamination rate exceeded the degradation potential of the bacterial community.

Concentration of available nitrogen in D3 soil was nearly ten times higher than in D5 and D6 soils, and similar to or even higher than that in common mineral microbiological media (437 ± 22 mg N kg^{-1} soil, Additional file 2: Table S1, is equivalent to 1680 mg NH$_4$Cl kg^{-1} soil, or 6.0 g NH$_4$Cl L^{-1} soil solution on the basis of 35 % water holding capacity and 80 % humidity of soil). It was obvious that D3 soil contained excess nitrogen and its availability did not limit bacterial growth. Therefore, atrazine seemed to stimulate the population growth of its degraders in D3 soil as a source of carbon and energy rather than of nitrogen. The excess nitrogen from the degraded atrazine could contribute to the pool of available nitrogen in soil. Thus, the specific genetic structure of the atrazine degrading community in D3 soil seemed to be caused mainly by its heavy contamination with atrazine and other components of the factory wastewater.

ERIC typing of atrazine degrading isolates from the maize rhizosphere revealed a narrow range of genotypes. And despite the possible presence of some minor genetic groups of atrazine degraders that were below the limit of detection, the contrast with the diversity observed in the industrially contaminated soils was striking. Bacteria of ERIC type B were the dominant or even the sole group of atrazine degraders detected in the maize rhizosphere at all agricultural sites investigated. This clearly indicated their prevalence on the large area of agricultural soils with a history of atrazine application. Taking into consideration that agricultural soils were treated with low doses of atrazine and only once a year, traits other than the ability to utilize atrazine seemed to contribute to the competence of ERIC type B bacteria in the maize rhizosphere.

Detection of genes for atrazine degradation and atrazine-degrading capacity in the isolates

The multiplex PCR mixture contained four primer pairs allowing selective amplification of the genes atzA, –B, –C and trzN to generate fragments with predicted sizes of 432, 275, 626, and 196 base pairs respectively. One of the reference strains was Arthrobacter sp. SD41, bearing the genes atzB, –C and trzN [34]. Later the genes atzA, –B and –C were detected in strain Pseudomonas sp. D3-1l isolated in this work. Multiplex PCRs with cell lysates of strains SD41 or D3-1l as templates produced respective fragments of the predicted sizes at the wide range of T_a (Additional file 5: Figure S4), indicating robust detection of the genes atzB, –C and trzN in Arthrobacter sp. SD41 and the genes atzA, –B and –C in strain D3-1l. No production of unexpected fragments was detected.

The fragments produced in the multiplex PCRs were sequenced to verify their identity to the genes targeted. The resulting nucleotide sequences of atzB and -C from Arthrobacter sp. SD41 and atzA from D3-1l were deposited in GenBank (http://www.ncbi.nlm.nih.gov/genbank) under accession Nos. KP994320 - KP994322. The nucleotide sequence of a short trzN fragment from Arthrobacter sp. SD41 is provided in Additional file 6. Additionally, an amplified with the common primers C190-10/C190-11 fragment of trzN gene from Arthrobacter sp. SD41 was sequenced (Accession No. KP994319). BLAST search results demonstrated that the determined nucleotide sequences shared high identity (98–100 %) with respective atrazine degradation genes from other bacteria, including the genes atzA, –B and -C from Pseudomonas sp. ADP and trzN, atzB and –C from Arthrobacter aurescens TC1.

Besides strain D3-1l, the gene atzA was detected in all ERIC type G bacteria (Table 4, Additional file 7: Figure S5). All isolates belonging to other ERIC types contained trzN gene. The genes atzB and atzC were detected in all isolates except 12 strains of ERIC type B and three isolates of ERIC type L. Also, atzB gene was not found in strains D6r1-2 (ERIC type A) and D3-2 (ERIC type G).

Atrazine-degrading capacity of isolates representing distinguished ERIC types was assessed in liquid medium SM25 with 25 mg L^{-1} atrazine as a sole nitrogen source. After 1 week incubation, HPLC-MS/MS analysis detected trace amounts of atrazine (from <0.003 to 0.09 μg L^{-1}) in cultural liquids of the strains D5r1-1 (ERIC type A), DnL1-1 (ERIC type B), TD1-5-1 (ERIC type C), GD1-3 (ERIC type D), D3-1l (ERIC type G), D3-3 (ERIC type H), D5-13 (ERIC type L), D6r2-4 (ERIC type P) and D5-1 (ERIC type Q), indicating over 99.99 % degradation. The concentration of atrazine in the cultural liquid of strain D3-7 (ERIC type F) was reduced to 3567 ± 312 μg L^{-1}, demonstrating nearly 85.7 % degradation. No significant degradation of atrazine was observed in the non-inoculated control.

Reported in numerous papers over the last 20 years, the dissolution of atrazine in liquid media or production of colonies with clearing halos on solid media with atrazine were always due to its rapid degradation by bacteria harboring atzA or trzN in combination with atzB and/or atzC [15]. The non-degradative microbial dissolution of atrazine and other chlorinated s-triazines has never been observed and seems to be impossible in principle due to the low solubility of these compounds in all known solvents. The results obtained in our study once again confirm that the production of clearing halos by growing colonies and the presence of the genes trzN and/or atzABC in the genome are robust indicators of the atrazine-degrading capacity in bacteria. Direct isolation of bacteria producing colonies with clearing halos on SM medium and detection of trzN and atzABC genes by multiplex PCR may be promising methods to facilitate ecological and biogeographical studies of atrazine degraders.

Table 4 Detection of atrazine degradation genes in the isolates and ribosomal diversity of the strains representing separate ERIC types

ERIC type	Genes for atrazine degradation	Strains (GenBank Accession No.)[a]	Nearest type strain (GenBank Accession No.) [Percent similarity][b]
A	*trzN atzBC*	**D5-2 (KF889364)**, D5-4, D5-9, D5-10, D5-14, **D5r1-1 (KF889366)**, D5r1-3, D5r1-4, D5r2-1, D5r2-3, D5rh3-1, D5r3-3, D5r3-4, D5r3-5, **D6-2 (KF889365)**, D6-4, D6-5, **D6r1-1 (KF889367)**, D6r1-3, D6r1-4, D6r2-1, D6r2-2, D6r2-3, **D6r2-5 (KF889368)**, D6r3-1, D6r3-2, D6r3-3, D6r3-4	*Arthrobacter nitroguajacolicus* G2-1[T] (NR_027199) [99.3–99.4 %]
	trzN atzC	D6r1-2	
B	*trzN atzBC*	**D5r1-2 (KF889369)**, TD1-1, TD1-2, TD1-3, TD1-4, TD2-2, **TD2-4 (KF889370)**, TD3-1, TD3-2, TD3-3, **TD4-1 (KF889371)**, TD 4–2, TD5-1, TD6-1, TD6-3; **DnW1-1 (KF889372)**, DnW1-2, DnW2-1, DnW2-2, DnW3-1, DnW3-2; **DnL1-1 (KF889373)**, DnL1-3, DnL2-1, DnL3-1, DnL3-3; **GD1-1 (KF889374)**, GD1-4, GD2-2, GD2-3, GD3-1, GD3-2, GD3-3, GD3-4, GD3-5; WS1-1, WS2-3	*Arthrobacter ureafaciens* NC[T] (NR_029281) [99.5– 99.6 %]
	trzN	D5r2-2, TD2-1, TD6-2, DnL1-2, DnL2-2, DnL3-2, WS1-2, WS2-1, **WS2-2 (KF889375)**, WS2-5, WS3-1, WS3-2.	
C	*trzN atzBC*	TD1-5-1, **TD1-5-2 (KF889379)**, **TD2-3 (KF889376)**, **D4-3 (KF889377)**, **TD4-4 (KF889378)**, TD5-2, TD5-3	*Nocardioides panacihumi* Gsoil 616[T] (NR_041518) [98.1–98.3 %]
D	*trzN atzBC*	GD1-2, **GD1-3 (KF889380)**, GD2-1, GD2-4, **WS1-3 (KF889381)**, WS2-4	*Nocardioides ganghwensis* JC2055 [T] (NR_025776) [99.8 %]
E	*trzN atzBC*	**D6rh3-5 (KF889382)**	*Arthrobacter sulfonivorans* ALL[T] (NR_025084) [99.4–99.6 %]
	trzN atzC	**D5rh3-2 (KF889383)**	
F	*trzN atzBC*	D3-7, **D3-8 (KF889384)**	*Gulosibacter molinativorax* ON4[T] (NR_025451) [100 %]
G	*atzABC*	**D3-1l (KF889385)**, D3-1s, D3-4, D3-5	*Pseudomonas alcaliphila* AL15-21[T] (NR_024734) [99.1–99.3 %]
	atzAC	**D3-2 (KF889386)**	
H	*trzN atzBC*	**D3-3 (KF889387)**, **D3-6 (KF889388)**	*Arthrobacter crystallopoietes* DSM 20117[T] (NR_026189) [99.8 %]
I	*trzN atzBC*	**D5-3 (KF889389)**	*Arthrobacter crystallopoietes* DSM 20117[T] (NR_026189) [99.6 %]
J	*trzN atzBC*	**D5-5 (KF889390)**	*Arthrobacter crystallopoietes* DSM 20117[T] (NR_026189) [99.8 %]
K	*trzN atzBC*	**D5-7 (KF889391)**	*Arthrobacter crystallopoietes* DSM 20117[T] (NR_026189) [99.7 %]
L	*trzN atzBC*	D5-12, **D5-13 (KF889392)**	*Arthrobacter oxydans* DSM 20119[T] (NR_026236) [99.7–99.8 %]
	trzN atzBC	**D5-8 (KF889393)**	
	trzN	**D6-1 (KF889394)**, D6-3, D6-3 s	
M	*trzN atzBC*	**D5-6 (KF889395)**	*Arthrobacter phenanthrenivorans* Sphe3[T] (NR_074770) [99.9 %]
N	*trzN atzBC*	**D5-11 (KF889396)**	*Arthrobacter subterraneus* CH7[T] (NR_043546) [99.8 %]
O	*trzN atzBC*	**D6r1-5 (KF889397)**	*Arthrobacter nitroguajacolicus* G2-1[T] (NR_027199) [100 %]
P	*trzN atzBC*	**D6r2-4 (KF889398)**	*Arthrobacter nitroguajacolicus* G2-1[T] (NR_027199) [99.8 %]
Q	*trzN atzBC*	**D5-1 (KJ010189)**	*Arthrobacter nitroguajacolicus* G2-1[T] (NR_027199) [99.6 %]

[a]Names of the strains selected for sequencing of 16S rRNA genes and GenBank accession numbers are typed in bold. Letters in the name of each strain and numbers after letter D represent sampling sites listed in the Table 1. Letter "r" following D5 and D6 means that the strain has been isolated from the rhizosphere. The first digit in the names of strains originated from TD, DnW, DnL, GD and WS sites indicates replicate samples. Site TD(a) was represented by replicate samples 1–3, and site TD(b) – by replicate samples 4–6

[b] According to BLAST, based on over 1235 bp nucleotide sequences of 16S rRNA genes

Our data demonstrated the presence of atrazine degraders harboring the gene *trzN* in all studied soils and clearly indicated their prevalence over *atzA*-harboring bacteria at all agricultural sites and at two industrially contaminated sites. Arbeli and Fuentes [39] reached a similar conclusion for Colombian agricultural soils based on the analysis of soil DNA isolated after previous "microcosm enrichment" of atrazine degraders. They hypothesized that atrazine degraders harboring *trzN* had ecological superiority over *atzA*-harboring bacteria, owing to the faster reaction rate of TrzN, its higher affinity for substrates and wider range of substrates degraded [39]. Interestingly, Arbeli and Fuentes did detect *atzA* gene in enrichment cultures derived from some of the soils and found that enrichment with atrazine as a sole carbon and nitrogen source was markedly more helpful for the detection of *atzA* than with atrazine as the only source of nitrogen. Because atrazine is a poor source of carbon and rich in nitrogen, the growth of degrading bacteria in media with atrazine as sole carbon and nitrogen source is limited by carbon [10]. In contrast to all other sites, bacteria harboring *atzA* gene were found to be a major or even dominating group in the atrazine degrading community of D3 soil, indicating that *trzN* had no ecological superiority over *atzA*. Unlike soils from all other sites investigated, D3 soil had the highest content of atrazine and an excess of available nitrogen, low content of organic carbon and contained no plant roots that could provide substances rich in carbon and energy. These conditions are similar to those in the cultures with atrazine as a sole carbon and nitrogen source, which favor the enrichment of *atzA*-bearing atrazine degraders. Thus, it is possible to hypothesize that the high contamination rate has influenced the structure of atrazine-degrading community in D3 soil by altering the selective advantage for the degraders harboring *trzN* or *atzA*.

Sequencing of 16S rRNA genes and phylogenetic analysis

The 16S rRNA genes of 36 strains representing all the 17 distinguished ERIC types were sequenced (Table 4). Selected strains of the same ERIC type represented all the sites where bacteria belonging to this ERIC type were isolated. BLAST search results gave evidence that 26 strains representing 12 ERIC types, including predominant types A and B, belonged to the genus *Arthrobacter*. Bacteria of two other major ERIC types C and D were found to belong to the genus *Nocardioides*. Besides *Arthrobacter* and *Nocardioides* no other genera were detected among strains isolated from the maize rhizosphere. Isolates from the industrial soils (D samples) were somewhat more diverse taxonomically. Strains of ERIC types F and G were affiliated to the genera *Gulosibacter* (phylum *Actinobacteria*, class *Actinobacteria*, subclass *Actinobacteridae*, order *Actinomycetales*, suborder *Micrococcineae*, family *Microbacteriaceae*) and *Pseudomonas* respectively.

To accurately define taxonomic positions of the isolates at the sub-genus level, a phylogenetic analysis was carried out based on the determined and reference nucleotide sequences of 16S rRNA genes. As a result, 26 representatives of 12 ERIC types belonging to the genus *Arthrobacter* were distributed into eight phylogenetic groups (Fig. 2a). It was found that all representative strains of ERIC type A constituted a tight cluster with known atrazine-degrading strains *Arthrobacter* sp. AD30 and *Arthrobacter* sp. T$_3$AB$_1$, which were isolated in China. *Arthrobacter* sp. T$_3$AB$_1$ was the only agricultural strain in the cluster and originated from maize field soil sampled in Nehe County, Heilongjiang Province [31], located about 1300 km north from sampling site D. The cluster contained no species type strains and was closely related to the phylogenetic group formed by strains D5-1 (ERIC type Q), D6r1-5 (ERIC type O), D6r2-4 (ERIC type P), the type strains of *A. aurescens*, *Arthrobacter nitroguajacolicus*, *Arthrobacter ilicis*, and the known atrazine-degrading bacterium *A. aurescens* TC1, isolated in the United States from soil heavily contaminated with atrazine due to an accidental spill [36]. The divergence between these two clusters exceeded the divergence between the species *A. aurescens*, *A. nitroguajacolicus* and *A. ilicis*, suggesting that bacteria of ERIC type A represented a separate genomospecies.

All seven strains selected from the largest group of ERIC type B isolates tightly clustered with the type strain of *Arthrobacter ureafaciens*, suggesting their affiliation with this species. This robust cluster also contained eight of the 12 Chinese atrazine-degrading strains of the genus *Arthrobacter* for which 16 rRNA gene nucleotide sequences of the proper length and quality have been determined and published [25, 27, 30, 33, 34, 40, 41], 1 isolate originating from India [42] and 3 Colombian atrazine degrading strains [39]. While all the Colombian *A. ureafaciens* strains and the Indian strain were isolated from agricultural soils, 5 of the Chinese strains were reported to originate from the wastewater of atrazine plants. The other 3 Chinese *A. ureafaciens* strains were agricultural isolates originated from Shandong Province (*Arthrobacter* sp. SD41 [34]) as well as Jiangsu (*Arthrobacter* sp. ADH-2 [30]) and Hebei (*Arthrobacter* sp. DAT1 [33]) provinces bordering Shandong to the south and north-west, respectively. Thus, atrazine-degrading bacteria belonging to the species *A. ureafaciens* appeared to have intercontinental distribution and to be the most frequently isolated atrazine degraders in China.

Strains of ERIC type E D5rh3-2 and D6rh3-5 tightly clustered with the type strain of *Arthrobacter sulfonivorans*. Strains D5-8, D5-13 and D6-1 (ERIC type L) were included in the cluster of species *Arthrobacter scleromae*, *Arthrobacter oxydans* and *Arthrobacter polychromogenes*. Strain D5-6 (ERIC type M) was closely related to *Arthrobacter phenantrenivorans* type strain. Strain D5-11 (ERIC type N) clustered with the type strain of *Arthrobacter subterraneus*. These 4 robust clusters did not contain known atrazine-degraders.

A highly robust cluster comprised the type strain of *Arthrobacter crystallopoietes* with two tight sub-clusters of

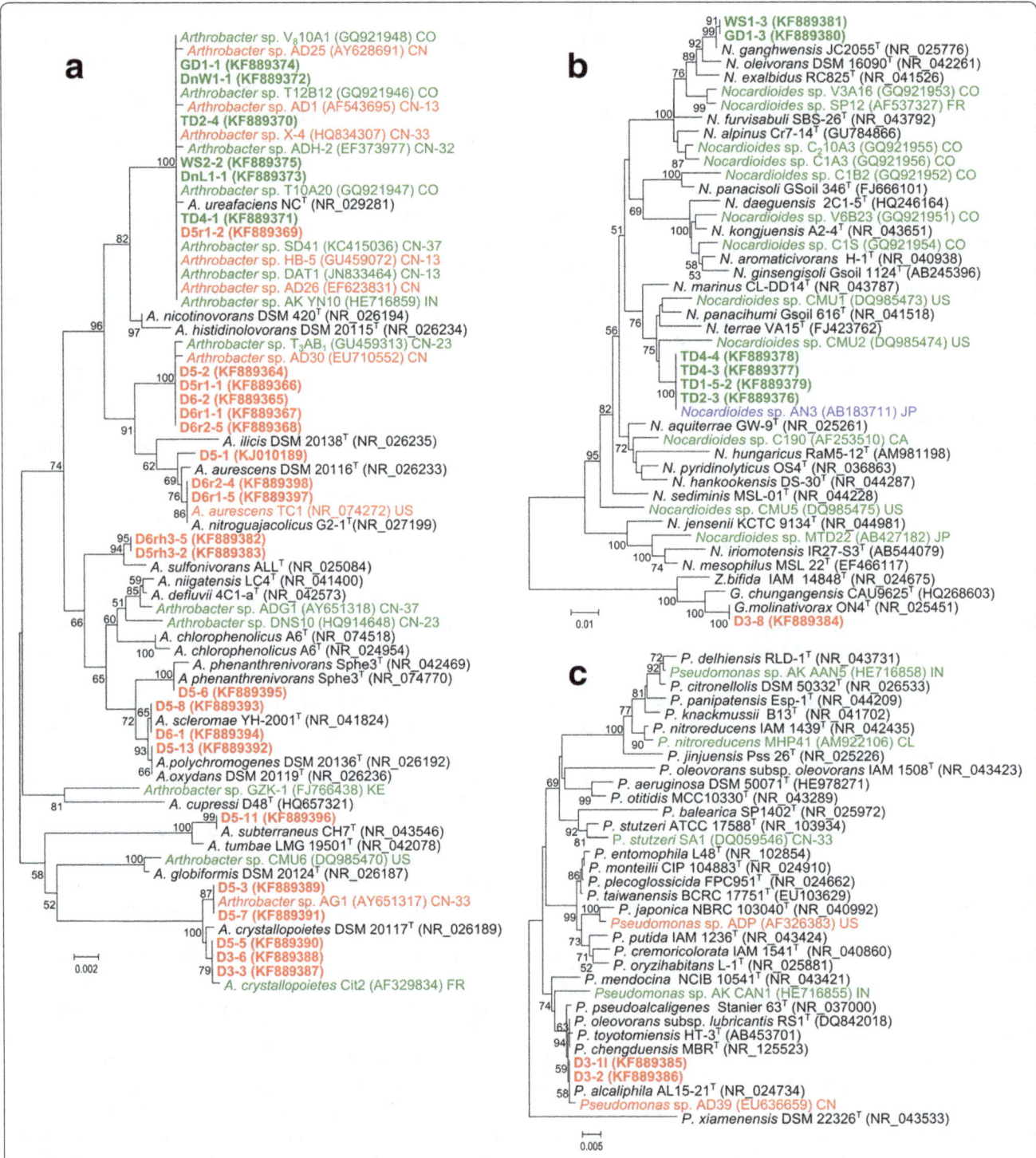

Fig 2 Neighbor-joining phylogenetic trees for atrazine-degrading bacteria belonging to the genera **a** *Arthrobacter*, **b** *Nocardioides* and *Gulosibacter*, **c** *Pseudomonas*, based on 16S rRNA gene sequences. Colors indicate atrazine-degrading bacteria isolated from industrial soils or sites of spill (*red*), agricultural soils (*green*), and riverbed sediment (*blue*). Names of the strains isolated in this work are printed in bold lettering. GenBank accession numbers are shown in parentheses. For known atrazine-degrading strains, countries of isolation, and for those isolated in China – provinces, are indicated by ISO 3166 codes. Bootstrap values (expressed as percentages of 1000 replications) greater than 50 % are shown at the branching points. There were a total of 1200 (**a**), 1206 (**b**) and 1258 (**c**) positions in the final datasets. Scale bars show substitutions per nucleotide position. Evolutionary analysis was conducted in MEGA5 [63]

atrazine degraders. One of the sub-clusters contained the atrazine-degrading bacterium *Arthrobacter* sp. AG1 isolated from the heavily contaminated industrial soil in China [26] and single representatives of ERIC type I (strain D5-3) and ERIC type J (strain D5-5). The other sub-cluster joined strains D3-3, D3-6 (ERIC type H), and D5-7 (ERIC type K) with atrazine-degrading strain *A. crystallopoietes* Cit2 isolated from French agricultural soils [43].

Isolates of ERIC types C and D were included into 2 distant phylogenetic groups of the genus *Nocardioides* (Fig. 2b). Strains of the ERIC type C and atrazine-degrading strain *Nocardioides* sp. AN3, isolated from riverbed sediment in Japan [44], formed a distinct subgroup within a robust cluster comprising the type strains of *Nocardioides marinus*, *Nocardioides panacihumi*, *Nocardioides terrae*, and the known atrazine-degrading isolates *Nocardioides* sp. CMU1 and CMU2 which originated from the United States [45]. ERIC type D isolates from GD and WS soils fitted in the large robust cluster formed by type strains of the species *Nocardioides alpinus*, *Nocardioides furvisabuli*, *Nocardioides exalbidus*, *Nocardioides oleivorans*, *Nocardioides ganghwensis*, 3 atrazine-degrading *Nocardioides* isolates from Colombian agricultural soils [39] and the French strain *Nocardioides* sp. SP12 reported to inhabit both maize rhizosphere and bulk soil [38]. Within this cluster, ERIC type D isolates were close to the type strain of *N. ganghwensis*. To our knowledge, atrazine-degrading *Nocardioides* spp. had not been previously isolated from Chinese soils.

Isolate D3-8 (ERIC type F) tightly clustered with the type strain of *Gulosibacter molinativorax* (Fig. 2b). The species was described based on characterization of a single strain, capable of transforming the herbicide molinate [46]. To our knowledge, this is the first report of an atrazine-degrading strain belonging to the genus *Gulosibacter*, and to the family *Microbacteriaceae*.

Bacteria D3-1l and D3-2 clustered with the type strain of *Pseudomonas alcaliphila* (Fig. 2c). This cluster also contained known atrazine-degrading bacterium *Pseudomonas* sp. AD39 isolated from the wastewater system of atrazine plant in China [28]. There were no differences between nucleotide sequences of 16S rRNA genes from D3-1l, D3-2 and *P. alcaliphila* AL15-21^T in the final dataset of the alignment. However, the sub-cluster had medium level of the bootstrap support, being included in a robust cluster together with closely related type strains *Pseudomonas oleovorans* ssp. *lubricantis* RS1^T, *Pseudomonas toyotomiensis* HT-3^T and *Pseudomonas pseudoalcaligenes* Stanier 63^T.

Despite the method of direct plating enabled us to discover a variety of atrazine degraders, the diversity revealed was limited to the genera *Arthrobacter*, *Nocardioides*, *Gulosibacter* and *Pseudomonas*. The diversity of atrazine-degrading *Arthrobacter* spp. originated from three adjacent sites of industrial soils exceeded taxonomic diversity of known atrazine-degrading *Arthrobacter* bacteria

isolated worldwide. At the same time, representatives of most other known atrazine-degrading genera, such as *Clavibacter*, *Nocardia* (phylum *Actinobacteria*), *Agrobacterium*, *Alcaligenes*, *Herbaspirillum*, *Pseudaminobacter*, *Pseudomonas*, *Polaromonas*, *Ralstonia*, *Rhizobium*, *Sinorhizobium*, *Stenotrophomonas* (phylum *Proteobacteria*) [15] were not found among the isolates, suggesting that they were either absent in the soils studied or were present as minor groups not directly detectable among the dominating atrazine degraders. Drastic differences between the genetic structures of the atrazine-degrading communities in industrially contaminated soils at different sites, and between those in the rhizosphere of plants growing in the industrially contaminated and agricultural soils clearly indicate the selection of bacteria that are better adapted to local environments. As a result, representatives of different species, genera, or even phyla became prevalent in the atrazine-degrading communities. However, in all sites except D3 where *Gulosibacter* strains were isolated, the adaptation did not extend the diversity of prevalent atrazine degraders over the range of known atrazine-degrading genera. Interestingly, that no *Pseudomonas* strains, considered to be excellent root-colonizing bacteria [47], or other atrazine degrading strains belonging to the genera of the phylum *Proteobacteria* [15], both culturable and unculturable representatives of which were proved to dominate in root-associated bacterial consortia [48, 49], were found among the rhizosphere isolates. This was especially notable for isolates from D5 and D6 sites where cogon grass and common reed thickets were close to the plantless D3 soil in which atrazine degrading bacteria of *P. alcaliphila* phylogenetic group were abundant. This fact indicates a failure of horizontal transfer of functional atrazine degradation genes to the rhizosphere-competent pseudomonads or other *Proteobacteria*. Alternatively, it demonstrates a low contribution of the associative plant-bacterial interaction [47, 50] to the rhizosphere competence of bacteria and their population growth compared to the advantages gained by *Arthrobacter* bacteria from the effective TrzN-mediated utilization of atrazine.

The observed limitation of atrazine degraders' diversity to various *Arthrobacter* spp. and some phylogenetic groups within a narrow range of other genera suggests that individual atrazine utilization is an uncommon capability of soil bacteria belonging to a limited number of lineages. Probable reasons for this may be a lack of efficient gene transfer systems or a physiological incompatibility of potential recipients with functional atrazine degradation genes. However, the genes *atzABC* are highly mobile, and their intergeneric transfer in the rhizosphere has been demonstrated [51]. Although horizontal transfer of *trzN* still has not been documented, location of the gene within transposon-like structures, on plasmids, and its presence in viral DNA suggest that *trzN* can spread and facilitate the adaptation of microbial communities to contamination by atrazine [52]. The high local diversity of bacteria harboring *trzN* in industrial soils found in this study also implies ecological significance of *trzN* mobility. Thus, the latter assumption looks more likely. Full or partial incompatibility of the pathways for fast atrazine degradation with physiological background of a host may interfere with utilization of atrazine or reduce soil or rhizosphere competence of bacteria acquired the atrazine degradation genes.

BOX-PCR genotyping of ERIC type B atrazine-degrading isolates

All related to *A. ureafaciens* atrazine-degrading isolates of ERIC type B exhibited identical ERIC-PCR patterns regardless their geographical origin. It was known that BOX-PCR genotyping (PCR targeting bacterial repetitive BOX element) revealed a higher diversity in Colombian atrazine-degrading strains of *A. ureafaciens* [39]. Aiming to examine their genetic uniformity, the ERIC type B isolates were genotyped by BOX-PCR. The type strain *A. ureafaciens* CGMCC 1.1897^T was included in the analysis in order to obtain more taxonomic information. Earlier, ERIC–PCR revealed no similarity between representative ERIC type B bacteria and *A. ureafaciens* type strain (Additional file 8: Figure S6). The amplicon patterns generated from BOX-PCR with DNAs of ERIC type B isolates and strain *A. ureafaciens* CGMCC 1.1897^T demonstrated marked identity (Additional file 9: Figure S7). Taking into account that Rep-PCR genotyping resolved genetic differences between strains within the same species [53], the BOX-PCR results provided evidence that ERIC type B atrazine degraders isolated in this work were a group of genetically similar bacteria belonging to the species *A. ureafaciens*.

Detection of atrazine degraders in soils by conventional PCR targeting genes for atrazine degradation

A narrow taxonomic range of atrazine degraders isolated in this work might be a result of limited culturability of some bacteria. Also, the transformation of atrazine in soils can be carried out by bacterial consortia whose members separately contain the genes *trzN*, *atzA*, *atzB* and *atzC* [15]. In order to better understand the role of the isolated atrazine degraders in atrazine-degrading communities in the industrial and agricultural soils, we analyzed the soil DNAs by conventional PCR targeting the genes *trzN*, *atzA*, *atzB* and *atzC*.

Selectivity of the primers for atrazine degradation genes was previously assessed in gradient PCRs with S1 soil DNA as a template. It was found that although no non-target nucleotide sequences with substantial complimentarily to the designed primers were found in the GenBank databases, most of the tested primer pairs yielded multiple unintended products at T_a values close to or even higher than the calculated melting midpoints (Additional file 10: Figure S8). This suggested that annealing of these primers to partly complementary non-target templates might interfere with annealing

to the perfect complements present in the soil DNA extracts at much lower concentrations, thus reducing the sensitivity of PCR detection. No amplification of unintended products was observed in PCRs with primer pairs *atzA*655f/*atzA*982r, *atzB*181f/*atzB*316r and *atzC*340f/*atzC*552r (Additional file 10: Figure S8) indicating a high specificity. Primer pair *trzN*1114f/*trzN*1271r produced unintended products at T_a range from 58 to 64 °C, but the visible yield was much lower than that detected in PCR with the commonly used pair C190-10/C190-11.

To evaluate the detection limits of PCRs with the selected primer pairs, aliquots of S1 soil were spiked with known titers of isolates *Pseudomonas* sp.D3-1l or *A. ureafaciens* DnL1-1 prior DNA extraction. Conventional PCRs with the selected primer pairs *atzA*655f/*atzA*982r and *atzB*181f/*atzB*316r and template DNA extracted from S1 soil supplemented with at least 10^3 CFU g^{-1} of *Pseudomonas* sp.D3-1l yielded products of the expected sizes near 0.37 kb and 0.18 kb, respectively (Fig. 3), demonstrating robust detection of this strain. Sensitivity of the PCR assay with primers *atzC*340f/*atzC*552r was even higher and a slight band of the expected 0.25 kb product indicated the detection of *Pseudomonas* sp.D3-1l population even at a density as low as 10^2 CFU g^{-1} soil. No amplification was detected with DNAs isolated from S1 soil to which *Pseudomonas* sp.D3-1l was not added or its density was nearly 10^1 CFU g^{-1} soil. In contrast to the strain *Pseudomonas* sp.D3-1l, the limits of *A. ureafaciens* DnL1-1 detection by PCR with primer pairs *trzN*1114f/*trzN*1271r and *atzB*181f/*atzB*316r were at the level of 10^5 CFU g^{-1} soil, and nearly 10^6 CFU g^{-1} soil with the pair *atzC*340f/*atzC*552r (Fig. 3). The band intensity in reactions with either of the primer pairs increased with CFU density, indicating that the assays could be used semi-quantitatively.

The fragments produced in PCRs with DNA extracted from S1 soil harboring *Pseudomonas* sp. D3-1l or *A. ureafaciens* DnL1-1 were sequenced to verify their identity to the targeted genes. The nucleotide sequences of the recovered *atzA* and *atzC* fragments were deposited in GenBank (Accession Nos. KP997248 - KP997250). The nucleotide sequences of short *trzN* fragment from *A. ureafaciens* DnL1-1 and *atzB* fragments from both strains are provided in Additional file 6. BLAST search results gave evidence that all the recovered fragments of the atrazine degradation genes shared 100 % identity with respective genes from known atrazine degrading bacteria, including *atzA*, –*B* and –*C* genes from *Pseudomonas* sp. ADP and *trzN*, *atzB* and –*C* genes from *A. aurescens* TC1.

A drastic difference between the limits of *Pseudomonas* sp.D3-1l and *A. ureafaciens* DnL1-1 detection can be explained by low extractability of DNA from the latter strain, rather than by actual selectivity and sensitivity of the reactions. A bias produced by DNA extraction method is known to affect composition and abundance of bacterial phylotypes which can be detected in soil [54, 55]. Feinstein

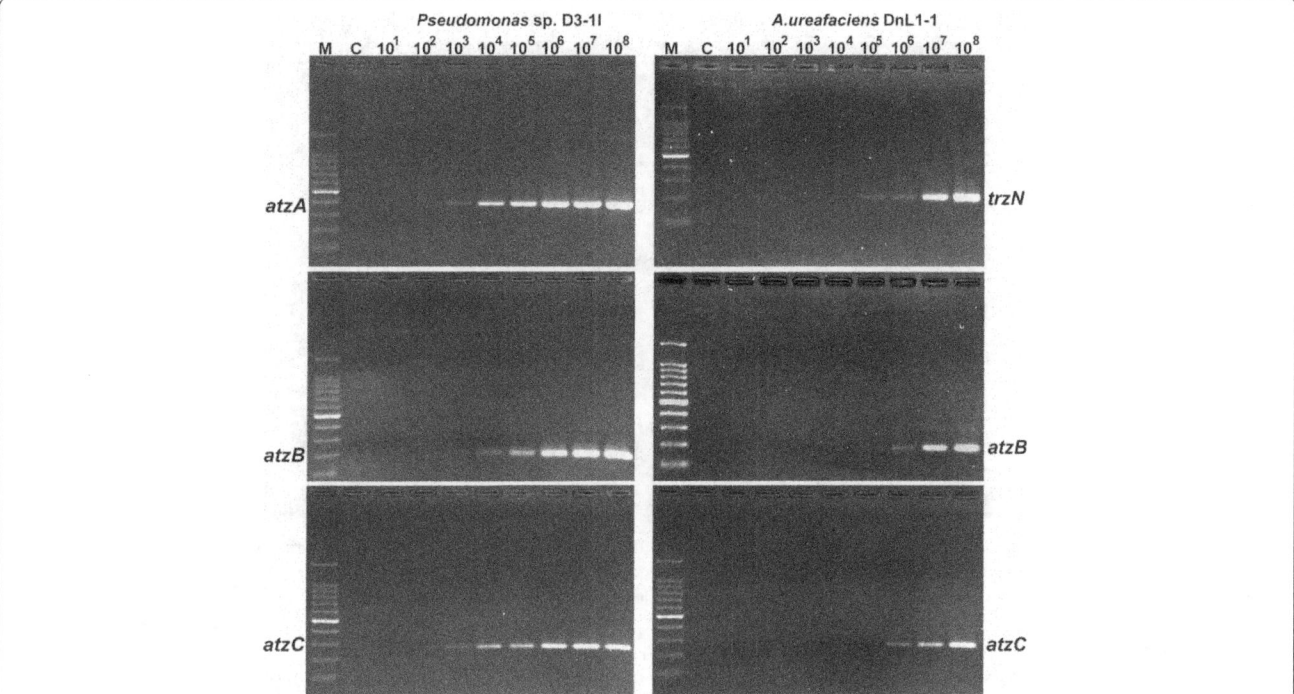

Fig. 3 The limits of *Pseudomonas* sp. D3-1l and *A. ureafaciens* DnL1-1 detection in soil by PCRs targeting atrazine degradation genes. Lanes are designated by titers of *Pseudomonas* sp.D3-1l (*left panel*) and *A. ureafaciens* DnL1-1 (*right panel*) CFU added per 1 g of S1 soil. Lane N and C are, respectively, no template control and a control in which DNA isolated from non-inoculated S1 soil was used as a template. The primer pairs for *trzN*, *atzA*, *atzB* and *atzC* were *trzN*1114f/*trzN*1271r, *atzA*655f/*atzA*982r, *atzB*181f/*atzB*316r and *atzC*340f/*atzC*552r, respectively. Lanes M contain a 100 bp DNA Ladder (Takara Biotechnology (Dalian) Co., Ltd., China)

et al. [56] found that the phylum *Actinobacteria* fell into the poorly lysed portion of soil bacterial communities. Our results demonstrate a possible link between the extraction bias and properties of the particular strains. Because *A. ureafaciens* bacteria appear to be predominant atrazine degraders in agricultural soils, the observed detection limit of 10^5 CFU g^{-1} soil can result in underestimation of the atrazine-degrading populations by culture-independent methods or even cause false negative results.

The attained sensitivity of *Pseudomonas* sp.D3-1l detection in soil significantly exceeded those of previously described assays exploiting conventional PCR. Thus, the reported detection limits of *E. coli* strain harboring *atzA* or *trzN* cloned in a high copy number plasmid pCR 2.1-TOPO (Invitrogen, USA) targeted by the commonly used primer pairs *atzA*f/*atzA*r [21] or C190-10/C190-11 [22] respectively were at the level of 10^4 CFU g^{-1} soil [39]. The same *atzA*f/*atzA*r primers allowed detecting of *Pseudomonas* sp. ADP at its density no less than 10^6 CFU g^{-1} soil, and additional purification or dilution of soil DNA did not improve the sensitivity of the assay [57]. The reported detection limit for *Nocardioides* sp.C190 targeted by primers C190-10/C190-11 was 10^8 CFU g^{-1} soil [22]. As a result, detection of atrazine degradation genes in soil DNA by conventional PCR required prior "microcosm enrichment" of atrazine degraders [39].

The experiments with strain *Pseudomonas* sp.D3-1l provided evidence that conventional PCR exploiting commercial high-yield *TaKaRa Ex Taq* polymerase kit and primers designed in this work is sensitive enough to detect atrazine degraders in the 10^3 CFU g^{-1} spiked soil sample. However, like methods of cultivation, direct PCR-detection of bacteria in soil has its own specific limitations, most likely caused by a poor lysability of many bacteria [56]. Thus, a combination of culturing and culture-independent methods seems to be a reasonable approach to extend the range of detectable atrazine degraders and to improve the detection sensitivity.

The analysis of DNAs extracted from bulk D3 and D5 soils clearly identified the genes *trzN*, *atzA*, *atzB*, and *atzC* (Fig. 4). No products of the expected sizes were amplified in reactions with S1 soil DNA that served "no template control". Reactions with D3 soil DNAs produced equally strong amplicon bands for all the targeted genes. At the same time, assays for the gene *atzA* with D5 template gave bands of slight intensity, indicating a lower density of *atzA*-harboring bacteria. PCRs with D6 soil DNAs clearly demonstrated amplification of the fragments for the genes *trzN*, *atzB*, and *atzC*, while a slight fragment of the expected size for the gene *atzA* was produced in reactions with only 1 of the 3 independent replicate templates.

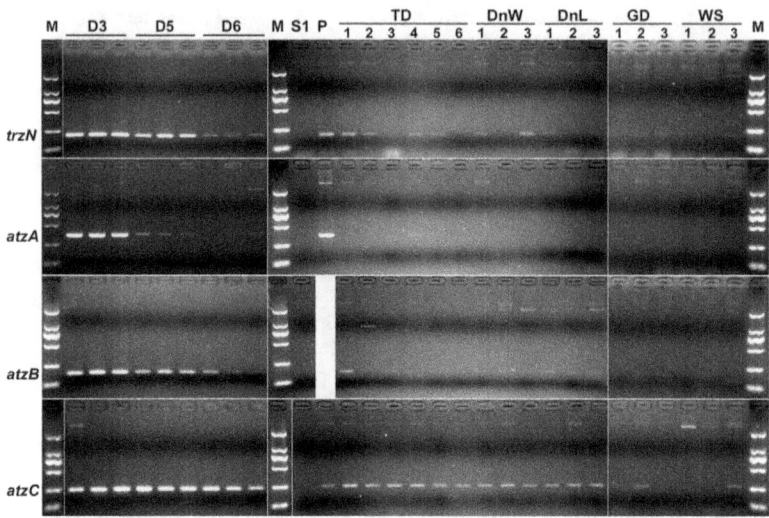

Fig 4 Detection of atrazine-degrading bacteria in soils by PCRs targeting the genes *trzN*, *atzA*, *atzB* and *atzC*. Lanes are designated by names of the sampling sites listed in Table 1. Numbers of replicate samples are given below the designations of sampling sites. Site TD(a) was represented by replicate samples 1–3, and site TD(b) by replicate samples 4–6. The primers for *trzN*, *atzA*, *atzB* and *atzC* were *trzN*1114f/*trzN*1271r, *atzA*655f/ *atzA*982r, *atzB*181f/*atzB*316r and *atzC*340f/*atzC*552r, respectively. Lanes P contain products of positive control reactions for which template DNAs were isolated from S1 soil supplemented with 10^4 CFU g^{-1} of *Pseudomonas* sp.D3-1I for *atzA* gene and with 10^6 CFU g^{-1} of *A. ureafaciens* DnL1-1 for *trzN* and *atzC*. Positive controls for the gene *atzB* were reactions with template DNAs isolated from S1 soil aliquots supplemented with known titers of *Pseudomonas* sp.D3-1I (Additional file 11: Figure S9). Lanes M contain DL 2000 DNA Marker (Takara Biotechnology (Dalian) Co., Ltd., China)

In agricultural soils, the gene *atzA* was detected only in replicate samples TD4 and DnW2. The genes *trzN*, *atzB* and *atzC* were found in the maize rhizosphere at all agricultural sampling sites (Fig. 4). However, the gene *trzN* was not detected in some of TD(b), DnL, DnW and WS replicate samples, and the genes *atzB* and *atzC* in some of GD and WS replicate samples.

The results of the PCR assays matched the results of the direct isolation of atrazine-degrading bacteria. Both methods demonstrated the striking prevalence of atrazine degraders possessing the gene *trzN* in all industrial and agricultural soils studied, except D3 soil where culturable *atzA*-bearing pseudomonads were one of the major groups in the atrazine-degrading community.

Weak amplicon bands produced in positive reactions and negative results obtained for some replicate samples indicated that the population size of atrazine degraders in agricultural soils was close to the limit of their detection by RCR. Direct plating on SM agar revealed *A. ureafaciens* bacteria of ERIC type B as a predominant group of atrazine degraders in all replicate samples from all agricultural sites tested. Population densities of culturable atrazine degraders exceeded the level of 10^4 CFU g^{-1} soil only at sites TD(a) and GD (Table 2), that was below the limit of PCR detection for *A. ureafaciens* DnL1-1. However, less than 40 % of viable *A. ureafaciens* cells could be recovered on SM agar at this population density (see Methods section), indicating that actual titers of the atrazine degrader in the maize

rhizosphere were significantly higher than those detected on SM plates. Adjustment based on the percent recovery gave the densities of atrazine-degrading *A. ureafaciens* in TD(a) and GD soils close to the limit of their detection by PCR assay (10^5 CFU g^{-1} soil), suggesting that *A. ureafaciens* bacteria might represent a substantial or even dominant fraction of atrazine-degrading communities in agricultural soils.

Conclusions

The present study provides the first example of direct detection and isolation of atrazine-degrading bacteria on a specially developed selective agar SM with atrazine as sole nitrogen source and Tween 80 as wetting agent and dispersant. This method overcomes the limitations of common enrichment protocols and facilitates enumeration and rapid isolation of atrazine degraders. Elimination of the enrichment bias enables analysis of the diversity and community structure of culturable atrazine degrading bacteria in soil. Advantages of the direct isolation make it a reasonable method for future investigations of the ecology and biogeography of atrazine degraders.

The highly specific primers designed in this work enabled PCR-detection of atrazine-degrading *Pseudomonas* sp. populations as low as 10^3 CFU g^{-1} soil. However, the assay was at least 100 times less sensitive for the detection of atrazine-degrading *A. ureafaciens*, indicating that its results could be greatly affected by the differential

DNA extractability from targeted soil bacteria. For this reason, the PCR assay is a useful supplemental method of the detection rather than an alternative to the direct plating.

Both the direct plating and culture-independent assays provided evidence that the atrazine degraders constituted a major component of microbial populations in industrially contaminated soils and a minor one in the maize rhizosphere. The industrial soils harbored communities of genetically distinct bacteria that were individually capable of degrading and utilizing atrazine. Genetic structures of the atrazine-degrading communities in soils differentially exposed to atrazine did not overlap, indicating intensive selection of bacteria better adapted to local environments. However, the range of the atrazine degraders was limited to a variety of phylotypes belonging to the genus *Arthrobacter*, 2 phylotypes within the genus *Nocardioides*, and single genotypes within phylogenetic groups of *Pseudomonas alcaliphila* and *Gulosibacter molinativorax*, suggesting that the individual atrazine utilization is a trait of soil bacteria belonging to a limited number of lineages.

Strains of *P. alcaliphila* phylogenetic group were the only harboring the gene *atzA*, while all other isolates possessed *trzN*. The strong prevalence of *trzN*-bearing atrazine degraders in all the industrial and agricultural soils, except that from the site where atrazine-degrading pseudomonads were isolated, was confirmed by PCR assay. Bacteria related to *P. alcaliphila*, *G. molinativorax* and *A. crystallopoietes* were major atrazine-degrading inhabitants of the heavily contaminated plantless soil. The rhizosphere of growing in the industrial soils cogon grass and common reed harbored an abundance of atrazine-degrading *Arthrobacter* spp. with a strong prevalence of a genomospecies closely related to, but distinct from, the species *A. aurescens*, *A. ilicis*, and *A. nitroguajacolicus*.

In contrast to the diversity of atrazine-degrading *Arthrobacter* spp. in industrial soils, genetically similar *A. ureafaciens* bacteria were the dominant culturable atrazine degraders and the only found atrazine degrading representatives of the genus in the maize rhizosphere at all agricultural sampling sites. The contribution of *A. ureafaciens* bacteria to the enhanced degradation of atrazine in agricultural soils and mechanisms causing their prevalence among atrazine degraders in the maize rhizosphere deserve further study.

Methods
Soil sampling and processing
The geographic coordinates of sampling sites were determined using GPS + GLONASS receiver Garmin eTrex 30 with an accuracy of 3 m. Cropping and herbicide histories for at least the past 5 years were obtained from specialists of the local agriculture bureaus and farm managers. At each field site, 3 replicate samples were collected 2.0 – 2.5 m apart and then processed independently. Soil cores 15 × 10 × 10 cm (length × width × depth) containing grass roots or the root system of an individual maize plant were carefully excavated with a surface-sterilized shovel avoiding mechanical disturbance to the samples in order to retain intact soil structure. Shoots of common reed and cogon grass were cut to a length of 15 cm, and stems of the maize plants were cut above the first node. The intact soil cores were placed into plastic boxes, delivered to the laboratory the day of sampling and kept at room temperature (23–25 °C) overnight till processing.

Soil samples were processed aseptically to prevent cross-contamination. S1 soil was sieved to 2 mm and thoroughly mixed. Triplicate 20 g sub-samples of the homogenized S1 soil were oven-dried at 105 °C to constant weight in order to determine moisture content.

For isolation of atrazine degraders, soil samples were processed the day after delivery (within 22–26 h after sampling). Plant roots were carefully retrieved from the cores and the loose soil was removed by hand using sterile latex gloves. Roots of common reed and cogon grass with tightly adhering soil were cut from the rhizomes, clipped into approximately 2.5 cm pieces, mixed in a Petri dish and then directly used in the isolation procedure. To recover rhizosphere soil of maize, the soil tightly associated with seminal and nodal roots was scraped off with the blunt side of a scalpel, placed into a Petri dish and mixed. Soil remaining in the box was considered to be bulk soil. The bulk soils were homogenized and the moisture contents were determined in the manner described for S1 soil. Because the high humidity prevented sieving the agricultural bulk soils, they were preliminary air-dried for 2 days. For this reason, dry matter contents required for normalizing CFU densities of detected atrazine-degraders were determined in 20 g sub-samples of agricultural bulk soils collected before their drying and further processing.

Samples of bulk soils reserved for analysis of texture and chemical characteristics were kept at –20 °C. Soil characteristics were analyzed in the Laboratory of Environmental Analysis of the Shandong Provincial Analysis and Test Center.

Atrazine extraction and quantification
Extraction of atrazine from soil samples was performed in the manner described by Krutz et al. [17]. The extracts were analyzed by high performance liquid chromatography-tandem mass spectrometry (HPLC-MS/MS) on an UltiMate 3000 HPLC system interfaced to a TSQ Vantage Triple Quadrupole Mass Spectrometer (Thermo Fisher Scientific, USA). The conditions

of HPLC and MS/MS parameters are provided in Additional file 12. Calibration standards of atrazine (# 45330 Sigma-Aldrich, USA) were prepared in HPLC grade methanol (Sinopharm, China) at the following concentrations: 0.01, 0.05, 0.1, 0.5, 1, 5, 10, 50 and 100 ng mL^{-1}. The typical calibration curve was Y = 2313.0 + 4073.8X. The curve displayed excellent linearity ($r^2 = 0.9995$) over the entire calibration range. Recovery of atrazine from fortified soil samples was 90.3–93.2 %. The method limit of atrazine quantification was 0.05 μg kg^{-1}.

Bacterial strains

The strain *Arthrobacter* sp. SD41 was previously isolated from the rhizosphere of wheat sampled in Yucheng County, Dezhou Prefecture of Shandong Province [34]. The species type strain *A. ureafaciens* CGMCC 1.1897^T was obtained from the China General Microbiological Culture Collection Center (www.cgmcc.net).

Growth media

Mineral media SM, SMY and nutrient medium TY were used for enumeration, isolation, purification and maintenance of atrazine-degrading bacteria according to the methods described. The medium SM contained (per liter of distilled water) 0.5 g K_2HPO_4, 0.2 g $MgSO_4 \cdot 7H_2O$, 0.1 g NaCl, 0.02 g $CaCl_2$, 2 g D-glucose, 10 mL atrazine stock solution, 5 mL ZnFe-citrate stock solution. The atrazine stock solution contained (per 100 mL of distilled water) 1 mL Tween 80 and 5 g atrazine powder (≥97 %, Shandong Dehao, China). ZnFe-citrate stock solution contained (per 100 mL of distilled water) 0.04 g $ZnSO_4 \cdot 7H_2O$, 0.4 g $FeSO_4 \cdot 7H_2O$ and 10 g trisodium citrate. SM25 medium was SM with concentration of atrazine reduced to 25 mg L^{-1}. SMY medium was SM amended with 0.1 g L^{-1} yeast extract (Oxoid, England). TY medium contained (per liter of distilled water) 10 g tryptone (Oxoid, England), 1 g yeast extract (Oxoid, England), and 0.02 g $CaCl_2$. The solid media SM, SMY and TY were supplemented with 13 g L^{-1} bacteriological agar (Beijing Dingguochangsheng Biotechnology Co., Ltd., China).

Recovery of a reference atrazine-degrading strain

The recovery of atrazine degraders was studied after inoculation of S1 soil with known titers of *Arthrobacter* sp. SD41. Previously, the presence of atrazine degraders in S1 soil was checked by enrichment in liquid SM medium performed in the manner described by Mandelbaum et al. [35]. No visible dissolution of atrazine was observed in the cultures during 3 enrichment cycles, and no colonies with clearing zones were detected after their plating on SM agar, indicating no culturable atrazine degraders.

To prepare inocula, twenty 3 day-old colonies of the strain *Arthrobacter* sp. SD41 were harvested from an SMY plate, suspended in 1 mL buffer (SM medium salt solution), twice washed by centrifugation (3000 × g, 2 min) and serially diluted. CFU titers were determined by plating the dilutions on SM agar and incubation at 28 °C for 4 days.

Raw S1 soil was weight (100 mg dry weight equivalents) to 2 mL polypropylene tubes. The samples were spiked with the dilutions of *Arthrobacter* sp. SD41 cell suspension (20 μL/ sample), incubated on the bench for 5 min., and then 1 mL buffer was added to each tube. The tubes were secured in a MO BIO Vortex Adapter assembled on a Vortex-Genie® 2 Vortex (MO BIO Laboratories, Inc., USA) and vortexed at maximum speed for 10 min. The resulting soil suspensions and their serial dilutions were plated on SM agar. Colonies with typical zones of atrazine dissolution were counted after 4-day incubation at 28 °C. Mean CFU numbers, percent recoveries and 95 % confidence intervals were calculated based on 3 replications for each dilution by using the descriptive statistics tool of MS Excel.

Direct plating of the inoculated S1 soil on SM agar enabled selective isolation of *Arthrobacter* sp. SD41 (Additional file 13: Figure S10). Nearly 1/3 of the bacterial cells were recovered at inoculation densities from 10^3 to 10^5 CFU g^{-1} soil (Table 5). No atrazine-degrading colonies were recovered from S1 soil spiked with 10^2 CFU g^{-1}, indicating the method limit of *Arthrobacter* sp. SD41 detection was 10^3 CFU g^{-1} soil. Values close to full recovery were obtained at inoculation density near 10^6 CFU g^{-1} soil, or about 1 % of the resident population, $(1.2 \pm 0.1) \times 10^8$ CFU g^{-1} soil, enumerated by plating on TY agar. Recoverability of *Arthrobacter* sp. SD41 from soil dilutions directly plated on SM agar was similar to that obtained for Tn5-marked *Pseudomonas putida* and *Rhizobium* spp. enumerated by the most-probable-number–DNA hybridization procedure [58]. Hence, the direct plating on SM agar was considered a promising method of detection, enumeration and

Table 5 Recovery of *Arthrobacter* sp.SD41 from S1 soil by direct plating on SM agar

Inoculation density, CFU g^{-1} soil	Recovery, CFU g^{-1} soil[a]	Percent recovery[a]
No inoculation (control)	<10^2	ND
1.6×10^2	<10^2	ND
1.6×10^3	$(6.0 \pm 2.4) \times 10^2$	37.5 ± 15.0
1.6×10^4	$(5.5 \pm 0.7) \times 10^3$	34.8 ± 4.8
1.6×10^5	$(6.3 \pm 0.8) \times 10^4$	39.4 ± 5.0
1.6×10^6	$(1.4 \pm 0.1) \times 10^6$	87.5 ± 6.3

[a] Means and 95 % confidence intervals. *ND* not determined

isolation of atrazine-degrading bacteria from industrially contaminated and agricultural soils.

Detection, enumeration and isolation of atrazine-degrading bacteria

Isolation of atrazine-degrading bacteria was commenced immediately after the processing of soil samples. To detect, enumerate and isolate atrazine degraders, 0.1 g of bulk or rhizosphere soil, or (for samples D5 and D6) 0.1 g of common reed or cogon grass root sections with tightly adhering soil were placed into 2 mL polypropylene tubes with 1 mL washing buffer (SM medium salt solution). The tubes were secured in a MO BIO Vortex Adapter assembled on a Vortex-Genie® 2 Vortex (MO BIO Laboratories, Inc., USA) and vortexed at maximum speed for 10 min. Washed root sections from D5 and D6 rhizosphere samples were removed, blotted and weighed to determine the exact amount of soil in the tubes. Serial dilutions of the resulting suspensions were plated onto solid media SM and TY. The plates were incubated at 28 °C. Colonies of soil bacteria growing on TY agar were counted after 5 days. Colonies of atrazine degraders, which produced clearing zones, were first counted after 3-day incubation. The appearance of additional atrazine-degrading colonies was checked daily during the following week. Bacterial densities and confidence intervals were calculated according to Koch [59].

For isolation of atrazine degraders, colonies were chosen based on their morphology, size of clearing zones and time required for their production. Several colonies within each distinguishable type were selected to ensure the most complete isolation of different bacteria. The colonies were repeatedly streaked on SMY agar until cultures of atrazine-degrading bacteria showing no presence of contaminating microorganisms on SMY, TY and R2A (Sigma-Aldrich, USA) agar media were obtained.

Evaluation of atrazine-degrading capacity of isolates

Strains of atrazine degraders were cultured in 25 mL of SM25 medium in 250 mL Erlenmeyer flasks without shaking at 28 °C for 7 days. After incubation, the cultures were centrifuged (4000 g, 2 min.). Clear cultural liquids were transferred to polypropylene tubes and kept at –20 °C until analysis. The medium SM25 kept at –20 °C from the beginning of the experiment, and the medium SM25 incubated under the same conditions as the cultures were used as controls. The liquids were analyzed by HPLC-MS/MS in the above described manner.

DNA extraction

Template DNAs for PCR were extracted from pure cultures of the isolates grown on SMY agar by thermal lysis in a 5 % suspension of Chelex 100 resin (BioRad, USA)

performed in the manner described by Mahenthiralingam et al. [60].

Soil DNA was extracted by using a Power Soil DNA Isolation Kit (MO BIO Laboratories, Inc., USA), according to the manufacturer's instructions, immediately after processing of the soil samples. The isolated soil DNAs were kept at –70 °C for analysis.

PCR primers

Characteristics of primers used in this research are summarized in Table 6. All primers were synthesized by Sangon Biotech (Shanghai) Co., Ltd., China.

New primers were designed by using the NCBI primer-BLAST tool, ensuring verification of their specificity to the target genes *atzABC* from *Pseudomonas* sp. ADP (GenBank accession no. U66917, regions 34964–36388, 44487–45932, and 70219–71430 respectively), *atzABC* from *Herbaspirillum huttiense* B601 (GenBank accession nos. DQ089655, AY965854, and AY965855 respectively), *trzN* from *Nocardioides* sp. C190 (GenBank accession no. AF416746, region 4490–5860). Since high sequence similarity was found between the melamine deaminase gene *triA* from *Pseudomonas* sp. strain NRRL B-12227 and known *atzA* genes [61], the *atzA*-targeting primers were designed so as to match *atzA*-specific positions by their 3' nucleotides.

To amplify 16S rRNA gene sequences of the diverse bacteria, the primer pair 63KWf/1389r was designed, based on the primers 63f and 1387r, described by Marchesi et al. [62] as useful ones for ecological and systematic studies. Alignment with 16S rRNA gene sequences from the GenBank database revealed G or T in the 3rd position, and A or T in the 8th position from the 5' end of the forward primer in many bacteria. Hence, these residues were synthesized as K and W respectively in the modified primer 63KWf. The 2nd position from the 3' end of the primer 1387r was also found to vary. Therefore, the primer 1389r was proposed instead of 1387r.

PCR-detection and sequencing of atrazine degradation genes

All PCRs were done in a BioRad Verity thermal cycler.

Specificity of the primers targeting genes for atrazine degradation was assessed in gradient PCR with DNA isolated from S1 soil as a template. Reactions were performed in a total volume of 20 μL containing: 2 μL of *TaKaRa* 10× *Ex Taq* Buffer, (Takara Biotechnology (Dalian) Co., Ltd., China), 2.0 mM MgCl$_2$, 200 μM of each dNTP, 0.5 μM of each primer, 0.5 U of *TaKaRa Ex Taq* polymerase (Takara Biotechnology (Dalian) Co., Ltd., China), and 0.5 μL of the soil DNA as a template. The temperature program was as follows: denaturation at

Table 6 DNA primers

Target gene	Primer[a]	Nucleotide sequences (5' → 3')	Reference
atzA	atzA250f	TCGCACGGGCGTCAAT	This study
	atzA655f	CGCTCCTGCCACTACCA	This study
	atzA650r	TGTCACCGCCGTGGTAG	This study
	atzA757r	GCGGGACTCATCCCATGAAT	This study
	atzA982r	TACGGAGTCATTACTATTCCCGTT	This study
atzB	B1f	AGGGTGTTGAGGTGGTGAAC	[51]
	B1r	CACCACTGTGCTGTGGTAGA	[51]
	atzB181f	GGGTGTTGAGGTGGTGAACT	This study
	atzB426f	ACCAGTACAACTACAGCCGC	This study
	atzB681f	TGATTGCCTACCCGGAAACC	This study
	atzB316r	TCTTCATCCACCAGGGCAAA	This study
	atzB662r	GGTTTCCGGGTAGGCAATCA	This study
	atzB919r	CTTCGGCACCCACCAGAAA	This study
atzC	atzCf	GCTCACATGCAGGTACTCCA	[21]
	atzCr	TGTACCATATCACCGTTGCCA	[21]
	Cf	GCTCACATGCAGGTACTCCA	[51]
	C1r	TCCCCCAACTAAATCACAGC	[51]
	atzC340f	TGTGATAGAACATGCTCACATGC	This study
	atzC552r	TAGCAGGATCAACTCCCCCA	This study
trzN	C190-10	CACCAGCACCTGTACGAAGG	[22]
	C190-11	GATTCGAACCATTCCAAACG	[22]
	trzN1114f	AATGGCAACCAGGGGATCAG	This study
	trzN1271f	GAGCACCTGACCATTCACGA	This study
rrs	63KWf	CAKGCCTWACACATGCAAGTC	[62], this study
	1389r	ACGGGCGGTGTGTACAAG	[62], this study
ERIC	ERIC2	AAGTAAGTGACTGGGGTGAGCG	[64]
BOX	BOXA1R	CTACGGCAAGGCGACGCTGACG	[65]

[a]Numbers in designations of the primers designed in this work are positions of their 3' nucleotides in coding direct sequences of the respective genes: atzA, atzB, atzC from Pseudomonas sp. ADP (GenBank accession no. U66917, regions 34964-36388, 44487-45932, 70219-71430 respectively), and trzN from Nocardioides sp. C190 (GenBank accession no. AF416746); f – forward, r - reverse

95 °C for 3 min; then 40 cycles consisting of 94 °C for 1 min, T_a range 58-68 °C with 2 °C step for 1 min, 72 °C for 1 min; and a final extension at 72 °C for 2 min. Conventional PCR to detect atrazine degradation genes in soil was performed in the same manner using hot start *TaKaRa Ex Taq* HS polymerase with the selected primer pairs *atzA655f/atzA982r*, *atzB181f/atzB316r*, *atzC340f/atzC552r* and *trzN1114f/trzN1271r* at T_a 68 °C. The number of cycles was reduced to 35.

Detection of the genes *atzA*, *atzB*, *atzC* and *trzN* in the isolated bacterial cultures was carried out by multiplex PCR using primer pairs *atzA250f/atzA650r*, *atzB426f/atzB662r*,

atzCf/atzCr, and *trzN1114/trzN1271r*. Reactions were done in a total reaction volume of 20 μL containing: 2 μL of *TaKaRa* 10× *Ex Taq* Buffer, (Takara Biotechnology (Dalian) Co., Ltd., China), 2.0 mM MgCl$_2$, 250 μM of each dNTP, 0.2 μM of each primer, 0.5 U of *TaKaRa Ex Taq* HS polymerase (Takara Biotechnology (Dalian) Co., Ltd., China) and 0.5 μL of a bacterial lysate as a template. The temperature program was as follows: denaturation at 95 °C for 3 min; then 25 cycles consisting of 94 °C for 1 min, 62° for 30 sec., 72 °C for 1 min; and a final extension at 72 °C for 2 min.

Fragments of the predicted sizes produced by the reference strains in the multiplex PCRs were cut out of the gel and purified using a *TaKaRa* MiniBEST Agarose Gel DNA Extraction Kit Ver. 4.0 (Takara Biotechnology (Dalian) Co., Ltd., China). Additionally, a 0.42 kb fragment of *trzN* from *Arthrobacter* sp. SD41 was amplified using a conventional primer pair C190-10/C190-11. The fragments were directly sequenced in both directions exploiting respective amplification primers. The sequencing reactions were performed using a BigDye Terminator v.3.1 Cycle Sequencing Kit (Applied Biosystems, United States) according to the manufacturer's recommendations. Automated sequencing was performed on a 3730xl DNA Analyzer (Applied Biosystems, United States) at the Sequencing Department of the Sangon Biotech (Shanghai) Co., Ltd. The resulting DNA traces and sequences were checked and corrected manually.

Sequencing of the fragments produced in the PCRs targeting the genes for atrazine degradation in soil DNA was carried out in a similar manner.

Genotyping

Genotyping of the isolates was performed by repetitive elements sequence-based PCR (Rep-PCR) with primers ERIC2 or BOXA1R (Table 6). The reaction mixture (20 μL) contained 2 μL of *TaKaRa* 10× *Ex Taq* Buffer, (Takara Biotechnology (Dalian) Co., Ltd., China), 2.0 mM MgCl$_2$, 250 μM of each dNTP, 1.0 μM of one of the primers, 0.5 U of *TaKaRa Ex Taq* polymerase (Takara Biotechnology (Dalian) Co., Ltd., China) and 0.5 μL of a bacterial lysate as a template. Rep-PCRs were started by denaturation at 95 °C for 3 min. followed by 4 cycles at 94 °C for 1 min., 40 °C (ERIC-PCR) or 55 °C (BOX-PCR) for 1 min., 68 °C for 8 min.; followed by 30 cycles: 94 °C for 1 min., 52 °C (ERIC-PCR) or 65 °C (BOX-PCR) for 1 min., 72 °C for 2 min. A final extension was performed at 72 °C for 5 min.

Products of the amplification were separated by electrophoresis on a 2.0 % agarose gel (Genview, China) in 0.5 × TBE. The gel was supplemented with 50 μL L^{-1} GoldView Nucleic Acid Stain (Beijing Dingguochangsheng Biotechnology Co., Ltd., China) in order to visualize DNA bands. The Rep-PCR banding patterns were analyzed visually, and similar ones were considered to belong to the same genotypic group.

Sequencing of 16S rRNA genes and phylogenetic analysis

The genes for 16S rRNA were amplified with the primer pair 63KWf/1387r. The PCRs were performed in a total volume of 50 μL containing 5 μL of *TaKaRa* 10× *Ex Taq* Buffer (Takara Biotechnology (Dalian) Co., Ltd., China), 2.0 mM $MgCl_2$, 250 μM of each dNTP, 0.5 μM of each primer, 0.5 U of *TaKaRa Ex Taq* polymerase (Takara Biotechnology (Dalian) Co., Ltd., China) and 1.25 μL of a bacterial lysate as a template. The PCRs were started by denaturation at 95 °C for 3 min.; and consisted of 30 cycles: 94 °C for 1 min., 55 °C for 1 min., 72 °C for 2 min.; followed by extension at 72 °C for 5 min. The amplification products were analyzed by electrophoresis on an agarose gel in 0.5 × TBE. Target fragments of about 1.3 kb were cut out from the gel, purified and sequenced in the manner described for the fragments of atrazine degradation genes.

The BLASTn similarity search was performed against 16S ribosomal RNA sequences database of GenBank. The phylogenetic analysis was performed using the MEGA5 software package [63]. 16S rRNA gene nucleotide sequences of known atrazine-degrading bacteria and species type strains sharing more than 98 % sequence similarity with the analyzed bacteria were included in the datasets. Multiple alignments were implemented using the CLUSTALW aligner of MEGA5 and then refined by hand. Phylogenies were inferred using the Neighbor-Joining algorithm with elimination of all positions containing gaps and missing data.

Additional files

Additional file 1: Figure S1. Map of Shandong Province and location of sampling sites. Location of sampling sites is indicated by the symbol ● with following designations: S - Jinan, Eastern Campus of the Shandong Academy of Sciences; D - Dajiawa Town, Weifang Prefecture; TD - Dongdaguan Village, Dawenkou Township, Tai'an Prefecture; DnW - Wangdian Village, Binhe Township, Dingtao County, Heze Prefecture; DnL - Liulou Village, Binhe Township, Dingtao County, Heze Prefecture; GD - Dawang Village, Guhe Township, Gaotang County, Liaocheng Prefecture; WS - Shadong Village, Guangyun Township, Wucheng County, Dezhou Prefecture.

Additional file 2: Table S1. Soil characteristics.

Additional file 3: Figure S2. Direct isolation of atrazine-degrading bacteria on SM agar. 1st dilution of soil suspension (a total of 5 mg soil) was spread on the medium surface. The total population of culturable bacteria was about 10^9 CFU g^{-1} soil. The incubation time is indicated near the plates.

Additional file 4: Figure S3. Discrimination of atrazine-degrading bacteria by ERIC-PCR banding patterns. ERIC types of the bacteria are designated by capital letters under the square brackets. Letters in the name of each strain are designations of the locations as indicated in Table 1. Numbers after letter D represent sampling sites at the location D as listed in Table 1. Letter "r" following D5 and D6 indicates the rhizosphere isolates. The first digit in the names of strains isolated from TD, DnW, DnL, GD and WS sites indicates replicate samples. Site TD(a) was represented by replicate samples 1–3, and site TD(b)

by replicate samples 4–6. Lanes M contain DL 2000 DNA Marker (Takara Biotechnology (Dalian) Co., Ltd., China).

Additional file 5: Figure S4. Amplification products obtained in multiplex gradient PCR targeting *trzN*, *atzA*, *atzB* and *atzC* with DNAs of the reference strains as templates. Lanes M contain DL 2000 DNA Marker (Takara Biotechnology (Dalian) Co., Ltd., China).

Additional file 6: Nucleotide sequences of short gene fragments.

Additional file 7: Figure S5. Detection of the genes for atrazine degradation in isolates by the multiplex PCR assay. Lanes are designated by the isolates' names. Lanes M1 contain a 100 bp DNA Ladder, lane M2 - DL 2000 DNA Marker (Takara Biotechnology (Dalian) Co., Ltd., China).

Additional file 8: Figure S6. ERIC-PCR patterns of representative ERIC type B isolates. Lanes are designated by the isolates' names. Lane T – pattern of the species type strain *A. ureafaciens* CGMCC 1.1897[T]. Lane M contains DL 2000 DNA Marker (Takara Biotechnology (Dalian) Co., Ltd., China).

Additional file 9: Figure S7. BOX-PCR typing of ERIC type B isolates. Lanes are designated by the strain names. Lane T – pattern of the species type strain *A. ureafaciens* CGMCC 1.1897[T]. Lanes M contain a 100 bp DNA Ladder (Takara Biotechnology (Dalian) Co., Ltd., China).

Additional file 10: Figure S8. Amplification of unintended products in PCRs with primers targeting the genes for atrazine degradation. Lanes are designated by values of T_a. Lanes designated by M contain DL 2000 DNA Marker (Takara Biotechnology (Dalian) Co., Ltd., China). The primer pairs are printed on respective gel zones under or below banding patterns.

Additional file 11: Figure S9. Detection of the gene *atzB* in PCRs with template DNAs isolated from S1 soil aliquots with known titers of *Pseudomonas* sp.D3-1l. (Positive controls for *atzB* detection results represented in Fig. 4). Lanes are designated by titers of *Pseudomonas* sp.D3-1l. Lane N and C are, respectively, no template control and a control in which DNA isolated from non-inoculated S1 soil was used as a template. Lanes M contain DL 2000 DNA Marker (Takara Biotechnology (Dalian) Co., Ltd., China).

Additional file 12: Conditions of HPLC and parameters of MS/MS.

Additional file 13: Figure S10. Recovery of *Arthrobacter* sp. SD41 from soil by direct plating on SM agar. S1 soil sample was inoculated with about 10^4 CFU g^{-1} of *Arthrobacter* sp.SD41. 1st (left dish) and 2nd (right dish) dilutions of soil suspension were plated on SM agar. Photos of the dishes were taken after 3 (left) and 5 (right) days of incubation at 28 °C.

Abbreviations

BOX-PCR: Polymerase chain reaction targeting bacterial repetitive BOX element; ERIC-PCR: Enterobacterial repetitive intergenic consensus PCR; HPLC-MS/MS: High performance liquid chromatography - tandem mass spectrometry; Rep-PCR: Repetitive elements sequence-based PCR

Acknowledgements

We are grateful to Xingwen Gao from Tai'an Agriculture Bureau, Yewu Cheng from Dingtao Agriculture Bureau, and Fengyan Wang from Dezhou Agriculture Bureau for locating the agricultural sampling sites and organizing samplings. We thank Maarten Ryder for reading this manuscript and providing helpful suggestions.

Funding

This work was supported by the grants WQ20123700079 from One Thousand Talents Plan of China (www.1000plan.org), tshw20120743 from Taishan Scholar Plan of Shandong Province, 2014QN021 from Youth Foundation of Shandong Academy of Sciences and BS2014SW030 from Science Foundation for Distinguished Young Scholars of Shandong Province (http://www.sdnsf.gov.cn).

www.ncbi.nlm.nih.gov/genbank) under Accession Nos. KP994319 - KP994322, KP997248 - KP997250. The nucleotide sequences of 16S rRNA genes were deposited in GenBank (http://www.ncbi.nlm.nih.gov/genbank) under Accession Nos. KF889364 - KF889398, KJ010189. Full sequence alignments and phylogenetic trees are available in TreeBASE (https://treebase.org/treebase-web/home.html), study ID: S20014, URL for programmatic access to the data using the PhyloWS API: http://purl.org/phylo/treebase/phylows/study/TB2:S20014

Authors' contributions

DPB CL XC HY and JL conceived and designed the experiments. CL HL DPB XC and HZ performed the experiments. DPB and CL drafted the manuscript. All authors read and approved the final manuscript.

Competing interests

The authors declare that they have no competing interest.

Author details

[1]Key Laboratory for Applied Microbiology of Shandong Province, Ecology Institute (Biotechnology Center) of Shandong Academy of Sciences, Jinan, Shandong Province, People's Republic of China. [2]Biology Institute of Shandong Academy of Sciences, Jinan, Shandong Province, People's Republic of China. [3]Shandong Provincial Analysis and Test Center of Shandong Academy of Sciences, Jinan, Shandong Province, People's Republic of China.

References

1. Müller G. History of the discovery and development of triazine herbicides. In: LeBaron HM, McFarland J, Burnside OC, editors. The triazine herbicides: 50 years revolutionizing agriculture. San Diego: Elsevier; 2008. p. 13–43.
2. Thelin GP, Stone WW. Estimation of annual agricultural pesticide use for counties of the conterminous United States, 1992–2009. Reston: U.S. Geological Survey; 2013. p. 54. http://101.96.10.62/pubs.usgs.gov/sir/2013/5009/pdf/sir20135009.pdf. Accessed 27 Oct 2016.
3. Ye C, Gong A, Wang X, Zheng H, Lei Z. Distribution of atrazine in a crop-soil-groundwater system at Baiyangdian lake area in China. J Environ Sci. 2001;13:148–52.
4. Ren J, Jiang K. Atrazine and its degradation products in surface and ground waters in Zhangjiakou district, China. Chin Sci Bull. 2002;47:1612–5.
5. Kookana RS, Baskaran S, Naidu R. Pesticide fate and behaviour in Australian soils in relation to contamination and management of soil and water: a review. Aust J Soil Res. 1998;36:715–64.
6. Cašić S, Budimir M, Brkić D, Nešković N. Residues of atrazine in agricultural areas of Serbia. Journal of Serbian Chemical Society. 2002;67:887–92.
7. Pick FE, van Dick LP, Botha E. Atrazine in ground and surface water in maize production areas of the Transvaal, South Africa. Chemosphere. 1992;25:335–41.
8. Shomar BH, Müller G, Yahya A. Occurrence of pesticides in groundwater and topsoil of the Gaza Strip. Water Air Soil Pollut. 2006;171:237–51.
9. Aslam M, Alam M, Rais S. Detection of atrazine and simazine in ground water of Delhi using high performance liquid chromatography with ultraviolet detector. Current World Environment. 2013;8:323–9.
10. Cook AM. Biodegradation of s-triazine xenobiotics. FEMS Microbiol Rev. 1987;46:93–116.
11. Smith AE. Herbicides and the soil environment in Canada. Can J Soil Sci. 1982;62:433–60.
12. Smith AE, Walker A. Prediction of the persistence of the triazine herbicides atrazine, cyanazine, and metribuzin in Regina heavy clay. Can J Soil Sci. 1989;69:587–95.
13. Tasli S, Patty L, Boetti H, Ravanel P, Vachaud G, Scharff C, et al. Persistence and leaching of atrazine in corn culture in the experimental site of La Côte

Saint André (Isère, France). Arch Environ Contam Toxicol. 1996;30:203–12.
14. Capriel P, Haisch A, Khan SU. Distribution and nature of bound (nonextractable) residues of atrazine in a mineral soil nine years after the herbicide application. J Agric Food Chem. 1985;33:567–9.
15. Krutz LJ, Shaner DL, Weaver MA, Webb RMT, Zablotowicz RM, Reddy KN, et al. Agronomic and environmental implications of enhanced s-triazine degradation. Pest Manag Sci. 2010;66:461–81.
16. Vonberg D, Hofmann D, Vanderborght J, Lelickens A, Köppchen S, Pütz T, Barauel P, Vereecken H. Atrazine soil core residues analysis from an agricultural field 21 years after its ban. J Environ Qual. 2014;43:1450–9. doi:10.2134/jeq2013.12.0497.
17. Krutz LJ, Shaner DL, Accinelli C, Zablotowicz RM, Henry WB. Atrazine dissipation in s-triazine-adapted and non-adapted soil from Colorado and Mississippi: implications of enhanced degradation on atrazine fate and transport parameters. J Environ Qual. 2008;37:848–57.
18. Krutz LJ, Zablotowicz RM, Reddy KN. Selection pressure, cropping system, and rhizosphere proximity affect atrazine degrader populations and activity in s-triazine–adapted soil. Weed Science. 2012;60:516–24.
19. Mandelbaum RT, Allan DL, Wackett LP. Isolation and characterization of a *Pseudomonas* sp. that mineralize the s-triazine herbicide atrazine. Appl Environ Microbiol. 1995;61:1451–7.
20. Shapir N, Mongodin EF, Sadowsky MJ, Daugherty SC, Nelson KE, Wackett LP. Evolution of catabolic pathways: genomic insights into microbial s-triazine metabolism. J Bacteriol. 2007;189:674–82.
21. de Souza ML, Seffernick J, Martinez B, Sadowsky MJ, Wackett LP. The atrazine catabolism genes *atzABC* are widespread and highly conserved. J Bacteriol. 1998;180:1951–4.
22. Mulbry WW, Zhu H, Nour SM, Topp E. The triazine hydrolase gene *trzN* from *Nocardioides* sp. strain C190: Cloning and construction of gene-specific primers. FEMS Microbiol Let. 2002;206:75–9.
23. Saldick J. Biodegradation of cyanuric acid. Appl Microbiol. 1974;28:1004–8.
24. Cook AM, Beilstein P, Grossenbacher H, Hutter R. Ring cleavage and degradative pathway of cyanuric acid in bacteria. Biochem J. 1985;231:25–30.
25. Cai B, Han Y, Liu B, Ren Y, Jiang S. Isolation and characterization of an atrazine-degrading bacterium from industrial wastewater in China. Lett Appl Microbiol. 2003;36:272–6.
26. Dai X, Jiang H, Jiang J, Wei H, Li S. Isolation and identification of in situ atrazine-degrading bacteria from contaminated soils. Acta Pedologica Sinica. 2006;43:467–72. (In Chinese) http://pedologica.issas.ac.cn/trxb/ch/reader/view_abstract.aspx?flag=1&file_no=20060316&journal_id=trxb. Accessed 27 Oct 2016.
27. Li Q, Li Y, Zhu X, Cai B. Isolation and characterization of atrazine-degrading *Arthrobacter* sp. AD26 and use of this strain in bioremediation of contaminated soil. J Environ Sci (China). 2008;20:1226–30.
28. Zheng L, Yuan B, Zhu X, Cai B. Isolation and characterization of atrazine-degrading strains and biotreatment experiment of industrial wastewater. Microbiol China. 2009;36:1099–114. (In Chinese) http://journals.im.ac.cn/wswxtbcn/ch/reader/view_abstract.aspx?file_no=tb09071099&flag=1. Accessed 27 Oct 2016.
29. Dai X, Jiang H, Gu L, Li R, Li S. Isolation and characterization of atrazine-degrading bacterium strain SA1. Acta Microbiologica Sinica. 2007;47:544–7. (In Chinese) http://journals.im.ac.cn/actamicrocn/ch/reader/view_abstract.aspx?file_no=07030544&flag=1. Accessed 27 Oct 2016.
30. Han P, Hong Q, He L, Yan Q, Li S. Isolation, identification and characterization of an atrazine-degrading bacteria ADH-2. J Agro-Environ Sci. 2009;28:406–10. (In Chinese) http://www.aes.org.cn/nyhjkxxb/ch/reader/view_abstract.aspx?file_no=10848&flag=1. Accessed 27 Oct 2016.
31. Liu C, Yang F, Lu X, Huang F, Liu L, Yang C. Isolation, identification and soil remediation of atrazine-degrading strain T3AB1. Acta Microbiologica Sinica. 2010;50:1642–50. (In Chinese) http://journals.im.ac.cn/actamicrocn/ch/reader/view_abstract.aspx?file_no=1012-10&flag=1. Accessed 27 Oct 2016.
32. Zhang Y, Jiang Z, Cao B, Hu M, Wang Z, Dong X. Metabolic ability and gene characteristics of *Arthrobacter* sp. strain DNS10, the sole atrazine-degrading strain in a consortium isolated from black soil. Int Biodeterior Biodegradation. 2011;65:1140–4.
33. Wang Q, Xie S. Isolation and characterization of a high-efficiency soil atrazine-degrading *Arthrobacter* sp. strain. Int Biodeterior Biodegradation. 2012;71:61–6.
34. Li H, Zhang X, Li J, Wei Y, Chen K, Bazhanov D, et al. Experimental study on atrazine degrading strain SD41: isolation, identification and soil remediation.

Environ Sci Technol. 2014;37(4):38–41. (In Chinese) http://fjks.chinajournal.net.cn/WKB2/WebPublication/paperDigest.aspx?paperID=1ae129d8-0eb9-4d00-aa72-6170d884548f. Accessed 27 Oct 2016.

35. Mandelbaum RT, Wackett LP, Allan DL. Mineralization of the s-triazine ring of atrazine by stable bacterial mixed cultures. Appl Environ Microbiol. 1993; 59:1695–701.

36. Strong LC, Rosendahl C, Johnson G, Sadowsky MJ, Wackett LP. *Arthrobacter aurescens* TC1 Metabolizes Diverse s-Triazine Ring Compounds. Appl Environ Microbiol. 2002;68:5973–80.

37. Devers M, El Azhari N, Kolic N-U, Martin-Laurent F. Detection and organization of atrazine-degrading genetic potential of seventeen bacterial isolates belonging to divergent taxa indicate a recent common origin of their catabolic functions. FEMS Microbiol Let. 2007;273:78–86.

38. Piutti S, Semon E, Landry D, Hartmann A, Dousset S, Lichtfouse E, et al. Isolation and characterisation of *Nocardioides* sp. SP12, an atrazine-degrading bacterial strain possessing the gene *trzN* from bulk and maize rhizosphere soil. FEMS Microbiol Let. 2003;221:111–7.

39. Arbeli Z, Fuentes C. Prevalence of the gene *trzN* and biogeographic patterns among atrazine-degrading bacteria isolated from 13 Colombian agricultural soils. FEMS Microbiol Ecol. 2010;73:611–23.

40. Wang J, Zhu L, Liu A, Ma T, Wang Q, Xie H, et al. Isolation and characterization of an *Arthrobacter* sp. strain HB-5 that transforms atrazine. Environ Geochem Health. 2011;33:259–66.

41. Yan C, Lou X, Hong Q, Li S. Isolation, identification and characterization of an atrazine-degrading strain. Microbiol China. 2011;38:493–7. (In Chinese) http://journals.im.ac.cn/wswxtbcn/ch/reader/view_abstract.aspx?file_no=tb11040493&flag=1. Accessed 27 Oct 2016.

42. Sagarcar S, Nousiainen A, Shaligram S, Björklöf K, Lindström K, Jørgensen KS, et al. Soil mesocosm studies on atrazine bioremediation. J Environ Manage. 2014; 139:208–16.

43. Rousseaux S, Hartmann A, Soulas G. Isolation and characterisation of new Gram-negative and Gram-positive atrazine degrading bacteria from different French soils. FEMS Microbiol Ecol. 2001;36:211–22.

44. Satsuma K. Characterization of new strains of atrazine degrading *Nocardioides* sp. isolated from Japanese riverbed sediment using naturally derived river ecosystem. Pest Manag Sci. 2006;62:340–9.

45. Vibber LL, Pressler MJ, Colores GM. Isolation and characterization of novel atrazine-degrading microorganisms from an agricultural soil. Appl Microbiol Biotechnol. 2007;75:921–8.

46. Manaia CM, Nogales B, Weiss N, Nunes OC. *Gulosibacter molinativorax* gen. nov., sp. nov., a molinate-degrading bacterium, and classification of 'Brevibacterium helvolum' DSM 20419 as *Pseudoclavibacter helvolus* gen. nov., sp. nov. Int J Syst Evol Microbiol. 2004;53:985–94.

47. Lugtenberg BJJ, Dekkers L, Bloemberg GV. Molecular determinants of rhizosphere colonization by *Pseudomonas*. Annu Rev Phytopathol. 2001;39:461–90.

48. Chelius MK, Triplett EW. The diversity of *Archaea* and *Bacteria* in association with the roots of *Zea mays* L. Microb Ecol. 2001;41:252–63.

49. Vandenkoornhuyse P, Mahé S, Ineson P, Staddon P, Ostle N, Cliquet J-B, et al. Active root-inhabiting microbes identified by rapid incorporation of plant-derived carbon into RNA. Proc Natl Acad Sci U S A. 2007;104:16970–5.

50. Huang CF, Chaparro JM, Reardon KF, Zhang R, Shen Q, Vivanco JM. Rhizosphere interactions: root exudates, microbes, and microbial communities. Botany. 2014;92:267–75.

51. Devers M, Henry S, Hartmann A, Martin-Laurent F. Horizontal gene transfer of atrazine-degrading genes (*atz*) from *Agrobacterium tumefaciens* St96-4 pADP1::Tn5 to bacteria of maize cultivated soil. Pest Manag Sci. 2005;61:870–80.

52. Liang B, Jiang J, Zhang J, Zhao Y, Li S. Horizontal transfer of dehalogenase genes involved in the catalysis of chlorinated compounds: evidence and ecological role. Crit Rev Microbiol. 2012;38:95–110.

53. Olive DM, Bean P. Principles and application of methods for DNA-based typing of microbial organisms. J Clin Microbiol. 1999;37:1661–9.

54. Frostegård Å, Courtois S, Ramisse V, Clerc S, Bernillon D, Le Gall F, et al. Quantification of bias related to the extraction of DNA directly from soils. Appl Environ Microbiol. 1999;65:5409–20.

55. Martin-Laurent F, Philippot L, Hallet S, Chaussod R, Germon JC, Soulas G, et al. DNA extraction from soils: old bias for new microbial analysis methods. Appl Environ Microbiol. 2001;67:2354–9.

56. Feinstein LM, Sul WJ, Blackwood CB. Assessment of bias associated with incomplete extraction of microbial DNA from soil. Appl Environ Microbiol. 2009;75:5428–33.

57. Arbeli Z, Fuentes CL. Improved purification and amplification of DNA from environmental samples. FEMS Microbiol Lett. 2007;272:269–75.

58. Fredrickson JK, Bezdicek DF, Brockman FJ, Li SW. Enumeration of Tn5 mutant bacteria in soil by using most-probable-number-DNA hybridization procedure and antibiotic resistance. Appl Environ Microbiol. 1988;54:446–53.

59. Koch AL. Growth measurement. In: Gerhardt P, Murray RGE, Costilow RN, Nester EW, Wood WA, Krieg NR, Phillips GH, editors. Manual of methods for general microbiology. Washington: American Society for Microbiology; 1981. p. 179–207.

60. Mahenthiralingam E, Marchbank A, Drevinek P, Garaiova I, Plummer S. Use of colony-based bacterial strain typing for tracking the fate of *Lactobacillus* strains during human consumption. BMC Microbiol. 2009;9:251. http://www.biomedcentral.com/1471-2180/9/251 . Accessed 27 Oct 2016.

61. Seffernick J, de Souza ML, Sadowsky MJ, Wackett LP. Melamine deaminase and atrazine chlorohydrolase: 98 percent identical but functionally different. J Bacteriol. 2001;183:2405–10.

62. Marchesi JR, Sato T, Weightman AJ, Martin TA, Fry JC, Hiom SJ, et al. Design and evaluation of useful bacterium-specific PCR primers that amplify genes coding for bacterial 16S rRNA. Appl Environ Microbiol. 1998;64:795–9.

63. Tamura K, Peterson D, Peterson N, Stecher G, Nei M, Kumar S. MEGA5: Molecular evolutionary genetics analysis using maximum likelihood, evolutionary distance, and maximum parsimony methods. Mol Biol Evol. 2011;28:2731–9.

64. Versalovic J, Koeuth T, Lupski JR. Distribution of repetitive DNA sequences in eubacteria and application to fingerprinting of bacterial genomes. Nucleic Acids Res. 1991;19:6823–31.

65. Koeuth T, Versalovic J, Lupski JR. Differential subsequence conservation of interspersed repetitive *Streptococcus pneumoniae* BOX elements in diverse bacteria. Genome Res. 1995;5:408–18.

PERMISSIONS

LIST OF CONTRIBUTORS

Feng Zhao
Department of Marine Organism Taxonomy and Phylogeny, Institute of Oceanology, Chinese Academy of Sciences, 7 Nanhai Road, Qingdao 266071, People's Republic of China
Department of Molecular Ecology, University of Kaiserslautern, 67663 Kaiserslautern, Germany

Sabine Filker
Department of Molecular Ecology, University of Kaiserslautern, 67663 Kaiserslautern, Germany

Thorsten Stoeck
Department of Ecology, University of Kaiserslautern, 67663 Kaiserslautern, Germany

Kuidong Xu
Department of Marine Organism Taxonomy and Phylogeny, Institute of Oceanology, Chinese Academy of Sciences, 7 Nanhai Road, Qingdao 266071, People's Republic of China
University of Chinese Academy of Sciences, Beijing 100049, China

Gislaine Aurelie Kemegne, Jean Justin Essia Ngang and Sylvain Leroy Sado Kamdem
Department of Microbiology, Faculty of Science, University of Yaoundé I, Yaoundé, Cameroon

Pierre Mkounga and Augustin Ephrem Nkengfack
Department of Organic Chemistry, Faculty of Science, University of Yaoundé I, Yaoundé, Cameroon

Bailin Cong, Nengfei Wang, Shenghao Liu, Feng Liu, Xiaofei Yin and Jihong Shen
The First Institute of Oceanography, State Oceanic Administration, Qingdao 266061, People's Republic of China

R. Doug Wagner and Shemedia J. Johnson
Microbiology Division, National Center for Toxicological Research, U.S. Food and Drug Administration, 3900 NCTR Rd, Jefferson, AR 72079, USA

Federica Briani and Gianni Dehò
Dipartimento di Bioscienze, Università degli Studi di Milano, via Celoria 26, Milan 20133, Italy

Thomas Carzaniga
Dipartimento di Bioscienze, Università degli Studi di Milano, via Celoria 26, Milan 20133, Italy
Dipartimento di Biotecnologie mediche e medicina traslazionale, Università degli Studi di Milano, via F.lli Cervi 93, Segrate, MI 20090, Italy

Giulia Sbarufatti
Dipartimento di Bioscienze, Università degli Studi di Milano, via Celoria 26, Milan 20133, Italy
Eurofins BioPharma Product Testing Italy, Eurofins Biolab srl, via Bruno Buozzi, 2, Vimodrone 20090, Italy

Lin-Li Han, Huan-Huan Shao, Yong-Cheng Liu, Gang Liu, Chao-Ying Xie, Xiao-Jie Cheng, Hai-Yan Wang, Xue-Mei Tan and Hong Feng
Key Laboratory of Bio-resources and Eco-environment, Ministry of Education, Sichuan Key Laboratory of Molecular Biology and Biotechnology, Sichuan University, Chengdu 610064, Sichuan, People's Republic of China
College of Life Sciences, Sichuan University, Chengdu 610064, Sichuan, People's Republic of China

Wilson H. Johnson, Marlis R. Douglas, Jeffrey A. Lewis, Tara N. Stuecker, Bradley J. Austin, Michelle A. Evans-White and Michael E. Douglas
Department of Biological Sciences, University of Arkansas, Fayetteville, AR, USA

Franck G. Carbonero
Department of Food Sciences, University of Arkansas, Fayetteville, AR, USA

Sally A. Entrekin
Department of Biology, University of Central Arkansas, Conway, AR 72035, USA

Ismail M. Al-Bulushi, Nejib S. Guizani and Mohammed K. Al-Khusaibi
Department of Food Science and Nutrition, College of Agricultural and Marine Sciences, Sultan Qaboos University, Al-Khod 123, Oman

Abdullah M. Al-Sadi
Department of Crop Sciences, College of Agricultural and Marine Sciences, Sultan Qaboos University, Al-Khod 123, Oman

Dan Wang, Fengqiu Zhu, Qian Wang, Peng Yu, Jing Gong and Gejiao Wang
State Key Laboratory of Agricultural Microbiology, College of Life Science and Technology, Huazhong Agricultural University, Wuhan, People's Republic of China

Christopher Rensing
College of Resources and Environment, Fujian Agriculture and Forestry University, Fuzhou, People's Republic of China

Yang Liu and Yanan Gao
State Key Laboratory of Bioreactor Engineering, East China University of Science and Technology, Shanghai 200237, China

Xiaohong Liu, Qin Liu, Yuanxing Zhang, Qiyao Wang and Jingfan Xiao
State Key Laboratory of Bioreactor Engineering, East China University of Science and Technology, Shanghai 200237, China
Shanghai Engineering Research Center of Maricultured Animal Vaccines, Shanghai, China
Shanghai Collaborative Innovation Center for Biomanufacturing, 130 Meilong Road, Shanghai 200237, China

Jorge S. Oliveira
Laboratório de Biologia Molecular e Genômica, Departamento de Biologia
Celular e Genética, Centro de Biociências, Universidade Federal do Rio Grande do Norte, Natal, RN, Brazil
INESC-ID/IST Instituto de Engenharia de Sistemas e Computadores/Instituto Superior Técnico, Universidade de Lisboa, Rua Alves Redol, 9, 1000-029 Lisbon, Portugal
Laboratório de Bioinformática, Laboratório Nacional de Computação Científica, Petrópolis, RJ, Brazil

Wydemberg J. Araújo
Laboratório de Biologia Molecular e Genômica, Departamento de Biologia
Celular e Genética, Centro de Biociências, Universidade Federal do Rio
Grande do Norte, Natal, RN, Brazil.
Laboratório de Bioinformática, Laboratório Nacional de Computação Científica, Petrópolis, RJ, Brazil

Rita C. B. Silva-Portela, Alaine de Brito Guerra, Sinara Carla da Silva Araújo, Carolina Minnicelli, Aline Cardoso Carlos and Lucymara F. Agnez-Lima
Laboratório de Biologia Molecular e Genômica, Departamento de Biologia Celular e Genética, Centro de Biociências, Universidade Federal do Rio Grande do Norte, Natal, RN, Brazil

Ana Teresa Freitas and Ricardo M. Figueiredo
INESC-ID/IST Instituto de Engenharia de Sistemas e Computadores/Instituto Superior Técnico, Universidade de Lisboa, Rua Alves Redol, 9, 1000-029 Lisbon, Portugal

Ana Tereza Ribeiro de Vasconcelos
Laboratório de Bioinformática, Laboratório Nacional de Computação Científica, Petrópolis, RJ, Brazil

Ali Atas, Alan M. Seddon and Andrey V. Karlyshev
School of Life Sciences, Pharmacy and Chemistry; Faculty of Science, Engineering and Computing, Kingston University, Penrhyn Road, Kingston upon Thames KT1 2EE, UK

Brendan W. Wren
Department of Pathogen Molecular Biology, London School of Hygiene and Tropical Medicine, Keppel Street, London WC1E 7HT, UK

Donna C. Ford, Petra C. F. Oyston and Ian A. Cooper
Biomedical Sciences, DSTL Porton Down, Salisbury, Wiltshire SP4 0JQ, UK

Surajit De Mandal and Nachimuthu Senthil Kumar
Department of Biotechnology, Mizoram University, Aizawl, Mizoram 796004, India

Raghunath Chatterjee
Human Genetics Unit, Indian Statistical Institute, Kolkata 700108, India
Xiao-Lan He and Bing-Cheng Gan, Wei-Hong Peng, Jie Zhou, Xue-Lian Cao, Di Wang, Zhong-Qian Huang and Wei Tan Soil and Fertilizer Institute, Sichuan Academy of Agricultural Sciences, Chengdu 610066, China

Qian Li, Yu Li
Soil and Fertilizer Institute, Sichuan Academy of Agricultural Sciences, Chengdu 610066, China

Jilin Agricultural University, Changchun 130118, China
Mianyang Institute of Agricultural Sciences, Mianyang 621023, China

Juliana Ferraz Rosa, Camila Rizek, Ana Paula Marchi, Thais Guimaraes and Anna S Levin
Department of Infectious Diseases, University of São Paulo, Laboratory of Medical Investigation 54 (LIM-54), Hospital Das Clínicas FMUSP, São Paulo, Brazil

Lourdes Miranda
Hospital de Itapecerica da Serra, Itapecerica da Serra, SP, Brazil

Claudia Carrilho
Hospital University of Londrina, Londrina, Paraná, Brazil

Silvia F Costa
LIM-54, Faculdade de Medicina da Universidade de São Paulo, São Paulo, Brazil

Katherine García and Diliana Pérez-Reytor
Centro de Investigación Biomédica, Facultad de Ciencias de la Salud, Instituto de Ciencias Biomédicas, Universidad Autónoma de Chile, Av El Llano Subercaseaux, 2801 Santiago, Chile

Cristian Yáñez and Romilio T. Espejo
Institute of Nutrition and Food Technology, Universidad de Chile, Av. El Líbano 5524, Macul, Santiago, Chile

Nicolás Plaza
Centro de Investigación Biomédica, Facultad de Ciencias de la Salud, Instituto de Ciencias Biomédicas, Universidad Autónoma de Chile, Av El Llano Subercaseaux, 2801 Santiago, Chile
Institute of Nutrition and Food Technology, Universidad de Chile, Av. El Líbano 5524, Macul, Santiago, Chile

Francisca Peña and Pedro Sepúlveda
School of Bioinformatics Engineering, University of Talca, Talca, Chile

Nicole Werner, Melissa Gómez, Marcelo Baeza, Víctor Cifuentes and Jennifer Alcaíno
Departamento de Ciencias Ecológicas y Centro de Biotecnología, Facultad de
Ciencias, Universidad de Chile, Las Palmeras 3425, Casilla 653, Ñuñoa, Santiago, Chile

Dmitry P. Bazhanov and Xinjian Zhang
Key Laboratory for Applied Microbiology of Shandong Province, Ecology Institute (Biotechnology Center) of Shandong Academy of Sciences, Jinan,Shandong Province, People's Republic of China

Chengyun Li, Hongmei Li, Jishun Li and Hetong Yang
Biology Institute of Shandong Academy of Sciences, Jinan, Shandong Province, People's Republic of China

Xiangfeng Chen
Shandong Provincial Analysis and Test Center of Shandong Academy of Sciences, Jinan, Shandong Province, People's Republic of China

Index